Two-Dimensional Two-Product Cubic Systems, Vol. II

Albert C. J. Luo

Two-Dimensional Two-Product Cubic Systems, Vol. II

Crossing-linear and Self-quadratic Product Vector Fields

Springer

Albert C. J. Luo
Department of Mechanical and Mechatronics Engineering
Southern Illinois University
Edwardsville, IL, USA

ISBN 978-3-031-48490-2 ISBN 978-3-031-48491-9 (eBook)
https://doi.org/10.1007/978-3-031-48491-9

This Springer imprint is published by the registered company Springer Nature Switzerland AG
The registered company address is: Gewerbestrasse 11, 6330 Cham, Switzerland

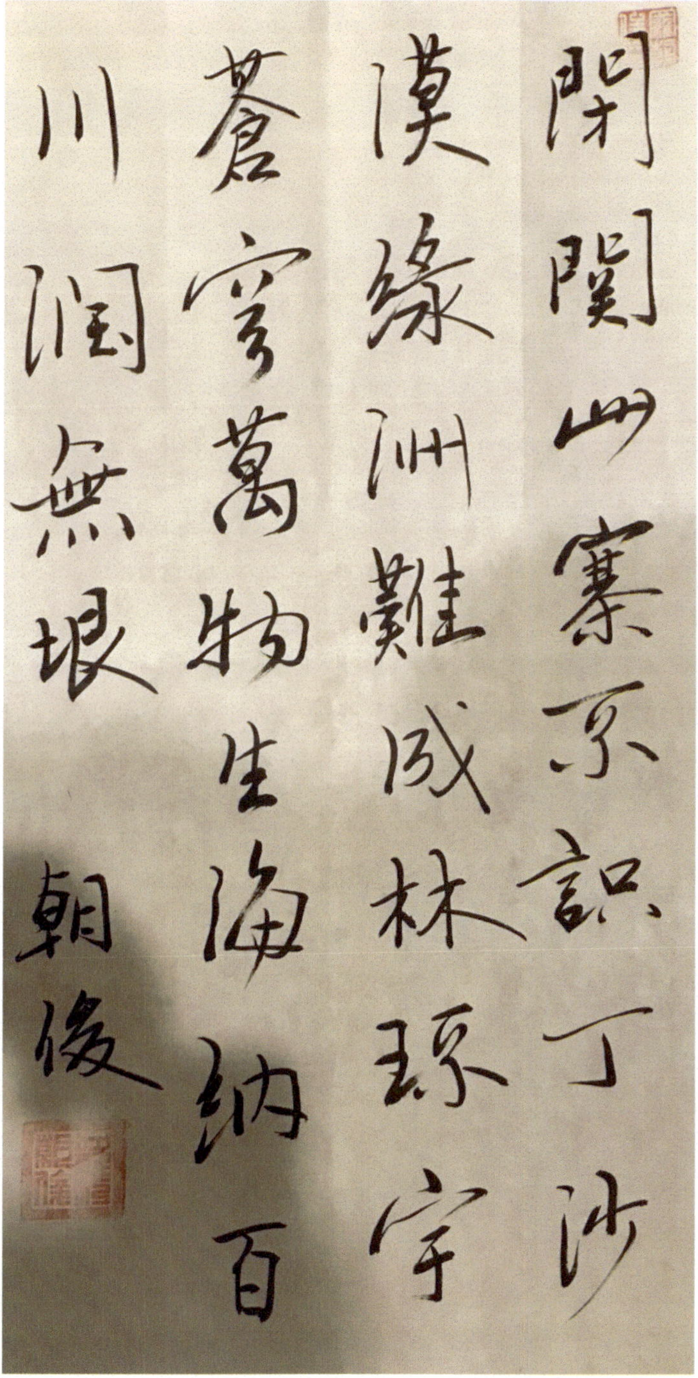

闭关山寨不识丁，沙漠绿洲难成林，环宇苍穹万物生，海纳百川润无垠。

—朝俊【海纳百川润无垠】辛丑冬

Preface

In this book, consider product-cubic nonlinear systems with two crossing-linear and self-quadratic products vector fields. Equilibrium singularity and bifurcations are discussed through appearing and switching bifurcations. Switching bifurcations are presented through infinite-equilibriums. Double-saddle equilibriums are the appearing bifurcations of the saddle-sources and saddle-sinks, and such double-saddle equilibriums are also the appearing bifurcations of the network of saddles, sink, and source. Infinite-equilibriums for switching bifurcations exist in such cubic nonlinear systems. Inflection upper-saddle and lower-saddle infinite-equilibriums are the switching bifurcation of saddle-sink and saddle-source with inflection sink and source flows. Hyperbolic and hyperbolic-secant, sink, and source infinite-equilibriums are for the switching bifurcation of saddle, source, and sink with inflection-sink and source flows. Up-down and down-up, asymptotic lower-saddle and upper-saddle infinite-equilibriums are the switching bifurcations of saddle-source and saddle-sink with hyperbolic and hyperbolic-secant flows. Inflection-source and sink infinite-equilibriums are the switching bifurcations of saddle, sink, and source with hyperbolic and hyperbolic-secant flows. Dynamic behaviors and singularity are presented through the first integral manifolds.

In Chap. 1, nonlinear dynamics and singularity of cubic dynamical systems possessing two cubic vector fields of different product structures are discussed through a theorem. Equilibrium singularity and bifurcations are presented through the appearing and switching bifurcations. Parabola-saddles and saddle-source (sink) singularities with hyperbolic-to-hyperbolic-secant flow and inflection-source (sink) singularities in two product-cubic systems are discussed, and the corresponding bifurcation dynamics are discussed. Corresponding switching bifurcations are discussed. Such nonlinear dynamics and singularity properties are presented through a theorem.

In Chap. 2, double-saddles and hyperbolic flow singularity in cubic systems with two crossing-linear and self-quadratic products vector fields are discussed. Double-saddle equilibriums are the appearing bifurcations of the saddle-source and saddle-sinks, and such double-saddle equilibriums are also the appearing bifurcations of the network of saddles, sink, and source. The hyperbolic flow singularity generates the appearing bifurcations of hyperbolic and hyperbolic-secant flows. Switching bifurcations are presented through infinite-equilibriums. Up-down and down-up infinite-equilibriums are the switching bifurcation. The hyperbolic lower-to-upper saddle is for the switching of positive (negative) saddle with paralleled hyperbolic-secant-to-hyperbolic flow. The hyperbolic-secant lower-to-upper saddle is for the switching of a center with paralleled hyperbolic-to-hyperbolic-secant flow. Up-down (down-up) saddles are for the switching bifurcation of double-saddles connected with a hyperbolic-to-hyperbolic-secant flow.

In Chap. 3, vertically paralleled saddle-source and saddle-sink with hyperbolic flow singularity exist in equilibrium networks, and the center or saddle with separated hyperbolic and hyperbolic-secant flows are paralleled. The hyperbolic flow singularity is for the appearing bifurcation of paralleled hyperbolic and hyperbolic-secant flows. Switching bifurcations for center (saddle) with paralleled hyperbolic (hyperbolic-secant) flow and saddle-source (sink) with connected hyperbolic singular flows are discussed. In addition, the switching bifurcations of the saddle-source (sink) with connected saddle-source (sink) and the saddle and center with paralleled hyperbolic singular flows are discussed through infinite-equilibriums.

In Chap. 4, horizontally paralleled saddle-source and saddle-sink with hyperbolic flow singularity exist in equilibrium networks, and the center or saddle with separated hyperbolic and hyperbolic-secant flows are paralleled. The hyperbolic flow singularity is for the appearing bifurcation of the paralleled hyperbolic and hyperbolic-secant flows. Switching bifurcations for center (saddle) with paralleled hyperbolic (hyperbolic-secant) flow and saddle-source (sink) with connected hyperbolic singular flows are discussed. In addition, the switching bifurcations of the saddle-source (sink) with connected saddle-source (sink) and the saddle and center with paralleled hyperbolic singular flows are discussed through infinite-equilibriums.

In Chap. 5, simple equilibrium networks are presented for the cubic product system with two self-linear and crossing-quadratic product vector fields, and the corresponding inflection-source and sink infinite-equilibriums are for hyperbolic flows with paralleled saddle and center and for hyperbolic flows with connected saddle, source, and sink. The (3×3)-simple equilibrium networks consist of negative and positive saddles, counter-clockwise and clockwise centers, saddles, sink and source, vertical and horizontal hyperbolic and hyperbolic-secant flows. In addition, double switching bifurcations are generated through inflection-saddles formed by two inflection-source (sink) infinite-equilibriums.

Finally, the author hopes materials presented herein can provide a better understanding of cubic nonlinear systems in science and engineering.

Edwardsville, IL, USA Albert C. J. Luo

The original version of the book was revised: A correction to this book can be found at https://doi.org/10.1007/978-3-031-48491-9_6

Contents

Chapter 1
Crossing-Linear and Self-Quadratic Product Systems

In this chapter, cubic product nonlinear systems with two crossing-linear and self-quadratic products vector fields are presented. The first integral manifolds are developed, and the equilibrium singularity and bifurcations are discussed through the appearing and switching bifurcations. The switching bifurcations are presented through the infinite-equilibriums. The double-saddle equilibriums are the appearing bifurcations of the saddle-sources and saddle-sinks, and such double-saddle equilibriums are also the appearing bifurcations of the network of saddles, sink, and source. The theory of the two-cubic product systems with crossing-linear and self-quadratic vector field is presented through a theorem and the corresponding proof of the theorem is given.

1.1 Crossing-Linear and Self-Quadratic Product Vector Fields

In this section, consider a cubic system with two crossing-linear and self-quadratic product vector fields. The corresponding singularity with appearing and switching bifurcations are presented through the following theorem, and the first integral manifolds are presented as well.

Theorem 1.1 *Consider a two-dimensional, cubic dynamical system with two crossing-linear and self-quadratic product cubic vector fields as*

$$\begin{aligned} \dot{x}_{j_2} &= a_{j_2j_20}(x_{j_1} - a_{j_2j_11})(x_{j_2}^2 + B_{j_2j_2}x_{j_2} + C_{j_2j_2}), \\ \dot{x}_{j_1} &= a_{j_1j_10}(x_{j_2} - a_{j_1j_21})(x_{j_1}^2 + B_{j_1j_1}x_{j_1} + C_{j_1j_1}) \end{aligned} \tag{1.1}$$

with

A. C. J. Luo, *Two-Dimensional Two-Product Cubic Systems, Vol. II*,
https://doi.org/10.1007/978-3-031-48491-9_1

$$\Delta_{j_1j_1} = B_{j_1j_1}^2 - 4C_{j_1j_1}, \Delta_{j_2j_2} = B_{j_2j_2}^2 - 4C_{j_2j_2}. \tag{1.2}$$

(i) For $\Delta_{j_1j_1} < 0$ *and* $\Delta_{j_2j_2} < 0$*, the standard form is*

$$\begin{aligned} \dot{x}_{j_2} &= a_{j_2j_20}(x_{j_1} - a_{j_2j_1}1)\left[(x_{j_2} - a_{j_2j_2})^2 + b_{j_2j_2}\right], \\ \dot{x}_{j_1} &= a_{j_1j_10}(x_{j_2} - a_{j_1j_2}1)\left[(x_{j_1} - a_{j_1j_1})^2 + b_{j_1j_1}\right] \end{aligned} \tag{1.3}$$

where

$$\begin{aligned} a_{j_1j_1} &= -\frac{1}{2}B_{j_1j_1}, b_{j_1j_1} = \frac{1}{4}(-\Delta_{j_1j_1}), \\ a_{j_2j_2} &= -\frac{1}{2}B_{j_2j_2}, b_{j_2j_2} = \frac{1}{4}(-\Delta_{j_2j_2}). \end{aligned} \tag{1.4}$$

The first integral manifold is

$$\begin{aligned} &\frac{1}{2}\ln\frac{\left|(x_{j_1} - a_{j_2j_1})^2 + b_{j_1j_1}\right|}{\left|(x_{j_10} - a_{j_1j_1})^2 + b_{j_1j_1}\right|} \\ &+\frac{(a_{j_1j_1} - a_{j_2j_1}1)}{\sqrt{b_{j_1j_1}}}\left(\arctan\frac{x_{j_1} - a_{j_1j_1}}{\sqrt{b_{j_1j_1}}} - \arctan\frac{x_{j_10} - a_{j_1j_1}}{\sqrt{b_{j_1j_1}}}\right) \\ &= \frac{a_{j_1j_10}}{a_{j_2j_20}}\left[\frac{1}{2}\ln\frac{\left|(x_{j_2} - a_{j_2j_2})^2 + b_{j_2j_2}\right|}{\left|(x_{j_20} - a_{j_2j_2})^2 + b_{j_2j_2}\right|}\right. \\ &\left.+\frac{(a_{j_2j_2} - a_{j_1j_2}1)}{\sqrt{b_{j_2j_2}}}\left(\arctan\frac{x_{j_2} - a_{j_2j_2}}{\sqrt{b_{j_2j_2}}} - \arctan\frac{x_{j_20} - a_{j_2j_2}}{\sqrt{b_{j_2j_2}}}\right)\right]. \end{aligned} \tag{1.5}$$

The equilibrium of $(x_{j_1}^*, x_{j_2}^*) = (a_{j_1j_2}1, a_{j_2j_1}1)$ *has the following properties:*

- *For* $a_{j_1j_10} > 0$ *and* $a_{j_2j_20} > 0$,

$$(a_{j_1j_2}1, a_{j_2j_1}1) = \underbrace{(\mathrm{UP}_+, \mathrm{UP}_+)}_{\text{positive saddle}}. \tag{1.6}$$

The equilibrium of $(x_{j_2}^*, x_{j_1}^*) = (a_{j_1j_2}1, a_{j_2j_1}1)$ *is a* (UP₊, UP₊)*-positive saddle.*

- *For* $a_{j_1j_10} < 0$ *and* $a_{j_2j_20} > 0$,

$$(a_{j_1j_2}1, a_{j_2j_1}1) = \underbrace{(\mathrm{DP}_+, \mathrm{DP}_-)}_{\text{CCW center}}. \tag{1.7}$$

The equilibrium of $(x_{j_2}^*, x_{j_1}^*) = (a_{j_1j_2}1, a_{j_2j_1}1)$ *is a* $(\mathrm{DP}_+, \mathrm{DP}_-)$*-counterclockwise center.*

- *For $a_{j_1j_10} > 0$ and $a_{j_2j_20} < 0$,*

$$(a_{j_1j_21}, a_{j_2j_11}) = \underbrace{(\mathrm{DP}_-, \mathrm{DP}_+)}_{\text{CW center}}. \tag{1.8}$$

The equilibrium of $(x^*_{j_2}, x^*_{j_1}) = (a_{j_1j_21}, a_{j_2j_11})$ is a (DP$_-$, DP$_+$)*-clockwise center.*

- *For $a_{j_1j_10} < 0$ and $a_{j_2j_20} < 0$,*

$$(a_{j_1j_21}, a_{j_2j_11}) = \underbrace{(\mathrm{UP}_-, \mathrm{UP}_-)}_{\text{negative saddle}}. \tag{1.9}$$

The equilibrium of $(x^*_{j_2}, x^*_{j_1}) = (a_{j_1j_21}, a_{j_2j_11})$ is a (UP$_-$,UP$_-$)*-negative saddle.*

(ii) For $\Delta_{j_1j_1} < 0$ and $\Delta_{j_2j_2} = 0$, the standard form is

$$\begin{aligned} \dot{x}_{j_2} &= a_{j_2j_20}(x_{j_1} - a_{j_2j_11})(x_{j_2} - a_{j_2j_21})^2, \\ \dot{x}_{j_1} &= a_{j_1j_10}(x_{j_2} - a_{j_1j_21})\left[(x_{j_1} - a_{j_1j_1})^2 + b_{j_1j_1}\right] \end{aligned} \tag{1.10}$$

where

$$\begin{aligned} a_{j_1j_1} &= -\frac{1}{2}B_{j_1j_1}, b_{j_1j_1} = \frac{1}{4}(-\Delta_{j_1j_1}), \\ a_{j_2j_21} &= -\frac{1}{2}B_{j_2j_2}. \end{aligned} \tag{1.11}$$

(ii$_1$) The first integral manifold for $a_{j_1j_21} \neq a_{j_2j_21}$ is

$$\begin{aligned} &\frac{1}{2}\ln\frac{|(x_{j_1} - a_{j_2j_1})^2 + b_{j_1j_1}|}{|(x_{j_10} - a_{j_1j_1})^2 + b_{j_1j_1}|} \\ &\quad + \frac{(a_{j_1j_1} - a_{j_2j_11})}{\sqrt{b_{j_1j_1}}}(\arctan\frac{x_{j_1} - a_{j_1j_1}}{\sqrt{b_{j_1j_1}}} - \arctan\frac{x_{j_10} - a_{j_1j_1}}{\sqrt{b_{j_1j_1}}}) \\ &= \frac{a_{j_1j_10}}{a_{j_2j_20}}\left[\ln\frac{|x_{j_2} - a_{j_2j_21}|}{|x_{j_20} - a_{j_2j_21}|} - (a_{j_2j_21} - a_{j_1j_21})(\frac{1}{x_{j_2} - a_{j_2j_21}} - \frac{1}{x_{j_20} - a_{j_2j_21}})\right]. \end{aligned} \tag{1.12}$$

(ii$_{1a}$) The equilibrium of $(x^*_{j_2}, x^*_{j_1}) = (a_{j_1j_21}, a_{j_2j_11})$ with $a_{j_1j_21} \neq a_{j_2j_21}$ has the following properties:

- *For $a_{j_1j_10} > 0$ and $a_{j_2j_20}(a_{j_1j_21} - a_{j_2j_21})^2 > 0$,*

$$(a_{j_1j_21}, a_{j_2j_11}) = \underbrace{(\mathrm{UP}_+, \mathrm{UP}_+)}_{\text{positive saddle}}. \tag{1.13}$$

The equilibrium of $(x^*_{j_2}, x^*_{j_1}) = (a_{j_1j_21}, a_{j_2j_11})$ is a (UP$_+$,UP$_+$)*-positive saddle.*

- *For $a_{j_1j_1 0} < 0$ and $a_{j_2j_2 0}(a_{j_1j_2 1} - a_{j_2j_2 1})^2 > 0$,*

$$(a_{j_1j_2 1}, a_{j_2j_1 1}) = \underbrace{(\text{DP}_+,\text{DP}_-)}_{\text{CCW center}}. \tag{1.14}$$

The equilibrium of $(x_{j_2}^*, x_{j_1}^*) = (a_{j_1j_2 1}, a_{j_2j_1 1})$ is a (DP$_+$,DP$_-$)*-counterclockwise center.*

- *For $a_{j_1j_1 0} > 0$ and $a_{j_2j_2 0}(a_{j_1j_2 1} - a_{j_2j_2 1})^2 < 0$,*

$$(a_{j_1j_2 1}, a_{j_2j_1 1}) = \underbrace{(\text{DP}_-,\text{DP}_+)}_{\text{CW center}}. \tag{1.15}$$

The equilibrium of $(x_{j_2}^*, x_{j_1}^*) = (a_{j_1j_2 1}, a_{j_2j_1 1})$ is a (DP$_-$, DP$_+$)*-clockwise center.*

- *For $a_{j_1j_1 0} < 0$ and $a_{j_2j_2 0}(a_{j_1j_2 1} - a_{j_2j_2 1})^2 < 0$,*

$$(a_{j_1j_2 1}, a_{j_2j_1 1}) = \underbrace{(\text{UP}_-,\text{UP}_-)}_{\text{negative saddle}}. \tag{1.16}$$

The equilibrium of $(x_{j_2}^*, x_{j_1}^*) = (a_{j_1j_2 1}, a_{j_2j_1 1})$ is a (UP$_-$,UP$_-$)*-negative saddle.*

(ii_{1b}) The equilibrium of $(x_{j_1}^*, x_{j_2}^*) = (a_{j_2j_1 1}, a_{j_2j_2 1})$ with $a_{j_2j_2 1} \neq a_{j_1j_2 1}$ has the following properties:

- *For $a_{j_1j_1 0}(a_{j_2j_2 1} - a_{j_1j_2 1}) > 0$ and $a_{j_2j_2 0} > 0$,*

$$(a_{j_2j_1 1}, a_{j_2j_2 1}) = \underbrace{(\text{pF,UP:UP})}_{\text{hyperbolic-secant-to-hyperbolic flow } (+)}. \tag{1.17}$$

The equilibrium of $(x_{j_1}^*, x_{j_2}^*) = (a_{j_2j_1 1}, a_{j_2j_2 1})$ is a (pF, UP:UP)*-positive hyperbolic-secant-to-hyperbolic flow.*

- *For $a_{j_1j_1 0}(a_{j_2j_2 1} - a_{j_1j_2 1}) < 0$ and $a_{j_2j_2 0} > 0$,*

$$(a_{j_2j_1 1}, a_{j_2j_2 1}) = \underbrace{(\text{nF,DP:DP})}_{\text{hyperbolic-to-hyperbolic-secant flow } (-)}. \tag{1.18}$$

The equilibrium of $(x_{j_1}^*, x_{j_2}^*) = (a_{j_2j_1 1}, a_{j_2j_2 1})$ is an (nF,DP:DP)*-negative hyperbolic-to-hyperbolic-secant flow.*

- *For $a_{j_1j_1 0}(a_{j_2j_2 1} - a_{j_1j_2 1}) > 0$ and $a_{j_2j_2 0} < 0$,*

$$(a_{j_2j_1 1}, a_{j_2j_2 1}) = \underbrace{(\text{pF,DP:DP})}_{\text{hyperbolic-to-hyperbolic-secant flow } (+)}. \tag{1.19}$$

The equilibrium of $(x^*_{j_1}, x^*_{j_2}) = (a_{j_2j_1}1, a_{j_2j_2}1)$ *is a* (pF,DP:DP)-*positive hyperbolic-to-hyperbolic-secant flow.*

- *For* $a_{j_1j_1}0(a_{j_2j_2}1 - a_{j_1j_2}1) < 0$ *and* $a_{j_2j_2}0 < 0$,

$$(a_{j_2j_1}1, a_{j_2j_2}1) = \underbrace{(\text{nF,UP:UP})}_{\text{hyperbolic-secant-to-hyperbolic flow }(-)}. \tag{1.20}$$

The equilibrium of $(x^*_{j_1}, x^*_{j_2}) = (a_{j_2j_1}1, a_{j_2j_2}1)$ *is an* (nF, UP:UP)-*negative hyperbolic secant-to-hyperbolic flow.*

– *The hyperbolic-to-hyperbolic-secant and hyperbolic-secant-to-hyperbolic flows are the appearing bifurcations of two separated hyperbolic and hyperbolic secant flows.*

(ii₂) The first integral manifold for $a_{j_1j_2}1 = a_{j_2j_2}1$ *is*

$$\begin{aligned}
&\frac{1}{2}\ln\frac{|(x_{j_1} - a_{j_2j_1})^2 + b_{j_1j_1}|}{|(x_{j_10} - a_{j_1j_1})^2 + b_{j_1j_1}|} \\
&\quad + \frac{(a_{j_1j_1} - a_{j_2j_1}1)}{\sqrt{b_{j_1j_1}}}(\arctan\frac{x_{j_1} - a_{j_1j_1}}{\sqrt{b_{j_1j_1}}} - \arctan\frac{x_{j_10} - a_{j_1j_1}}{\sqrt{b_{j_1j_1}}}) \\
&= \frac{a_{j_1j_10}}{a_{j_2j_20}}\ln\frac{|x_{j_2} - a_{j_2j_2}1|}{|x_{j_20} - a_{j_2j_2}1|}.
\end{aligned} \tag{1.21}$$

(ii₂ₐ) The infinite-equilibrium of $x^*_{j_2} = a_{j_1j_2}1 = a_{j_2j_2}1$ *and* $\bar{x}_{j_1} \neq a_{j_2j_1}1$ *has the following properties:*

- *For* $a_{j_1j_1}0 > 0$ *and* $a_{j_2j_2}0(\bar{x}_{j_1} - a_{j_2j_1}1) > 0$,

$$(\bar{x}_{j_1}, a_{j_2j_2}1) = \underbrace{(\text{DU,US})}_{\text{down-up upper-saddle}}. \tag{1.22}$$

The infinite-equilibrium of $x^*_{j_2} = a_{j_1j_2}1 = a_{j_2j_2}1$ *is a* (DU,US)-*down-up upper saddle.*

- *For* $a_{j_1j_1}0 < 0$ *and* $a_{j_2j_2}0(\bar{x}_{j_1} - a_{j_2j_1}1) > 0$,

$$(\bar{x}_{j_1}, a_{j_2j_2}1) = \underbrace{(\text{UD,US})}_{\text{up-down upper-saddle}}. \tag{1.23}$$

The infinite-equilibrium of $x^*_{j_2} = a_{j_1j_2}1 = a_{j_2j_2}1$ *is a* (UD,US)-*up-down upper saddle.*

- *For* $a_{j_1j_1}0 > 0$ *and* $a_{j_2j_2}0(\bar{x}_{j_1} - a_{j_2j_1}1) < 0$,

$$(\bar{x}_{j_1}, a_{j_2j_21}) = \underbrace{(\text{UD,LS})}_{\text{up-down lower-saddle}}. \tag{1.24}$$

The infinite-equilibrium of $x_{j_2}^* = a_{j_1j_21} = a_{j_2j_21}$ *is a* (UD,LS)*-up-down-lower saddle.*

- *For* $a_{j_1j_10} < 0$ *and* $a_{j_2j_20}(\bar{x}_{j_1} - a_{j_2j_11}) < 0$,

$$(\bar{x}_{j_1} a_{j_2j_21}) = \underbrace{(\text{DU,LS})}_{\text{down-up lower-saddle}}. \tag{1.25}$$

The infinite-equilibrium of $x_{j_2}^* = a_{j_1j_21} = a_{j_2j_21}$ *is a* (DU,LS)*-down-up lower saddle.*

(ii_{2b}) *The equilibrium of* $(x_{j_1}^*, x_{j_2}^*) = (a_{j_2j_11}, a_{j_2j_21})$ *has the following properties:*

- *For* $a_{j_1j_10} > 0$ *and* $a_{j_2j_20} > 0$,

$$(a_{j_2j_11}, a_{j_2j_21}) = \underbrace{({}_{\text{UD}}\text{LS:}{}_{\text{DU}}\text{US,DP}_-\text{:UP}_+)}_{\text{hyperbolic lower-to-upper saddle}}. \tag{1.26}$$

The equilibrium of $(x_{j_1}^*, x_{j_2}^*) = (a_{j_2j_11}, a_{j_2j_21})$ *is an* (${}_{\text{UD}}$LS:${}_{\text{DU}}$US, DP$_-$:UP$_+$)*-hyperbolic lower-to-upper saddle.*

- *For* $a_{j_1j_10} < 0$ *and* $a_{j_2j_20} > 0$,

$$(a_{j_2j_11}, a_{j_2j_21}) = \underbrace{({}_{\text{DU}}\text{LS:}{}_{\text{UD}}\text{US,UP}_+\text{:DP}_-)}_{\text{hyperbolic-secant lower-to-upper saddle}}. \tag{1.27}$$

The equilibrium of $(x_{j_1}^*, x_{j_2}^*) = (a_{j_2j_11}, a_{j_2j_21})$ *is an* (${}_{\text{DU}}$LS:${}_{\text{UD}}$US, UP$_+$:DP$_-$)-hyperbolic-secant lower-to-upper saddle.

- *For* $a_{j_1j_10} > 0$ *and* $a_{j_2j_20} < 0$,

$$(a_{j_2j_11}, a_{j_2j_21}) = \underbrace{({}_{\text{DU}}\text{US:}{}_{\text{UD}}\text{LS,UP}_-\text{:DP}_+)}_{\text{hyperbolic-secant upper-to-lower saddle}}. \tag{1.28}$$

The equilibrium of $(x_{j_1}^*, x_{j_2}^*) = (a_{j_2j_11}, a_{j_2j_21})$ *is a* (${}_{\text{DU}}$US:${}_{\text{UD}}$LS, UP$_-$:DP$_+$)-hyperbolic-secant upper-to-lower saddle.

- *For* $a_{j_1j_10} < 0$ *and* $a_{j_2j_20} < 0$,

$$(a_{j_2j_11}, a_{j_2j_21}) = \underbrace{({}_{\text{UD}}\text{US:}{}_{\text{DU}}\text{LS,DP}_+\text{:UP}_-)}_{\text{hyperbolic upper-to-lower saddle}}. \tag{1.29}$$

The equilibrium of $(x^*_{j_1}, x^*_{j_2}) = (a_{j_2 j_1 1}, a_{j_2 j_2 1})$ *is a* ($_{\text{UD}}$US:$_{\text{DU}}$LS,DP$_+$:UP$_-$)-hyperbolic upper-to-lower saddle.

(iii) For $\Delta_{j_1 j_1} < 0$ *and* $\Delta_{j_2 j_2} > 0$, *the standard form is*

$$\begin{aligned} \dot{x}_{j_2} &= a_{j_2 j_2 0}(x_{j_1} - a_{j_2 j_1 1})(x_{j_2} - a_{j_2 j_2 1})(x_{j_2} - a_{j_2 j_2 2}), \\ \dot{x}_{j_1} &= a_{j_1 j_1 0}\left[(x_{j_1} - a_{j_1 j_1})^2 + b_{j_1 j_1}\right](x_{j_2} - a_{j_1 j_2 1}) \end{aligned} \tag{1.30}$$

where

$$\begin{aligned} a_{j_1 j_1} &= -\frac{1}{2}B_{j_1 j_1}, b_{j_1 j_1} = \frac{1}{4}(-\Delta_{j_1 j_1}), \\ b_{j_2 j_2 1}, b_{j_2 j_2 2} &= -\frac{1}{2}(B_{j_2 j_2} \pm \sqrt{\Delta_{j_2 j_2}}), \\ \{a_{j_2 j_2 1}, a_{j_2 j_2 2}\} &= \text{sort}\{b_{j_2 j_2 1}, b_{j_2 j_2 2}\}, a_{j_2 j_2 1} < a_{j_2 j_2 2}. \end{aligned} \tag{1.31}$$

(iii$_1$) The first integral manifold $(a_{j_1 j_2 1} \neq a_{j_2 j_2 s_1}, a_{j_2 j_2 s_2}, s_1, s_2 \in \{1, 2\}, s_1 \neq s_2)$ *is*

$$\begin{aligned} &\frac{1}{2}\ln\frac{|(x_{j_1} - a_{j_2 j_1})^2 + b_{j_1 j_1}|}{|(x_{j_1 0} - a_{j_1 j_1})^2 + b_{j_1 j_1}|} \\ &\quad + \frac{(a_{j_1 j_1} - a_{j_2 j_1 1})}{\sqrt{b_{j_1 j_1}}}\left(\arctan\frac{x_{j_1} - a_{j_1 j_1}}{\sqrt{b_{j_1 j_1}}} - \arctan\frac{x_{j_1 0} - a_{j_1 j_1}}{\sqrt{b_{j_1 j_1}}}\right) \\ &= \frac{a_{j_1 j_1 0}}{a_{j_2 j_2 0}}\sum\nolimits_{s_1 = 1}^{2}\frac{(a_{j_2 j_2 s_1} - a_{j_1 j_2 1})}{(a_{j_2 j_2 s_1} - a_{j_2 j_2 s_2})}\ln\frac{|x_{j_2} - a_{j_2 j_2 s_1}|}{|x_{j_2 0} - a_{j_2 j_2 s_1}|}. \end{aligned} \tag{1.32}$$

(iii$_{1a}$) The equilibrium of $x^*_{j_2} = a_{j_1 j_2 1} \neq a_{j_2 j_2 s_1}, a_{j_2 j_2 s_2}$ $(s_1, s_2 \in \{1, 2\}, s_1 \neq s_2)$ *has the following properties:*

- *For* $a_{j_1 j_1 0} > 0$ *and* $a_{j_2 j_2 0}\prod_{s_1 = 1}^{2}(a_{j_1 j_2 1} - a_{j_2 j_2 s_1}) > 0$,

$$(a_{j_1 j_2 1}, a_{j_2 j_1 1}) = \underbrace{(\text{UP}_+, \text{UP}_+)}_{\text{positive saddle}}. \tag{1.33}$$

  The equilibrium of $(x^*_{j_1}, x^*_{j_2}) = (a_{j_1 j_2 1}, a_{j_2 j_1 1})$ *is a* (UP$_+$, UP$_+$)*-positive saddle.*

- *For* $a_{j_1 j_1 0} < 0$ *and* $a_{j_2 j_2 0}\prod_{s_1 = 1}^{2}(a_{j_1 j_2 1} - a_{j_2 j_2 s_1}) > 0$,

$$(a_{j_1 j_2 1}, a_{j_2 j_1 1}) = \underbrace{(\text{DP}_+, \text{DP}_-)}_{\text{CCW center}}. \tag{1.34}$$

  The equilibrium of $(x^*_{j_1}, x^*_{j_2}) = (a_{j_1 j_2 1}, a_{j_2 j_1 1})$ *is a* (DP$_+$,DP$_-$)*-counterclockwise center.*

- *For* $a_{j_1 j_1 0} > 0$ *and* $a_{j_2 j_2 0}\prod_{s_1 = 1}^{2}(a_{j_1 j_2 1} - a_{j_2 j_2 s_1}) < 0$,

$$(a_{j_1j_2 1}, a_{j_2j_1 1}) = \underbrace{(\text{DP}_-, \text{DP}_+)}_{\text{CW center}}. \tag{1.35}$$

The equilibrium of $(x^*_{j_1}, x^*_{j_2}) = (a_{j_1j_2 1}, a_{j_2j_1 1})$ *is a* (DP$_-$, DP$_+$)*-clockwise center.*

- *For* $a_{j_1j_1 0} < 0$ *and* $a_{j_2j_2 0}\prod_{s_1=1}^{2}(a_{j_1j_2 1} - a_{j_2j_2 s_1}) < 0$,

$$(a_{j_1j_2 1}, a_{j_2j_1 1}) = \underbrace{(\text{UP}_-, \text{UP}_-)}_{\text{negative saddle}}. \tag{1.36}$$

The equilibrium of $(x^*_{j_1}, x^*_{j_2}) = (a_{j_1j_2 1}, a_{j_2j_1 1})$ *is a* (UP$_-$,UP$_-$)*-negative saddle.*

(iii$_{1b}$) The equilibrium of $(x^*_{j_1}, x^*_{j_2}) = (a_{j_2j_1 1}, a_{j_2j_2 s_1})$ $(s_1, s_2 \in \{1, 2\}, s_1 \neq s_2)$ *with* $a_{j_2j_2 s_1} \neq a_{j_1j_2 1}$ *has the following properties:*

- *For* $a_{j_1j_1 0}(a_{j_2j_2 s_1} - a_{j_1j_2 1}) > 0$ *and* $a_{j_2j_2 0}(a_{j_2j_2 s_1} - a_{j_2j_2 s_2}) > 0$,

$$(a_{j_2j_1 1}, a_{j_2j_2 s_1}) = \underbrace{(\text{pF}, \text{DP:UP})}_{\text{hyperbolic flow }(+)}. \tag{1.37}$$

The equilibrium of $(x^*_{j_1}, x^*_{j_2}) = (a_{j_2j_1 1}, a_{j_2j_2 s_1})$ *is a* (pF, DP:UP)*-positive hyperbolic flow.*

- *For* $a_{j_1j_1 0}(a_{j_2j_2 s_1} - a_{j_1j_2 1}) < 0$ *and* $a_{j_2j_2 0}(a_{j_2j_2 s_1} - a_{j_2j_2 s_2}) > 0$,

$$(a_{j_2j_1 1}, a_{j_2j_2 s_1}) = \underbrace{(\text{nF}, \text{UP:DP})}_{\text{hyperbolic-secant flow }(-)}. \tag{1.38}$$

The equilibrium of $(x^*_{j_1}, x^*_{j_2}) = (a_{j_2j_1 1}, a_{j_2j_2 s_1})$ *is an* (nF, UP:DP)*-negative hyperbolic secant flow.*

- *For* $a_{j_1j_1 0}(a_{j_2j_2 s_1} - a_{j_1j_2 1}) > 0$ *and* $a_{j_2j_2 0}(a_{j_2j_2 s_1} - a_{j_2j_2 s_2}) < 0$,

$$(a_{j_2j_1 1}, a_{j_2j_2 s_1}) = \underbrace{(\text{pF}, \text{UP:DP})}_{\text{hyperbolic-secant flow }(+)}. \tag{1.39}$$

The equilibrium of $(x^*_{j_1}, x^*_{j_2}) = (a_{j_2j_1 1}, a_{j_2j_2 s_1})$ *is a* (pF, UP:DP)*-positive hyperbolic secant flow.*

- *For* $a_{j_1j_1 0}(a_{j_2j_2 s_1} - a_{j_1j_2 1}) < 0$ *and* $a_{j_2j_2 0}(a_{j_2j_2 s_1} - a_{j_2j_2 s_2}) < 0$,

$$(a_{j_2j_1 1}, a_{j_2j_2 s_1}) = \underbrace{(\text{nF}, \text{DP:UP})}_{\text{hyperbolic flow }(-)}. \tag{1.40}$$

The equilibrium of $(x^*_{j_1}, x^*_{j_2}) = (a_{j_2j_1 1}, a_{j_2j_2 s_1})$ *is an* (nF, DP:UP)*-negative hyperbolic flow.*

(iii_2) The first integral manifold for $a_{j_1j_21} = a_{j_2j_2s_1}$ ($s_1, s_2 \in \{1,2\}$, $s_1 \neq s_2$) *is*

$$\begin{aligned}
&\frac{1}{2}\ln\frac{|(x_{j_1}-a_{j_2j_1})^2+b_{j_1j_1}|}{|(x_{j_10}-a_{j_1j_1})^2+b_{j_1j_1}|}\\
&+\frac{(a_{j_1j_1}-a_{j_2j_11})}{\sqrt{b_{j_1j_1}}}(\arctan\frac{x_{j_1}-a_{j_1j_1}}{\sqrt{b_{j_1j_1}}}-\arctan\frac{x_{j_10}-a_{j_1j_1}}{\sqrt{b_{j_1j_1}}})\\
&=\frac{a_{j_1j_10}}{a_{j_2j_20}}\ln\frac{|x_{j_2}-a_{j_2j_2s_2}|}{|x_{j_20}-a_{j_2j_2s_2}|}.
\end{aligned}\tag{1.41}$$

(iii_{2a}) The infinite-equilibrium of $x^*_{j_2} = a_{j_1j_21} = a_{j_2j_2s_1}$ *has the following properties:*

- *For* $a_{j_1j_10} > 0$ *and* $a_{j_2j_20}(\bar{x}_{j_1} - a_{j_2j_11})(a_{j_2j_2s_1} - a_{j_2j_2s_2}) > 0$,

$$(\bar{x}_{j_1}, a_{j_2j_2s_1}) = \underbrace{(\text{II,SO})}_{\text{increasing-inflection source}}. \tag{1.42}$$

  The infinite-equilibrium of $x^*_{j_2} = a_{j_1j_21} = a_{j_2j_2s_1}$ *is an* (II,SO)*-increasing inflection source.*

- *For* $a_{j_1j_10} < 0$ *and* $a_{j_2j_20}(\bar{x}_{j_1} - a_{j_2j_11})(a_{j_2j_2s_1} - a_{j_2j_2s_2}) > 0$,

$$(\bar{x}_{j_1}, a_{j_2j_2s_1}) = \underbrace{(\text{DI,SO})}_{\text{decreasing-inflection source}}. \tag{1.43}$$

  The infinite-equilibrium of $x^*_{j_2} = a_{j_1j_21} = a_{j_2j_2s_1}$ *is a* (DI,SO)*-down-parabola source.*

- *For* $a_{j_1j_10} > 0$ *and* $a_{j_2j_20}(\bar{x}_{j_1} - a_{j_2j_11})(a_{j_2j_2s_1} - a_{j_2j_2s_2}) < 0$,

$$(\bar{x}_{j_1}, a_{j_2j_2s_1}) = \underbrace{(\text{DI,SI})}_{\text{decreasing-inflection sink}}. \tag{1.44}$$

  The infinite-equilibrium of $x^*_{j_2} = a_{j_1j_21} = a_{j_2j_2s_1}$ *is a* (DI,SI)*-decreasing inflection sink.*

- *For* $a_{j_1j_10} < 0$ *and* $a_{j_2j_20}(\bar{x}_{j_1} - a_{j_2j_11})(a_{j_2j_2s_1} - a_{j_2j_2s_2}) < 0$,

$$(\bar{x}_{j_1}, a_{j_2j_2s_1}) = \underbrace{(\text{II,SI})}_{\text{increasing-inflection sink}}. \tag{1.45}$$

  The infinite-equilibrium of $x^*_{j_2} = a_{j_1j_21} = a_{j_2j_2s_1}$ *is an* (II,SI)*-increasing infection sink.*

(iii_{2b}) The equilibrium of $(x^*_{j_2}, x^*_{j_1}) = (a_{j_1j_21}, a_{j_2j_11})$ *with* $a_{j_1j_21} = a_{j_2j_2s_1}$ *has the following properties:*

- *For* $a_{j_1 j_1 0} > 0$ *and* $a_{j_2 j_2 0}(a_{j_2 j_2 s_1} - a_{j_2 j_2 s_2}) > 0$,

$$(a_{j_1 j_2 1}, a_{j_2 j_1 1}) = \underbrace{({}_{\mathrm{SI:SO}}\mathrm{US}, {}_{\mathrm{DI:II}}\mathrm{UP})}_{\text{up-parabola upper-saddle}}. \tag{1.46}$$

The equilibrium of $(x_{j_1}^*, x_{j_2}^*) = (a_{j_2 j_1 1}, a_{j_2 j_2 s_1})$ *is an* $({}_{\mathrm{SI:SO}}\mathrm{US}, {}_{\mathrm{DI:II}}\mathrm{UP})$*-up-parabola upper-saddle.*

- *For* $a_{j_1 j_1 0} < 0$ *and* $a_{j_2 j_2 0}(a_{j_2 j_2 s_1} - a_{j_2 j_2 s_2}) > 0$,

$$(a_{j_1 j_2 1}, a_{j_2 j_1 1}) = \underbrace{({}_{\mathrm{SI:SO}}\mathrm{US}, {}_{\mathrm{II:DI}}\mathrm{DP})}_{\text{down-parabola upper-saddle}}. \tag{1.47}$$

The equilibrium of $(x_{j_1}^*, x_{j_2}^*) = (a_{j_2 j_1 1}, a_{j_2 j_2 s_1})$ *is an* $({}_{\mathrm{SI:SO}}\mathrm{US}, {}_{\mathrm{II:DI}}\mathrm{DP})$*-down-parabola upper-saddle.*

- *For* $a_{j_1 j_1 0} > 0$ *and* $a_{j_2 j_2 0}(a_{j_2 j_2 s_1} - a_{j_2 j_2 s_2}) < 0$,

$$(a_{j_1 j_2 1}, a_{j_2 j_1 1}) = \underbrace{({}_{\mathrm{SO:SI}}\mathrm{LS}, {}_{\mathrm{II:DI}}\mathrm{DP})}_{\text{down-parabola lower-saddle}}. \tag{1.48}$$

The equilibrium of $(x_{j_1}^*, x_{j_2}^*) = (a_{j_2 j_1 1}, a_{j_2 j_2 s_1})$ *is an* $({}_{\mathrm{SO:SI}}\mathrm{LS}, {}_{\mathrm{II:DI}}\mathrm{DP})$*-down-parabola lower-saddle.*

- *For* $a_{j_1 j_1 0} < 0$ *and* $a_{j_2 j_2 0}(a_{j_2 j_2 s_1} - a_{j_2 j_2 s_2}) < 0$,

$$(a_{j_1 j_2 1}, a_{j_2 j_1 1}) = \underbrace{({}_{\mathrm{SO:SI}}\mathrm{LS}, {}_{\mathrm{DI:II}}\mathrm{UP})}_{\text{up-parabola lower-saddle}}. \tag{1.49}$$

The equilibrium of $(x_{j_1}^*, x_{j_2}^*) = (a_{j_2 j_1 1}, a_{j_2 j_2 s_1})$ *is an* $({}_{\mathrm{SO:SI}}\mathrm{LS}, {}_{\mathrm{DI:II}}\mathrm{UP})$*-up-parabola lower-saddle.*

(iv) For $\Delta_{j_1 j_1} = 0$ *and* $\Delta_{j_2 j_2} < 0$*, the standard form is*

$$\begin{aligned} \dot{x}_{j_2} &= a_{j_2 j_2 0}(x_{j_1} - a_{j_2 j_1 1})\left[(x_{j_2} - a_{j_2 j_2})^2 + b_{j_2 j_2}\right], \\ \dot{x}_{j_1} &= a_{j_1 j_1 0}(x_{j_1} - a_{j_1 j_1 1})^2 (x_{j_2} - a_{j_1 j_2 1}) \end{aligned} \tag{1.50}$$

where

$$\begin{aligned} a_{j_1 j_1 1} &= -\frac{1}{2} B_{j_1 j_1}, \\ a_{j_2 j_2} &= -\frac{1}{2} B_{j_2 j_2}, b_{j_2 j_2} = \frac{1}{4}(-\Delta_{j_2 j_2}). \end{aligned} \tag{1.51}$$

(iv$_1$*) The first integral manifold for* $a_{j_1 j_1 1} \neq a_{j_2 j_1 1}$ *is*

$$\begin{aligned}
&\ln\frac{|x_{j_1}-a_{j_1j_11}|}{|x_{j_10}-a_{j_1j_11}|}-(a_{j_1j_11}-a_{j_2j_11})(\frac{1}{x_{j_1}-a_{j_1j_11}}-\frac{1}{x_{j_10}-a_{j_1j_11}})\\
&=\frac{a_{j_1j_10}}{a_{j_2j_20}}\left[\frac{1}{2}\ln\frac{|(x_{j_2}-a_{j_2j_2})^2+b_{j_2j_2}|}{|(x_{j_20}-a_{j_2j_2})^2+b_{j_2j_2}|}\right.\\
&\left.+\frac{(a_{j_2j_2}-a_{j_1j_21})}{\sqrt{b_{j_2j_2}}}(\arctan\frac{x_{j_2}-a_{j_2j_2}}{\sqrt{b_{j_2j_2}}}-\arctan\frac{x_{j_20}-a_{j_2j_2}}{\sqrt{b_{j_2j_2}}})\right].
\end{aligned}\tag{1.52}$$

(iv_{1a}) The equilibrium of $(x^*_{j_2},x^*_{j_1})=(a_{j_1j_21},a_{j_2j_11})$ *has the following properties:*

- *For* $a_{j_1j_10}(a_{j_2j_11}-a_{j_1j_11})^2>0$ *and* $a_{j_2j_20}>0$,

$$(a_{j_1j_21},a_{j_2j_11})=\underbrace{(\mathrm{UP}_+,\mathrm{UP}_+)}_{\text{positive saddle}}.\tag{1.53}$$

  The equilibrium of $(x^*_{j_2},x^*_{j_1})=(a_{j_1j_21},a_{j_2j_11})$ *is a* $(\mathrm{UP}_+,\ \mathrm{UP}_+)$*-positive saddle.*

- *For* $a_{j_1j_10}(a_{j_2j_11}-a_{j_1j_11})^2<0$ *and* $a_{j_2j_20}>0$,

$$(a_{j_1j_21},a_{j_2j_11})=\underbrace{(\mathrm{DP}_+,\mathrm{DP}_-)}_{\text{CCW center}}.\tag{1.54}$$

  The equilibrium of $(x^*_{j_2},x^*_{j_1})=(a_{j_1j_21},a_{j_2j_11})$ *is a* $(\mathrm{DP}_+,\mathrm{DP}_-)$*-counterclockwise center.*

- *For* $a_{j_1j_10}(a_{j_2j_11}-a_{j_1j_11})^2>0$ *and* $a_{j_2j_20}<0$,

$$(a_{j_1j_21},a_{j_2j_11})=\underbrace{(\mathrm{DP}_-,\mathrm{DP}_+)}_{\text{CW center}}.\tag{1.55}$$

  The equilibrium of $(x^*_{j_2},x^*_{j_1})=(a_{j_1j_21},a_{j_2j_11})$ *is a* $(\mathrm{DP}_-,\ \mathrm{DP}_+)$*-clockwise center.*

- *For* $a_{j_1j_10}(a_{j_2j_11}-a_{j_1j_11})^2<0$ *and* $a_{j_2j_20}<0$,

$$(a_{j_1j_21},a_{j_2j_11})=\underbrace{(\mathrm{UP}_-,\mathrm{UP}_-)}_{\text{negative saddle}}.\tag{1.56}$$

  The equilibrium of $(x^*_{j_2},x^*_{j_1})=(a_{j_1j_21},a_{j_2j_11})$ *is a* $(\mathrm{UP}_-,\mathrm{UP}_-)$*-negative saddle.*

(iv_{1b}) The equilibrium of $(x^*_{j_1},x^*_{j_2})=(a_{j_1j_11},a_{j_1j_21})$ *has the following properties:*

- *For* $a_{j_1j_10}>0$ *and* $a_{j_2j_20}(a_{j_1j_11}-a_{j_2j_11})>0$,

$$(a_{j_1j_11}, a_{j_1j_21}) = \underbrace{\text{(UP:UP,pF)}}_{\text{hyperbolic-secant to hyperbolic flow }(+)}. \tag{1.57}$$

The equilibrium of $(x^*_{j_1}, x^*_{j_2}) = (a_{j_1j_11}, a_{j_1j_21})$ *is a* (UP:UP, pF)*-positive hyperbolic secant-to-hyperbolic flow.*

- *For* $a_{j_1j_10} < 0$ *and* $a_{j_2j_20}(a_{j_1j_11} - a_{j_2j_11}) > 0$,

$$(a_{j_1j_11}, a_{j_1j_21}) = \underbrace{\text{(DP:DP,pF)}}_{\text{hyperbolic to hyperbolic-secant flow }(+)}. \tag{1.58}$$

The equilibrium of $(x^*_{j_1}, x^*_{j_2}) = (a_{j_1j_11}, a_{j_1j_21})$ *is a* (DP:DP,pF)*-positive hyperbolic-to-hyperbolic secant flow.*

- *For* $a_{j_1j_10} > 0$ *and* $a_{j_2j_20}(a_{j_1j_11} - a_{j_2j_11}) < 0$,

$$(a_{j_1j_11}, a_{j_1j_21}) = \underbrace{\text{(DP:DP,nF)}}_{\text{hyperbolic to hyperbolic-secant flow }(-)}. \tag{1.59}$$

The equilibrium of $(x^*_{j_1}, x^*_{j_2}) = (a_{j_1j_11}, a_{j_1j_21})$ *is a* (DP:DP, nF)*-negative hyperbolic secant-to-hyperbolic flow.*

- *For* $a_{j_1j_10} < 0$ *and* $a_{j_2j_20}(a_{j_1j_11} - a_{j_2j_11}) < 0$,

$$(a_{j_1j_11}, a_{j_1j_21}) = \underbrace{\text{(UP:UP,nF)}}_{\text{hyperbolic to hyperbolic-secant flow }(-)}. \tag{1.60}$$

The equilibrium of $(x^*_{j_1}, x^*_{j_2}) = (a_{j_1j_11}, a_{j_1j_21})$ *is a* (UP:UP,nF)*-negative hyperbolic-to-hyperbolic secant flow.*

(iv$_2$) For $a_{j_2j_11} = a_{j_1j_11}$, *the first integral manifold is*

$$\begin{aligned}\ln\frac{|x_{j_1} - a_{j_1j_11}|}{|x_{j_10} - a_{j_1j_11}|} = \frac{a_{j_1j_10}}{a_{j_2j_20}}\bigg[&\frac{1}{2}\ln\frac{|(x_{j_2} - a_{j_2j_2})^2 + b_{j_2j_2}|}{|(x_{j_20} - a_{j_2j_2})^2 + b_{j_2j_2}|}\\ &+\frac{(a_{j_2j_2} - a_{j_1j_21})}{\sqrt{b_{j_2j_2}}}\Big(\arctan\frac{x_{j_2} - a_{j_2j_2}}{\sqrt{b_{j_2j_2}}} - \arctan\frac{x_{j_20} - a_{j_2j_2}}{\sqrt{b_{j_2j_2}}}\Big)\bigg].\end{aligned} \tag{1.61}$$

(iv$_{2a}$) The infinite equilibrium of $x^*_{j_1} = a_{j_2j_11} = a_{j_1j_11}$ *with* $\overline{x}_{j_2} \neq a_{j_1j_21}$ *has the following properties:*

- *For* $a_{j_1j_10}(\overline{x}_{j_2} - a_{j_1j_21}) > 0$ *and* $a_{j_2j_20} > 0$,

$$(a_{j_1j_1}1, \bar{x}_{j_2}) = \underbrace{(\text{US,DU})}_{\text{down-up upper-saddle}}. \tag{1.62}$$

The infinite equilibrium of $x^*_{j_1} = a_{j_2j_1}1 = a_{j_1j_1}1$ *is a* (US,DU)*-down-up upper-saddle.*

- *For* $a_{j_1j_1}0(\bar{x}_{j_2} - a_{j_1j_2}1) < 0$ *and* $a_{j_2j_2}0 > 0$,

$$(a_{j_1j_1}1, \bar{x}_{j_2}) = \underbrace{(\text{LS,UD})}_{\text{up-down lower-saddle}}. \tag{1.63}$$

The infinite equilibrium of $x^*_{j_1} = a_{j_2j_1}1 = a_{j_1j_1}1$ *is an* (LS,UD)*-up-down lower-saddle.*

- *For* $a_{j_1j_1}0(\bar{x}_{j_2} - a_{j_1j_2}1) > 0$ *and* $a_{j_2j_2}0 < 0$,

$$(a_{j_1j_1}1, \bar{x}_{j_2}) = \underbrace{(\text{UD,US})}_{\text{up-down upper-saddle}}. \tag{1.64}$$

The infinite equilibrium of $x^*_{j_1} = a_{j_2j_1}1 = a_{j_1j_1}1$ *is a* (US,UD)*-up-down upper-saddle.*

- *For* $a_{j_1j_1}0(\bar{x}_{j_2} - a_{j_1j_2}1) < 0$ *and* $a_{j_2j_2}0 < 0$,

$$(a_{j_1j_1}1, \bar{x}_{j_2}) = \underbrace{(\text{LS,DU})}_{\text{down-up lower-saddle}}. \tag{1.65}$$

The infinite equilibrium of $x^*_{j_1} = a_{j_2j_1}1 = a_{j_1j_1}1$ *is an* (LS,DU)*-down-up lower-saddle.*

(iv$_{2b}$) The equilibrium of $(x^*_{j_1}, x^*_{j_2}) = (a_{j_1j_1}1, a_{j_1j_2}1)$ *has the following properties:*

- *For* $a_{j_1j_1}0 > 0$ *and* $a_{j_2j_2}0 > 0$,

$$(a_{j_1j_1}1, a_{j_1j_2}1) = \underbrace{(\text{DP}_-\text{:UP}_+, {}_{\text{UD}}\text{LS:}_{\text{DU}}\text{US})}_{\text{hyperbolic lower-to-upper saddle}}. \tag{1.66}$$

The equilibrium of $(x^*_{j_1}, x^*_{j_2}) = (a_{j_1j_1}1, a_{j_1j_2}1)$ *is a* (DP_-:UP_+,$_{\text{UD}}$LS:$_{\text{DU}}$US)*-hyperbolic-lower-to-upper saddle.*

- *For* $a_{j_1j_1}0 < 0$ *and* $a_{j_2j_2}0 > 0$,

$$(a_{j_1j_1}1, a_{j_1j_2}1) = \underbrace{(\text{UP}_-\text{:DP}_+, {}_{\text{DU}}\text{US:}_{\text{UD}}\text{LS})}_{\text{hyperbolic-secant upper-to-lower saddle}}. \tag{1.67}$$

The equilibrium of $(x^*_{j_1}, x^*_{j_2}) = (a_{j_1j_1}1, a_{j_1j_2}1)$ *is a* (UP_-:DP_+,$_{\text{DU}}$US:$_{\text{UD}}$LS)*-hyperbolic secant upper-to-lower saddle.*

- *For $a_{j_1j_10}>0$ and $a_{j_2j_20}<0$,*

$$(a_{j_1j_11},a_{j_1j_21})=\underbrace{(\mathrm{UP}_{+}\text{:}\mathrm{DP}_{-},{}_{\mathrm{DU}}\mathrm{LS}\text{:}_{\mathrm{UD}}\mathrm{US})}_{\text{hyperbolic-secant lower-to-upper saddle}}. \tag{1.68}$$

The equilibrium of $(x_{j_1}^*,x_{j_2}^*)=(a_{j_1j_11},a_{j_1j_21})$ is a (UP$_+$:DP$_-$, $_{\mathrm{DU}}$LS:$_{\mathrm{UD}}$US)*-hyperbolic-secant lower-to-upper saddle.*

- *For $a_{j_1j_10}<0$ and $a_{j_2j_20}<0$,*

$$(a_{j_1j_11},a_{j_1j_21})=\underbrace{(\mathrm{DP}_{+}\text{:}\mathrm{UP}_{-},{}_{\mathrm{UD}}\mathrm{US}\text{:}_{\mathrm{DU}}\mathrm{LS})}_{\text{hyperbolic upper-to-lower saddle}}. \tag{1.69}$$

The equilibrium of $(x_{j_2}^*,x_{j_1}^*)=(a_{j_1j_21},a_{j_2j_11})$ is a (DP$_+$:UP$_-$, $_{\mathrm{UD}}$US:$_{\mathrm{DU}}$LS)*-hyperbolic upper-to-lower saddle.*

(v) For $\Delta_{j_1j_1}=0$ and $\Delta_{j_2j_2}=0$, the standard form is

$$\begin{aligned}\dot{x}_{j_2}&=a_{j_2j_20}(x_{j_1}-a_{j_2j_11})(x_{j_2}-a_{j_2j_21})^2,\\ \dot{x}_{j_1}&=a_{j_1j_10}(x_{j_1}-a_{j_1j_11})^2(x_{j_2}-a_{j_1j_21})\end{aligned} \tag{1.70}$$

where

$$a_{j_1j_11}=-\frac{1}{2}B_{j_1j_1},\, a_{j_2j_21}=-\frac{1}{2}B_{j_2j_2}. \tag{1.71}$$

(v_1) The first integral manifold for $a_{j_2j_21}\neq a_{j_1j_21}$ is

$$\begin{aligned}&\ln\frac{|x_{j_1}-a_{j_1j_11}|}{|x_{j_10}-a_{j_1j_11}|}-(a_{j_1j_11}-a_{j_2j_11})(\frac{1}{x_{j_1}-a_{j_1j_11}}-\frac{1}{x_{j_10}-a_{j_1j_11}})\\ &=\frac{a_{j_1j_10}}{a_{j_2j_20}}\left[\ln\frac{|x_{j_2}-a_{j_2j_21}|}{|x_{j_20}-a_{j_2j_21}|}-(a_{j_2j_21}-a_{j_1j_21})(\frac{1}{x_{j_2}-a_{j_2j_21}}-\frac{1}{x_{j_20}-a_{j_2j_21}})\right].\end{aligned} \tag{1.72}$$

(v_{1a}) The equilibrium of $(x_{j_1}^*,x_{j_2}^*)=(a_{j_1j_21},a_{j_2j_11})$ with $a_{j_1j_21}\neq a_{j_2j_21}$ and $a_{j_1j_11}\neq a_{j_2j_11}$ *has the following properties:*

- *For $a_{j_1j_10}(a_{j_2j_11}-a_{j_1j_11})^2>0$ and $a_{j_2j_20}(a_{j_1j_21}-a_{j_2j_21})^2>0$,*

$$(a_{j_1j_21},a_{j_2j_11})=\underbrace{(\mathrm{UP}_{+},\mathrm{UP}_{+})}_{\text{positive saddle}}. \tag{1.73}$$

The equilibrium of $(x_{j_2}^*,x_{j_1}^*)=(a_{j_1j_21},a_{j_2j_11})$ is a (UP$_+$, UP$_+$)*-positive saddle.*

- *For $a_{j_1j_10}(a_{j_2j_11}-a_{j_1j_11})^2<0$ and $a_{j_2j_20}(a_{j_1j_21}-a_{j_2j_21})^2>0$,*

$$(a_{j_1j_2 1}, a_{j_2j_1 1}) = \underbrace{(\mathrm{DP}_+, \mathrm{DP}_-)}_{\text{CCW center}}. \tag{1.74}$$

The equilibriumv of $(x^*_{j_2}, x^*_{j_1}) = (a_{j_1j_2 1}, a_{j_2j_1 1})$ *is a* (DP₊,DP₋)-*counterclockwise center.*

- *For* $a_{j_1j_1 0}(a_{j_2j_1 1} - a_{j_1j_1 1})^2 > 0$ *and* $a_{j_2j_2 0}(a_{j_1j_2 1} - a_{j_2j_2 1})^2 < 0$,

$$(a_{j_1j_2 1}, a_{j_2j_1 1}) = \underbrace{(\mathrm{DP}_-, \mathrm{DP}_+)}_{\text{CW center}}. \tag{1.75}$$

The equilibrium of $(x^*_{j_2}, x^*_{j_1}) = (a_{j_1j_2 1}, a_{j_2j_1 1})$ *is a* (DP_-, DP_+)-*clockwise center.*

- *For* $a_{j_1j_1 0}(a_{j_2j_1 1} - a_{j_1j_1 1})^2 < 0$ *and* $a_{j_2j_2 0}(a_{j_1j_2 1} - a_{j_2j_2 1})^2 < 0$,

$$(a_{j_1j_2 1}, a_{j_2j_1 1}) = \underbrace{(\mathrm{UP}_-, \mathrm{UP}_-)}_{\text{negative saddle}}. \tag{1.76}$$

The equilibrium of $(x^*_{j_2}, x^*_{j_1}) = (a_{j_1j_2 1}, a_{j_2j_1 1})$ *is a* (UP_-,UP_-)-negative saddle.

(v_{1b}) The equilibrium of $(x^*_{j_1}, x^*_{j_2}) = (a_{j_1j_1 1}, a_{j_2j_2 1})$ *with* $a_{j_1j_2 1} \neq a_{j_2j_2 1}$ and $a_{j_1j_1 1} \neq a_{j_2j_1 1}$ *has the following properties:*

- *For* $a_{j_1j_1 0}(a_{j_2j_2 1} - a_{j_1j_2 1}) > 0$ *and* $a_{j_2j_2 0}(a_{j_1j_1 1} - a_{j_2j_1 1}) > 0$,

$$(a_{j_1j_1 1}, a_{j_2j_2 1}) = \underbrace{(\mathrm{US}, \mathrm{US})}_{\text{double-saddle}}. \tag{1.77}$$

The equilibrium of $(x^*_{j_1}, x^*_{j_2}) = (a_{j_1j_1 1}, a_{j_2j_2 1})$ *is a* (US,US)-*double-saddle.*

- *For* $a_{j_1j_1 0}(a_{j_2j_2 1} - a_{j_1j_2 1}) < 0$ *and* $a_{j_2j_2 0}(a_{j_1j_1 1} - a_{j_2j_1 1}) > 0$,

$$(a_{j_1j_1 1}, a_{j_2j_2 1}) = \underbrace{(\mathrm{LS}, \mathrm{US})}_{\text{double-saddle}}. \tag{1.78}$$

The equilibrium of $(x^*_{j_1}, x^*_{j_2}) = (a_{j_1j_1 1}, a_{j_2j_2 1})$ *is an* (LS,US)-*double-saddle.*

- *For* $a_{j_1j_1 0}(a_{j_2j_2 1} - a_{j_1j_2 1}) > 0$ *and* $a_{j_2j_2 0}(a_{j_1j_1 1} - a_{j_2j_1 1}) < 0$,

$$(a_{j_1j_1 1}, a_{j_2j_2 1}) = \underbrace{(\mathrm{US}, \mathrm{LS})}_{\text{double-saddle}}. \tag{1.79}$$

The equilibrium of $(x^*_{j_1}, x^*_{j_2}) = (a_{j_1j_1 1}, a_{j_2j_2 1})$ *is a* (US,LS)-*double-saddle.*

- *For* $a_{j_1j_1 0}(a_{j_2j_2 1} - a_{j_1j_2 1}) < 0$ *and* $a_{j_2j_2 0}(a_{j_1j_1 1} - a_{j_2j_1 1}) < 0$,

$$(a_{j_1j_1 1}, a_{j_2j_2 1}) = \underbrace{(\text{LS,LS})}_{\text{double-saddle}}. \tag{1.80}$$

The equilibrium of $(x_{j_1}^*, x_{j_2}^*) = (a_{j_1j_1 1}, a_{j_2j_2 1})$ *is an* (LS,LS)-*double-saddle.*

– *The double-saddles are the appearing bifurcations of saddle source and saddle sink, and such bifurcations are also the appearing bifurcations of the network of saddles, source, and sink.*

(v_{1c}) *The equilibrium of* $(x_{j_1}^*, x_{j_2}^*) = (a_{j_1j_1 1}, a_{j_1j_2 1})$ *with* $a_{j_1j_2 1} \neq a_{j_2j_2 1}$ and $a_{j_1j_1 1} \neq a_{j_2j_1 1}$ *has the following properties:*

- *For* $a_{j_1j_1 0} > 0$ *and* $a_{j_2j_2 0}(a_{j_1j_1 1} - a_{j_2j_1 1})(a_{j_1j_2 1} - a_{j_2j_2 1})^2 > 0$,

$$(a_{j_1j_1 1}, a_{j_1j_2 1}) = \underbrace{(\text{UP:UP,pF})}_{\text{hyperbolic-secant to hyperbolic flow } (+)}. \tag{1.81}$$

  The equilibrium of $(x_{j_1}^*, x_{j_2}^*) = (a_{j_1j_1 1}, a_{j_1j_2 1})$ *is a* (UP:UP, pF)-*positive hyperbolic-secant-to-hyperbolic flow.*

- *For* $a_{j_1j_1 0} < 0$ *and* $a_{j_2j_2 0}(a_{j_1j_1 1} - a_{j_2j_1 1})(a_{j_1j_2 1} - a_{j_2j_2 1})^2 > 0$,

$$(a_{j_1j_1 1}, a_{j_1j_2 1}) = \underbrace{(\text{DP:DP,pF})}_{\text{hyperbolic to hyperbolic-secant flow } (+)}. \tag{1.82}$$

  The equilibrium of $(x_{j_1}^*, x_{j_2}^*) = (a_{j_1j_1 1}, a_{j_1j_2 1})$ *is a* (DP:DP, pF)-*positive hyperbolic-to-hyperbolic-secant flow.*

- *For* $a_{j_1j_1 0} > 0$ *and* $a_{j_2j_2 0}(a_{j_1j_1 1} - a_{j_2j_1 1})(a_{j_1j_2 1} - a_{j_2j_2 1})^2 < 0$,

$$(a_{j_1j_1 1}, a_{j_1j_2 1}) = \underbrace{(\text{DP:DP,nF})}_{\text{hyperbolic to hyperbolic-secant flow } (-)}. \tag{1.83}$$

  The equilibrium of $(x_{j_1}^*, x_{j_2}^*) = (a_{j_1j_1 1}, a_{j_1j_2 1})$ *is a* (DP:DP, nF)-*negative hyperbolic-to-hyperbolic-secant flow.*

- *For* $a_{j_1j_1 0} < 0$ *and* $a_{j_2j_2 0}(a_{j_1j_1 1} - a_{j_2j_1 1})(a_{j_1j_2 1} - a_{j_2j_2 1})^2 < 0$,

$$(a_{j_1j_1 1}, a_{j_1j_2 1}) = \underbrace{(\text{UP:UP,nF})}_{\text{hyperbolic-secant to hyperbolic flow } (-)}. \tag{1.84}$$

  The equilibrium of $(x_{j_1}^*, x_{j_2}^*) = (a_{j_1j_1 1}, a_{j_1j_2 1})$ *is a* (UP:UP, nF)-*negative hyperbolic-secant-to-hyperbolic flow.*

(v_{1d}) *The equilibrium of* $(x_{j_1}^*, x_{j_2}^*) = (a_{j_2j_1 1}, a_{j_2j_2 1})$ *with* $a_{j_1j_2 1} \neq a_{j_2j_2 1}$ and $a_{j_1j_1 1} \neq a_{j_2j_1 1}$ *has the following properties:*

- *For* $a_{j_1j_10}(a_{j_2j_11}-a_{j_1j_11})(a_{j_2j_21}-a_{j_1j_21})^2>0$ *and* $a_{j_2j_20}>0$,

$$(a_{j_2j_11},a_{j_2j_21})=\underbrace{\text{(pF,UP:UP)}}_{\text{hyperbolic-secant to hyperbolic flow }(+)}. \tag{1.85}$$

The equilibrium of $(x_{j_1}^*,x_{j_2}^*)=(a_{j_2j_11},a_{j_2j_21})$ *is a* (pF,UP:UP)*-positive hyperbolic-secant-to-hyperbolic flow.*

- *For* $a_{j_1j_10}(a_{j_2j_11}-a_{j_1j_11})(a_{j_2j_21}-a_{j_1j_21})^2<0$ *and* $a_{j_2j_20}>0$,

$$(a_{j_2j_11},a_{j_2j_21})=\underbrace{\text{(nF,DP:DP)}}_{\text{hyperbolic to hyperbolic-secant flow }(-)}. \tag{1.86}$$

The equilibrium of $(x_{j_1}^*,x_{j_2}^*)=(a_{j_2j_11},a_{j_2j_21})$ *is an* (nF, DP:DP)*-negative hyperbolic-to-hyperbolic-secant flow.*

- *For* $a_{j_1j_10}(a_{j_2j_11}-a_{j_1j_11})(a_{j_2j_21}-a_{j_1j_21})^2>0$ *and* $a_{j_2j_20}<0$,

$$(a_{j_2j_11},a_{j_2j_21})=\underbrace{\text{(pF,DP:DP)}}_{\text{hyperbolic to hyperbolic-secant flow }(+)}. \tag{1.87}$$

The equilibrium of $(x_{j_1}^*,x_{j_2}^*)=(a_{j_2j_11},a_{j_2j_21})$ *is a* (pF, DP:DP)*-negative hyperbolic-to-hyperbolic-secant flow.*

- *For* $a_{j_1j_10}(a_{j_2j_11}-a_{j_1j_11})(a_{j_2j_21}-a_{j_1j_21})^2<0$ *and* $a_{j_2j_20}<0$,

$$(a_{j_2j_11},a_{j_2j_21})=\underbrace{\text{(nF,UP:UP)}}_{\text{hyperbolic-secant to hyperbolic flow }(-)}. \tag{1.88}$$

The equilibrium of $(x_{j_1}^*,x_{j_2}^*)=(a_{j_2j_11},a_{j_2j_21})$ *is an* (nF,UP:UP)*-negative hyperbolic-secant-to-hyperbolic flow.*

(v_2) The first integral manifold for $a_{j_2j_21}=a_{j_1j_21}$ and $a_{j_1j_11}\neq a_{j_2j_11}$ *is*

$$\begin{aligned}&\ln\frac{|x_{j_1}-a_{j_1j_11}|}{|x_{j_10}-a_{j_1j_11}|}-(a_{j_1j_11}-a_{j_2j_11})(\frac{1}{x_{j_1}-a_{j_1j_11}}-\frac{1}{x_{j_10}-a_{j_1j_11}})\\&=\frac{a_{j_1j_10}}{a_{j_2j_20}}\ln\frac{|x_{j_2}-a_{j_2j_21}|}{|x_{j_20}-a_{j_2j_21}|}.\end{aligned} \tag{1.89}$$

(v_{2a}) The infinite equilibrium of $x_{j_2}^*=a_{j_2j_21}=a_{j_1j_21}$ *with* $\overline{x}_1\neq a_{j_1j_11},a_{j_2j_11}$ and $a_{j_1j_11}\neq a_{j_2j_11}$ *has the following properties:*

- *For* $a_{j_1j_10}(\overline{x}_{j_1}-a_{j_1j_11})^2>0$ *and* $a_{j_2j_20}(\overline{x}_{j_1}-a_{j_2j_11})>0$,

$$(\bar{x}_{j_1}, a_{j_2j_2 1}) = \underbrace{\text{(DU,US)}}_{\text{down-up upper-saddle}} . \tag{1.90}$$

The infinite equilibrium of $x^*_{j_2} = a_{j_2j_2 1} = a_{j_1j_2 1}$ *is a* (DU,US)*-down-up upper saddle.*

- *For* $a_{j_1j_1 0}(\bar{x}_{j_1} - a_{j_1j_1 1})^2 < 0$ *and* $a_{j_2j_2 0}(\bar{x}_{j_1} - a_{j_2j_1 1}) > 0$,

$$(\bar{x}_{j_1}, a_{j_2j_2 1}) = \underbrace{\text{(UD,US)}}_{\text{up-down upper-saddle}} . \tag{1.91}$$

The infinite equilibrium of $x^*_{j_2} = a_{j_2j_2 1} = a_{j_1j_2 1}$ *is a* (UD,US)*-up-down upper saddle.*

- *For* $a_{j_1j_1 0}(\bar{x}_{j_1} - a_{j_1j_1 1})^2 > 0$ *and* $a_{j_2j_2 0}(\bar{x}_{j_1} - a_{j_2j_1 1}) < 0$,

$$(\bar{x}_{j_1}, a_{j_2j_2 1}) = \underbrace{\text{(UD,LS)}}_{\text{up-down lower-saddle}} . \tag{1.92}$$

The infinite equilibrium of $x^*_{j_2} = a_{j_2j_2 1} = a_{j_1j_2 1}$ *is a* (UD,LS)*-up-down lower saddle.*

- *For* $a_{j_1j_1 0}(\bar{x}_{j_1} - a_{j_1j_1 1})^2 < 0$ *and* $a_{j_2j_2 0}(\bar{x}_{j_1} - a_{j_2j_1 1}) < 0$,

$$(\bar{x}_{j_1}, a_{j_2j_2 1}) = \underbrace{\text{(DU,LS)}}_{\text{down-up lower-saddle}} . \tag{1.93}$$

The infinite equilibrium of $x^*_{j_2} = a_{j_2j_2 1} = a_{j_1j_2 1}$ *is a* (DU,LS)*-down-up lower saddle.*

(v_{2b}) The equilibrium of $(x^*_{j_1}, x^*_{j_2}) = (a_{j_1j_1 1}, a_{j_2j_2 1})$ *with* $a_{j_2j_2 1} = a_{j_1j_2 1}$ and $a_{j_1j_1 1} \neq a_{j_2j_1 1}$ *has the following properties:*

- *For* $a_{j_1j_1 0} > 0$ *and* $a_{j_2j_2 0}(a_{j_1j_1 1} - a_{j_2j_1 1}) > 0$,

$$(a_{j_1j_1 1}, a_{j_2j_2 1}) = \underbrace{\text{(DU:DU,US)}}_{\text{down-up upper-saddle}} . \tag{1.94}$$

The equilibrium of $(x^*_{j_1}, x^*_{j_2}) = (a_{j_1j_1 1}, a_{j_2j_2 1})$ *is a* (DU:DU,US)*-double down-up upper saddle.*

- *For* $a_{j_1j_1 0} < 0$ *and* $a_{j_2j_2 0}(a_{j_1j_1 1} - a_{j_2j_1 1}) > 0$,

$$(a_{j_1j_1 1}, a_{j_2j_2 1}) = \underbrace{\text{(UD:UD,US)}}_{\text{up-down upper-saddle}} . \tag{1.95}$$

The equilibrium of $(x^*_{j_1}, x^*_{j_2}) = (a_{j_1j_1}1, a_{j_2j_2}1)$ *is a* (UD:UD,US)-*double up-down upper saddle.*

- *For* $a_{j_1j_10} > 0$ *and* $a_{j_2j_20}(a_{j_1j_1}1 - a_{j_2j_1}1) < 0$,

$$(a_{j_1j_1}1, a_{j_2j_2}1) = \underbrace{(\text{UD:UD,LS})}_{\text{up-down lower-saddle}}. \tag{1.96}$$

The equilibrium of $(x^*_{j_1}, x^*_{j_2}) = (a_{j_1j_1}1, a_{j_2j_2}1)$ *is a* (UD:UD,LS)-*double up-down lower saddle.*

- *For* $a_{j_1j_10} < 0$ *and* $a_{j_2j_20}(a_{j_1j_1}1 - a_{j_2j_1}1) < 0$,

$$(a_{j_1j_1}1, a_{j_2j_2}1) = \underbrace{(\text{DU:DU,LS})}_{\text{down-up upper-saddle}}. \tag{1.97}$$

The equilibrium of $(x^*_{j_1}, x^*_{j_2}) = (a_{j_1j_1}1, a_{j_2j_2}1)$ *is a* (DU:DU,LS)-*double down-up lower saddle.*

(v_{2c}) *The equilibrium of* $(x^*_{j_1}, x^*_{j_2}) = (a_{j_2j_1}1, a_{j_2j_2}1)$ *with* $a_{j_2j_2}1 = a_{j_1j_2}1$ and $a_{j_1j_1}1 \neq a_{j_2j_1}1$ *has the following properties:*

- *For* $a_{j_1j_10}(a_{j_2j_1}1 - a_{j_1j_1}1)^2 > 0$ *and* $a_{j_2j_20} > 0$,

$$(a_{j_2j_1}1, a_{j_2j_2}1) = \underbrace{({}_{\text{DU}}\text{LS}:{}_{\text{UD}}\text{US},\text{DP}_-:\text{UP}_+)}_{\text{hyperbolic lower-to-upper saddle}}. \tag{1.98}$$

The equilibrium of $(x^*_{j_1}, x^*_{j_2}) = (a_{j_2j_1}1, a_{j_2j_2}1)$ *is a* (${}_{\text{DU}}$LS:${}_{\text{UD}}$US, DP$_-$:UP$_+$)-*hyperbolic lower-to-upper saddle.*

- *For* $a_{j_1j_10}(a_{j_2j_1}1 - a_{j_1j_1}1)^2 < 0$ *and* $a_{j_2j_20} > 0$,

$$(a_{j_2j_1}1, a_{j_2j_2}1) = \underbrace{({}_{\text{UD}}\text{US}:{}_{\text{DU}}\text{LS},\text{DP}_+:\text{UP}_-)}_{\text{hyperbolic-secant upper-to-lower saddle}}. \tag{1.99}$$

The equilibrium of $(x^*_{j_1}, x^*_{j_2}) = (a_{j_2j_1}1, a_{j_2j_2}1)$ *is a* (${}_{\text{UD}}$US:${}_{\text{DU}}$LS, UP$_+$:DP$_-$)-*hyperbolic-secant upper-to-lower saddle.*

- *For* $a_{j_1j_10}(a_{j_2j_1}1 - a_{j_1j_1}1)^2 > 0$ *and* $a_{j_2j_20} < 0$,

$$(a_{j_2j_1}1, a_{j_2j_2}1) = \underbrace{({}_{\text{UD}}\text{LS}:{}_{\text{DU}}\text{US},\text{UP}_-:\text{DP}_+)}_{\text{hyperbolic-secant lower-to-upper saddle}}. \tag{1.100}$$

The equilibrium of $(x^*_{j_1}, x^*_{j_2}) = (a_{j_2j_1}1, a_{j_2j_2}1)$ *is a* (${}_{\text{UD}}$LS:${}_{\text{DU}}$US, UP$_-$:DP$_+$)-*hyperbolic-secant lower-to-upper saddle.*

- *For* $a_{j_1j_10}(a_{j_2j_1}1 - a_{j_1j_1}1)^2 < 0$ *and* $a_{j_2j_20} < 0$,

$$(a_{j_2j_11}, a_{j_2j_21}) = \underbrace{({}_{\mathrm{DU}}\mathrm{US}{:}_{\mathrm{UD}}\mathrm{LS},\mathrm{DP}_+{:}\mathrm{UP}_-)}_{\text{hyperbolic upper-to-lower saddle}}. \tag{1.101}$$

The equilibrium of $(x_{j_1}^*, x_{j_2}^*) = (a_{j_2j_11}, a_{j_2j_21})$ *is a* $({}_{\mathrm{DU}}\mathrm{LS}{:}_{\mathrm{UD}}\mathrm{US},\mathrm{DP}_+{:}\mathrm{UP}_-)$*-hyperbolic upper-to-lower saddle.*

(v_3) *The first integral manifold for* $a_{j_1j_11} = a_{j_2j_11}$ and $a_{j_2j_21} \neq a_{j_1j_21}$ *is*

$$\begin{aligned}\ln\frac{|x_{j_1} - a_{j_1j_11}|}{|x_{j_10} - a_{j_1j_11}|} = \frac{a_{j_1j_10}}{a_{j_2j_20}}\Big[&\ln\frac{|x_{j_2} - a_{j_2j_21}|}{|x_{j_20} - a_{j_2j_21}|}\\ &- (a_{j_2j_21} - a_{j_1j_21})(\frac{1}{x_{j_2} - a_{j_2j_21}} - \frac{1}{x_{j_20} - a_{j_2j_21}})\Big].\end{aligned} \tag{1.102}$$

(v_{3a}) *The infinite equilibrium of* $x_{j_1}^* = a_{j_1j_11} = a_{j_2j_11}$ *and* $\bar{x}_{j_1} \in (-\infty, \infty)$ *with* $a_{j_2j_21} \neq a_{j_1j_21}$ *has the following properties:*

- *For* $a_{j_1j_10}(\bar{x}_{j_2} - a_{j_1j_21}) > 0$ *and* $a_{j_2j_20}(\bar{x}_{j_2} - a_{j_2j_21})^2 > 0$,

$$(a_{j_1j_11}, \bar{x}_{j_2}) = \underbrace{(\mathrm{US},\mathrm{DU})}_{\text{down-up upper-saddle}}. \tag{1.103}$$

  The infinite equilibrium of $x_{j_1}^* = a_{j_1j_11} = a_{j_2j_11}$ *is a* (US,DU)*-down-up upper saddle.*

- *For* $a_{j_1j_10}(\bar{x}_{j_2} - a_{j_1j_21}) < 0$ *and* $a_{j_2j_20}(\bar{x}_{j_2} - a_{j_2j_21})^2 > 0$,

$$(a_{j_1j_11}, \bar{x}_{j_2}) = \underbrace{(\mathrm{LS},\mathrm{UD})}_{\text{up-down lower-saddle}}. \tag{1.104}$$

  The infinite equilibrium of $x_{j_1}^* = a_{j_1j_11} = a_{j_2j_11}$ *is an* (LS,UD)*-up-down lower saddle.*

- *For* $a_{j_1j_10}(\bar{x}_{j_2} - a_{j_1j_21}) > 0$ *and* $a_{j_2j_20}(\bar{x}_{j_2} - a_{j_2j_21})^2 < 0$,

$$(a_{j_1j_11}, \bar{x}_{j_2}) = \underbrace{(\mathrm{US},\mathrm{UD})}_{\text{up-down upper-saddle}}. \tag{1.105}$$

  The infinite equilibrium of $x_{j_1}^* = a_{j_1j_11} = a_{j_2j_11}$ *is a* (US.UD)*-up-down upper saddle.*

- *For* $a_{j_1j_10}(\bar{x}_{j_2} - a_{j_1j_21}) < 0$ *and* $a_{j_2j_20}(\bar{x}_{j_2} - a_{j_2j_21})^2 < 0$,

$$(a_{j_1j_11}, \bar{x}_{j_2}) = \underbrace{(\mathrm{LS},\mathrm{DU})}_{\text{down-up lower-saddle}}. \tag{1.106}$$

The infinite equilibrium of $x^*_{j_1} = a_{j_1 j_1 1} = a_{j_2 j_1 1}$ *is an* (LS,DU)-*down-up lower saddle.*

(v_{3b}) The equilibrium of $(x^*_{j_1}, x^*_{j_2}) = (a_{j_1 j_1 1}, a_{j_2 j_2 1})$ *with* $a_{j_1 j_1 1} = a_{j_2 j_1 1}$ and $a_{j_2 j_2 1} \neq a_{j_1 j_2 1}$ *has the following properties:*

- *For* $a_{j_1 j_1 0}(a_{j_2 j_2 1} - a_{j_1 j_2 1}) > 0$ *and* $a_{j_2 j_2 0} > 0$,

$$(a_{j_1 j_1 1}, a_{j_2 j_2 1}) = \underbrace{(\text{US,DU:DU})}_{\text{down-up upper-saddle}} . \tag{1.107}$$

The equilibrium of $(x^*_{j_1}, x^*_{j_2}) = (a_{j_1 j_1 1}, a_{j_2 j_2 1})$ *is a* (US,DU:DU)-*double down-up upper saddle.*

- *For* $a_{j_1 j_1 0}(a_{j_2 j_2 1} - a_{j_1 j_2 1}) < 0$ *and* $a_{j_2 j_2 0} > 0$,

$$(a_{j_1 j_1 1}, a_{j_2 j_2 1}) = \underbrace{(\text{LS,UD:UD})}_{\text{up-down lower-saddle}} . \tag{1.108}$$

The equilibrium of $(x^*_{j_1}, x^*_{j_2}) = (a_{j_1 j_1 1}, a_{j_2 j_2 1})$ *is an* (LS,UD:UD)-*double up-down lower saddle.*

- *For* $a_{j_1 j_1 0}(a_{j_2 j_2 1} - a_{j_1 j_2 1}) > 0$ *and* $a_{j_2 j_2 0} < 0$,

$$(a_{j_1 j_1 1}, a_{j_2 j_2 1}) = \underbrace{(\text{US,UD:UD})}_{\text{up-down upper-saddle}} . \tag{1.109}$$

The equilibrium of $(x^*_{j_1}, x^*_{j_2}) = (a_{j_1 j_1 1}, a_{j_2 j_2 1})$ *is a* (US,UD:UD)-*double up-down upper saddle.*

- *For* $a_{j_1 j_1 0}(a_{j_2 j_2 1} - a_{j_1 j_2 1}) < 0$ *and* $a_{j_2 j_2 0} < 0$,

$$(a_{j_1 j_1 1}, a_{j_2 j_2 1}) = \underbrace{(\text{LS,DU:DU})}_{\text{down-up lower-saddle}} . \tag{1.110}$$

The equilibrium of $(x^*_{j_1}, x^*_{j_2}) = (a_{j_1 j_1 1}, a_{j_2 j_2 1})$ *is an* (LS,DU:DU)-*double down-up lower saddle.*

(v_{3c}) The equilibrium of $(x^*_{j_1}, x^*_{j_2}) = (a_{j_1 j_1 1}, a_{j_1 j_2 1})$ *has the following properties:*

- *For* $a_{j_1 j_1 0} > 0$ *and* $a_{j_2 j_2 0}(a_{j_1 j_2 1} - a_{j_2 j_2 1})^2 > 0$,

$$(a_{j_1 j_1 1}, a_{j_1 j_2 1}) = \underbrace{(\text{DP}_-\text{:UP}_+, {}_{\text{DU}}\text{LS:}_{\text{UD}}\text{US})}_{\text{hyperbolic lower-to-upper saddle}} . \tag{1.111}$$

The equilibrium of $(x^*_{j_1}, x^*_{j_2}) = (a_{j_1 j_1 1}, a_{j_1 j_2 1})$ *is a* (DP_-:UP_+, ${}_{\text{DU}}$LS:${}_{\text{UD}}$US)-hyperbolic lower-to-upper saddle.

- *For $a_{j_1j_10}<0$ and $a_{j_2j_20}(a_{j_1j_21}-a_{j_2j_21})^2>0$,*

$$(a_{j_1j_11},a_{j_1j_21})=\underbrace{(\mathrm{UP}_{+}:\mathrm{DP}_{-},{}_{\mathrm{UD}}\mathrm{US}:{}_{\mathrm{DU}}\mathrm{LS})}_{\text{hyperbolic-secant upper-to-lower saddle}}. \tag{1.112}$$

The equilibrium of $(x^*_{j_1},x^*_{j_2})=(a_{j_1j_11},a_{j_1j_21})$ is a (UP$_+$:DP$_-$, $_{\mathrm{UD}}$US:$_{\mathrm{DU}}$LS)-*hyper-secant upper-to-lower saddle.*

- *For $a_{j_1j_10}>0$ and $a_{j_2j_20}(a_{j_1j_21}-a_{j_2j_21})^2<0$,*

$$(a_{j_1j_11},a_{j_1j_21})=\underbrace{(\mathrm{UP}_{+}:\mathrm{DP}_{-},{}_{\mathrm{UD}}\mathrm{LS}:{}_{\mathrm{DU}}\mathrm{US})}_{\text{hyperbolic-secant lower-to-upper saddle}}. \tag{1.113}$$

The equilibrium of $(x^*_{j_1},x^*_{j_2})=(a_{j_1j_11},a_{j_1j_21})$ is an (UP$_+$:DP$_-$, $_{\mathrm{UD}}$LS:$_{\mathrm{DU}}$US)-*hyper-secant lower-to-upper saddle.*

- *For $a_{j_1j_10}<0$ and $a_{j_2j_20}(a_{j_1j_21}-a_{j_2j_21})^2<0$,*

$$(a_{j_1j_11},a_{j_1j_21})=\underbrace{(\mathrm{DP}_{+}:\mathrm{UP}_{-},{}_{\mathrm{DU}}\mathrm{US}:{}_{\mathrm{UD}}\mathrm{LS})}_{\text{hyperbolic upper-to-lower saddle}}. \tag{1.114}$$

The equilibrium of $(x^*_{j_1},x^*_{j_2})=(a_{j_1j_11},a_{j_1j_21})$ is a (DP$_+$:UP$_-$, $_{\mathrm{DU}}$US:$_{\mathrm{UD}}$LS)-*hyperbolic upper-to-lower saddle.*

(v_4) The first integral manifold for $a_{j_1j_11}=a_{j_2j_11}$ and *$a_{j_2j_21}=a_{j_1j_21}$ is*

$$\ln\frac{|x_{j_1}-a_{j_1j_11}|}{|x_{j_10}-a_{j_1j_11}|}=\frac{a_{j_1j_10}}{a_{j_2j_20}}\ln\frac{|x_{j_2}-a_{j_2j_21}|}{|x_{j_20}-a_{j_2j_21}|}. \tag{1.115}$$

The equilibrium of $(x^*_{j_2},x^*_{j_1})=(a_{j_1j_21},a_{j_2j_11})$ with $a_{j_1j_11}=a_{j_2j_11}$ and *$a_{j_2j_21}=a_{j_1j_21}$ has the following properties:*

- *For $a_{j_1j_10}>0$ and $a_{j_2j_20}>0$,*

$$(a_{j_1j_21},a_{j_2j_11})=\underbrace{(\mathrm{DU}_{+},\mathrm{DU}_{+})}_{\text{positive down-up saddle}}. \tag{1.116}$$

The equilibrium of $(x^*_{j_2},x^*_{j_1})=(a_{j_1j_21},a_{j_2j_11})$ is an (DU$_+$,DU$_+$)-*positive down-up saddle.*

- *For $a_{j_1j_10}<0$ and $a_{j_2j_20}>0$,*

$$(a_{j_1j_21},a_{j_2j_11})=\underbrace{(\mathrm{UD}_{+},\mathrm{UD}_{-})}_{\text{CCW up-down saddle}}. \tag{1.117}$$

The equilibrium of $(x_{j_2}^*, x_{j_1}^*) = (a_{j_1j_2 1}, a_{j_2j_1 1})$ *is a* (UD$_+$,UD$_-$)*-counterclockwise up-down saddle.*

- *For* $a_{j_1j_1 0} < 0$ *and* $a_{j_2j_2 0} < 0$,

$$(a_{j_1j_2 1}, a_{j_2j_1 1}) = \underbrace{(\mathrm{UD}_-, \mathrm{UD}_+)}_{\text{CW up-down saddle}}, \tag{1.118}$$

The equilibrium of $(x_{j_2}^*, x_{j_1}^*) = (a_{j_1j_2 1}, a_{j_2j_1 1})$ is a (UD$_-$, UD$_+$)-clockwise up-down saddle.

- *For* $a_{j_1j_1 0} < 0$ *and* $a_{j_2j_2 0} < 0$,

$$(a_{j_1j_2 1}, a_{j_2j_1 1}) = \underbrace{(\mathrm{DU}_-, \mathrm{DU}_-)}_{\text{negative down-up saddle}}. \tag{1.119}$$

The equilibrium of $(x_{j_2}^*, x_{j_1}^*) = (a_{j_1j_2 1}, a_{j_2j_1 1})$ *is* a (DU$_-$, DU$_-$)-negative down-up saddle.

(vi) For $\Delta_{j_1j_1} = 0$ *and* $\Delta_{j_2j_2} > 0$, *the standard form is*

$$\begin{aligned} \dot{x}_{j_2} &= a_{j_2j_2 0}(x_{j_1} - a_{j_2j_1 1})(x_{j_2} - a_{j_2j_2 1})(x_{j_2} - a_{j_2j_2 2}), \\ \dot{x}_{j_1} &= a_{j_1j_1 0}(x_{j_1} - a_{j_1j_1 1})^2(x_{j_2} - a_{j_1j_2 1}) \end{aligned} \tag{1.120}$$

where

$$\begin{aligned} a_{j_1j_1 1} &= -\frac{1}{2}B_{j_1j_1}, \\ b_{j_2j_2 1}, b_{j_2j_2 2} &= -\frac{1}{2}(B_{j_2j_2} \pm \sqrt{\Delta_{j_2j_2}}), \\ \{a_{j_2j_2 1}, a_{j_2j_2 2}\} &= \mathrm{sort}\{b_{j_2j_2 1}, b_{j_2j_2 2}\}, a_{j_2j_2 1} < a_{j_2j_2 2}. \end{aligned} \tag{1.121}$$

(vi$_1$) The first integral manifold $(a_{j_1j_2 1} \neq a_{j_2j_2 s_1}, a_{j_1j_1 1} \neq a_{j_2j_1 1})$ *is*

$$\begin{aligned} &\ln\frac{|x_{j_1} - a_{j_1j_1 1}|}{|x_{j_1 0} - a_{j_1j_1 1}|} - (a_{j_1j_1 1} - a_{j_2j_1 1})\left(\frac{1}{x_{j_1} - a_{j_1j_1 1}} - \frac{1}{x_{j_1 0} - a_{j_1j_1 1}}\right) \\ &= \frac{a_{j_1j_1 0}}{a_{j_2j_2 0}}\sum\nolimits_{s_1=1}^{2}\frac{(a_{j_2j_2 s_1} - a_{j_1j_2 1})}{(a_{j_2j_2 s_1} - a_{j_2j_2 s_2})}\ln\frac{|x_{j_2} - a_{j_2j_2 s_1}|}{|x_{j_2 0} - a_{j_2j_2 s_1}|}. \end{aligned} \tag{1.122}$$

(vi$_{1a}$) The equilibrium of $(x_{j_2}^*, x_{j_1}^*) = (a_{j_1j_2 1}, a_{j_2j_1 1})$ *with* $a_{j_1j_2 1} \neq a_{j_2j_2 s_1}$ and $a_{j_1j_1 1} \neq a_{j_2j_1 1}$ *has the following properties:*

- *For* $a_{j_1j_1 0}(a_{j_2j_1 1} - a_{j_1j_1 1})^2 > 0$ *and* $a_{j_2j_2 0}\prod_{s_1=1}^{2}(a_{j_1j_2 1} - a_{j_2j_2 s_1}) > 0$,

$$(a_{j_1j_2 1}, a_{j_2j_1 1}) = \underbrace{(\mathrm{UP}_+,\mathrm{UP}_+)}_{\text{positive saddle}}. \tag{1.123}$$

The equilibrium of $(x^*_{j_2}, x^*_{j_1}) = (a_{j_1j_2 1}, a_{j_2j_1 1})$ *is a* (UP$_+$, UP$_+$)*-positive saddle.*

- *For* $a_{j_1j_1 0}(a_{j_2j_1 1} - a_{j_1j_1 1})^2 < 0$ *and* $a_{j_2j_2 0}\prod_{s_1=1}^{2}(a_{j_1j_2 1} - a_{j_2j_2 s_1}) > 0$,

$$(a_{j_1j_2 1}, a_{j_2j_1 1}) = \underbrace{(\mathrm{DP}_+,\mathrm{DP}_-)}_{\text{CCW center}}. \tag{1.124}$$

The equilibrium of $(x^*_{j_2}, x^*_{j_1}) = (a_{j_1j_2 1}, a_{j_2j_1 1})$ *is a* (DP$_+$,DP$_-$)*-counterclockwise center.*

- *For* $a_{j_1j_1 0}(a_{j_2j_1 1} - a_{j_1j_1 1})^2 > 0$ *and* $a_{j_2j_2 0}\prod_{s_1=1}^{2}(a_{j_1j_2 1} - a_{j_2j_2 s_1}) < 0$,

$$(a_{j_1j_2 1}, a_{j_2j_1 1}) = \underbrace{(\mathrm{DP}_-,\mathrm{DP}_+)}_{\text{CW center}}. \tag{1.125}$$

The equilibrium of $(x^*_{j_2}, x^*_{j_1}) = (a_{j_1j_2 1}, a_{j_2j_1 1})$ *is a* (DP$_-$, DP$_+$)*-clockwise center.*

- *For* $a_{j_1j_1 0}(a_{j_2j_1 1} - a_{j_1j_1 1})^2 < 0$ *and* $a_{j_2j_2 0}\prod_{s_1=1}^{2}(a_{j_1j_2 1} - a_{j_2j_2 s_1}) < 0$,

$$(a_{j_1j_2 1}, a_{j_2j_1 1}) = \underbrace{(\mathrm{UP}_-,\mathrm{UP}_-)}_{\text{negative saddle}}. \tag{1.126}$$

The equilibrium of $(x^*_{j_2}, x^*_{j_1}) = (a_{j_1j_2 1}, a_{j_2j_1 1})$ *is a* (UP$_-$,UP$_-$)*-negative saddle.*

(vi$_{1b}$) The equilibrium of $(x^*_{j_1}, x^*_{j_2}) = (a_{j_1j_1 1}, a_{j_2j_2 s_1})$ $(s_1, s_2 \in \{1,2\};\ s_1 \neq s_2)$ *has the following properties:*

- *For* $a_{j_1j_1 0}(a_{j_2j_2 s_1} - a_{j_1j_2 1}) > 0$ *and* $a_{j_2j_2 0}(a_{j_1j_1 1} - a_{j_2j_1 1})(a_{j_2j_2 s_1} - a_{j_2j_2 s_2}) > 0$,

$$(a_{j_1j_1 1}, a_{j_2j_2 s_1}) = \underbrace{(\mathrm{US},\mathrm{SO})}_{\text{upper-saddle source}}. \tag{1.127}$$

The equilibrium of $(x^*_{j_1}, x^*_{j_2}) = (a_{j_1j_1 1}, a_{j_2j_2 s_1})$ *is a* (US,SO)*-upper saddle source.*

- *For* $a_{j_1j_1 0}(a_{j_2j_2 s_1} - a_{j_1j_2 1}) < 0$ *and* $a_{j_2j_2 0}(a_{j_1j_1 1} - a_{j_2j_1 1})(a_{j_2j_2 s_1} - a_{j_2j_2 s_2}) > 0$,

$$(a_{j_1j_1 1}, a_{j_2j_2 s_1}) = \underbrace{(\mathrm{LS},\mathrm{SO})}_{\text{lower-saddle source}}. \tag{1.128}$$

The equilibrium of $(x^*_{j_1}, x^*_{j_2}) = (a_{j_1j_1 1}, a_{j_2j_2 s_1})$ *is an* (LS,SO)*-lower saddle source.*

- *For* $a_{j_1j_1 0}(a_{j_2j_2 s_1} - a_{j_1j_2 1}) > 0$ *and* $a_{j_2j_2 0}(a_{j_1j_1 1} - a_{j_2j_1 1})(a_{j_2j_2 s_1} - a_{j_2j_2 s_2}) < 0$,

$$(a_{j_1 j_1 1}, a_{j_2 j_2 s_1}) = \underbrace{\text{(US,SI)}}_{\text{upper-saddle sink}}. \tag{1.129}$$

The equilibrium of $(x_{j_1}^*, x_{j_2}^*) = (a_{j_1 j_1 1}, a_{j_2 j_2 s_1})$ *is a* (US,SI)-*upper saddle sink.*

- *For* $a_{j_1 j_1 0}(a_{j_2 j_2 s_1} - a_{j_1 j_2 1}) < 0$ *and* $a_{j_2 j_2 0}(a_{j_1 j_1 1} - a_{j_2 j_1 1})(a_{j_2 j_2 s_1} - a_{j_2 j_2 s_2}) < 0$,

$$(a_{j_1 j_1 1}, a_{j_2 j_2 s_1}) = \underbrace{\text{(LS,SI)}}_{\text{lower-saddle sink}}. \tag{1.130}$$

The equilibrium of $(x_{j_1}^*, x_{j_2}^*) = (a_{j_1 j_1 1}, a_{j_2 j_2 s_1})$ *is an* (LS,SI)-*lower saddle sink.*

(vi$_{1c}$) The equilibrium of $(x_{j_1}^*, x_{j_2}^*) = (a_{j_1 j_1 1}, a_{j_1 j_2 1})$ *with* $a_{j_1 j_2 1} \neq a_{j_2 j_2 1}, a_{j_2 j_2 2}$ *has the following properties:*

- *For* $a_{j_1 j_1 0} > 0$ *and* $a_{j_2 j_2 0}(a_{j_1 j_1 1} - a_{j_2 j_1 1})\prod_{s_1=1}^{2}(a_{j_1 j_2 1} - a_{j_2 j_2 s_1}) > 0$,

$$(a_{j_1 j_1 1}, a_{j_1 j_2 1}) = \underbrace{\text{(UP:UP,pF)}}_{\text{hyperbolic-secant to hyperbolic flow }(+)}. \tag{1.131}$$

The equilibrium of $(x_{j_1}^*, x_{j_2}^*) = (a_{j_1 j_1 1}, a_{j_1 j_2 1})$ *is a* (UP:UP, pF)-*positive hyperbolic secant-to-hyperbolic flow.*

- *For* $a_{j_1 j_1 0} < 0$ *and* $a_{j_2 j_2 0}(a_{j_1 j_1 1} - a_{j_2 j_1 1})\prod_{s_1=1}^{2}(a_{j_1 j_2 1} - a_{j_2 j_2 s_1}) > 0$,

$$(a_{j_1 j_1 1}, a_{j_1 j_2 1}) = \underbrace{\text{(DP:DP,pF)}}_{\text{hyperbolic to hyperbolic-secant flow }(+)}. \tag{1.132}$$

The equilibrium of $(x_{j_1}^*, x_{j_2}^*) = (a_{j_1 j_1 1}, a_{j_1 j_2 1})$ *is a* (DP:DP, pF)-*positive hyperbolic-to-hyperbolic secant flow.*

- *For* $a_{j_1 j_1 0} > 0$ *and* $a_{j_2 j_2 0}(a_{j_1 j_1 1} - a_{j_2 j_1 1})\prod_{s_1=1}^{2}(a_{j_1 j_2 1} - a_{j_2 j_2 s_1}) < 0$,

$$(a_{j_1 j_1 1}, a_{j_1 j_2 1}) = \underbrace{\text{(DP:DP,nF)}}_{\text{hyperbolic to hyperbolic-secant flow }(-)}. \tag{1.133}$$

The equilibrium of $(x_{j_1}^*, x_{j_2}^*) = (a_{j_1 j_1 1}, a_{j_1 j_2 1})$ *is a* (DP:DP, nF)-*negative hyperbolic-to-hyperbolic secant flow.*

- *For* $a_{j_1 j_1 0} < 0$ *and* $a_{j_2 j_2 0}(a_{j_1 j_1 1} - a_{j_2 j_1 1})\prod_{s_1=1}^{2}(a_{j_1 j_2 1} - a_{j_2 j_2 s_1}) < 0$,

$$(a_{j_1 j_1 1}, a_{j_1 j_2 1}) = \underbrace{\text{(UP:UP,nF)}}_{\text{hyperbolic-secant to hyperbolic flow }(-)}. \tag{1.134}$$

The equilibrium of $(x_{j_1}^*, x_{j_2}^*) = (a_{j_1 j_1 1}, a_{j_1 j_2 1})$ *is a* (UP:UP, nF)-*negative hyperbolic secant-to-hyperbolic flow.*

(vi_{1d}) The equilibrium of $(x^*_{j_1}, x^*_{j_2}) = (a_{j_2 j_1 1}, a_{j_2 j_2 s_1})$ $(s_1,\ s_2 \in \{1,2\},\ s_1 \neq s_2)$ *with* $a_{j_2 j_2 s_1} \neq a_{j_1 j_2 1}$ *has the following properties:*

- *For* $a_{j_1 j_1 0}(a_{j_2 j_2 s_1} - a_{j_1 j_2 1})(a_{j_2 j_1 1} - a_{j_1 j_1 1})^2 > 0$ *and* $a_{j_2 j_2 0}(a_{j_2 j_2 s_1} - a_{j_2 j_2 s_2}) > 0$,

$$(a_{j_2 j_1 1}, a_{j_2 j_2 s_1}) = \underbrace{(\text{pF,DP:UP})}_{\text{hyperbolic flow }(+)}. \tag{1.135}$$

The equilibrium of $(x^*_{j_1}, x^*_{j_2}) = (a_{j_2 j_1 1}, a_{j_2 j_2 s_1})$ *is a* (pF,DP:UP)*-positive hyperbolic flow.*

- *For* $a_{j_1 j_1 0}(a_{j_2 j_2 s_1} - a_{j_1 j_2 1})(a_{j_2 j_1 1} - a_{j_1 j_1 1})^2 < 0$ *and* $a_{j_2 j_2 0}(a_{j_2 j_2 s_1} - a_{j_2 j_2 s_2}) > 0$,

$$(a_{j_2 j_1 1}, a_{j_2 j_2 s_1}) = \underbrace{(\text{nF,UP:DP})}_{\text{hyperbolic-secant flow }(-)}. \tag{1.136}$$

The equilibrium of $(x^*_{j_1}, x^*_{j_2}) = (a_{j_2 j_1 1}, a_{j_2 j_2 s_1})$ *is an* (nF,UP:DP)*-negative hyperbolic secant flow.*

- *For* $a_{j_1 j_1 0}(a_{j_2 j_2 s_1} - a_{j_1 j_2 1})(a_{j_2 j_1 1} - a_{j_1 j_1 1})^2 > 0$ *and* $a_{j_2 j_2 0}(a_{j_2 j_2 s_1} - a_{j_2 j_2 s_2}) < 0$,

$$(a_{j_2 j_1 1}, a_{j_2 j_2 s_1}) = \underbrace{(\text{pF,UP:DP})}_{\text{hyperbolic-secant flow }(+)}. \tag{1.137}$$

The equilibrium of $(x^*_{j_1}, x^*_{j_2}) = (a_{j_2 j_1 1}, a_{j_2 j_2 s_1})$ *is a* (pF,UP:DP)*-positive hyperbolic secant flow.*

- *For* $a_{j_1 j_1 0}(a_{j_2 j_2 s_1} - a_{j_1 j_2 1})(a_{j_2 j_1 1} - a_{j_1 j_1 1})^2 < 0$ *and* $a_{j_2 j_2 0}(a_{j_2 j_2 s_1} - a_{j_2 j_2 s_2}) < 0$,

$$(a_{j_2 j_1 1}, a_{j_2 j_2 s_1}) = \underbrace{(\text{nF,DP:UP})}_{\text{hyperbolic flow }(-)}. \tag{1.138}$$

The equilibrium of $(x^*_{j_1}, x^*_{j_2}) = (a_{j_2 j_1 1}, a_{j_2 j_2 s_1})$ *is an* (nF,DP:UP)*-negative hyperbolic flow.*

(vi_2) The first integral manifold $(a_{j_1 j_2 1} = a_{j_2 j_2 s_1},\ a_{j_1 j_1 1} \neq a_{j_2 j_1 1})$ *is*

$$\begin{aligned} &\ln\frac{|x_{j_1} - a_{j_1 j_1 1}|}{|x_{j_1 0} - a_{j_1 j_1 1}|} - (a_{j_1 j_1 1} - a_{j_2 j_1 1})\Big(\frac{1}{x_{j_1} - a_{j_1 j_1 1}} - \frac{1}{x_{j_1 0} - a_{j_1 j_1 1}}\Big) \\ &= \frac{a_{j_1 j_1 0}}{a_{j_2 j_2 0}} \ln\frac{|x_{j_2} - a_{j_2 j_2 s_2}|}{|x_{j_2} - a_{j_2 j_2 s_2}|}. \end{aligned} \tag{1.139}$$

(vi_{2a}) The infinite equilibrium of $x^*_{j_2} = a_{j_1 j_2 1} = a_{j_2 j_2 s_1}$ *with* $\bar{x}_{j_1} \neq a_{j_1 j_1 1}, a_{j_2 j_1 1}$ *has the following properties:*

- *For* $a_{j_1j_10}(\bar{x}_{j_1}-a_{j_1j_11})^2>0$ *and* $a_{j_2j_20}(\bar{x}_{j_1}-a_{j_2j_11})(a_{j_2j_2s_1}-a_{j_2j_2s_2})>0$,

$$(\bar{x}_{j_1},a_{j_2j_21})=\underbrace{\text{(II,SO)}}_{\text{increasing-inflection source}}. \tag{1.140}$$

The infinite equilibrium of $x^*_{j_2}=a_{j_1j_21}=a_{j_2j_2s_1}$ *is an* (II,SO)-*increasing inflection source.*

- *For* $a_{j_1j_10}(\bar{x}_{j_1}-a_{j_1j_11})^2<0$ *and* $a_{j_2j_20}(\bar{x}_{j_1}-a_{j_2j_11})(a_{j_2j_2s_1}-a_{j_2j_2s_2})>0$,

$$(\bar{x}_{j_1},a_{j_2j_2s_1})=\underbrace{\text{(DI,SO)}}_{\text{decreasing-inflection source}}. \tag{1.141}$$

The infinite equilibrium of $x^*_{j_2}=a_{j_1j_21}=a_{j_2j_2s_1}$ *is a* (DI,SO)-*decreasing inflection source.*

- *For* $a_{j_1j_10}(\bar{x}_{j_1}-a_{j_1j_11})^2>0$ *and* $a_{j_2j_20}(\bar{x}_{j_1}-a_{j_2j_11})(a_{j_2j_2s_1}-a_{j_2j_2s_2})<0$,

$$(\bar{x}_{j_1},a_{j_2j_2s_1})=\underbrace{\text{(DI,SI)}}_{\text{decreasing-inflection sink}}. \tag{1.142}$$

The infinite equilibrium of $x^*_{j_2}=a_{j_1j_21}=a_{j_2j_2s_1}$ *is a* (DI,SI)-*decreasing inflection sink.*

- *For* $a_{j_1j_10}(\bar{x}_{j_1}-a_{j_1j_11})^2<0$ *and* $a_{j_2j_20}(\bar{x}_{j_1}-a_{j_2j_11})(a_{j_2j_2s_1}-a_{j_2j_2s_2})<0$,

$$(\bar{x}_{j_1},a_{j_2j_2s_1})=\underbrace{\text{(II,SI)}}_{\text{increasing-inflection sink}}. \tag{1.143}$$

The infinite equilibrium of $x^*_{j_2}=a_{j_1j_21}=a_{j_2j_2s_1}$ *is an* (II,SI)-*increasing inflection sink.*

(vi_{2b}) The equilibrium of $(x^*_{j_1},x^*_{j_2})=(a_{j_1j_11},a_{j_2j_2s_1})$ *with* $a_{j_1j_21}=a_{j_2j_2s_1}$ and $a_{j_1j_11}\neq a_{j_2j_11}$ *has the following properties:*

- *For* $a_{j_1j_10}>0$ *and* $a_{j_2j_20}(a_{j_1j_11}-a_{j_2j_11})(a_{j_2j_2s_1}-a_{j_2j_2s_2})>0$,

$$(a_{j_1j_11},a_{j_2j_2s_1})=\underbrace{\text{(II:II,SO)}}_{\text{increasing-inflection source}}. \tag{1.144}$$

The equilibrium of $(x^*_{j_1},x^*_{j_2})=(a_{j_1j_11},a_{j_2j_2s_1})$ *is an* (II:II,SO)-*increasing inflection source of the second kind.*

- *For* $a_{j_1j_10}<0$ *and* $a_{j_2j_20}(a_{j_1j_11}-a_{j_2j_11})(a_{j_2j_2s_1}-a_{j_2j_2s_2})>0$,

$$(a_{j_1j_11},a_{j_2j_2s_1})=\underbrace{\text{(DI:DI,SO)}}_{\text{decreasing-inflection source}}. \tag{1.145}$$

The equilibrium of $(x_{j_1}^*, x_{j_2}^*) = (a_{j_1 j_1 1}, a_{j_2 j_2 s_1})$ *is a* (DI:DI,SO)-*decreasing inflection source of the second kind.*

- *For* $a_{j_1 j_1 0} > 0$ *and* $a_{j_2 j_2 0}(a_{j_1 j_1 1} - a_{j_2 j_1 1})(a_{j_2 j_2 s_1} - a_{j_2 j_2 s_2}) < 0$,

$$(a_{j_1 j_1 1}, a_{j_2 j_2 s_1}) = \underbrace{\text{(DI:DI,SI)}}_{\text{decreasing-inflection sink}}. \tag{1.146}$$

The equilibrium of $(x_{j_1}^*, x_{j_2}^*) = (a_{j_1 j_1 1}, a_{j_2 j_2 s_1})$ *is a* (DI:DI,SI)-*decreasing inflection sink of the second kind.*

- *For* $a_{j_1 j_1 0} < 0$ *and* $a_{j_2 j_2 0}(a_{j_1 j_1 1} - a_{j_2 j_1 1})(a_{j_2 j_2 s_1} - a_{j_2 j_2 s_2}) < 0$,

$$(a_{j_1 j_1 1}, a_{j_2 j_2 s_1}) = \underbrace{\text{(II:II,SI)}}_{\text{increasing-inflection sink}}. \tag{1.147}$$

The equilibrium of $(x_{j_1}^*, x_{j_2}^*) = (a_{j_1 j_1 1}, a_{j_2 j_2 s_1})$ *is an* (II:II,SI)-*increasing inflection sink of the second kind.*

(vi$_{2c}$*) The equilibrium of* $(x_{j_2}^*, x_{j_1}^*) = (a_{j_1 j_2 1}, a_{j_2 j_1 1})$ *with* $a_{j_1 j_2 1} = a_{j_2 j_2 s_1}$ and $a_{j_1 j_1 1} \neq a_{j_2 j_1 1}$ *has the following properties:*

- *For* $a_{j_1 j_1 0}(a_{j_2 j_1 1} - a_{j_1 j_1 1})^2 > 0$ *and* $a_{j_2 j_2 0}(a_{j_2 j_2 s_1} - a_{j_2 j_2 s_2}) > 0$,

$$(a_{j_1 j_2 1}, a_{j_2 j_1 1}) = \underbrace{({}_{\text{SI:SO}}\text{US}, {}_{\text{DI:II}}\text{UP})}_{\text{up-parabola upper-saddle}}. \tag{1.148}$$

The equilibrium of $(x_{j_2}^*, x_{j_1}^*) = (a_{j_1 j_2 1}, a_{j_2 j_1 1})$ *is a* $({}_{\text{SI:SO}}\text{US}, {}_{\text{DI:II}}\text{UP})$-*up-parabola upper-saddle.*

- *For* $a_{j_1 j_1 0}(a_{j_2 j_1 1} - a_{j_1 j_1 1})^2 < 0$ *and* $a_{j_2 j_2 0}(a_{j_2 j_2 s_1} - a_{j_2 j_2 s_2}) > 0$,

$$(a_{j_1 j_2 1}, a_{j_2 j_1 1}) = \underbrace{({}_{\text{SI:SO}}\text{US}, {}_{\text{II:DI}}\text{DP})}_{\text{down-parabola upper-saddle}}. \tag{1.149}$$

The equilibrium of $(x_{j_2}^*, x_{j_1}^*) = (a_{j_1 j_2 1}, a_{j_2 j_1 1})$ *is an* $({}_{\text{SI:SO}}\text{US}, {}_{\text{II:DI}}\text{DP})$-*down-parabola upper-saddle.*

- *For* $a_{j_1 j_1 0}(a_{j_2 j_1 1} - a_{j_1 j_1 1})^2 > 0$ *and* $a_{j_2 j_2 0}(a_{j_2 j_2 s_1} - a_{j_2 j_2 s_2}) < 0$,

$$(a_{j_1 j_2 1}, a_{j_2 j_1 1}) = \underbrace{({}_{\text{SO:SI}}\text{LS}, {}_{\text{II:DI}}\text{DP})}_{\text{down-parabola lower-saddle}}. \tag{1.150}$$

The equilibrium of $(x_{j_2}^*, x_{j_1}^*) = (a_{j_1 j_2 1}, a_{j_2 j_1 1})$ *is an* $({}_{\text{SO:SI}}\text{LS}, {}_{\text{II:DI}}\text{DP})$-*down-parabola lower-saddle.*

- *For* $a_{j_1j_10}(a_{j_2j_11}-a_{j_1j_11})^2<0$ *and* $a_{j_2j_20}(a_{j_2j_2s_1}-a_{j_2j_2s_2})<0$,

$$(a_{j_1j_21},a_{j_2j_11})=\underbrace{({}_{\text{SO:SI}}\text{LS},{}_{\text{DI:II}}\text{UP})}_{\text{up-parabola lower-saddle}}. \tag{1.151}$$

The equilibrium of $(x_{j_2}^*,x_{j_1}^*)=(a_{j_1j_21},a_{j_2j_11})$ *is an* $({}_{\text{SO:SI}}\text{LS},{}_{\text{DI:II}}\text{UP})$*-up-parabola lower-saddle.*

(vi₃) The first integral manifold $(a_{j_1j_21}\neq a_{j_2j_2s_1},\ a_{j_1j_11}=a_{j_2j_11})$ *is*

$$\ln\frac{|x_{j_1}-a_{j_1j_11}|}{|x_{j_10}-a_{j_1j_11}|}=\frac{a_{j_1j_10}}{a_{j_2j_20}}\sum\nolimits_{s_1=1}^{2}\frac{(a_{j_2j_2s_1}-a_{j_1j_21})}{(a_{j_2j_2s_1}-a_{j_2j_2s_2})}\ln\frac{|x_{j_2}-a_{j_2j_2s_1}|}{|x_{j_20}-a_{j_2j_2s_1}|}. \tag{1.152}$$

(vi_{3a}) The infinite-equilibrium of $x_{j_1}^*=a_{j_1j_11}=a_{j_2j_11}$ *and* $\bar{x}_{j_2}\neq a_{j_1j_21},a_{j_2j_21},a_{j_2j_22}$ *with has the following properties:*

- *For* $a_{j_1j_10}(\bar{x}_{j_2}-a_{j_1j_21})>0$ *and* $a_{j_2j_20}\prod_{s_1=1}^{2}(\bar{x}_{j_2}-a_{j_2j_2s_1})>0$,

$$(a_{j_1j_11},\bar{x}_{j_2})=\underbrace{(\text{US,DU})}_{\text{down-up upper-saddle}}. \tag{1.153}$$

The infinite-equilibrium of $x_{j_1}^*=a_{j_1j_11}=a_{j_2j_11}$ *is a* (US, DU)*-down-up upper-saddle.*

- *For* $a_{j_1j_10}(\bar{x}_{j_2}-a_{j_1j_21})<0$ *and* $a_{j_2j_20}\prod_{s_1=1}^{2}(\bar{x}_{j_2}-a_{j_2j_2s_1})>0$,

$$(a_{j_1j_11},\bar{x}_{j_2})=\underbrace{(\text{LS,UD})}_{\text{up-down lower-saddle}}. \tag{1.154}$$

The infinite-equilibrium of $x_{j_1}^*=a_{j_1j_11}=a_{j_2j_11}$ *is an* (LS, UD)*-up-down lower-saddle.*

- *For* $a_{j_1j_10}(\bar{x}_{j_2}-a_{j_1j_21})>0$ *and* $a_{j_2j_20}\prod_{s_1=1}^{2}(\bar{x}_{j_2}-a_{j_2j_2s_1})<0$,

$$(a_{j_1j_11},\bar{x}_{j_2})=\underbrace{(\text{US,UD})}_{\text{up-down upper-saddle}}. \tag{1.155}$$

The infinite-equilibrium of $x_{j_1}^*=a_{j_1j_11}=a_{j_2j_11}$ *is a* (US, UD)*-up-down upper-saddle.*

- *For* $a_{j_1j_10}(\bar{x}_{j_2}-a_{j_1j_21})<0$ *and* $a_{j_2j_20}\prod_{s_1=1}^{2}(\bar{x}_{j_2}-a_{j_2j_2s_1})<0$,

$$(a_{j_1j_11},\bar{x}_{j_2})=\underbrace{(\text{LS,DU})}_{\text{down-up lower-saddle}}. \tag{1.156}$$

The infinite-equilibrium of $x^*_{j_1} = a_{j_1 j_1 1} = a_{j_2 j_1 1}$ *is an* (LS, DU)-*down-up lower-saddle.*

(*vi*$_{3b}$) *The equilibrium of* $(x^*_{j_1}, x^*_{j_2}) = (a_{j_1 j_1 1}, a_{j_2 j_2 s_1})$ *with* $a_{j_1 j_1 1} = a_{j_2 j_1 1}$ and $a_{j_2 j_2 s_1} \neq a_{j_1 j_2 1}$ *has the following properties:*

- *For* $a_{j_1 j_1 0}(a_{j_2 j_2 s_1} - a_{j_1 j_2 1}) > 0$ *and* $a_{j_2 j_2 0}(a_{j_2 j_2 s_1} - a_{j_2 j_2 s_2}) > 0$

$$(a_{j_1 j_1 1}, a_{j_2 j_2 s_1}) = \underbrace{(\text{US,UD:DU})}_{\text{up-down-to-down-up upper-saddle}}. \tag{1.157}$$

The equilibrium of $(x^*_{j_1}, x^*_{j_2}) = (a_{j_1 j_1 1}, a_{j_2 j_2 s_1})$ *is a* (US,UD:DU)-*up-down-to-up-down upper-saddle.*

- *For* $a_{j_1 j_1 0}(a_{j_2 j_2 s_1} - a_{j_1 j_2 1}) < 0$ *and* $a_{j_2 j_2 0}(a_{j_2 j_2 s_1} - a_{j_2 j_2 s_2}) > 0$

$$(a_{j_1 j_1 1}, a_{j_2 j_2 s_1}) = \underbrace{(\text{LS,DU:UD})}_{\text{down-up-to-up-down lower-saddle}}. \tag{1.158}$$

The equilibrium of $(x^*_{j_1}, x^*_{j_2}) = (a_{j_1 j_1 1}, a_{j_2 j_2 s_1})$ *is an* (LS,DU:UD)-*down-up-to-up-down lower-saddle.*

- *For* $a_{j_1 j_1 0}(a_{j_2 j_2 s_1} - a_{j_1 j_2 1}) > 0$ *and* $a_{j_2 j_2 0}(a_{j_2 j_2 s_1} - a_{j_2 j_2 s_2}) < 0$

$$(a_{j_1 j_1 1}, a_{j_2 j_2 s_1}) = \underbrace{(\text{US,DU:UD})}_{\text{down-up-to-up-down upper-saddle}}. \tag{1.159}$$

The equilibrium of $(x^*_{j_1}, x^*_{j_2}) = (a_{j_1 j_1 1}, a_{j_2 j_2 s_1})$ *is a* (US,DU:UD)-*down-up-to-up-down upper-saddle.*

- *For* $a_{j_1 j_1 0}(a_{j_2 j_2 s_1} - a_{j_1 j_2 1}) < 0$ *and* $a_{j_2 j_2 0}(a_{j_2 j_2 s_1} - a_{j_2 j_2 s_2}) < 0$

$$(a_{j_1 j_1 1}, a_{j_2 j_2 s_1}) = \underbrace{(\text{LS,UD:DU})}_{\text{up-down-to-down-up lower-saddle}}. \tag{1.160}$$

The equilibrium of $(x^*_{j_1}, x^*_{j_2}) = (a_{j_1 j_1 1}, a_{j_2 j_2 s_1})$ *is an* (LS,UD:DU)-*up-down-to-down-up lower-saddle.*

(*vi*$_{3c}$) *The equilibrium of* $(x^*_{j_1}, x^*_{j_2}) = (a_{j_1 j_1 1}, a_{j_1 j_2 1})$ *with* $a_{j_1 j_1 1} = a_{j_2 j_1 1}$ *has the following properties:*

- *For* $a_{j_1 j_1 0} > 0$ *and* $a_{j_2 j_2 0}\prod_{s_1=1}^{2}(a_{j_1 j_2 1} - a_{j_2 j_2 s_1}) > 0$,

$$(a_{j_1 j_1 1}, a_{j_1 j_2 1}) = \underbrace{(\text{DP}_{-}\text{:UP}_{+}, {}_{\text{UD}}\text{LS:}{}_{\text{DU}}\text{US})}_{\text{hyperbolic lower-to-upper saddle}}. \tag{1.161}$$

The equilibrium of $(x^*_{j_1}, x^*_{j_2}) = (a_{j_1j_1}1, a_{j_1j_2}1)$ *is a* (DP$_-$:UP$_+$, $_{\rm UD}$LS:$_{\rm DU}$US)*-hyperbolic lower-to-upper saddle.*

- *For* $a_{j_1j_1}0 < 0$ *and* $a_{j_2j_2}0\prod_{s_1=1}^{2}(a_{j_1j_2}1 - a_{j_2j_2s_1}) > 0$,

$$(a_{j_1j_1}1, a_{j_1j_2}1) = \underbrace{(\text{UP}_-\text{:DP}_+, {}_{\text{DU}}\text{US:}{}_{\text{UD}}\text{LS})}_{\text{hyperbolic-secant upper-to-lower saddle}}. \tag{1.162}$$

The equilibrium of $(x^*_{j_1}, x^*_{j_2}) = (a_{j_1j_1}1, a_{j_1j_2}1)$ *is a* (UP$_-$:DP$_+$,$_{\rm DU}$US:$_{\rm UD}$LS)*-hyperbolic-secant upper-to-lower saddle.*

- *For* $a_{j_1j_1}0 > 0$ *and* $a_{j_2j_2}0\prod_{s_1=1}^{2}(a_{j_1j_2}1 - a_{j_2j_2s_1}) < 0$,

$$(a_{j_1j_1}1, a_{j_1j_2}1) = \underbrace{(\text{UP}_+\text{:DP}_-, {}_{\text{DU}}\text{LS:}{}_{\text{UD}}\text{US})}_{\text{hyperbolic-secant lower-to-upper saddle}}. \tag{1.163}$$

The equilibrium of $(x^*_{j_1}, x^*_{j_2}) = (a_{j_1j_1}1, a_{j_1j_2}1)$ *is an* (UP$_+$:DP$_-$, $_{\rm DU}$LS:$_{\rm UD}$US)*-hyperbolic-secant lower-to-upper saddle.*

- *For* $a_{j_1j_1}0 < 0$ *and* $a_{j_2j_2}0\prod_{s_1=1}^{2}(a_{j_1j_2}1 - a_{j_2j_2s_1}) < 0$,

$$(a_{j_1j_1}1, a_{j_1j_2}1) = \underbrace{(\text{DP}_+\text{:UP}_-, {}_{\text{DU}}\text{US:}{}_{\text{UD}}\text{LS})}_{\text{hyperbolic upper-to-lower saddle}}. \tag{1.164}$$

The equilibrium of $(x^*_{j_1}, x^*_{j_2}) = (a_{j_1j_1}1, a_{j_1j_2}1)$ *is a* (DP$_+$:UP$_-$, $_{\rm DU}$US:$_{\rm UD}$LS)*-hyperbolic upper-to-lower saddle.*

(vi$_4$) The first integral manifold $(a_{j_1j_2}1 = a_{j_2j_2s_1},\ a_{j_1j_1}1 = a_{j_2j_1}1)$ *is*

$$\ln\frac{|x_{j_1} - a_{j_1j_1}1|}{|x_{j_1}0 - a_{j_1j_1}1|} = \frac{a_{j_1j_1}0}{a_{j_2j_2}0}\ln\frac{|x_{j_2} - a_{j_2j_2s_2}|}{|x_{j_2}0 - a_{j_2j_2s_2}|}. \tag{1.165}$$

The equilibrium of $(x^*_{j_2}, x^*_{j_1}) = (a_{j_1j_2}1, a_{j_2j_1}1)$ *with* $a_{j_1j_2}1 = a_{j_2j_2s_1}$ and $a_{j_1j_1}1 = a_{j_2j_1}1$ *has the following properties:*

- *For* $a_{j_1j_1}0 > 0$ *and* $a_{j_2j_2}0(a_{j_2j_2s_1} - a_{j_2j_2s_2}) > 0$,

$$(a_{j_1j_2}1, a_{j_2j_1}1) = \underbrace{({}_{\text{LS:US}}\text{DU}, {}_{\text{DI}}\text{SI:}{}_{\text{II}}\text{SO})}_{\text{down-up inflection sink-to-source}}. \tag{1.166}$$

The equilibrium of $(x^*_{j_2}, x^*_{j_1}) = (a_{j_1j_2}1, a_{j_2j_1}1)$ *is an* ($_{\rm LS:US}$DU,$_{\rm DI}$SI:$_{\rm II}$SO)- down-up inflection sink to source.

- *For* $a_{j_1j_1}0 < 0$ *and* $a_{j_2j_2}0(a_{j_2j_2s_1} - a_{j_2j_2s_2}) > 0$,

$$(a_{j_1j_2 1}, a_{j_2j_1 1}) = \underbrace{({}_{\text{US:LS}}\text{UD}, {}_{\text{II}}\text{SI:}{}_{\text{DI}}\text{SO})}_{\text{up-down inflection sink-to-source}}. \tag{1.167}$$

The equilibrium of $(x^*_{j_2}, x^*_{j_1}) = (a_{j_1j_2 1}, a_{j_2j_1 1})$ *is a* (${}_{\text{US:LS}}$UD,${}_{\text{II}}$SI:${}_{\text{DI}}$SO)-up-down inflection sink to source.

- *For* $a_{j_1j_1 0} > 0$ *and* $a_{j_2j_2 0}(a_{j_2j_2 s_1} - a_{j_2j_2 s_2}) < 0$,

$$(a_{j_1j_2 1}, a_{j_2j_1 1}) = \underbrace{({}_{\text{LS:US}}\text{UD}, {}_{\text{II}}\text{SO:}{}_{\text{DI}}\text{SI})}_{\text{up-down inflection source-to-sink}}. \tag{1.168}$$

The equilibrium of $(x^*_{j_2}, x^*_{j_1}) = (a_{j_1j_2 1}, a_{j_2j_1 1})$ *is an* (${}_{\text{LS:US}}$UD,${}_{\text{II}}$SO:${}_{\text{DI}}$SI)-up-down inflection source to sink.

- *For* $a_{j_1j_1 0} < 0$ *and* $a_{j_2j_2 0}(a_{j_2j_2 s_1} - a_{j_2j_2 s_2}) < 0$,

$$(a_{j_1j_2 1}, a_{j_2j_1 1}) = \underbrace{({}_{\text{US:LS}}\text{DU}, {}_{\text{DI}}\text{SO:}{}_{\text{II}}\text{SI})}_{\text{down-up inflection source-to-sink}}. \tag{1.169}$$

The equilibrium of $(x^*_{j_2}, x^*_{j_1}) = (a_{j_1j_2 1}, a_{j_2j_1 1})$ *is a* (${}_{\text{US:LS}}$DU,${}_{\text{DI}}$SO:${}_{\text{II}}$SI)- down-up inflection source to sink.

(vii) For $\Delta_{j_1j_1} > 0$ *and* $\Delta_{j_2j_2} < 0$*, the standard form is*

$$\begin{aligned} \dot{x}_{j_2} &= a_{j_2j_2 0}(x_{j_1} - a_{j_2j_1 1})\left[(x_{j_2} - a_{j_2j_2})^2 + b_{j_2j_2}\right], \\ \dot{x}_{j_1} &= a_{j_1j_1 0}(x_{j_1} - a_{j_1j_1 1})(x_{j_1} - a_{j_1j_1 2})(x_{j_2} - a_{j_1j_2 1}) \end{aligned} \tag{1.170}$$

where

$$\begin{aligned} b_{j_1j_1 1}, b_{j_1j_1 2} &= -\frac{1}{2}(B_{j_1j_1} \pm \sqrt{\Delta_{j_1j_1}}), \\ \{a_{j_1j_1 1}, a_{j_1j_1 2}\} &= \text{sort}\{b_{j_1j_1 1}, b_{j_1j_1 2}\}, a_{j_1j_1 1} < a_{j_1j_1 2}; \\ a_{j_2j_2} &= -\frac{1}{2}B_{j_2j_2}, b_{j_2j_2} = \frac{1}{4}(-\Delta_{j_2j_2}). \end{aligned} \tag{1.171}$$

The first integral manifold is

$$
\begin{aligned}
&\frac{a_{j_1j_1 2}-a_{j_2j_1 1}}{a_{j_1j_1 2}-a_{j_1j_1 1}}\ln\frac{|x_{j_1}-a_{j_1j_1 2}|}{|x_{j_1 0}-a_{j_1j_1 2}|}+\frac{a_{j_1j_1 1}-a_{j_2j_1 1}}{a_{j_1j_1 1}-a_{j_1j_1 2}}\ln\frac{|x_{j_1}-a_{j_1j_1 1}|}{|x_{j_1 0}-a_{j_1j_1 1}|}\\
&\quad=\frac{a_{j_1j_1 0}}{a_{j_2j_2 0}}\left[\frac{1}{2}\ln\frac{|(x_{j_2}-a_{j_2j_2})^2+b_{j_2j_2}|}{|(x_{j_2 0}-a_{j_2j_2})^2+b_{j_2j_2}|}\right.\\
&\qquad\left.+\frac{a_{j_2j_2}-a_{j_1j_2 1}}{\sqrt{b_{j_2j_2}}}\left(\arctan\frac{x_{j_2}-a_{j_2j_2}}{\sqrt{b_{j_2j_2}}}-\arctan\frac{x_{j_2 0}-a_{j_2j_2}}{\sqrt{b_{j_2j_2}}}\right)\right].
\end{aligned}
\tag{1.172}
$$

(vii$_{1a}$) The equilibrium of $(x^*_{j_2}, x^*_{j_1}) = (a_{j_1j_2 1}, a_{j_2j_1 1})$ *has the following properties:*

- *For* $a_{j_1j_1 0}\prod_{s_1=1}^{2}(a_{j_2j_1 1}-a_{j_1j_1 s_1})>0$ *and* $a_{j_2j_2 0}>0$,

$$(a_{j_1j_2 1}, a_{j_2j_1 1}) = \underbrace{(\mathrm{UP}_+, \mathrm{UP}_+)}_{\text{positive saddle}}. \tag{1.173}$$

  The equilibrium of $(x^*_{j_2}, x^*_{j_1}) = (a_{j_1j_2 1}, a_{j_2j_1 1})$ *is a* (UP_+, UP_+)*-positive saddle.*
- *For* $a_{j_1j_1 0}\prod_{s_1=1}^{2}(a_{j_2j_1 1}-a_{j_1j_1 s_1})<0$ *and* $a_{j_2j_2 0}>0$,

$$(a_{j_1j_2 1}, a_{j_2j_1 1}) = \underbrace{(\mathrm{DP}_+, \mathrm{DP}_-)}_{\text{CCW center}}. \tag{1.174}$$

  The equilibrium of $(x^*_{j_2}, x^*_{j_1}) = (a_{j_1j_2 1}, a_{j_2j_1 1})$ *is a* (DP_+,DP_-)*-counterclockwise center.*
- *For* $a_{j_1j_1 0}\prod_{s_1=1}^{2}(a_{j_2j_1 1}-a_{j_1j_1 s_1})>0$ *and* $a_{j_2j_2 0}<0$,

$$(a_{j_1j_2 1}, a_{j_2j_1 1}) = \underbrace{(\mathrm{DP}_-, \mathrm{DP}_+)}_{\text{CW center}}. \tag{1.175}$$

  The equilibrium of $(x^*_{j_2}, x^*_{j_1}) = (a_{j_1j_2 1}, a_{j_2j_1 1})$ *is a* (DP_-, DP_+)*-clockwise center.*
- *For* $a_{j_1j_1 0}\prod_{s_1=1}^{2}(a_{j_2j_1 1}-a_{j_1j_1 s_1})<0$ *and* $a_{j_2j_2 0}<0$,

$$(a_{j_1j_2 1}, a_{j_2j_1 1}) = \underbrace{(\mathrm{UP}_-, \mathrm{UP}_-)}_{\text{negative saddle}}. \tag{1.176}$$

  The equilibrium of $(x^*_{j_2}, x^*_{j_1}) = (a_{j_1j_2 1}, a_{j_2j_1 1})$ *is a* (UP_-,UP_-)*-negative saddle.*

(vii$_{1b}$) The equilibrium of $(x^*_{j_1}, x^*_{j_2}) = (a_{j_1j_1 s_1}, a_{j_1j_2 1})$ (s_1, $s_2 \in \{1, 2\}$, $s_1 \neq s_2$) *has the following properties:*

- *For* $a_{j_1j_1 0}(a_{j_1j_1 s_1}-a_{j_1j_1 s_2})>0$ *and* $a_{j_2j_2 0}(a_{j_1j_1 s_1}-a_{j_2j_1 1})>0$,

$$(a_{j_1j_1s_1}, a_{j_1j_2 1}) = \underbrace{(\text{DP:UP,pF})}_{\text{hyperbolic flow } (+)}. \quad (1.177)$$

The equilibrium of $(x^*_{j_1}, x^*_{j_2}) = (a_{j_1j_1s_1}, a_{j_1j_2 1})$ *is a* (DP:UP, pF)*-positive hyperbolic flow.*

- *For* $a_{j_1j_1 0}(a_{j_1j_1s_1} - a_{j_1j_1s_2}) < 0$ *and* $a_{j_2j_2 0}(a_{j_1j_1s_1} - a_{j_2j_1 1}) > 0$,

$$(a_{j_1j_1s_1}, a_{j_1j_2 1}) = \underbrace{(\text{UP:DP,pF})}_{\text{hyperbolic-secant flow } (+)}. \quad (1.178)$$

The equilibrium of $(x^*_{j_1}, x^*_{j_2}) = (a_{j_1j_1s_1}, a_{j_1j_2 1})$ *is a* (UP:DP, pF)*-positive hyperbolic-secant flow.*

- *For* $a_{j_1j_1 0}(a_{j_1j_1s_1} - a_{j_1j_1s_2}) > 0$ *and* $a_{j_2j_2 0}(a_{j_1j_1s_1} - a_{j_2j_1 1}) < 0$,

$$(a_{j_1j_1s_1}, a_{j_1j_2 1}) = \underbrace{(\text{UP:DP,nF})}_{\text{hyperbolic-secant flow } (-)}. \quad (1.179)$$

The equilibrium of $(x^*_{j_1}, x^*_{j_2}) = (a_{j_1j_1s_1}, a_{j_1j_2 1})$ *is a* (UP:DP, nF)*-negative hyperbolic-secant flow.*

- *For* $a_{j_1j_1 0}(a_{j_1j_1s_1} - a_{j_1j_1s_2}) < 0$ *and* $a_{j_2j_2 0}(a_{j_1j_1s_1} - a_{j_2j_1 1}) < 0$,

$$(a_{j_1j_1s_1}, a_{j_1j_2 1}) = \underbrace{(\text{DP:UP,nF})}_{\text{hyperbolic flow } (-)}. \quad (1.180)$$

The equilibrium of $(x^*_{j_1}, x^*_{j_2}) = (a_{j_1j_1s_1}, a_{j_1j_2 1})$ *is a* (DP:UP, nF)*-negative hyperbolic flow.*

(vii₂) For $a_{j_2j_1 1} = a_{j_1j_1s_1}$, the first integral manifold is

$$\ln \frac{|x_{j_1} - a_{j_1j_1 2}|}{|x_{j_1 0} - a_{j_1j_1 2}|} = \frac{a_{j_1j_1 0}}{a_{j_2j_2 0}} \left[\frac{1}{2} \ln \frac{|(x_{j_2} - a_{j_2j_2})^2 + b_{j_2j_2}|}{|(x_{j_2 0} - a_{j_2j_2})^2 + b_{j_2j_2}|} + \frac{a_{j_2j_2} - a_{j_1j_2 1}}{\sqrt{b_{j_2j_2}}} \left(\arctan \frac{x_{j_2} - a_{j_2j_2}}{\sqrt{b_{j_2j_2}}} - \arctan \frac{x_{j_2 0} - a_{j_2j_2}}{\sqrt{b_{j_2j_2}}} \right) \right]. \quad (1.181)$$

(vii$_{2a}$) The infinite-equilibrium of $x^*_{j_1} = a_{j_2j_1 1} = a_{j_1j_1s_1}$ *with* $\overline{x}_{j_2} \neq a_{j_1j_2 1}$ *has the following properties:*

- *For* $a_{j_1j_1 0}(\overline{x}_{j_2} - a_{j_1j_2 1})(a_{j_2j_1 1} - a_{j_1j_1s_2}) > 0$ *and* $a_{j_2j_2 0} > 0$,

$$(a_{j_1j_1s_1}, \overline{x}_{j_2}) = \underbrace{(\text{SO,II})}_{\text{increasing-inflection source}}. \tag{1.182}$$

The infinite-equilibrium of $x^*_{j_1} = a_{j_2j_11} = a_{j_1j_1s_1}$ *is an* (SO,II)*-increasing inflection source.*

- *For* $a_{j_1j_10}(\overline{x}_{j_2} - a_{j_1j_21})(a_{j_2j_11} - a_{j_1j_1s_2}) < 0$ *and* $a_{j_2j_20} > 0$,

$$(a_{j_1j_1s_1}, \overline{x}_{j_2}) = \underbrace{(\text{SI,DI})}_{\text{decreasing-inflection sink}}. \tag{1.183}$$

The infinite-equilibrium of $x^*_{j_1} = a_{j_2j_11} = a_{j_1j_1s_1}$ *is an* (SI,DI)*-decreasing inflection sink.*

- *For* $a_{j_1j_10}(\overline{x}_{j_2} - a_{j_1j_21})(a_{j_2j_11} - a_{j_1j_1s_2}) > 0$ *and* $a_{j_2j_20} < 0$,

$$(a_{j_1j_1s_1}, \overline{x}_{j_2}) = \underbrace{(\text{SO,DI})}_{\text{decreasing-inflection source}}. \tag{1.184}$$

The infinite-equilibrium of $x^*_{j_1} = a_{j_2j_11} = a_{j_1j_1s_1}$ *is an* (SO,DI)*-decreasing inflection source.*

- *For* $a_{j_1j_10}(\overline{x}_{j_2} - a_{j_1j_21})(a_{j_2j_11} - a_{j_1j_1s_2}) < 0$ *and* $a_{j_2j_20} < 0$,

$$(a_{j_1j_1s_1}, \overline{x}_{j_2}) = \underbrace{(\text{SI,II})}_{\text{increasing-inflection sink}}. \tag{1.185}$$

The infinite-equilibrium of $x^*_{j_1} = a_{j_2j_11} = a_{j_1j_1s_1}$ *is an* (SI, II)*-increasing inflection sink.*

(*vii*$_{2b}$) *The equilibrium of* $(x^*_{j_2}, x^*_{j_1}) = (a_{j_1j_21}, a_{j_2j_11})$ *has the following properties:*

- *For* $a_{j_1j_10}(a_{j_2j_11} - a_{j_1j_1s_2}) > 0$ *and* $a_{j_2j_20} > 0$,

$$(a_{j_1j_21}, a_{j_2j_11}) = \underbrace{({}_{\text{DI:II}}\text{UP}, {}_{\text{SI:SO}}\text{US})}_{\text{inflection up-parabola upper-saddle}}. \tag{1.186}$$

The equilibrium of $(x^*_{j_2}, x^*_{j_1}) = (a_{j_1j_21}, a_{j_2j_11})$ *is a* (${}_{\text{DI:II}}$UP, ${}_{\text{SI:SO}}$US)*-inflection up-parabola upper saddle.*

- *For* $a_{j_1j_10}(a_{j_2j_11} - a_{j_1j_1s_2}) < 0$ *and* $a_{j_2j_20} > 0$,

$$(a_{j_1j_21}, a_{j_2j_11}) = \underbrace{({}_{\text{II:DI}}\text{DP}, {}_{\text{SO:SI}}\text{LS})}_{\text{inflection down-parabola lower-saddle}}. \tag{1.187}$$

The equilibrium of $(x^*_{j_2}, x^*_{j_1}) = (a_{j_1j_21}, a_{j_2j_11})$ *is an* (${}_{\text{II:DI}}$DP: ${}_{\text{SO:SI}}$LS)-inflection down-parabola lower saddle.

- *For* $a_{j_1j_1 0}(a_{j_2j_1 1}-a_{j_1j_1 s_2})>0$ *and* $a_{j_2j_2 0}<0$,

$$(a_{j_1j_2 1},a_{j_2j_1 1})=\underbrace{({}_{\text{II:DI}}\text{DP},\ {}_{\text{SI:SO}}\text{US})}_{\text{inflection down-parabola upper-saddle}}. \tag{1.188}$$

The equilibrium of $(x_{j_2}^*,x_{j_1}^*)=(a_{j_1j_2 1},a_{j_2j_1 1})$ *is an* $({}_{\text{II:DI}}\text{DP},\ {}_{\text{SI:SO}}\text{US})$*-inflection down-parabola upper saddle.*

- *For* $a_{j_1j_1 0}(a_{j_2j_1 1}-a_{j_1j_1 s_2})<0$ *and* $a_{j_2j_2 0}<0$,

$$(a_{j_1j_2 1},a_{j_2j_1 1})=\underbrace{({}_{\text{DI:II}}\text{UP},\ {}_{\text{SO:SI}}\text{LS})}_{\text{inflection up-parabola lower-saddle}}. \tag{1.189}$$

The equilibrium of $(x_{j_2}^*,x_{j_1}^*)=(a_{j_1j_2 1},a_{j_2j_1 1})$ *is a* $({}_{\text{DI:II}}\text{UP},\ {}_{\text{SO:SI}}\text{LS})$*-inflection up-parabola lower saddle.*

(viii) For $\Delta_{j_1j_1}>0$ *and* $\Delta_{j_2j_2}=0$*, the standard form is*

$$\begin{aligned}\dot{x}_{j_2}&=a_{j_2j_2 0}(x_{j_1}-a_{j_2j_1 1})(x_{j_2}-a_{j_2j_2 1})^2,\\ \dot{x}_{j_1}&=a_{j_1j_1 0}(x_{j_1}-a_{j_1j_1 1})(x_{j_1}-a_{j_1j_1 2})(x_{j_2}-a_{j_1j_2 1})\end{aligned} \tag{1.190}$$

where

$$\begin{aligned}&b_{j_1j_1 1},b_{j_1j_1 2}=-\frac{1}{2}(B_{j_1j_1}\pm\sqrt{\Delta_{j_1j_1}}),\\ &\{a_{j_1j_1 1},a_{j_1j_1 2}\}=\text{sort}\{b_{j_1j_1 1},b_{j_1j_1 2}\},a_{j_1j_1 1}<a_{j_1j_1 2};\\ &a_{j_2j_2 1}=-\frac{1}{2}B_{j_2j_2}.\end{aligned} \tag{1.191}$$

(viii$_1$) The first integral manifold for $a_{j_2j_2 1}\neq a_{j_1j_2 1},a_{j_1j_2 2}$ *is*

$$\begin{aligned}&\frac{a_{j_1j_1 2}-a_{j_2j_1 1}}{a_{j_1j_1 2}-a_{j_1j_1 1}}\ln\frac{|x_{j_1}-a_{j_1j_1 2}|}{|x_{j_1 0}-a_{j_1j_1 2}|}+\frac{a_{j_1j_1 1}-a_{j_2j_1 1}}{a_{j_1j_1 1}-a_{j_1j_1 2}}\ln\frac{|x_{j_1}-a_{j_1j_1 1}|}{|x_{j_1 0}-a_{j_1j_1 1}|})\\ &=\frac{a_{j_1j_1 0}}{a_{j_2j_2 0}}\left[\ln\frac{|x_{j_2}-a_{j_2j_2 1}|}{|x_{j_2 0}-a_{j_2j_2 1}|}-(a_{j_2j_2 1}-a_{j_1j_2 1})(\frac{1}{x_{j_2}-a_{j_2j_2 1}}-\frac{1}{x_{j_2 0}-a_{j_2j_2 1}})\right].\end{aligned} \tag{1.192}$$

(viii$_{1a}$) The equilibrium of $(x_{j_2}^*,x_{j_1}^*)=(a_{j_1j_2 1},a_{j_2j_1 1})$ *with* $a_{j_1j_2 1}\neq a_{j_2j_2 1}$ and $a_{j_1j_1 1}\neq a_{j_2j_1 1}$ *has the following properties:*

- *For* $a_{j_1j_1 0}\prod_{s_1=1}^{2}(a_{j_2j_1 1}-a_{j_1j_1 s_1})>0$ *and* $a_{j_2j_2 0}(a_{j_1j_2 1}-a_{j_2j_2 1})^2>0$,

$$(a_{j_1j_2 1}, a_{j_2j_1 1}) = \underbrace{(\mathrm{UP}_+, \mathrm{UP}_+)}_{\text{positive saddle}}. \tag{1.193}$$

The equilibrium of $(x^*_{j_2}, x^*_{j_1}) = (a_{j_1j_2 1}, a_{j_2j_1 1})$ *is a* (UP$_+$, UP$_+$)*-positive saddle.*

- *For* $a_{j_1j_1 0}\prod_{s_1=1}^{2}(a_{j_2j_1 1} - a_{j_1j_1 s_1}) < 0$ *and* $a_{j_2j_2 0}(a_{j_1j_2 1} - a_{j_2j_2 1})^2 > 0$,

$$(a_{j_1j_2 1}, a_{j_2j_1 1}) = \underbrace{(\mathrm{DP}_+, \mathrm{DP}_-)}_{\text{CCW center}}. \tag{1.194}$$

The equilibrium of $(x^*_{j_2}, x^*_{j_1}) = (a_{j_1j_2 1}, a_{j_2j_1 1})$ *is a* (DP$_+$,DP$_-$)*-counterclockwise center.*

- *For* $a_{j_1j_1 0}\prod_{s_1=1}^{2}(a_{j_2j_1 1} - a_{j_1j_1 s_1}) > 0$ *and* $a_{j_2j_2 0}(a_{j_1j_2 1} - a_{j_2j_2 1})^2 < 0$,

$$(a_{j_1j_2 1}, a_{j_2j_1 1}) = \underbrace{(\mathrm{DP}_-, \mathrm{DP}_+)}_{\text{CW center}}. \tag{1.195}$$

The equilibrium of $(x^*_{j_2}, x^*_{j_1}) = (a_{j_1j_2 1}, a_{j_2j_1 1})$ *is a* (DP$_-$, DP$_+$)*-clockwise center.*

- *For* $a_{j_1j_1 0}\prod_{s_1=1}^{2}(a_{j_2j_1 1} - a_{j_1j_1 s_1}) < 0$ *and* $a_{j_2j_2 0}(a_{j_1j_2 1} - a_{j_2j_2 1})^2 < 0$,

$$(a_{j_1j_2 1}, a_{j_2j_1 1}) = \underbrace{(\mathrm{UP}_-, \mathrm{UP}_-)}_{\text{negative saddle}}. \tag{1.196}$$

The equilibrium of $(x^*_{j_2}, x^*_{j_1}) = (a_{j_1j_2 1}, a_{j_2j_1 1})$ *is a* (UP$_-$,UP$_-$)*-negative saddle.*

(viii$_{1b}$) The equilibrium of $(x^*_{j_1}, x^*_{j_2}) = (a_{j_1j_1 s_1}, a_{j_2j_2 1})$ *has the following properties:*

- *For* $a_{j_1j_1 0}(a_{j_1j_1 s_1} - a_{j_1j_1 s_2})(a_{j_2j_2 1} - a_{j_1j_2 1}) > 0$ *and* $a_{j_2j_2 0}(a_{j_1j_1 s_1} - a_{j_2j_1 1}) > 0$,

$$(a_{j_1j_1 s_1}, a_{j_2j_2 1}) = \underbrace{(\mathrm{SO}, \mathrm{US})}_{\text{upper-saddle source}}. \tag{1.197}$$

The equilibrium of $(x^*_{j_1}, x^*_{j_2}) = (a_{j_1j_1 s_1}, a_{j_2j_2 1})$ *is an* (SO,US)*-upper-saddle source.*

- *For* $a_{j_1j_1 0}(a_{j_1j_1 s_1} - a_{j_1j_1 s_2})(a_{j_2j_2 1} - a_{j_1j_2 1}) < 0$ *and* $a_{j_2j_2 0}(a_{j_1j_1 s_1} - a_{j_2j_1 1}) > 0$,

$$(a_{j_1j_1 s_1}, a_{j_2j_2 1}) = \underbrace{(\mathrm{SI}, \mathrm{US})}_{\text{upper-saddle-sink}}. \tag{1.198}$$

The equilibrium of $(x^*_{j_1}, x^*_{j_2}) = (a_{j_1j_1 s_1}, a_{j_2j_2 1})$ *is an* (SI,US)*-upper-saddle sink.*

- *For* $a_{j_1j_1 0}(a_{j_1j_1 s_1} - a_{j_1j_1 s_2})(a_{j_2j_2 1} - a_{j_1j_2 1}) > 0$ *and* $a_{j_2j_2 0}(a_{j_1j_1 s_1} - a_{j_2j_1 1}) < 0$,

$$(a_{j_1j_1s_1}, a_{j_2j_21}) = \underbrace{(\text{SO,LS})}_{\text{lower-saddle source}}. \quad (1.199)$$

The equilibrium of $(x^*_{j_1}, x^*_{j_2}) = (a_{j_1j_1s_1}, a_{j_2j_21})$ *is an* (SO,LS)-*lower-saddle source.*

- *For* $a_{j_1j_10}(a_{j_1j_1s_1} - a_{j_1j_1s_2})(a_{j_2j_21} - a_{j_1j_21}) < 0$ *and* $a_{j_2j_20}(a_{j_1j_1s_1} - a_{j_2j_11}) < 0$,

$$(a_{j_1j_1s_1}, a_{j_2j_21}) = \underbrace{(\text{SI,LS})}_{\text{lower-saddle source}}. \quad (1.200)$$

The equilibrium of $(x^*_{j_1}, x^*_{j_2}) = (a_{j_1j_1s_1}, a_{j_2j_21})$ *is an* (SI,LS)-*lower-saddle sink.*

– *The saddle-source and saddle-sink are the appearing bifurcations of saddle-source and saddle-sink.*

(viii$_{1c}$) The equilibrium of $(x^*_{j_1}, x^*_{j_2}) = (a_{j_1j_1s_1}, a_{j_1j_21})$ *with* $a_{j_1j_21} \neq a_{j_2j_21}$ $(s_1, s_2 \in \{1,2\}, s_1 \neq s_2)$ *has the following properties:*

- *For* $a_{j_1j_10}(a_{j_1j_1s_1} - a_{j_1j_1s_2}) > 0$ *and* $a_{j_2j_20}(a_{j_1j_1s_1} - a_{j_2j_11})(a_{j_1j_21} - a_{j_2j_21})^2 > 0$,

$$(a_{j_1j_1s_1}, a_{j_1j_21}) = \underbrace{(\text{DP:UP,pF})}_{\text{hyperbolic flow }(+)}. \quad (1.201)$$

The equilibrium of $(x^*_{j_1}, x^*_{j_2}) = (a_{j_1j_1s_1}, a_{j_1j_21})$ *is a* (DP:UP, pF)-*positive hyperbolic flow.*

- *For* $a_{j_1j_10}(a_{j_1j_1s_1} - a_{j_1j_1s_2}) < 0$ *and* $a_{j_2j_20}(a_{j_1j_1s_1} - a_{j_2j_11})(a_{j_1j_21} - a_{j_2j_21})^2 > 0$,

$$(a_{j_1j_1s_1}, a_{j_1j_21}) = \underbrace{(\text{UP:DP,pF})}_{\text{hyperbolic-secant flow }(+)}. \quad (1.202)$$

The equilibrium of $(x^*_{j_1}, x^*_{j_2}) = (a_{j_1j_1s_1}, a_{j_1j_21})$ *is a* (UP:DP, pF)-*positive hyperbolic secant flow*

- *For* $a_{j_1j_10}(a_{j_1j_1s_1} - a_{j_1j_1s_2}) > 0$ *and* $a_{j_2j_20}(a_{j_1j_1s_1} - a_{j_2j_11})(a_{j_1j_2s_1} - a_{j_2j_21})^2 < 0$,

$$(a_{j_1j_1s_1}, a_{j_1j_21}) = \underbrace{(\text{UP:DP,nF})}_{\text{hyperbolic-secant flow }(-)}. \quad (1.203)$$

The equilibrium of $(x^*_{j_1}, x^*_{j_2}) = (a_{j_1j_1s_1}, a_{j_1j_21})$ *is a* (UP:DP, nF)-*negative hyperbolic secant flow.*

- *For* $a_{j_1j_10}(a_{j_1j_1s_1} - a_{j_1j_1s_2}) < 0$ *and* $a_{j_2j_20}(a_{j_1j_1s_1} - a_{j_2j_11})(a_{j_1j_2s_1} - a_{j_2j_21})^2 < 0$,

$$(a_{j_1j_1s_1}, a_{j_1j_2 1}) = \underbrace{(\text{DP:UP,nF})}_{\text{hyperbolic flow }(-)}. \tag{1.204}$$

The equilibrium of $(x_{j_1}^*, x_{j_2}^*) = (a_{j_1j_1s_1}, a_{j_1j_2 1})$ *is a* (DP:UP, nF)*-negative hyperbolic flow.*

(viii$_{1d}$) The equilibrium of $(x_{j_1}^*, x_{j_2}^*) = (a_{j_2j_1 1}, a_{j_2j_2 1})$ *with* $a_{j_1j_2 1} \neq a_{j_2j_2 1}$ *and* $a_{j_1j_1 1} \neq a_{j_2j_1 1}$ *has the following properties:*

- *For* $a_{j_1j_1 0}(a_{j_2j_2 1} - a_{j_1j_2 1}) \prod_{s_1=1}^{2} (a_{j_2j_1 1} - a_{j_1j_1s_1}) > 0$ *and* $a_{j_2j_2 0} > 0$,

$$(a_{j_2j_1 1}, a_{j_2j_2 1}) = \underbrace{(\text{pF,UP:UP})}_{\text{hyperbolic-secant to hyperbolic flow }(+)}. \tag{1.205}$$

The equilibrium of $(x_{j_1}^*, x_{j_2}^*) = (a_{j_2j_1 1}, a_{j_2j_2 1})$ *is a* (pF,UP:UP)*-positive hyperbolic secant-to-hyperbolic flow.*

- *For* $a_{j_1j_1 0}(a_{j_2j_2 1} - a_{j_1j_2 1}) \prod_{s_1=1}^{2} (a_{j_2j_1 1} - a_{j_1j_1s_1}) < 0$ *and* $a_{j_2j_2 0} > 0$,

$$(a_{j_2j_1 1}, a_{j_2j_2 1}) = \underbrace{(\text{nF,DP:DP})}_{\text{hyperbolic to hyperbolic-secant flow }(-)}. \tag{1.206}$$

The equilibrium of $(x_{j_1}^*, x_{j_2}^*) = (a_{j_2j_1 1}, a_{j_2j_2 1})$ *is an* (nF,DP:DP)*-negative hyperbolic-to-hyperbolic secant flow.*

- *For* $a_{j_1j_1 0}(a_{j_2j_2 1} - a_{j_1j_2 1}) \prod_{s_1=1}^{2} (a_{j_2j_1 1} - a_{j_1j_1s_1}) > 0$ *and* $a_{j_2j_2 0} < 0$,

$$(a_{j_2j_1 1}, a_{j_2j_2 1}) = \underbrace{(\text{pF,DP:DP})}_{\text{hyperbolic to hyperbolic-secant flow }(+)}. \tag{1.207}$$

The equilibrium of $(x_{j_1}^*, x_{j_2}^*) = (a_{j_2j_1 1}, a_{j_2j_2 1})$ *is a* (pF,DP:DP)*-negative hyperbolic-to-hyperbolic secant flow.*

- *For* $a_{j_1j_1 0}(a_{j_2j_2 1} - a_{j_1j_2 1}) \prod_{s_1=1}^{2} (a_{j_2j_1 1} - a_{j_1j_1s_1}) < 0$ *and* $a_{j_2j_2 0} < 0$,

$$(a_{j_2j_1 1}, a_{j_2j_2 1}) = \underbrace{(\text{nF,UP:UP})}_{\text{hyperbolic-secant to hyperbolic flow }(-)}. \tag{1.208}$$

The equilibrium of $(x_{j_1}^*, x_{j_2}^*) = (a_{j_2j_1 1}, a_{j_2j_2 1})$ *is an* (nF,UP:UP)*-negative hyperbolic secant-to-hyperbolic flow.*

(viii$_2$) The first integral manifold for $a_{j_2j_2 1} = a_{j_1j_2 1}$ and $a_{j_1j_1s_1} \neq a_{j_2j_1 1}$ *is*

$$\begin{aligned}&\frac{a_{j_1j_12}-a_{j_2j_11}}{a_{j_1j_12}-a_{j_1j_11}}\ln\frac{|x_{j_1}-a_{j_1j_12}|}{|x_{j_10}-a_{j_1j_12}|}+\frac{a_{j_1j_11}-a_{j_2j_11}}{a_{j_1j_11}-a_{j_1j_12}}\ln\frac{|x_{j_1}-a_{j_1j_11}|}{|x_{j_10}-a_{j_1j_11}|}\\&\quad=\frac{a_{j_1j_10}}{a_{j_2j_20}}\ln\frac{|x_{j_2}-a_{j_2j_21}|}{|x_{j_20}-a_{j_2j_21}|}.\end{aligned}\tag{1.209}$$

(viii$_{2a}$) The infinite-equilibrium of $x^*_{j_2}=a_{j_1j_21}=a_{j_2j_21}$ with $\bar{x}_{j_1}\neq a_{j_1j_1s_1},a_{j_2j_11}$ $(s_1=1,2)$ has the following properties:

- *For $a_{j_1j_10}\prod_{s_1=1}^{2}(\bar{x}_{j_1}-a_{j_1j_1s_1})>0$ and $a_{j_2j_20}(\bar{x}_{j_1}-a_{j_2j_11})>0$,*

$$(\bar{x}_{j_1},a_{j_2j_21})=\underbrace{(\mathrm{DU,US})}_{\text{down-up upper-saddle}}.\tag{1.210}$$

The infinite-equilibrium of $x^*_{j_2}=a_{j_1j_21}=a_{j_2j_21}$ is an (DU,US)*-down-up upper saddle.*

- *For $a_{j_1j_10}\prod_{s_1=1}^{2}(\bar{x}_{j_1}-a_{j_1j_1s_1})<0$ and $a_{j_2j_20}(\bar{x}_{j_1}-a_{j_2j_11})>0$,*

$$(\bar{x}_{j_1},a_{j_2j_21})=\underbrace{(\mathrm{UD,US})}_{\text{up-down upper-saddle}}.\tag{1.211}$$

The infinite-equilibrium of $x^*_{j_2}=a_{j_1j_21}=a_{j_2j_21}$ is a (UD,US)*-up-down upper saddle.*

- *For $a_{j_1j_10}\prod_{s_1=1}^{2}(\bar{x}_{j_1}-a_{j_1j_1s_1})>0$ and $a_{j_2j_20}(\bar{x}_{j_1}-a_{j_2j_11})<0$,*

$$(\bar{x}_{j_1},a_{j_2j_21})=\underbrace{(\mathrm{UD,LS})}_{\text{up-down lower-saddle}}.\tag{1.212}$$

The infinite-equilibrium of $x^*_{j_2}=a_{j_1j_21}=a_{j_2j_21}$ is a (UD,LS)*-up-down lower saddle.*

- *For $a_{j_1j_10}\prod_{s_1=1}^{2}(\bar{x}_{j_1}-a_{j_1j_1s_1})<0$ and $a_{j_2j_20}(\bar{x}_{j_1}-a_{j_2j_11})<0$,*

$$(\bar{x}_{j_1},a_{j_2j_21})=\underbrace{(\mathrm{DU,LS})}_{\text{down-up lower-saddle}}.\tag{1.213}$$

The infinite-equilibrium of $x^*_{j_2}=a_{j_1j_21}=a_{j_2j_21}$ is a (DU,LS)*-down-up lower saddle.*

(viii$_{2b}$) The equilibrium of $(x^*_{j_1},x^*_{j_2})=(a_{j_1j_1s_1},a_{j_2j_21})$ with $a_{j_2j_21}=a_{j_1j_21}$ and $a_{j_1j_1s_1}\neq a_{j_2j_11}$ has the following properties:

- *For $a_{j_1j_10}(a_{j_1j_1s_1}-a_{j_1j_1s_2})>0$ and $a_{j_2j_20}(a_{j_1j_1s_1}-a_{j_2j_11})>0$,*

$$(a_{j_1j_1s_1}, a_{j_2j_2 1}) = \underbrace{(\text{UD:DU,US})}_{\text{up-down-to-down-up upper-saddle}}. \tag{1.214}$$

The equilibrium of $(x_{j_1}^*, x_{j_2}^*) = (a_{j_1j_1s_1}, a_{j_2j_2 1})$ *is a* (UD:DU,US)-*up-down-to-down-up upper-saddle.*

- *For* $a_{j_1j_1 0}(a_{j_1j_1s_1} - a_{j_1j_1s_2}) < 0$ *and* $a_{j_2j_2 0}(a_{j_1j_1s_1} - a_{j_2j_1 1}) > 0$,

$$(a_{j_1j_1s_1}, a_{j_2j_2 1}) = \underbrace{(\text{DU:UD,US})}_{\text{down-up-to-up-down upper-saddle}}. \tag{1.215}$$

The equilibrium of $(x_{j_1}^*, x_{j_2}^*) = (a_{j_1j_1s_1}, a_{j_2j_2 1})$ *is a* (DU:UD,US)-*down-up-to-up-down upper-saddle.*

- *For* $a_{j_1j_1 0}(a_{j_1j_1s_1} - a_{j_1j_1s_2}) > 0$ *and* $a_{j_2j_2 0}(a_{j_1j_1s_1} - a_{j_2j_1 1}) < 0$,

$$(a_{j_1j_1s_1}, a_{j_2j_2 1}) = \underbrace{(\text{DU:UD,LS})}_{\text{down-up-to-up-down lower-saddle}}. \tag{1.216}$$

The equilibrium of $(x_{j_1}^*, x_{j_2}^*) = (a_{j_1j_1s_1}, a_{j_2j_2 1})$ *is a* (DU:UD,LS)-*down-up-to-up-down lower-saddle.*

- *For* $a_{j_1j_1 0}(a_{j_1j_1s_1} - a_{j_1j_1s_2}) < 0$ *and* $a_{j_2j_2 0}(a_{j_1j_1s_1} - a_{j_2j_1 1}) < 0$,

$$(a_{j_1j_1s_1}, a_{j_2j_2 1}) = \underbrace{(\text{UD:DU,LS})}_{\text{up-down-to-down-up lower-saddle}}. \tag{1.217}$$

The equilibrium of $(x_{j_1}^*, x_{j_2}^*) = (a_{j_1j_1s_1}, a_{j_2j_2 1})$ *is a* (UD:DU,LS)-*up-down-to-down-up lower-saddle.*

(viii$_{2c}$) The equilibrium of $(x_{j_1}^*, x_{j_2}^*) = (a_{j_2j_1 1}, a_{j_2j_2 1})$ *with* $a_{j_2j_2 1} = a_{j_1j_2 1}$ *and* $a_{j_1j_1s_1} \neq a_{j_2j_1 1}$ *has the following properties:*

- *For* $a_{j_1j_1 0}(a_{j_1j_1s_1} - a_{j_1j_1s_2}) > 0$ *and* $a_{j_2j_2 0}(a_{j_1j_1s_1} - a_{j_2j_1 1}) > 0$,

$$(a_{j_2j_1 1}, a_{j_2j_2 1}) = \underbrace{({}_{\text{UD}}\text{LS:}{}_{\text{DU}}\text{US,DP}_{-}\text{:UP}_{+})}_{\text{hyperbolic lower-to-upper saddle}}. \tag{1.218}$$

The equilibrium of $(x_{j_1}^*, x_{j_2}^*) = (a_{j_2j_1 1}, a_{j_2j_2 1})$ *is a* (${}_{\text{UD}}$LS:${}_{\text{DU}}$US, DP$_-$:UP$_+$)-*hyperbolic lower-to-upper saddle.*

- *For* $a_{j_1j_1 0}(a_{j_1j_1s_1} - a_{j_1j_1s_2}) < 0$ *and* $a_{j_2j_2 0}(a_{j_1j_1s_1} - a_{j_2j_1 1}) > 0$,

$$(a_{j_2j_11}, a_{j_2j_21}) = \underbrace{({}_{\mathrm{DU}}\mathrm{LS}{:}{}_{\mathrm{UD}}\mathrm{US}, \mathrm{UP}_+{:}\mathrm{DP}_-)}_{\text{hyperbolic-secant lower-to-upper saddle}}. \tag{1.219}$$

The equilibrium of $(x_{j_1}^*, x_{j_2}^*) = (a_{j_2j_11}, a_{j_2j_21})$ *is a* (${}_{\mathrm{DU}}$LS:${}_{\mathrm{UD}}$US, UP$_+$:DP$_-$)-*hyperbolic secant lower-to-upper saddle.*

- *For* $a_{j_1j_10}(a_{j_1j_1s_1} - a_{j_1j_1s_2}) > 0$ *and* $a_{j_2j_20}(a_{j_1j_1s_1} - a_{j_2j_11}) < 0$,

$$(a_{j_2j_11}, a_{j_2j_21}) = \underbrace{({}_{\mathrm{DU}}\mathrm{US}{:}{}_{\mathrm{UD}}\mathrm{LS}, \mathrm{UP}_-{:}\mathrm{DP}_+)}_{\text{hyperbolic-secant upper-to-lower saddle}}. \tag{1.220}$$

The equilibrium of $(x_{j_1}^*, x_{j_2}^*) = (a_{j_2j_11}, a_{j_2j_21})$ *is a* (${}_{\mathrm{DU}}$US:${}_{\mathrm{UD}}$LS, UP$_-$:DP$_+$)-*hyperbolic secant upper-to-lower saddle.*

- *For* $a_{j_1j_10}(a_{j_1j_1s_1} - a_{j_1j_1s_2}) < 0$ *and* $a_{j_2j_20}(a_{j_1j_1s_1} - a_{j_2j_11}) < 0$,

$$(a_{j_2j_11}, a_{j_2j_21}) = \underbrace{({}_{\mathrm{UD}}\mathrm{US}{:}{}_{\mathrm{DU}}\mathrm{LS}, \mathrm{DP}_+{:}\mathrm{UP}_-)}_{\text{hyperbolic upper-to-lower saddle}}. \tag{1.221}$$

The equilibrium of $(x_{j_1}^*, x_{j_2}^*) = (a_{j_2j_11}, a_{j_2j_21})$ *is a* (${}_{\mathrm{UD}}$US:${}_{\mathrm{DU}}$LS,DP$_+$:UP$_-$)-*hyperbolic upper-to-lower saddle.*

(viii$_3$) The first integral manifold for $a_{j_1j_1s_1} = a_{j_2j_11}$ and $a_{j_1j_21} \neq a_{j_2j_21}$ *is*

$$\begin{aligned} \ln\frac{|x_{j_1} - a_{j_1j_1s_2}|}{|x_{j_10} - a_{j_1j_1s_2}|} = \frac{a_{j_1j_10}}{a_{j_2j_20}}\Bigg[&\ln\frac{|x_{j_2} - a_{j_2j_21}|}{|x_{j_20} - a_{j_2j_21}|} \\ &- (a_{j_2j_21} - a_{j_1j_21})(\frac{1}{x_{j_2} - a_{j_2j_21}} - \frac{1}{x_{j_20} - a_{j_2j_21}})\Bigg]. \end{aligned} \tag{1.222}$$

(viii$_{3a}$) The infinite-equilibrium of $x_{j_1}^* = a_{j_1j_1s_1} = a_{j_2j_11}$ $(s_1 = 1, 2)$ *with* $\bar{x}_{j_2} \neq a_{j_2j_21}\ a_{j_1j_21}$ has the following properties:

- *For* $a_{j_1j_10}(\bar{x}_{j_2} - a_{j_1j_21})(a_{j_1j_1s_1} - a_{j_1j_1s_2}) > 0$ *and* $a_{j_2j_20}(\bar{x}_{j_2} - a_{j_2j_21})^2 > 0$,

$$(a_{j_1j_1s_1}, \bar{x}_{j_2}) = \underbrace{(\mathrm{SO}, \mathrm{II})}_{\text{increasing-inflection source}}. \tag{1.223}$$

The infinite-equilibrium of $x_{j_1}^* = a_{j_2j_11} = a_{j_1j_1s_1}$ *is an* (SO, II)-*increasing inflection source.*

- *For* $a_{j_1j_10}(\bar{x}_{j_2} - a_{j_1j_21})(a_{j_1j_1s_1} - a_{j_1j_1s_2}) < 0$ *and* $a_{j_2j_20}(\bar{x}_{j_2} - a_{j_2j_21})^2 > 0$,

$$(a_{j_1j_1s_1}, \bar{x}_{j_2}) = \underbrace{(\text{SI,DI})}_{\text{decreasing-inflection sink}}. \tag{1.224}$$

The infinite-equilibrium of $x_{j_1}^* = a_{j_2j_1 1} = a_{j_1j_1s_1}$ *is an* (SI,DI)-*decreasing inflection sink.*

- *For* $a_{j_1j_1 0}(\bar{x}_{j_2} - a_{j_1j_2 1})(a_{j_1j_1s_1} - a_{j_1j_1s_2}) > 0$ *and* $a_{j_2j_2 0}(\bar{x}_{j_2} - a_{j_2j_2 1})^2 < 0$,

$$(a_{j_1j_1s_1}, \bar{x}_{j_2}) = \underbrace{(\text{SO,DI})}_{\text{decreasing-inflection source}}. \tag{1.225}$$

The infinite-equilibrium of $x_{j_1}^* = a_{j_2j_1 1} = a_{j_1j_1s_1}$ *is an* (SO,DI)-*decreasing inflection source.*

- *For* $a_{j_1j_1 0}(\bar{x}_{j_2} - a_{j_1j_2 1})(a_{j_1j_1s_1} - a_{j_1j_1s_2}) < 0$ *and* $a_{j_2j_2 0}(\bar{x}_{j_2} - a_{j_2j_2 1})^2 < 0$,

$$(a_{j_1j_1s_1}, \bar{x}_{j_2}) = \underbrace{(\text{SI,II})}_{\text{increasing-inflection sink}}. \tag{1.226}$$

The infinite-equilibrium of $x_{j_1}^* = a_{j_2j_1 1} = a_{j_1j_1s_1}$ *is an* (SI, II)-*increasing inflection sink.*

(viii$_{3b}$) The equilibrium of $(x_{j_2}^*, x_{j_1}^*) = (a_{j_1j_2 1}, a_{j_2j_1 1})$ *for* $a_{j_1j_1s_1} = a_{j_2j_1 1}$ and $a_{j_1j_2 1} \neq a_{j_2j_2 1}$ *has the following properties:*

- *For* $a_{j_1j_1 0}(a_{j_2j_1 1} - a_{j_1j_1s_2}) > 0$ *and* $a_{j_2j_2 0}(a_{j_1j_2 1} - a_{j_2j_2 1})^2 > 0$,

$$(a_{j_1j_2 1}, a_{j_2j_1 1}) = \underbrace{({}_{\text{DI:II}}\text{UP},\ {}_{\text{SI:SO}}\text{US})}_{\text{inflection up-parabola upper-saddle}}. \tag{1.227}$$

The equilibrium of $(x_{j_2}^*, x_{j_1}^*) = (a_{j_1j_2 1}, a_{j_2j_1 1})$ *is a* ($_{\text{DI:II}}$UP, $_{\text{SI:SO}}$US)-*inflection up-parabola upper-saddle.*

- *For* $a_{j_1j_1 0}(a_{j_2j_1 1} - a_{j_1j_1s_2}) < 0$ *and* $a_{j_2j_2 0}(a_{j_1j_2 1} - a_{j_2j_2 1})^2 > 0$,

$$(a_{j_1j_2 1}, a_{j_2j_1 1}) = \underbrace{({}_{\text{II:DI}}\text{DP}, {}_{\text{SO:SI}}\text{LS})}_{\text{inflection down-parabola lower-saddle}}. \tag{1.228}$$

The equilibrium of $(x_{j_2}^*, x_{j_1}^*) = (a_{j_1j_2 1}, a_{j_2j_1 1})$ *is an* ($_{\text{II:DI}}$DP, $_{\text{SO:SI}}$LS)-*inflection down-parabola lower-saddle.*

- *For* $a_{j_1j_1 0}(a_{j_2j_1 1} - a_{j_1j_1s_2}) > 0$ *and* $a_{j_2j_2 0}(a_{j_1j_2 1} - a_{j_2j_2 1})^2 < 0$,

$$(a_{j_1j_2 1}, a_{j_2j_1 1}) = \underbrace{({}_{\text{II:DI}}\text{DP}, {}_{\text{SI:SO}}\text{US})}_{\text{inflection down-parabola upper-saddle}}. \tag{1.229}$$

The equilibrium of $(x_{j_2}^*, x_{j_1}^*) = (a_{j_1 j_2 1}, a_{j_2 j_1 1})$ *is an* ($_{\text{II:DI}}$DP,$_{\text{SI:SO}}$US)*-inflection down-parabola upper-saddle.*

- *For* $a_{j_1 j_1 0}(a_{j_2 j_1 1} - a_{j_1 j_1 s_2}) < 0$ *and* $a_{j_2 j_2 0}(a_{j_1 j_2 1} - a_{j_2 j_2 1})^2 < 0$,

$$(a_{j_1 j_2 1}, a_{j_2 j_1 1}) = \underbrace{({}_{\text{DI:II}}\text{UP}, {}_{\text{SO:SI}}\text{LS})}_{\text{inflection up-parabola lower-saddle}}. \quad (1.230)$$

The equilibrium of $(x_{j_2}^*, x_{j_1}^*) = (a_{j_1 j_2 1}, a_{j_2 j_1 1})$ *is a* ($_{\text{DI:II}}$UP, $_{\text{SI:SO}}$LS)*-inflection up-parabola lower-saddle.*

(viii$_{3c}$) The equilibrium of $(x_{j_1}^*, x_{j_2}^*) = (a_{j_1 j_1 s_1}, a_{j_2 j_2 1})$ *for* $a_{j_1 j_1 s_1} = a_{j_2 j_1 1}$ and $a_{j_1 j_2 1} \neq a_{j_2 j_2 1}$ *has the following properties:*

- *For* $a_{j_1 j_1 0}(a_{j_2 j_2 1} - a_{j_1 j_2 1})(a_{j_1 j_1 s_1} - a_{j_1 j_1 s_2}) > 0$ *and* $a_{j_2 j_2 0} > 0$,

$$(a_{j_1 j_1 s_1}, a_{j_2 j_2 1}) = \underbrace{(\text{SO,II:II})}_{\text{increasing-inflection source}}. \quad (1.231)$$

The equilibrium of $(x_{j_1}^*, x_{j_2}^*) = (a_{j_1 j_1 s_1}, a_{j_2 j_2 1})$ *is an* (SO,II:II)*-increasing-inflection source of the second kind.*

- *For* $a_{j_1 j_1 0}(a_{j_2 j_2 1} - a_{j_1 j_2 1})(a_{j_1 j_1 s_1} - a_{j_1 j_1 s_2}) < 0$ *and* $a_{j_2 j_2 0} > 0$,

$$(a_{j_1 j_1 s_1}, a_{j_2 j_2 1}) = \underbrace{(\text{SI,DI:DI})}_{\text{decreasing-inflection sink}}. \quad (1.232)$$

The equilibrium of $(x_{j_1}^*, x_{j_2}^*) = (a_{j_1 j_1 s_1}, a_{j_2 j_2 1})$ *is an* (SI:DI:DI)*-decreasing-inflection sink of the second kind.*

- *For* $a_{j_1 j_1 0}(a_{j_2 j_2 1} - a_{j_1 j_2 1})(a_{j_1 j_1 s_1} - a_{j_1 j_1 s_2}) > 0$ *and* $a_{j_2 j_2 0} < 0$,

$$(a_{j_1 j_1 s_1}, a_{j_2 j_2 1}) = \underbrace{(\text{SO,DI:DI})}_{\text{decreasing-inflection source}}. \quad (1.233)$$

The equilibrium of $(x_{j_1}^*, x_{j_2}^*) = (a_{j_1 j_1 s_1}, a_{j_2 j_2 1})$ *is an* (SO,DI:DI)*-decreasing-inflection source of the second kind.*

- *For* $a_{j_1 j_1 0}(a_{j_2 j_2 1} - a_{j_1 j_2 1})(a_{j_1 j_1 s_1} - a_{j_1 j_1 s_2}) < 0$ *and* $a_{j_2 j_2 0} < 0$,

$$(a_{j_1 j_1 s_1}, a_{j_2 j_2 1}) = \underbrace{(\text{SI,II:II})}_{\text{increasing-inflection source}}. \quad (1.234)$$

The equilibrium of $(x_{j_1}^*, x_{j_2}^*) = (a_{j_1 j_1 s_1}, a_{j_2 j_2 1})$ *is an* (SI, II:II)*-increasing-inflection sink* of the second kind.

(viii$_4$) The first integral manifold for $a_{j_1 j_1 s_1} = a_{j_2 j_1 1}$ and $a_{j_1 j_2 1} = a_{j_2 j_2 1}$ *is*

$$\ln\frac{|x_{j_1}-a_{j_1j_1s_2}|}{|x_{j_10}-a_{j_1j_1s_2}|}=\frac{a_{j_1j_10}}{a_{j_2j_20}}\ln\frac{|x_{j_2}-a_{j_2j_21}|}{|x_{j_20}-a_{j_2j_21}|}. \tag{1.235}$$

The equilibrium of $(x_{j_2}^*,x_{j_1}^*)=(a_{j_1j_21},a_{j_2j_11})$ *with* $a_{j_1j_1s_1}=a_{j_2j_11}$ and $a_{j_1j_21}=a_{j_2j_21}$ *has the following properties:*

- *For* $a_{j_1j_10}(a_{j_1j_1s_1}-a_{j_1j_1s_2})>0$ *and* $a_{j_2j_20}>0$,

$$(a_{j_1j_21},a_{j_2j_11})=\underbrace{({}_{\text{DI}}\text{SI:}{}_{\text{II}}\text{SO},{}_{\text{LS:US}}\text{DU})}_{\text{down-up inflection sink-to-source}}. \tag{1.236}$$

  The equilibrium of $(x_{j_2}^*,x_{j_1}^*)=(a_{j_1j_1s_1},a_{j_2j_21})$ *is a* (${}_{\text{DI}}$SI:${}_{\text{II}}$SO, ${}_{\text{LS:US}}$DU)- *down-up inflection sink to source.*
- *For* $a_{j_1j_10}(a_{j_2j_11}-a_{j_1j_1s_2})<0$ *and* $a_{j_2j_20}>0$,

$$(a_{j_1j_21},a_{j_2j_11})=\underbrace{({}_{\text{II}}\text{SO:}{}_{\text{DI}}\text{SI},{}_{\text{LS:US}}\text{UD})}_{\text{up-down inflection source-to-sink}}. \tag{1.237}$$

  The equilibrium of $(x_{j_2}^*,x_{j_1}^*)=(a_{j_1j_1s_1},a_{j_2j_21})$ *is an* (${}_{\text{II}}$SO:${}_{\text{DI}}$SI, ${}_{\text{LS:US}}$UD)*-up-down inflection source to sink.*
- *For* $a_{j_1j_10}(a_{j_2j_11}-a_{j_1j_1s_2})>0$ *and* $a_{j_2j_20}<0$,

$$(a_{j_1j_21},a_{j_2j_11})=\underbrace{({}_{\text{II}}\text{SI:}{}_{\text{DI}}\text{SO},{}_{\text{US:LS}}\text{UD})}_{\text{up-down inflection sink-to-source}}. \tag{1.238}$$

  The equilibrium of $(x_{j_2}^*,x_{j_1}^*)=(a_{j_1j_1s_1},a_{j_2j_21})$ *is a* (${}_{\text{II}}$SI:${}_{\text{DI}}$SO, ${}_{\text{US:LS}}$UD)*-up-down inflection sink to source.*
- *For* $a_{j_1j_10}(a_{j_2j_11}-a_{j_1j_1s_2})<0$ *and* $a_{j_2j_20}<0$,

$$(a_{j_1j_21},a_{j_2j_11})=\underbrace{({}_{\text{DI}}\text{SO:}{}_{\text{II}}\text{SI},{}_{\text{US:LS}}\text{DU})}_{\text{down-up inflection source-to-sink}}. \tag{1.239}$$

  The equilibrium of $(x_{j_2}^*,x_{j_1}^*)=(a_{j_1j_1s_1},a_{j_2j_21})$ *is a* (${}_{\text{DI}}$SO:${}_{\text{II}}$SI, ${}_{\text{US:LS}}$DU)- *down-up inflection source-to-sink.*

(ix) For $\Delta_{j_1j_1}>0$ *and* $\Delta_{j_2j_2}>0$, the standard form is

$$\begin{aligned}\dot{x}_{j_2}&=a_{j_2j_20}(x_{j_1}-a_{j_2j_11})(x_{j_2}-a_{j_2j_21})(x_{j_2}-a_{j_2j_22}),\\ \dot{x}_{j_1}&=a_{j_1j_10}(x_{j_1}-a_{j_1j_11})(x_{j_1}-a_{j_1j_12})(x_{j_2}-a_{j_1j_21})\end{aligned} \tag{1.240}$$

where

$$\begin{aligned}
&b_{j_1j_1 1}, b_{j_1j_1 2} = -\frac{1}{2}(B_{j_1j_1} \pm \sqrt{\Delta_{j_1j_1}}),\\
&\{a_{j_1j_1 1}, a_{j_1j_1 2}\} = \text{sort}\{b_{j_1j_1 1}, b_{j_1j_1 2}\}, a_{j_1j_1 1} < a_{j_1j_1 2};\\
&b_{j_2j_2 1}, b_{j_2j_2 2} = -\frac{1}{2}(B_{j_2j_2} \pm \sqrt{\Delta_{j_2j_2}}),\\
&\{a_{j_2j_2 1}, a_{j_2j_2 2}\} = \text{sort}\{b_{j_2j_2 1}, b_{j_2j_2 2}\}, a_{j_2j_2 1} < a_{j_2j_2 2}.
\end{aligned} \tag{1.241}$$

(ix$_1$) The first integral manifold ($a_{j_2j_2s_1} \neq a_{j_1j_21}$) is

$$\begin{aligned}
&\sum\nolimits_{s_1=1, s_2\neq s_1}^{2} \frac{a_{j_1j_1s_1} - a_{j_2j_11}}{a_{j_1j_1s_1} - a_{j_1j_1s_1}} \ln \frac{|x_{j_1} - a_{j_1j_1s_1}|}{|x_{j_10} - a_{j_1j_1s_1}|}\\
&= \frac{a_{j_1j_10}}{a_{j_2j_20}} \sum\nolimits_{l_1=1, l_2\neq l_1}^{2} \frac{a_{j_2j_2l_1} - a_{j_1j_21}}{a_{j_2j_2l_1} - a_{j_2j_2l_2}} \ln \frac{|x_{j_2} - a_{j_2j_2l_1}|}{|x_{j_20} - a_{j_2j_2l_1}|}.
\end{aligned} \tag{1.242}$$

(ix$_{1a}$) The equilibrium of $(x_{j_2}^*, x_{j_1}^*) = (a_{j_1j_21}, a_{j_2j_11})$ with $a_{j_1j_21} \neq a_{j_2j_21}$ and *$a_{j_1j_11} \neq a_{j_2j_11}$ has the following properties:*

- *For $a_{j_1j_10}\prod_{s_1=1}^{2}(a_{j_2j_11} - a_{j_1j_1s_1}) > 0$ and $a_{j_2j_20}\prod_{l_1=1}^{2}(a_{j_1j_21} - a_{j_2j_2l_1}) > 0$,*

$$(a_{j_1j_21}, a_{j_2j_11}) = \underbrace{(\text{UP}_+, \text{UP}_+)}_{\text{positive saddle}}. \tag{1.243}$$

The equilibrium of $(x_{j_2}^*, x_{j_1}^*) = (a_{j_1j_21}, a_{j_2j_11})$ is a (UP$_+$, UP$_+$)*-positive saddle.*

- *For $a_{j_1j_10}\prod_{s_1=1}^{2}(a_{j_2j_11} - a_{j_1j_1s_1}) < 0$ and $a_{j_2j_20}\prod_{l_1=1}^{2}(a_{j_1j_21} - a_{j_2j_2l_1}) > 0$,*

$$(a_{j_1j_21}, a_{j_2j_11}) = \underbrace{(\text{DP}_+, \text{DP}_-)}_{\text{CCW center}}. \tag{1.244}$$

The equilibrium of $(x_{j_2}^*, x_{j_1}^*) = (a_{j_1j_21}, a_{j_2j_11})$ is a (DP$_+$,DP$_-$)*-counterclockwise center.*

- *For $a_{j_1j_10}\prod_{s_1=1}^{2}(a_{j_2j_11} - a_{j_1j_1s_1}) > 0$ and $a_{j_2j_20}\prod_{l_1=1}^{2}(a_{j_1j_21} - a_{j_2j_2l_1}) < 0$,*

$$(a_{j_1j_21}, a_{j_2j_11}) = \underbrace{(\text{DP}_-, \text{DP}_+)}_{\text{CW center}}. \tag{1.245}$$

The equilibrium of $(x_{j_2}^*, x_{j_1}^*) = (a_{j_1j_21}, a_{j_2j_11})$ is a (DP$_-$, DP$_+$)*-clockwise center.*

- *For $a_{j_1j_10}\prod_{s_1=1}^{2}(a_{j_2j_11} - a_{j_1j_1s_1}) < 0$ and $a_{j_2j_20}\prod_{l_1=1}^{2}(a_{j_1j_21} - a_{j_2j_2l_1}) < 0$,*

$$(a_{j_1 j_2 1}, a_{j_2 j_1 1}) = \underbrace{(\text{UP}_-,\text{UP}_-)}_{\text{negative saddle}}. \tag{1.246}$$

The equilibrium of $(x^*_{j_2}, x^*_{j_1}) = (a_{j_1 j_2 1}, a_{j_2 j_1 1})$ *is a* (UP$_-$,UP$_-$)*-negative saddle.*

(ix_{Ib}*) The equilibrium of* $(x^*_{j_1}, x^*_{j_2}) = (a_{j_1 j_1 s_1}, a_{j_2 j_2 l_1})$ $(s_1, s_2, l_1, l_2 \in \{1,2\}, s_1 \neq s_2, l_1 \neq l_2)$ *has the following properties:*

- *For* $a_{j_1 j_1 0}(a_{j_2 j_2 l_1} - a_{j_1 j_2 1})(a_{j_1 j_1 s_1} - a_{j_1 j_1 s_2}) > 0$ *and* $a_{j_2 j_2 0}(a_{j_1 j_1 s_1} - a_{j_2 j_1 1})(a_{j_2 j_2 l_1} - a_{j_2 j_2 l_2}) > 0$,

$$(a_{j_1 j_1 s_1}, a_{j_2 j_2 l_1}) = \underbrace{(\text{SO,SO})}_{\text{source}}. \tag{1.247}$$

  The equilibrium of $(x^*_{j_1}, x^*_{j_2}) = (a_{j_1 j_1 s_1}, a_{j_2 j_2 l_1})$ *is an* (SO,SO) *source.*

- *For* $a_{j_1 j_1 0}(a_{j_2 j_2 l_1} - a_{j_1 j_2 1})(a_{j_1 j_1 s_1} - a_{j_1 j_1 s_2}) < 0$ *and* $a_{j_2 j_2 0}(a_{j_1 j_1 s_1} - a_{j_2 j_1 1})(a_{j_2 j_2 l_1} - a_{j_2 j_2 l_2}) > 0$,

$$(a_{j_1 j_1 1}, a_{j_2 j_2 s_1}) = \underbrace{(\text{SI,SO})}_{\text{saddle}}. \tag{1.248}$$

  The equilibrium of $(x^*_{j_1}, x^*_{j_2}) = (a_{j_1 j_1 s_1}, a_{j_2 j_2 l_1})$ *is an* (SI,SO) *saddle.*

- *For* $a_{j_1 j_1 0}(a_{j_2 j_2 l_1} - a_{j_1 j_2 1})(a_{j_1 j_1 s_1} - a_{j_1 j_1 s_2}) > 0$ *and* $a_{j_2 j_2 0}(a_{j_1 j_1 s_1} - a_{j_2 j_1 1})(a_{j_2 j_2 l_1} - a_{j_2 j_2 l_2}) < 0$,

$$(a_{j_1 j_1 1}, a_{j_2 j_2 s_1}) = \underbrace{(\text{SO,SI})}_{\text{saddle}}. \tag{1.249}$$

  The equilibrium of $(x^*_{j_1}, x^*_{j_2}) = (a_{j_1 j_1 s_1}, a_{j_2 j_2 l_1})$ *is an* (SO,SI) *saddle.*

- *For* $a_{j_1 j_1 0}(a_{j_2 j_2 l_1} - a_{j_1 j_2 1})(a_{j_1 j_1 s_1} - a_{j_1 j_1 s_2}) < 0$ *and* $a_{j_2 j_2 0}(a_{j_1 j_1 s_1} - a_{j_2 j_1 1})(a_{j_2 j_2 l_1} - a_{j_2 j_2 l_2}) < 0$,

$$(a_{j_1 j_1 1}, a_{j_2 j_2 s_1}) = \underbrace{(\text{SI,SI})}_{\text{sink}}. \tag{1.250}$$

  The equilibrium of $(x^*_{j_1}, x^*_{j_2}) = (a_{j_1 j_1 s_1}, a_{j_2 j_2 l_1})$ *is an* (SI,SI) *sink.*

(ix_{1c}) The equilibrium of $(x^*_{j_1}, x^*_{j_2}) = (a_{j_1 j_1 s_1}, a_{j_1 j_2 1})$ $(s_1,\ s_2 \in \{1,2\},\ s_1 \neq s_2)$ *with* $a_{j_1 j_2 1} \neq a_{j_2 j_2 1}, a_{j_2 j_2 2}$ *has the following properties:*

- *For* $a_{j_1 j_1 0}(a_{j_1 j_1 s_1} - a_{j_1 j_1 s_2}) > 0$ *and* $a_{j_2 j_2 0}(a_{j_1 j_1 s_1} - a_{j_2 j_1 1})\prod_{l_1=1}^{2}(a_{j_1 j_2 1} - a_{j_2 j_2 l_1}) > 0$,

$$(a_{j_1 j_1 s_1}, a_{j_1 j_2 1}) = \underbrace{\text{(DP:UP,pF)}}_{\text{hyperbolic flow }(+)}. \tag{1.251}$$

  The equilibrium of $(x^*_{j_1}, x^*_{j_2}) = (a_{j_1 j_1 s_1}, a_{j_1 j_2 1})$ *is a* (DP:UP,pF)*-positive hyperbolic flow.*

- *For* $a_{j_1 j_1 0}(a_{j_1 j_1 s_1} - a_{j_1 j_1 s_2}) < 0$ *and* $a_{j_2 j_2 0}(a_{j_1 j_1 s_1} - a_{j_2 j_1 1})\prod_{l_1=1}^{2}(a_{j_1 j_2 1} - a_{j_2 j_2 l_1}) > 0$,

$$(a_{j_1 j_1 s_1}, a_{j_1 j_2 1}) = \underbrace{\text{(UP:DP,pF)}}_{\text{hyperbolic-secant flow }(+)}. \tag{1.252}$$

  The equilibrium of $(x^*_{j_1}, x^*_{j_2}) = (a_{j_1 j_1 s_1}, a_{j_1 j_2 1})$ *is a* (UP:DP,pF)*-positive hyperbolic secant flow.*

- *For* $a_{j_1 j_1 0}(a_{j_1 j_1 s_1} - a_{j_1 j_1 s_2}) > 0$ *and* $a_{j_2 j_2 0}(a_{j_1 j_1 s_1} - a_{j_2 j_1 1})\prod_{l_1=1}^{2}(a_{j_1 j_2 1} - a_{j_2 j_2 l_1}) < 0$,

$$(a_{j_1 j_1 s_1}, a_{j_1 j_2 1}) = \underbrace{\text{(UP:DP,nF)}}_{\text{hyperbolic-secant flow }(-)}. \tag{1.253}$$

  The equilibrium of $(x^*_{j_1}, x^*_{j_2}) = (a_{j_1 j_1 s_1}, a_{j_1 j_2 1})$ *is a* (UP:DP,nF)*-negative hyperbolic secant flow.*

- *For* $a_{j_1 j_1 0}(a_{j_1 j_1 s_1} - a_{j_1 j_1 s_2}) < 0$ *and* $a_{j_2 j_2 0}(a_{j_1 j_1 s_1} - a_{j_2 j_1 1})\prod_{l_1=1}^{2}(a_{j_1 j_2 1} - a_{j_2 j_2 l_1}) < 0$,

$$(a_{j_1 j_1 s_1}, a_{j_1 j_2 1}) = \underbrace{\text{(DP:UP,nF)}}_{\text{hyperbolic flow }(-)}. \tag{1.254}$$

  The equilibrium of $(x^*_{j_1}, x^*_{j_2}) = (a_{j_1 j_1 s_1}, a_{j_1 j_2 1})$ *is a* (DP:UP,nF)*-negative hyperbolic flow.*

(ix_{1d}) The equilibrium of $(x^*_{j_1}, x^*_{j_2}) = (a_{j_2 j_1 1}, a_{j_2 j_2 l_1})$ $(l_1,\ l_2 \in \{1,2\},\ l_1 \neq l_2)$ *with* $a_{j_2 j_1 1} \neq a_{j_1 j_1 1}, a_{j_1 j_1 2}$ *has the following properties:*

- *For* $a_{j_1 j_1 0}(a_{j_2 j_2 l_1} - a_{j_1 j_2 1})\prod_{s_1=1}^{2}(a_{j_2 j_1 1} - a_{j_1 j_1 s_1}) > 0$ *and* $a_{j_2 j_2 0}(a_{j_1 j_2 l_1} - a_{j_2 j_2 l_2}) > 0$,

$$(a_{j_2 j_1 1}, a_{j_2 j_2 l_1}) = \underbrace{\text{(pF,DP:UP)}}_{\text{hyperbolic flow }(+)}. \tag{1.255}$$

The equilibrium of $(x^*_{j_1}, x^*_{j_2}) = (a_{j_2 j_1 1}, a_{j_2 j_2 l_1})$ *is a* (pF,DP:UP)-*positive hyperbolic flow.*

- *For* $a_{j_1 j_1 0}(a_{j_2 j_2 l_1} - a_{j_1 j_2 1})\prod_{s_1=1}^{2}(a_{j_2 j_1 1} - a_{j_1 j_1 s_1}) < 0$ *and* $a_{j_2 j_2 0}(a_{j_1 j_2 l_1} - a_{j_2 j_2 l_2}) > 0$,

$$(a_{j_1 j_1 s_1}, a_{j_1 j_2 1}) = \underbrace{(\text{nF,UP:DP})}_{\text{hyperbolic-secant flow }(-)}. \tag{1.256}$$

The equilibrium of $(x^*_{j_1}, x^*_{j_2}) = (a_{j_1 j_1 s_1}, a_{j_1 j_2 1})$ *is an* (nF, UP:DP)-*negative hyperbolic secant flow.*

- *For* $a_{j_1 j_1 0}(a_{j_2 j_2 l_1} - a_{j_1 j_2 1})\prod_{s_1=1}^{2}(a_{j_2 j_1 1} - a_{j_1 j_1 s_1}) > 0$ *and* $a_{j_2 j_2 0}(a_{j_1 j_2 l_1} - a_{j_2 j_2 l_2}) < 0$,

$$(a_{j_1 j_1 s_1}, a_{j_1 j_2 1}) = \underbrace{(\text{pF,UP:DP})}_{\text{hyperbolic-secant flow }(+)}. \tag{1.257}$$

The equilibrium of $(x^*_{j_1}, x^*_{j_2}) = (a_{j_1 j_1 s_1}, a_{j_1 j_2 1})$ *is a* (pF, UP:DP)-*positive hyperbolic secant flow.*

- *For* $a_{j_1 j_1 0}(a_{j_2 j_2 l_1} - a_{j_1 j_2 1})\prod_{s_1=1}^{2}(a_{j_2 j_1 1} - a_{j_1 j_1 s_1}) < 0$ *and* $a_{j_2 j_2 0}(a_{j_1 j_2 l_1} - a_{j_2 j_2 l_2}) < 0$,

$$(a_{j_1 j_1 s_1}, a_{j_1 j_2 1}) = \underbrace{(\text{nF,DP:UP})}_{\text{hyperbolic flow }(-)}. \tag{1.258}$$

The equilibrium of $(x^*_{j_1}, x^*_{j_2}) = (a_{j_1 j_1 s_1}, a_{j_1 j_2 1})$ *is an* (nF, DP:UP)-*negative hyperbolic flow.*

(ix$_2$) If $a_{j_1 j_2 1} = a_{j_2 j_2 l_1}$ *and* $a_{j_2 j_1 1} \neq a_{j_1 j_1 s_1}$ $(s_1, s_2, l_1, l_2 \in \{1, 2\},\ s_1 \neq s_2,\ l_1 \neq l_2)$, *the first integral manifold is*

$$\sum_{s_1=1, s_2 \neq s_1}^{2} \frac{a_{j_1 j_1 s_1} - a_{j_2 j_1 1}}{a_{j_1 j_1 s_1} - a_{j_1 j_1 s_1}} \ln \frac{|x_{j_1} - a_{j_1 j_1 s_1}|}{|x_{j_1 0} - a_{j_1 j_1 s_1}|} = \frac{a_{j_1 j_1 0}}{a_{j_2 j_2 0}} \ln \frac{|x_{j_2} - a_{j_2 j_2 l_2}|}{|x_{j_2 0} - a_{j_2 j_2 l_2}|}. \tag{1.259}$$

(ix$_{2a}$) The infinite equilibrium of $x^*_{j_2} = a_{j_1 j_2 1} = a_{j_2 j_2 l_1}$ *with* $\overline{x}_{j_1} \neq a_{j_1 j_1 s_1}, a_{j_1 j_1 s_2}, a_{j_2 j_1 1}$ *has the following properties:*

- *For* $a_{j_1 j_1 0}\prod_{s_1=1}^{2}(\overline{x}_{j_1} - a_{j_1 j_1 s_1}) > 0$ *and* $a_{j_2 j_2 0}(\overline{x}_{j_1} - a_{j_2 j_1 1})(a_{j_2 j_2 l_1} - a_{j_2 j_2 l_2}) > 0$,

$$(\bar{x}_{j_1}, a_{j_2 j_2 l_1}) = \underbrace{(\mathrm{II,SO})}_{\text{increasing-inflection source}} . \tag{1.260}$$

The infinite equilibrium of $x^*_{j_2} = a_{j_1 j_2 1} = a_{j_2 j_2 l_1}$ *is an* (II,SO)-*increasing inflection source.*

- *For* $a_{j_1 j_1 0} \prod_{s_1=1}^{2} (\bar{x}_{j_1} - a_{j_1 j_1 s_1}) < 0$ *and* $a_{j_2 j_2 0}(\bar{x}_{j_1} - a_{j_2 j_1 1})(a_{j_2 j_2 l_1} - a_{j_2 j_2 l_2}) > 0$,

$$(\bar{x}_{j_1}, a_{j_2 j_2 l_1}) = \underbrace{(\mathrm{DI,SO})}_{\text{decreasing-inflection source}} . \tag{1.261}$$

The infinite equilibrium of $x^*_{j_2} = a_{j_1 j_2 1} = a_{j_2 j_2 l_1}$ *is a* (DI,SO)-*decreasing inflection source.*

- *For* $a_{j_1 j_1 0} \prod_{s_1=1}^{2} (\bar{x}_{j_1} - a_{j_1 j_1 s_1}) > 0$ *and* $a_{j_2 j_2 0}(\bar{x}_{j_1} - a_{j_2 j_1 1})(a_{j_2 j_2 l_1} - a_{j_2 j_2 l_2}) < 0$,

$$(\bar{x}_{j_1}, a_{j_2 j_2 l_1}) = \underbrace{(\mathrm{DI,SI})}_{\text{decreasing-inflection sink}} . \tag{1.262}$$

The infinite equilibrium of $x^*_{j_2} = a_{j_1 j_2 1} = a_{j_2 j_2 l_1}$ *is a* (DI,SI)-*decreasing inflection sink.*

- *For* $a_{j_1 j_1 0} \prod_{s_1=1}^{2} (\bar{x}_{j_1} - a_{j_1 j_1 s_1}) < 0$ *and* $a_{j_2 j_2 0}(\bar{x}_{j_1} - a_{j_2 j_1 1})(a_{j_2 j_2 l_1} - a_{j_2 j_2 l_2}) < 0$,

$$(\bar{x}_{j_1}, a_{j_2 j_2 l_1}) = \underbrace{(\mathrm{II,SI})}_{\text{increasing-inflection sink}} . \tag{1.263}$$

The infinite equilibrium of $x^*_{j_2} = a_{j_1 j_2 1} = a_{j_2 j_2 l_1}$ *is an* (II,SI)-*increasing inflection sink.*

(ix$_{2b}$*) The equilibrium of* $(x^*_{j_1}, x^*_{j_2}) = (a_{j_1 j_1 s_1}, a_{j_2 j_2 l_1})$ *with* $a_{j_1 j_2 1} = a_{j_2 j_2 l_1}$ and $a_{j_2 j_1 1} \neq a_{j_1 j_1 s_1}$ *has the following properties:*

- *For* $a_{j_1 j_1 0}(a_{j_1 j_1 s_1} - a_{j_1 j_1 s_2}) > 0$ *and* $a_{j_2 j_2 0}(a_{j_1 j_1 s_1} - a_{j_2 j_1 1})(a_{j_2 j_2 l_1} - a_{j_2 j_2 l_2}) > 0$,

$$(a_{j_1 j_1 s_1}, a_{j_2 j_2 l_1}) = \underbrace{(\mathrm{DI{:}II,SO})}_{\text{inflection source}} . \tag{1.264}$$

The equilibrium of $(x^*_{j_1}, x^*_{j_2}) = (a_{j_1 j_1 s_1}, a_{j_2 j_2 l_1})$ *is a* (DI:II,SO)-*inflection source.*

- *For* $a_{j_1 j_1 0}(a_{j_1 j_1 s_1} - a_{j_1 j_1 s_2}) < 0$ *and* $a_{j_2 j_2 0}(a_{j_1 j_1 s_1} - a_{j_2 j_1 1})(a_{j_2 j_2 l_1} - a_{j_2 j_2 l_2}) > 0$,

$$(a_{j_1j_1s_1}, a_{j_2j_2l_1}) = \underbrace{(\text{II:DI,SO})}_{\text{inflection source}}. \tag{1.265}$$

The equilibrium of $(x^*_{j_1}, x^*_{j_2}) = (a_{j_1j_1s_1}, a_{j_2j_2l_1})$ *is an* (II:DI,SO)-*inflection source.*

- *For* $a_{j_1j_10}(a_{j_1j_1s_1} - a_{j_1j_1s_2}) > 0$ *and* $a_{j_2j_20}(a_{j_1j_1s_1} - a_{j_2j_11})(a_{j_2j_2l_1} - a_{j_2j_2l_2}) < 0$,

$$(a_{j_1j_1s_1}, a_{j_2j_2l_1}) = \underbrace{(\text{II:DI,SI})}_{\text{inflection sink}}. \tag{1.266}$$

The equilibrium of $(x^*_{j_1}, x^*_{j_2}) = (a_{j_1j_1s_1}, a_{j_2j_2l_1})$ *is an* (II:DI,SI)-*inflection sink.*

- *For* $a_{j_1j_10}(a_{j_1j_1s_1} - a_{j_1j_1s_2}) < 0$ *and* $a_{j_2j_20}(a_{j_1j_1s_1} - a_{j_2j_11})(a_{j_2j_2l_1} - a_{j_2j_2l_2}) < 0$,

$$(a_{j_1j_1s_1}, a_{j_2j_2l_1}) = \underbrace{(\text{DI:II,SI})}_{\text{inflection sink}}. \tag{1.267}$$

The equilibrium of $(x^*_{j_1}, x^*_{j_2}) = (a_{j_1j_1s_1}, a_{j_2j_2l_1})$ *is a* (DI:II,SI)-*inflection sink.*

(ix$_{2c}$) The equilibrium of $(x^*_{j_2}, x^*_{j_1}) = (a_{j_1j_21}, a_{j_2j_11})$ *with* $a_{j_1j_21} = a_{j_2j_2l_1}$ and $a_{j_2j_11} \neq a_{j_1j_1s_1}$ *has the following properties:*

- *For* $a_{j_1j_10}\prod_{s_1=1}^{2}(a_{j_2j_11} - a_{j_1j_1s_1}) > 0$ *and* $a_{j_2j_20}(a_{j_2j_2l_1} - a_{j_2j_2l_2}) > 0$,

$$(a_{j_1j_21}, a_{j_2j_11}) = \underbrace{({}_{\text{SI:SO}}\text{US}, {}_{\text{DI:II}}\text{UP})}_{\text{up-parabola upper-saddle}}. \tag{1.268}$$

The equilibrium of $(x^*_{j_2}, x^*_{j_1}) = (a_{j_1j_21}, a_{j_2j_11})$ *is an* (${}_{\text{SI:SO}}$US, ${}_{\text{DI:II}}$UP)-*up-parabola upper-saddle.*

- *For* $a_{j_1j_10}\prod_{s_1=1}^{2}(a_{j_2j_11} - a_{j_1j_1s_1}) < 0$ *and* $a_{j_2j_20}(a_{j_2j_2l_1} - a_{j_2j_2l_2}) > 0$,

$$(a_{j_1j_21}, a_{j_2j_11}) = \underbrace{({}_{\text{SI:SO}}\text{US}, {}_{\text{II:DI}}\text{DP})}_{\text{down-parabola upper-saddle}}. \tag{1.269}$$

The equilibrium of $(x^*_{j_2}, x^*_{j_1}) = (a_{j_1j_21}, a_{j_2j_11})$ *is an* (${}_{\text{SI:SO}}$US, ${}_{\text{II:DI}}$DP)-*down-parabola upper-saddle.*

- *For* $a_{j_1j_10}\prod_{s_1=1}^{2}(a_{j_2j_11} - a_{j_1j_1s_1}) > 0$ *and* $a_{j_2j_20}(a_{j_2j_2l_1} - a_{j_2j_2l_2}) < 0$,

$$(a_{j_1j_21}, a_{j_2j_11}) = \underbrace{({}_{\text{SO:SI}}\text{LS}, {}_{\text{II:DI}}\text{DP})}_{\text{down-parabola lower-saddle}}. \tag{1.270}$$

The equilibrium of $(x^*_{j_2}, x^*_{j_1}) = (a_{j_1j_21}, a_{j_2j_11})$ *is an* (${}_{\text{SO:SI}}$LS, ${}_{\text{II:DI}}$DP)-*down-parabola lower-saddle.*

- *For* $a_{j_1j_10}\prod_{s_1=1}^{2}(a_{j_2j_11} - a_{j_1j_1s_1}) < 0$ *and* $a_{j_2j_20}(a_{j_2j_2l_1} - a_{j_2j_2l_2}) < 0$,

$$(a_{j_1j_21}, a_{j_2j_11}) = \underbrace{({}_{\mathrm{SO:SI}}\mathrm{LS}, {}_{\mathrm{DI:II}}\mathrm{UP})}_{\text{up-parabola lower-saddle}}. \tag{1.271}$$

The equilibrium of $(x_{j_2}^*, x_{j_1}^*) = (a_{j_1j_21}, a_{j_2j_11})$ *is an* (${}_{\mathrm{SO:SI}}$LS, ${}_{\mathrm{DI:II}}$UP)*-up-parabola lower-saddle.*

(ix₃) The first integral manifold for $a_{j_1j_1s_1} = a_{j_2j_11}$ *and* $a_{j_1j_21} \neq a_{j_2j_2l_1}$ $(s_1, s_2, l_1, l_2 \in \{1,2\}, s_1 \neq s_2, l_1 \neq l_2)$ *is*

$$\ln\frac{|x_{j_1} - a_{j_1j_1s_2}|}{|x_{j_10} - a_{j_1j_1s_2}|} = \frac{a_{j_1j_10}}{a_{j_2j_20}}\sum\nolimits_{l_1=1, l_2\neq l_1}^{2}\frac{a_{j_2j_2l_1} - a_{j_1j_21}}{a_{j_2j_2l_1} - a_{j_2j_2l_2}}\ln\frac{|x_{j_2} - a_{j_2j_2l_1}|}{|x_{j_20} - a_{j_2j_2l_1}|}. \tag{1.272}$$

(ix₃ₐ) The infinite equilibrium of $x_{j_1}^* = a_{j_1j_1s_1} = a_{j_2j_11}$ $(s_1, s_2 \in \{1,2\})$ *with* $\bar{x}_{j_2} \neq a_{j_2j_21}, a_{j_2j_22}, a_{j_1j_21}$ *has the following properties:*

- *For* $a_{j_1j_10}(\bar{x}_{j_2} - a_{j_1j_21})(a_{j_1j_1s_1} - a_{j_1j_1s_2}) > 0$ *and* $a_{j_2j_20}\prod_{l_1=1}^{2}(\bar{x}_{j_2} - a_{j_2j_2l_1}) > 0$,

$$(a_{j_1j_1s_1}, \bar{x}_{j_2}) = \underbrace{(\mathrm{SO,II})}_{\text{increasing-inflection source}}. \tag{1.273}$$

The infinite equilibrium of $x_{j_1}^* = a_{j_2j_11} = a_{j_1j_1s_1}$ *is an* (SO, II)*-increasing-inflection source.*

- *For* $a_{j_1j_10}(\bar{x}_{j_2} - a_{j_1j_21})(a_{j_1j_1s_1} - a_{j_1j_1s_2}) < 0$ *and* $a_{j_2j_20}\prod_{l_1=1}^{2}(\bar{x}_{j_2} - a_{j_2j_2l_1}) > 0$,

$$(a_{j_1j_1s_1}, \bar{x}_{j_2}) = \underbrace{(\mathrm{SI,DI})}_{\text{decreasing-inflection sink}}. \tag{1.274}$$

The infinite equilibrium of $x_{j_1}^* = a_{j_2j_11} = a_{j_1j_1s_1}$ *is an* (SI, DI)*-decreasing-inflection sink.*

- *For* $a_{j_1j_10}(\bar{x}_{j_2} - a_{j_1j_21})(a_{j_1j_1s_1} - a_{j_1j_1s_2}) > 0$ *and* $a_{j_2j_20}\prod_{l_1=1}^{2}(\bar{x}_{j_2} - a_{j_2j_2l_1}) < 0$,

$$(a_{j_1j_1s_1}, \bar{x}_{j_2}) = \underbrace{(\mathrm{SO,DI})}_{\text{decreasing-inflection source}}. \tag{1.275}$$

The infinite equilibrium of $x_{j_1}^* = a_{j_2j_11} = a_{j_1j_1s_1}$ *is an* (SO, DI)*-decreasing-inflection source.*

- *For* $a_{j_1j_10}(\bar{x}_{j_2} - a_{j_1j_21})(a_{j_1j_1s_1} - a_{j_1j_1s_2}) < 0$ *and* $a_{j_2j_20}\prod_{l_1=1}^{2}(\bar{x}_{j_2} - a_{j_2j_2l_1}) < 0$,

$$(a_{j_1j_1s_1}, \bar{x}_{j_2}) = \underbrace{(\mathrm{SI,II})}_{\text{increasing-inflection sink}}. \tag{1.276}$$

The infinite equilibrium of $x_{j_1}^* = a_{j_2j_11} = a_{j_1j_1s_1}$ *is an* (SI,II)*-increasing-inflection sink.*

(ix_{3b}) The equilibrium of $(x_{j_2}^*, x_{j_1}^*) = (a_{j_1 j_2 1}, a_{j_2 j_1 1})$ *for* $a_{j_1 j_1 s_1} = a_{j_2 j_1 1}$ and $a_{j_1 j_2 1} \neq a_{j_2 j_2 l_1}$ *has the following properties:*

- *For* $a_{j_1 j_1 0}(a_{j_2 j_1 1} - a_{j_1 j_1 s_2}) > 0$ *and* $a_{j_2 j_2 0} \prod_{l_1=1}^{2} (a_{j_1 j_2 1} - a_{j_2 j_2 l_1}) > 0$,

$$(a_{j_1 j_2 1}, a_{j_2 j_1 1}) = \underbrace{({}_{\text{DI:II}}\text{UP}, {}_{\text{SI:SO}}\text{US})}_{\text{up-parabola upper-saddle}}. \tag{1.277}$$

  The equilibrium of $(x_{j_2}^*, x_{j_1}^*) = (a_{j_1 j_2 1}, a_{j_2 j_1 1})$ *is a* (${}_{\text{DI:II}}$UP, ${}_{\text{SI:SO}}$US)*-up-parabola upper-saddle.*

- *For* $a_{j_1 j_1 0}(a_{j_2 j_1 1} - a_{j_1 j_1 s_2}) < 0$ *and* $a_{j_2 j_2 0} \prod_{l_1=1}^{2} (a_{j_1 j_2 1} - a_{j_2 j_2 l_1}) > 0$,

$$(a_{j_1 j_2 1}, a_{j_2 j_1 1}) = \underbrace{({}_{\text{II:DI}}\text{DP}, {}_{\text{SO:SI}}\text{LS})}_{\text{down-parabola lower-saddle}}. \tag{1.278}$$

  The equilibrium of $(x_{j_2}^*, x_{j_1}^*) = (a_{j_1 j_2 1}, a_{j_2 j_1 1})$ *is a* (${}_{\text{II:DI}}$DP:${}_{\text{SO:SI}}$LS)*-down-parabola lower-saddle.*

- *For* $a_{j_1 j_1 0}(a_{j_2 j_1 1} - a_{j_1 j_1 s_2}) > 0$ *and* $a_{j_2 j_2 0} \prod_{l_1=1}^{2} (a_{j_1 j_2 1} - a_{j_2 j_2 l_1}) < 0$,

$$(a_{j_1 j_2 1}, a_{j_2 j_1 1}) = \underbrace{({}_{\text{II:DI}}\text{DP}, {}_{\text{SI:SO}}\text{US})}_{\text{inflection down-parabola upper-saddle}}. \tag{1.279}$$

  The equilibrium of $(x_{j_2}^*, x_{j_1}^*) = (a_{j_1 j_2 1}, a_{j_2 j_1 1})$ *is a* (${}_{\text{II:DI}}$DP,${}_{\text{SO:SI}}$US)-down-parabola lower-saddle.

- *For* $a_{j_1 j_1 0}(a_{j_2 j_1 1} - a_{j_1 j_1 s_2}) < 0$ *and* $a_{j_2 j_2 0} \prod_{l_1=1}^{2} (a_{j_1 j_2 1} - a_{j_2 j_2 l_1}) < 0$,

$$(a_{j_1 j_2 1}, a_{j_2 j_1 1}) = \underbrace{({}_{\text{DI:II}}\text{UP}, {}_{\text{SO:SI}}\text{LS})}_{\text{up-parabola lower-saddle}}. \tag{1.280}$$

  The equilibrium of $(x_{j_2}^*, x_{j_1}^*) = (a_{j_1 j_2 1}, a_{j_2 j_1 1})$ *is a* (${}_{\text{DI:II}}$UP,${}_{\text{SO:SI}}$LS)*-up-parabola lower-saddle.*

(ix_{3c}) The equilibrium of $(x_{j_1}^*, x_{j_2}^*) = (a_{j_1 j_1 s_1}, a_{j_2 j_2 l_1})$ *for* $a_{j_1 j_1 s_1} = a_{j_2 j_1 1}$ and $a_{j_1 j_2 1} \neq a_{j_2 j_2 1}$ *has the following properties:*

- *For* $a_{j_1 j_1 0}(a_{j_2 j_2 l_1} - a_{j_1 j_2 1})(a_{j_1 j_1 s_1} - a_{j_1 j_1} s_2) > 0$ *and* $a_{j_2 j_2 0}(a_{j_2 j_2 l_1} - a_{j_2 j_2 l_2}) > 0$,

$$(a_{j_1 j_1 s_1}, a_{j_2 j_2 l_1}) = \underbrace{(\text{SO,DI:II})}_{\text{inflection source}}. \tag{1.281}$$

  The equilibrium of $(x_{j_1}^*, x_{j_2}^*) = (a_{j_1 j_1 s_1}, a_{j_2 j_2 l_1})$ *is an* (SO,DI:II)*-inflection source.*

- *For* $a_{j_1 j_1 0}(a_{j_2 j_2 l_1} - a_{j_1 j_2 1})(a_{j_1 j_1 s_1} - a_{j_1 j_1} s_2) < 0$ *and* $a_{j_2 j_2 0}(a_{j_2 j_2 l_1} - a_{j_2 j_2 l_2}) > 0$,

$$(a_{j_1j_1s_1}, a_{j_2j_2l_1}) = \underbrace{(\text{SI,II:DI})}_{\text{inflection sink}}. \tag{1.282}$$

The equilibrium of $(x^*_{j_1}, x^*_{j_2}) = (a_{j_1j_1s_1}, a_{j_2j_2l_1})$ *is an* (SI:II:DI)-*inflection sink.*

- *For* $a_{j_1j_1 0}(a_{j_2j_2l_1} - a_{j_1j_2 1})(a_{j_1j_1s_1} - a_{j_1j_1s_2}) > 0$ *and* $a_{j_2j_2 0}(a_{j_2j_2l_1} - a_{j_2j_2l_2}) < 0$,

$$(a_{j_1j_1s_1}, a_{j_2j_2l_1}) = \underbrace{(\text{SO,II:DI})}_{\text{inflection source}}. \tag{1.283}$$

The equilibrium of $(x^*_{j_1}, x^*_{j_2}) = (a_{j_1j_1s_1}, a_{j_2j_2 1})$ *is an* (SO,II:DI)-*inflection source.*

- *For* $a_{j_1j_1 0}(a_{j_2j_2l_1} - a_{j_1j_2 1})(a_{j_2j_1 1} - a_{j_1j_1s_2}) < 0$ *and* $a_{j_2j_2 0}(a_{j_2j_2l_1} - a_{j_2j_2l_2}) < 0$,

$$(a_{j_1j_1s_1}, a_{j_2j_2l_1}) = \underbrace{(\text{SI,DI:II})}_{\text{inflection sink}}. \tag{1.284}$$

The equilibrium of $(x^*_{j_1}, x^*_{j_2}) = (a_{j_1j_1s_1}, a_{j_2j_2l_1})$ *is an* (SI,DI:II)-*inflection sink.*

(ix$_4$) The first integral manifold for $a_{j_1j_1s_1} = a_{j_2j_1 1}$ *and* $a_{j_1j_2 1} = a_{j_2j_2l_1}$ $(s_1, s_2, l_1, l_2 \in \{1,2\},$ $s_1 \neq s_2,\ l_1 \neq l_2)$ *is*

$$\ln \frac{|x_{j_1} - a_{j_1j_1s_2}|}{|x_{j_1 0} - a_{j_1j_1s_2}|} = \frac{a_{j_1j_1 0}}{a_{j_2j_2 0}} \ln \frac{|x_{j_2} - a_{j_2j_2l_2}|}{|x_{j_2 0} - a_{j_2j_2l_2}|} \tag{1.285}$$

The equilibrium of $(x^*_{j_1}, x^*_{j_2}) = (a_{j_1j_1s_1}, a_{j_2j_2l_1})$ *for* $a_{j_1j_1s_1} = a_{j_2j_1 1}$ *and* $a_{j_1j_2 1} = a_{j_2j_2l_1}$ *has the following properties:*

- *For* $a_{j_1j_1 0}(a_{j_1j_1s_1} - a_{j_1j_1s_2}) > 0$ *and* $a_{j_2j_2 0}(a_{j_2j_2l_1} - a_{j_2j_2l_2}) > 0$,

$$(a_{j_1j_1s_1}, a_{j_2j_2l_1}) = \underbrace{({}_{\text{SI:SO}}\text{II}, {}_{\text{SI:SO}}\text{II})}_{\text{increasing-inflection diagonal-saddle}}. \tag{1.286}$$

The equilibrium of $(x^*_{j_1}, x^*_{j_2}) = (a_{j_1j_1s_1}, a_{j_2j_2l_1})$ *is an* $({}_{\text{SI:SI}}\text{II}{:}{}_{\text{SI:SO}}\text{II})$-*increasing-inflection diagonal-saddle.*

- *For* $a_{j_1j_1 0}(a_{j_1j_1s_1} - a_{j_1j_1s_2}) < 0$ *and* $a_{j_2j_2 0}(a_{j_2j_2l_1} - a_{j_2j_2l_2}) > 0$,

$$(a_{j_1j_1s_1}, a_{j_2j_2l_1}) = \underbrace{({}_{\text{SI:SO}}\text{DI}, {}_{\text{SO:SI}}\text{DI})}_{\text{decreasing-inflection diagonal-saddle}}. \tag{1.287}$$

The equilibrium of $(x^*_{j_1}, x^*_{j_2}) = (a_{j_1j_1s_1}, a_{j_2j_2l_1})$ *is an* $({}_{\text{SI:SO}}\text{DI}{:}{}_{\text{SO:SI}}\text{DI})$-*decreasing-inflection diagonal-saddle.*

- *For* $a_{j_1j_1 0}(a_{j_1j_1s_1} - a_{j_1j_1s_2}) > 0$ *and* $a_{j_2j_2 0}(a_{j_2j_2l_1} - a_{j_2j_2l_2}) < 0$,

$$(a_{j_1j_1s_1}, a_{j_2j_2l_1}) = \underbrace{({}_{\text{SO:SI}}\text{DI}, {}_{\text{SI:SO}}\text{DI})}_{\text{decreasing-inflection diagonal-saddle}}. \tag{1.288}$$

The equilibrium of $(x^*_{j_1}, x^*_{j_2}) = (a_{j_1j_1s_1}, a_{j_2j_2l_1})$ *is an* $({}_{\text{SO:SI}}\text{DI:}{}_{\text{SI:SO}}\text{DI})$*-decreasing-inflection diagonal-saddle.*

- *For* $a_{j_1j_10}(a_{j_1j_1s_1} - a_{j_1j_1s_2}) < 0$ *and* $a_{j_2j_20}(a_{j_2j_2l_1} - a_{j_2j_2l_2}) < 0$,

$$(a_{j_1j_1s_1}, a_{j_2j_2l_1}) = \underbrace{({}_{\text{SO:SI}}\text{II}, {}_{\text{SO:SI}}\text{II})}_{\text{increasing-inflection diagonal-saddle}}. \tag{1.289}$$

The equilibrium of $(x^*_{j_1}, x^*_{j_2}) = (a_{j_1j_1s_1}, a_{j_2j_2l_1})$ *is an* $({}_{\text{SO:SI}}\text{II}, {}_{\text{SO:SI}}\text{II})$*-increasing-inflection diagonal-saddle.*

1.2 Proof of Theorem 1.1

Consider a two-dimensional, nonlinear dynamical system as

$$\begin{aligned}
\dot{x}_{j_2} &= a_{j_2j_20}(x_{j_1} - a_{j_2j_11})(x^2_{j_2} + B_{j_2j_2}x_{j_2} + C_{j_2j_2}),\\
\dot{x}_{j_1} &= a_{j_1j_10}(x_{j_2} - a_{j_1j_21})(x^2_{j_1} + B_{j_1j_1}x_{j_1} + C_{j_1j_1})
\end{aligned}$$

with

$$\begin{aligned}
\Delta_{j_2j_2} &= B^2_{j_2j_2} - 4C_{j_2j_2},\\
\Delta_{j_1j_1} &= B^2_{j_1j_1} - 4C_{j_1j_1}.
\end{aligned}$$

(i) For $\Delta_{j_1j_1} < 0$ and $\Delta_{j_2j_2} < 0$, the standard form is

$$\begin{aligned}
\dot{x}_{j_2} &= a_{j_2j_20}(x_{j_1} - a_{j_2j_11})\left[(x_{j_2} - a_{j_2j_2})^2 + b_{j_2j_2}\right],\\
\dot{x}_{j_1} &= a_{j_1j_10}\left[(x_{j_1} - a_{j_1j_1})^2 + b_{j_1j_1}\right](x_{j_2} - a_{j_1j_21})
\end{aligned}$$

where

$$\begin{aligned}
a_{j_2j_2} &= -\frac{1}{2}B_{j_2j_2}, b_{j_2j_2} = \frac{1}{4}(-\Delta_{j_1j_2});\\
a_{j_1j_1} &= -\frac{1}{2}B_{j_1j_1}, b_{j_1j_1} = \frac{1}{4}(-\Delta_{j_1j_1}).
\end{aligned}$$

In phase space,

$$\frac{dx_{j_1}}{dx_{j_2}}=\frac{a_{j_1j_10}}{a_{j_2j_20}}\frac{(x_{j_2}-a_{j_1j_21})\big[(x_{j_1}-a_{j_1j_1})^2+b_{j_1j_1}\big]}{(x_{j_1}-a_{j_2j_11})\big[(x_{j_2}-a_{j_2j_2})^2+b_{j_2j_2}\big]},$$

and the deformation of the foregoing equation is

$$\begin{aligned}&\left[\frac{(x_{j_1}-a_{j_1j_1})}{(x_{j_1}-a_{j_1j_1})^2+b_{j_1j_1}}+\frac{(a_{j_1j_1}-a_{j_2j_11})}{(x_{j_1}-a_{j_1j_1})^2+b_{j_1j_1}}\right]dx_{j_1}\\&=\frac{a_{j_1j_10}}{a_{j_2j_20}}\left[\frac{(x_{j_2}-a_{j_2j_2})}{(x_{j_2}-a_{j_2j_2})^2+b_{j_2j_2}}+\frac{(a_{j_2j_2}-a_{j_1j_21})}{(x_{j_2}-a_{j_2j_2})^2+b_{j_2j_2}}\right]dx_{j_2}.\end{aligned}$$

With an initial condition of (x_{j_10},x_{j_20}) at $t=t_0$, the integration of the above equation gives

$$\begin{aligned}&\frac{1}{2}\ln\frac{|(x_{j_1}-a_{j_1j_1})^2+b_{j_1j_1}|}{|(x_{j_10}-a_{j_1j_1})^2+b_{j_1j_1}|}\\&\quad+\frac{a_{j_1j_1}-a_{j_2j_11}}{\sqrt{b_{j_1j_1}}}\Big(\arctan\frac{x_{j_1}-a_{j_1j_1}}{\sqrt{b_{j_1j_1}}}-\arctan\frac{x_{j_10}-a_{j_1j_1}}{\sqrt{b_{j_1j_1}}}\Big)\\&=\frac{a_{j_1j_10}}{a_{j_2j_20}}\left[\frac{1}{2}\ln\frac{|(x_{j_2}-a_{j_2j_2})^2+b_{j_2j_2}|}{|(x_{j_20}-a_{j_2j_2})^2+b_{j_2j_2}|}\right.\\&\quad\left.+\frac{a_{j_2j_2}-a_{j_1j_21}}{\sqrt{b_{j_2j_2}}}\Big(\arctan\frac{x_{j_2}-a_{j_2j_2}}{\sqrt{b_{j_2j_2}}}-\arctan\frac{x_{j_20}-a_{j_2j_2}}{\sqrt{b_{j_2j_2}}}\Big)\right].\end{aligned}$$

Consider two cases (I) $x^*_{j_2}=a_{j_1j_21}$ and (II) $x^*_{j_1}=a_{j_2j_11}$.

(I) At $x^*_{j_2}=a_{j_1j_21}$ with $\overline{x}_{j_1}=a_{j_2j_11}$, in phase space,

$$\frac{dx_{j_1}}{dx_{j_2}}\Big|_{x^*_{j_2}=a_{j_1j_21}}=\frac{a_{j_1j_10}}{a_{j_2j_20}}\frac{(x_{j_2}-a_{j_1j_21})\big[(x_{j_1}-a_{j_1j_1})^2+b_{j_1j_1}\big]}{(\overline{x}_{j_1}-a_{j_2j_11})\big[(x_{j_2}-a_{j_2j_2})^2+b_{j_2j_2}\big]}\Big|_{x^*_{j_2}=a_{j_1j_21}}=0.$$

If

$$\frac{d^2x_{j_1}}{dx^2_{j_2}}\Big|_{x^*_{j_2}=a_{j_1j_21}}=\frac{a_{j_1j_10}}{a_{j_2j_20}}\frac{(\overline{x}_{j_1}-a_{j_1j_1})^2+b_{j_1j_1}}{(\overline{x}_{j_1}-a_{j_2j_11})\big[(a_{j_1j_21}-a_{j_2j_2})^2+b_{j_2j_2}\big]}>0,$$

there is an up-parabola flow. If

$$\left.\frac{d^2x_{j_1}}{dx_{j_2}^2}\right|_{x_{j_2}^*=a_{j_1j_21}} = \frac{a_{j_1j_10}}{a_{j_2j_20}} \frac{(\bar{x}_{j_1}-a_{j_1j_1})^2+b_{j_1j_1}}{(\bar{x}_{j_1}-a_{j_2j_11})\left[(a_{j_1j_21}-a_{j_2j_2})^2+b_{j_2j_2}\right]} < 0,$$

there is a down-parabola flow. Let

$$\dot{x}_{j_2} = a_{j_2j_20}(\bar{x}_{j_1}-a_{j_2j_11})\left[(a_{j_1j_21}-a_{j_2j_2})^2+b_{j_2j_2}\right].$$

Because

$$(a_{j_1j_21}-a_{j_2j_2})^2+b_{j_2j_2} > 0,$$

the parabola flows at $x_{j_1}^* = a_{j_1j_11}$ are positive and negative, for $a_{j_2j_20} > 0$ and $a_{j_2j_20} < 0$, respectively. In addition,

$$(\bar{x}_{j_1}-a_{j_1j_1})^2+b_{j_1j_1} > 0.$$

Thus, the equilibrium of $x_{j_2}^* = a_{j_1j_21}$ with $\bar{x}_{j_1} = a_{j_2j_11}$ has the following properties:

- For $a_{j_1j_10} > 0$ and $a_{j_2j_20}(\bar{x}_{j_1}-a_{j_2j_11}) > 0$,

$$(a_{j_1j_21}, \dot{x}_{j_2}) = \underbrace{\text{(UP,pF)}}_{\text{up-parabola flow }(+)}.$$

- For $a_{j_1j_10} < 0$ and $a_{j_2j_20}(\bar{x}_{j_1}-a_{j_2j_11}) > 0$,

$$(a_{j_1j_21}, \dot{x}_{j_2}) = \underbrace{\text{(DP,pF)}}_{\text{down-parabola flow }(+)}.$$

- For $a_{j_1j_10} > 0$ and $a_{j_2j_20}(\bar{x}_{j_1}-a_{j_2j_11}) < 0$,

$$(a_{j_1j_21}, \dot{x}_{j_2}) = \underbrace{\text{(DP,nF)}}_{\text{down-parabola flow }(-)}.$$

- For $a_{j_1j_10} < 0$ and $a_{j_2j_20}(\bar{x}_{j_1}-a_{j_2j_11}) < 0$,

$$(a_{j_1j_21}, \dot{x}_{j_2}) = \underbrace{\text{(UP,nF)}}_{\text{up-parabola flow }(-)}.$$

(II) Similarly, at $x_{j_1}^* = a_{j_2j_11}$ with $\overline{x}_{j_2} = a_{j_1j_21}$, in phase space,

$$\frac{dx_{j_2}}{dx_{j_1}}\bigg|_{x_{j_1}^*=a_{j_2j_11}} = \frac{a_{j_2j_20}}{a_{j_1j_10}}\frac{(x_{j_1} - a_{j_2j_11})\left[(\overline{x}_{j_2} - a_{j_2j_2})^2 + b_{j_2j_2}\right]}{(\overline{x}_{j_2} - a_{j_1j_21})\left[(x_{j_1} - a_{j_1j_1})^2 + b_{j_1j_1}\right]}\bigg|_{x_{j_1}^*=a_{j_2j_11}} = 0.$$

If

$$\frac{d^2x_{j_2}}{dx_{j_1}^2}\bigg|_{x_{j_1}^*=a_{j_2j_11}} = \frac{a_{j_2j_20}}{a_{j_1j_10}}\frac{(\overline{x}_{j_2} - a_{j_2j_2})^2 + b_{j_2j_2}}{(\overline{x}_{j_2} - a_{j_1j_21})\left[(a_{j_2j_11} - a_{j_1j_1})^2 + b_{j_1j_1}\right]} > 0,$$

there is an up-parabola flow. If

$$\frac{d^2x_{j_2}}{dx_{j_1}^2}\bigg|_{x_{j_1}^*=a_{j_2j_11}} = \frac{a_{j_2j_20}}{a_{j_1j_10}}\frac{(\overline{x}_{j_2} - a_{j_2j_2})^2 + b_{j_2j_2}}{(\overline{x}_{j_2} - a_{j_1j_21})\left[(a_{j_2j_11} - a_{j_1j_1})^2 + b_{j_1j_1}\right]} < 0,$$

there is a down-parabola flow. Let

$$\dot{x}_{j_1} = a_{j_1j_10}(\overline{x}_{j_2} - a_{j_1j_21})\left[(a_{j_2j_11} - a_{j_1j_1})^2 + b_{j_1j_1}\right].$$

Because

$$(a_{j_2j_11} - a_{j_1j_1})^2 + b_{j_1j_1} > 0,$$

the parabola flows at $x_{j_1}^* = a_{j_1j_11}$ are positive and negative, for $a_{j_1j_10} > 0$ and $a_{j_1j_10} < 0$, respectively. In addition,

$$(\overline{x}_{j_2} - a_{j_2j_2})^2 + b_{j_2j_2} > 0.$$

Thus, the equilibrium of $x_{j_1}^* = a_{j_2j_11}$ with $\overline{x}_{j_2} = a_{j_1j_21}$ has the following properties:

- For $a_{j_1j_10}(\overline{x}_{j_2} - a_{j_1j_21}) > 0$ and $a_{j_2j_20} > 0$,

$$(\dot{x}_{j_1}, a_{j_2j_11}) = \underbrace{(\text{pF,UP})}_{\text{up-parabola flow } (+)}.$$

- For $a_{j_1j_10}(\overline{x}_{j_2} - a_{j_1j_21}) < 0$ and $a_{j_2j_20} > 0$,

$$(\dot{x}_{j_1}, a_{j_2j_11}) = \underbrace{(\text{nF,DP})}_{\text{down-parabola flow } (-)}.$$

- For $a_{j_1j_10}(\overline{x}_{j_2} - a_{j_1j_21}) > 0$ and $a_{j_2j_20} < 0$,

$$(\dot{x}_{j_1}, a_{j_2j_11}) = \underbrace{\text{(pF,DP)}}_{\text{down-parabola flow }(+)}.$$

- For $a_{j_1j_10}(\overline{x}_{j_2} - a_{j_1j_21}) < 0$ and $a_{j_2j_20} < 0$,

$$(\dot{x}_{j_1}, a_{j_2j_11}) = \underbrace{\text{(nF,UP)}}_{\text{up-parabola flow }(-)}.$$

Therefore, from the analysis, the equilibrium of $(x^*_{j_2}, x^*_{j_1}) = (a_{j_1j_21}, a_{j_2j_11})$ has the following properties as in Eqs. (1.6)–(1.9):

- For $a_{j_1j_10} > 0$ and $a_{j_2j_20} > 0$,

$$(a_{j_1j_21}, a_{j_2j_11}) = \underbrace{(\text{UP}_+,\text{UP}_+)}_{\text{positive saddle}}.$$

  The equilibrium of $(x^*_{j_2}, x^*_{j_1}) = (a_{j_1j_21}, a_{j_2j_11})$ is a (UP$_+$, UP$_+$)-positive saddle.
- For $a_{j_1j_10} < 0$ and $a_{j_2j_20} > 0$,

$$(a_{j_1j_21}, a_{j_2j_11}) = \underbrace{(\text{DP}_+,\text{DP}_-)}_{\text{CCW center}}.$$

  The equilibrium of $(x^*_{j_2}, x^*_{j_1}) = (a_{j_1j_21}, a_{j_2j_11})$ is a (DP$_+$,DP$_-$)-counterclockwise center.
- For $a_{j_1j_10} > 0$ and $a_{j_2j_20} < 0$,

$$(a_{j_1j_21}, a_{j_2j_11}) = \underbrace{(\text{DP}_-,\text{DP}_+)}_{\text{CW center}}.$$

  The equilibrium of $(x^*_{j_2}, x^*_{j_1}) = (a_{j_1j_21}, a_{j_2j_11})$ is a (DP$_-$, DP$_+$)-clockwise center.
- For $a_{j_1j_10} < 0$ and $a_{j_2j_20} < 0$,

$$(a_{j_1j_21}, a_{j_2j_11}) = \underbrace{(\text{UP}_-,\text{UP}_-)}_{\text{negative saddle}}.$$

  The equilibrium of $(x^*_{j_2}, x^*_{j_1}) = (a_{j_1j_21}, a_{j_2j_11})$ is a (UP$_-$,UP$_-$)-negative saddle.

(ii) For $\Delta_{j_1j_2} < 0$ and $\Delta_{j_2j_2} = 0$, the standard form is

$$\begin{aligned}\dot{x}_{j_2} &= a_{j_2j_20}(x_{j_1} - a_{j_2j_11})(x_{j_2} - a_{j_2j_21})^2,\\ \dot{x}_{j_1} &= a_{j_1j_10}(x_{j_2} - a_{j_1j_21})\left[(x_{j_1} - a_{j_1j_1})^2 + b_{j_1j_1}\right]\end{aligned}$$

where

$$\begin{aligned}a_{j_1j_1} &= -\frac{1}{2}B_{j_1j_1}, b_{j_1j_1} = \frac{1}{4}(-\Delta_{j_1j_1}),\\ a_{j_2j_21} &= -\frac{1}{2}B_{j_2j_2}.\end{aligned}$$

In phase space,

$$\frac{dx_{j_1}}{dx_{j_2}} = \frac{a_{j_1j_10}}{a_{j_2j_20}} \frac{(x_{j_2} - a_{j_1j_21})\left[(x_{j_1} - a_{j_1j_1})^2 + b_{j_1j_1}\right]}{(x_{j_1} - a_{j_2j_11})(x_{j_2} - a_{j_2j_21})^2},$$

and the deformation of the foregoing equation is

$$\begin{aligned}&\left[\frac{(x_{j_1} - a_{j_jj_1})}{(x_{j_1} - a_{j_1j_1})^2 + b_{j_1j_1}} + \frac{(a_{j_jj_1} - a_{j_2j_11})}{(x_{j_1} - a_{j_1j_1})^2 + b_{j_1j_1}}\right] dx_{j_1}\\ &\quad = \frac{a_{j_1j_10}}{a_{j_2j_20}}\left[\frac{1}{(x_{j_2} - a_{j_2j_21})} + \frac{(a_{j_2j_21} - a_{j_1j_21})}{(x_{j_2} - a_{j_2j_21})^2}\right] dx_{j_2}.\end{aligned}$$

With an initial condition of (x_{j_10}, x_{j_20}) at $t = t_0$, the integration of the above equation gives

$$\begin{aligned}&\frac{1}{2}\ln\frac{|(x_{j_1} - a_{j_1j_1})^2 + b_{j_1j_1}|}{|(x_{j_10} - a_{j_1j_1})^2 + b_{j_1j_1}|}\\ &\quad + \frac{a_{j_1j_1} - a_{j_2j_11}}{\sqrt{b_{j_1j_1}}}(\arctan\frac{x_{j_1} - a_{j_1j_1}}{\sqrt{b_{j_1j_1}}} - \arctan\frac{x_{j_10} - a_{j_1j_1}}{\sqrt{b_{j_1j_1}}})\\ &= \frac{a_{j_1j_10}}{a_{j_2j_20}}\left[\ln\frac{|x_{j_2} - a_{j_2j_21}|}{|x_{j_20} - a_{j_2j_21}|} + (a_{j_2j_21} - a_{j_1j_21})(\frac{1}{x_{j_2} - a_{j_2j_21}} - \frac{1}{x_{j_20} - a_{j_2j_21}})\right].\end{aligned}$$

(ii_{1a}) Consider two cases (I) $x^*_{j_2} = a_{j_1j_21}$ and (II) $x^*_{j_1} = a_{j_2j_11}$.
(I) At $x^*_{j_2} = a_{j_1j_21} \neq a_{j_2j_21}$, similar to the proof (I) of case (i), in phase space,

$$\frac{dx_{j_1}}{dx_{j_2}}\Big|_{x^*_{j_2} = a_{j_1j_21}} = \frac{a_{j_1j_10}}{a_{j_2j_20}} \frac{(x_{j_2} - a_{j_1j_21})\left[(x_{j_1} - a_{j_1j_1})^2 + b_{j_1j_1}\right]}{(\overline{x}_{j_1} - a_{j_2j_11})(x_{j_2} - a_{j_2j_21})^2}\Big|_{x^*_{j_2} = a_{j_1j_21}} = 0.$$

If

$$\frac{d^2 x_{j_1}}{dx_{j_2}^2}\Big|_{x_{j_2}^*=a_{j_1j_2 1}} = \frac{a_{j_1j_1 0}}{a_{j_2j_2 0}} \frac{(\overline{x}_{j_1} - a_{j_1j_1})^2 + b_{j_1j_1}}{(\overline{x}_{j_1} - a_{j_2j_1 1})(a_{j_1j_2 1} - a_{j_2j_2 1})^2} > 0,$$

there is an up-parabola flow. If

$$\frac{d^2 x_{j_1}}{dx_{j_2}^2}\Big|_{x_{j_2}^*=a_{j_1j_2 1}} = \frac{a_{j_1j_1 0}}{a_{j_2j_2 0}} \frac{(\overline{x}_{j_1} - a_{j_1j_1})^2 + b_{j_1j_1}}{(\overline{x}_{j_1} - a_{j_2j_1 1})(a_{j_1j_2 1} - a_{j_2j_2 1})^2} < 0,$$

there is a down-parabola flow. Let

$$\dot{x}_{j_2} = a_{j_2j_2 0}(\overline{x}_{j_1} - a_{j_2j_1 1})(a_{j_1j_2 1} - a_{j_2j_2 1})^2 \neq 0.$$

The parabola flows at $x_{j_1}^* = a_{j_1j_1 1}$ in the x_{j_2}-direction are positive and negative, for $a_{j_2j_2 0}(a_{j_1j_2 1} - a_{j_2j_2 1})^2 > 0$ and $a_{j_2j_2 0}(a_{j_1j_2 1} - a_{j_2j_2 1})^2 < 0$, respectively. In addition,

$$(\overline{x}_{j_1} - a_{j_1j_1})^2 + b_{j_1j_1} > 0.$$

Thus, the equilibrium of $x_{j_2}^* = a_{j_1j_2 1} \neq a_{j_2j_2 1}$ has the following properties:

- For $a_{j_1j_1 0} > 0$ and $a_{j_2j_2 0}(\overline{x}_{j_1} - a_{j_2j_1 1})(a_{j_1j_2 1} - a_{j_2j_2 1})^2 > 0$,
$$(a_{j_1j_2 1}, \dot{x}_{j_2}) = \underbrace{(\text{UP,pF})}_{\text{up-parabola flow } (+)}.$$
- For $a_{j_1j_1 0} < 0$ and $a_{j_2j_2 0}(\overline{x}_{j_1} - a_{j_2j_1 1})(a_{j_1j_2 1} - a_{j_2j_2 1})^2 > 0$,
$$(a_{j_1j_2 1}, \dot{x}_{j_2}) = \underbrace{(\text{DP,pF})}_{\text{down-parabola flow } (+)}.$$
- For $a_{j_1j_1 0} > 0$ and $a_{j_2j_2 0}(\overline{x}_{j_1} - a_{j_2j_1 1})(a_{j_1j_2 1} - a_{j_2j_2 1})^2 < 0$,
$$(a_{j_1j_2 1}, \dot{x}_{j_2}) = \underbrace{(\text{DP,nF})}_{\text{down-parabola flow } (-)}.$$
- For $a_{j_1j_1 0} < 0$ and $a_{j_2j_2 0}(\overline{x}_{j_1} - a_{j_2j_1 1})(a_{j_1j_2 1} - a_{j_2j_2 1})^2 < 0$,
$$(a_{j_1j_2 1}, \dot{x}_{j_2}) = \underbrace{(\text{UP,nF})}_{\text{up-parabola flow } (-)}.$$

(II) Similarly, at $x_{j_1}^* = a_{j_2j_1 1}$ with $\overline{x}_{j_2} = a_{j_1j_2 1}$, in phase space,

$$\frac{dx_{j_2}}{dx_{j_1}}\bigg|_{x_{j_1}^*=a_{j_2j_11}} = \frac{a_{j_2j_20}}{a_{j_1j_10}}\frac{(x_{j_1}-a_{j_2j_11})(\overline{x}_{j_2}-a_{j_2j_21})^2}{(\overline{x}_{j_2}-a_{j_1j_21})\left[(x_{j_1}-a_{j_1j_1})^2+b_{j_1j_1}\right]}\bigg|_{x_{j_1}^*=a_{j_2j_11}}=0.$$

If

$$\frac{d^2x_{j_2}}{dx_{j_1}^2}\bigg|_{x_{j_1}^*=a_{j_2j_11}} = \frac{a_{j_2j_20}}{a_{j_1j_10}}\frac{(\overline{x}_{j_2}-a_{j_2j_21})^2}{(\overline{x}_{j_2}-a_{j_1j_21})\left[(a_{j_2j_11}-a_{j_1j_1})^2+b_{j_1j_1}\right]}>0,$$

there is an up-parabola flow. If

$$\frac{d^2x_{j_2}}{dx_{j_1}^2}\bigg|_{x_{j_1}^*=a_{j_2j_11}} = \frac{a_{j_2j_20}}{a_{j_1j_10}}\frac{(\overline{x}_{j_2}-a_{j_2j_21})^2}{(\overline{x}_{j_2}-a_{j_1j_21})\left[(a_{j_2j_11}-a_{j_1j_1})^2+b_{j_1j_1}\right]}<0,$$

there is a down-parabola flow. Let

$$\dot{x}_{j_1} = a_{j_1j_10}(\overline{x}_{j_2}-a_{j_1j_21})\left[(a_{j_2j_11}-a_{j_1j_1})^2+b_{j_1j_1}\right].$$

Because

$$(a_{j_2j_11}-a_{j_1j_1})^2+b_{j_1j_1}>0,$$

the parabola flows at $x_{j_1}^*=a_{j_1j_11}$ are positive and negative, for $a_{j_1j_10}>0$ and $a_{j_1j_10}<0$, respectively. In addition, for $\overline{x}_{j_2}\neq a_{j_2j_21}$,

$$(\overline{x}_{j_2}-a_{j_2j_21})^2>0.$$

Thus, the equilibrium of $x_{j_1}^*=a_{j_2j_11}$ with $\overline{x}_{j_2}=a_{j_1j_21}$ has the following properties:

- For $a_{j_1j_10}(\overline{x}_{j_2}-a_{j_1j_21})>0$ and $a_{j_2j_20}(\overline{x}_{j_2}-a_{j_2j_21})^2>0$,

$$(\dot{x}_{j_1}, a_{j_2j_11}) = \underbrace{(\text{pF,UP})}_{\text{up-parabola flow }(+)}.$$

- For $a_{j_1j_10}(\overline{x}_{j_2}-a_{j_1j_21})<0$ and $a_{j_2j_20}(\overline{x}_{j_2}-a_{j_2j_21})^2>0$,

$$(\dot{x}_{j_1}, a_{j_2j_11}) = \underbrace{(\text{nF,DP})}_{\text{down-parabola flow }(-)}.$$

- For $a_{j_1j_10}(\overline{x}_{j_2}-a_{j_1j_21})>0$ and $a_{j_2j_20}(\overline{x}_{j_2}-a_{j_2j_21})^2<0$,

$$(\dot{x}_{j_1}, a_{j_2 j_1 1}) = \underbrace{(\text{pF,DP})}_{\text{down-parabola flow }(+)}.$$

- For $a_{j_1 j_1 0}(\overline{x}_{j_2} - a_{j_1 j_2 1}) < 0$ and $a_{j_2 j_2 0}(\overline{x}_{j_2} - a_{j_2 j_2 1})^2 < 0$,

$$(\dot{x}_{j_1}, a_{j_2 j_1 1}) = \underbrace{(\text{nF,UP})}_{\text{up-parabola flow }(-)}.$$

Therefore, from the analysis, the equilibrium of $(x_{j_2}^*, x_{j_1}^*) = (a_{j_1 j_2 1}, a_{j_2 j_1 1})$ has the following properties as in Eqs. (1.13)–(1.16):

- For $a_{j_1 j_1 0} > 0$ and $a_{j_2 j_2 0}(a_{j_1 j_2 1} - a_{j_2 j_2 1})^2 > 0$,

$$(a_{j_1 j_2 1}, a_{j_2 j_1 1}) = \underbrace{(\text{UP}_+, \text{UP}_+)}_{\text{positive saddle}}.$$

  The equilibrium of $(x_{j_2}^*, x_{j_1}^*) = (a_{j_1 j_2 1}, a_{j_2 j_1 1})$ is a (UP$_+$, UP$_+$)-positive saddle.

- For $a_{j_1 j_1 0} < 0$ and $a_{j_2 j_2 0}(a_{j_1 j_2 1} - a_{j_2 j_2 1})^2 > 0$,

$$(a_{j_1 j_2 1}, a_{j_2 j_1 1}) = \underbrace{(\text{DP}_+, \text{DP}_-)}_{\text{CCW center}}.$$

  The equilibrium of $(x_{j_2}^*, x_{j_1}^*) = (a_{j_1 j_2 1}, a_{j_2 j_1 1})$ is a (DP$_+$,DP$_-$)-counterclockwise center.

- For $a_{j_1 j_1 0} > 0$ and $a_{j_2 j_2 0}(a_{j_1 j_2 1} - a_{j_2 j_2 1})^2 < 0$,

$$(a_{j_1 j_2 1}, a_{j_2 j_1 1}) = \underbrace{(\text{DP}_-, \text{DP}_+)}_{\text{CW center}}.$$

  The equilibrium of $(x_{j_2}^*, x_{j_1}^*) = (a_{j_1 j_2 1}, a_{j_2 j_1 1})$ is a (DP$_-$, DP$_+$)-clockwise center.

- For $a_{j_1 j_1 0} < 0$ and $a_{j_2 j_2 0}(a_{j_1 j_2 1} - a_{j_2 j_2 1})^2 < 0$,

$$(a_{j_1 j_2 1}, a_{j_2 j_1 1}) = \underbrace{(\text{UP}_-, \text{UP}_-)}_{\text{negative saddle}}.$$

  The equilibrium of $(x_{j_2}^*, x_{j_1}^*) = (a_{j_1 j_2 1}, a_{j_2 j_1 1})$ is a (UP$_-$,UP$_-$)-negative saddle.

(ii$_{1b}$) At $x_{j_2}^* = a_{j_2 j_2 1} \neq a_{j_1 j_2 1}$, the variational equation is

$$\Delta \dot{x}_{j_2} = a_{j_2 j_2 0}(\overline{x}_{j_1} - a_{j_2 j_1 1})(\Delta x_{j_2})^2.$$

The flows at $x_{j_2}^* = a_{j_2j_2 1} \neq a_{j_1j_2 1}$ in the x_{j_2}-direction are upper-saddle and lower-saddle for $a_{j_2j_2 0}(\overline{x}_{j_1} - a_{j_2j_1 1}) > 0$ and $a_{j_2j_2 0}(\overline{x}_{j_1} - a_{j_2j_1 1}) < 0$, respectively. Let

$$\dot{x}_{j_1} = a_{j_1j_1 0}(a_{j_2j_2 1} - a_{j_1j_1 1})\left[(a_{j_2j_1 1} - a_{j_1j_1})^2 + b_{j_1j_1}\right],$$

with

$$\left[(\overline{x}_{j_1} - a_{j_1j_1})^2 + b_{j_1j_1}\right] > 0.$$

Thus, the equilibrium of $x_{j_2}^* = a_{j_2j_2 1} \neq a_{j_1j_2 1}$ has the following properties:

- For $a_{j_1j_1 0}(a_{j_2j_2 1} - a_{j_1j_2 1}) > 0$ and $a_{j_2j_2 0}(\overline{x}_{j_1} - a_{j_2j_1 1}) > 0$,

$$(\dot{x}_{j_1}, a_{j_2j_2 1}) = \underbrace{(\text{pF,US})}_{\text{upper-saddle flow } (+)}.$$

- For $a_{j_1j_1 0}(a_{j_2j_2 1} - a_{j_1j_2 1}) < 0$ and $a_{j_2j_2 0}(\overline{x}_{j_1} - a_{j_2j_1 1}) > 0$,

$$(\dot{x}_{j_1}, a_{j_2j_2 1}) = \underbrace{(\text{nF,US})}_{\text{upper-saddle flow } (-)}.$$

- For $a_{j_1j_1 0}(a_{j_2j_2 1} - a_{j_1j_2 1}) > 0$ and $a_{j_2j_2 0}(\overline{x}_{j_1} - a_{j_2j_1 1}) < 0$,

$$(\dot{x}_{j_1}, a_{j_2j_2 1}) = \underbrace{(\text{pF,LS})}_{\text{lower-saddle flow } (+)}.$$

- For $a_{j_1j_1 0}(a_{j_2j_2 1} - a_{j_1j_2 1}) < 0$ and $a_{j_2j_2 0}(\overline{x}_{j_1} - a_{j_2j_1 1}) < 0$,

$$(\dot{x}_{j_1}, a_{j_2j_2 1}) = \underbrace{(\text{nF,LS})}_{\text{lower-saddle flow } (-)}.$$

From the above analysis, with case (II) of (i), therefore, the equilibrium of $(x_{j_1}^*, x_{j_2}^*) = (a_{j_2j_1 1}, a_{j_2j_2 1})$ with $a_{j_2j_2 1} \neq a_{j_1j_2 1}$ has the following properties as in, Eqs. (1.17)–(1.20):

- For $a_{j_1j_1 0}(a_{j_2j_2 1} - a_{j_1j_2 1}) > 0$ and $a_{j_2j_2 0} > 0$,

$$(a_{j_2j_1 1}, a_{j_2j_2 1}) = \underbrace{(\text{pF,UP:UP})}_{\text{hyperbolic-secant-to-hyperbolic flow } (+)}.$$

  The equilibrium of $(x_{j_1}^*, x_{j_2}^*) = (a_{j_2j_1 1}, a_{j_2j_2 1})$ is a (pF, UP:UP)-positive hyperbolic-secant-to-hyperbolic flow.
- For $a_{j_1j_1 0}(a_{j_2j_2 1} - a_{j_1j_2 1}) < 0$ and $a_{j_2j_2 0} > 0$,

$$(a_{j_2j_1 1}, a_{j_2j_2 1}) = \underbrace{(\text{nF,DP:DP})}_{\text{hyperbolic-to-hyperbolic-secant flow }(-)}.$$

The equilibrium of $(x^*_{j_1}, x^*_{j_2}) = (a_{j_2j_1 1}, a_{j_2j_2 1})$ is an (nF,DP:DP)-negative hyperbolic-to-hyperbolic-secant flow.

- For $a_{j_1j_1 0}(a_{j_2j_2 1} - a_{j_1j_2 1}) > 0$ and $a_{j_2j_2 0} < 0$,

$$(a_{j_2j_1 1}, a_{j_2j_2 1}) = \underbrace{(\text{pF,DP:DP})}_{\text{hyperbolic-to-hyperbolic-secant flow }(+)}.$$

The equilibrium of $(x^*_{j_1}, x^*_{j_2}) = (a_{j_2j_1 1}, a_{j_2j_2 1})$ is a (pF,DP:DP)-positive hyperbolic-to-hyperbolic-secant flow.

- For $a_{j_1j_1 0}(a_{j_2j_2 1} - a_{j_1j_2 1}) < 0$ and $a_{j_2j_2 0} < 0$,

$$(a_{j_2j_1 1}, a_{j_2j_2 1}) = \underbrace{(\text{nF,UP:UP})}_{\text{hyperbolic-secant-to-hyperbolic flow }(-)}.$$

The equilibrium of $(x^*_{j_1}, x^*_{j_2}) = (a_{j_2j_1 1}, a_{j_2j_2 1})$ is an (nF, UP:UP)-negative hyperbolic-secant-to-hyperbolic flow.

– The hyperbolic-to-hyperbolic-secant and hyperbolic-secant-to-hyperbolic flows are the appearing bifurcations of two separated hyperbolic and hyperbolic secant flows.

(ii$_2$) For $a_{j_1j_2 1} = a_{j_2j_2 1}$, in phase space,

$$\frac{dx_{j_1}}{dx_{j_2}} = \frac{a_{j_1j_1 0}}{a_{j_2j_2 0}} \frac{\left[(x_{j_1} - a_{j_1j_1})^2 + b_{j_1j_1}\right]}{(x_{j_1} - a_{j_2j_1 1})(x_{j_2} - a_{j_2j_2 1})},$$

and the deformation of the foregoing equation is

$$\left[\frac{(x_{j_1} - a_{j_1j_1})}{(x_{j_1} - a_{j_1j_1})^2 + b_{j_1j_1}} + \frac{(a_{j_1j_1} - a_{j_2j_1 1})}{(x_{j_1} - a_{j_1j_1})^2 + b_{j_1j_1}}\right] dx_{j_1} = \frac{a_{j_1j_1 0}}{a_{j_2j_2 0}} \frac{dx_{j_2}}{x_{j_2} - a_{j_2j_2 1}}.$$

With an initial condition of $(x_{j_1 0}, x_{j_2 0})$ at $t = t_0$, the integration of the above equation gives the first integral manifold for $a_{j_1j_2 1} = a_{j_2j_2 1}$ as

$$\begin{aligned}
&\frac{1}{2}\ln\frac{|(x_{j_1}-a_{j_1j_1})^2+b_{j_1j_1}|}{|(x_{j_10}-a_{j_1j_1})^2+b_{j_1j_1}|}\\
&\quad+\frac{a_{j_1j_1}-a_{j_2j_11}}{\sqrt{b_{j_1j_1}}}(\arctan\frac{x_{j_1}-a_{j_1j_1}}{\sqrt{b_{j_1j_1}}}-\arctan\frac{x_{j_10}-a_{j_1j_1}}{\sqrt{b_{j_1j_1}}})\\
&=\frac{a_{j_1j_10}}{a_{j_2j_20}}\ln\frac{|x_{j_2}-a_{j_2j_21}|}{|x_{j_20}-a_{j_2j_21}|}.
\end{aligned}$$

($\mathrm{ii}_{2\mathrm{a}}$) In phase space, at $x^*_{j_2}=a_{j_1j_21}=a_{j_2j_21}$ with $\overline{x}_{j_1}\neq a_{j_2j_11}$,

$$\frac{dx_{j_1}}{dx_{j_2}}\Big|_{x^*_{j_2}=a_{j_2j_21}\pm\varepsilon}=\pm\frac{a_{j_1j_10}}{a_{j_2j_20}}\frac{[(\overline{x}_{j_1}-a_{j_1j_1})^2+b_{j_1j_1}]}{(\overline{x}_{j_1}-a_{j_2j_11})\varepsilon}.$$

If

$$\frac{a_{j_1j_10}}{a_{j_2j_20}}\frac{[(\overline{x}_{j_1}-a_{j_1j_1})^2+b_{j_1j_1}]}{(\overline{x}_{j_1}-a_{j_2j_11})}>0,$$

there is a down-up flow with a separatrix of $x^*_{j_2}=a_{j_2j_21}$. If

$$\frac{a_{j_1j_10}}{a_{j_2j_20}}\frac{[(\overline{x}_{j_1}-a_{j_1j_1})^2+b_{j_1j_1}]}{(\overline{x}_{j_1}-a_{j_2j_11})}<0,$$

there is an up-down flow with a separatrix of $x^*_{j_2}=a_{j_2j_21}$. The variational equation is

$$\Delta\dot{x}_{j_2}=a_{j_2j_20}(\overline{x}_{j_1}-a_{j_2j_11})(\Delta x_{j_2})^2.$$

The up-down and down-up flows at $x^*_{j_1}=a_{j_1j_11}$ in the x_{j_2}-direction are upper-saddle and lower-saddle for $a_{j_2j_20}(\overline{x}_{j_1}-a_{j_2j_11})>0$ and $a_{j_2j_20}(\overline{x}_{j_1}-a_{j_2j_11})<0$, respectively. In addition,

$$(\overline{x}_{j_1}-a_{j_1j_1})^2+b_{j_1j_1}>0.$$

Therefore, the infinite equilibrium of $x^*_{j_2}=a_{j_1j_21}=a_{j_2j_21}$ with $\overline{x}_{j_1}\neq a_{j_2j_11}$ has the following properties as in Eqs. (1.22)–(1.25):

- For $a_{j_1j_10}>0$ and $a_{j_2j_20}(\overline{x}_{j_1}-a_{j_2j_11})>0$,

$$(\overline{x}_{j_1},a_{j_2j_21})=\underbrace{(\mathrm{DU,US})}_{\text{down-up upper-saddle}}.$$

  The infinite-equilibrium of $x^*_{j_2}=a_{j_1j_21}=a_{j_2j_21}$ is a (DU,US)-down-up upper-saddle.

- For $a_{j_1j_10} < 0$ and $a_{j_2j_20}(\overline{x}_{j_1} - a_{j_2j_11}) > 0$,

$$(\overline{x}_{j_1}, a_{j_2j_21}) = \underbrace{(\text{UD,US})}_{\text{up-down upper-saddle}}.$$

The infinite-equilibrium of $x^*_{j_2} = a_{j_1j_21} = a_{j_2j_21}$ is a (UD,US)-up-down upper-saddle.

- For $a_{j_1j_10} > 0$ and $a_{j_2j_20}(\overline{x}_{j_1} - a_{j_2j_11}) < 0$,

$$(\overline{x}_{j_1}, a_{j_2j_21}) = \underbrace{(\text{UD,LS})}_{\text{up-down lower-saddle}}.$$

The infinite equilibrium of $x^*_{j_2} = a_{j_1j_21} = a_{j_2j_21}$ is a (UD,LS)-up-down lower-saddle.

- For $a_{j_1j_10} < 0$ and $a_{j_2j_20}(\overline{x}_{j_1} - a_{j_2j_11}) < 0$,

$$(\overline{x}_{j_1}, a_{j_2j_21}) = \underbrace{(\text{DU,LS})}_{\text{down-up lower-saddle}}.$$

The infinite equilibrium of $x^*_{j_2} = a_{j_1j_21} = a_{j_2j_21}$ is a (DU,US)-down-up lower-saddle.

($\text{ii}_{2\text{b}}$) From case ($\text{ii}_{2\text{a}}$), for $\overline{x}_{j_1} = a_{j_2j_11}$, with case (II) of ($\text{ii}_{2\text{a}}$), therefore, the equilibrium of $(x^*_{j_1}, x^*_{j_2}) = (a_{j_2j_11}, a_{j_2j_21})$ has the following properties as in Eqs. (1.26)–(1.29):

- For $a_{j_1j_10} > 0$ and $a_{j_2j_20} > 0$,

$$(a_{j_2j_11}, a_{j_2j_21}) = \underbrace{({}_{\text{UD}}\text{LS:}{}_{\text{DU}}\text{US,DP}_-\text{:UP}_+)}_{\text{hyperbolic lower-to-upper saddle}}.$$

The equilibrium of $(x^*_{j_1}, x^*_{j_2}) = (a_{j_2j_11}, a_{j_2j_21})$ is an (${}_{\text{UD}}\text{LS:}{}_{\text{DU}}\text{US}$, $\text{DP}_-\text{:UP}_+$)-hyperbolic lower-to-upper saddle.

- For $a_{j_1j_10} < 0$ and $a_{j_2j_20} > 0$,

$$(a_{j_2j_11}, a_{j_2j_21}) = \underbrace{({}_{\text{DU}}\text{LS:}{}_{\text{UD}}\text{US,UP}_+\text{:DP}_-)}_{\text{hyperbolic-secant lower-to-upper saddle}}.$$

The equilibrium of $(x^*_{j_1}, x^*_{j_2}) = (a_{j_2j_11}, a_{j_2j_21})$ is an (${}_{\text{DU}}\text{LS:}{}_{\text{UD}}\text{US}$, $\text{UP}_+\text{:DP}_-$)-hyper-secant lower-to-upper saddle.

- For $a_{j_1j_10} > 0$ and $a_{j_2j_20} < 0$,

$$(a_{j_2j_11}, a_{j_2j_21}) = \underbrace{({}_{\mathrm{DU}}\mathrm{US}{:}_{\mathrm{UD}}\mathrm{LS}, \mathrm{UP}_{-}{:}\mathrm{DP}_{+})}_{\text{hyperbolic-secant upper-to-lower saddle}}.$$

The equilibrium of $(x_{j_1}^*, x_{j_2}^*) = (a_{j_2j_11}, a_{j_2j_21})$ is a (${}_{\mathrm{DU}}$US:${}_{\mathrm{UD}}$LS, UP$_-$:DP$_+$)-hyper-secant upper-to-lower saddle.

- For $a_{j_1j_10} < 0$ and $a_{j_2j_20} < 0$,

$$(a_{j_2j_11}, a_{j_2j_21}) = \underbrace{({}_{\mathrm{UD}}\mathrm{US}{:}_{\mathrm{DU}}\mathrm{LS}, \mathrm{DP}_{+}{:}\mathrm{UP}_{-})}_{\text{hyperbolic upper-to-lower saddle}}.$$

The equilibrium of $(x_{j_1}^*, x_{j_2}^*) = (a_{j_2j_11}, a_{j_2j_21})$ is a (${}_{\mathrm{UD}}$US:${}_{\mathrm{DU}}$LS,DP$_+$:UP$_-$)-hyperbolic upper-to-lower saddle.

(iii) For $\Delta_{j_1j_2} < 0$ and $\Delta_{j_2j_2} > 0$, the standard form is

$$\begin{aligned}
\dot{x}_{j_2} &= a_{j_2j_20}(x_{j_1} - a_{j_2j_11})(x_{j_2} - a_{j_2j_21})(x_{j_2} - a_{j_2j_22}),\\
\dot{x}_{j_1} &= a_{j_1j_10}\left[(x_{j_1} - a_{j_1j_1})^2 + b_{j_1j_1}\right](x_{j_2} - a_{j_1j_21})
\end{aligned}$$

where

$$\begin{aligned}
a_{j_1j_1} &= -\frac{1}{2}B_{j_1j_1}, b_{j_1j_1} = \frac{1}{4}(-\Delta_{j_1j_1}),\\
b_{j_2j_21}, b_{j_2j_22} &= -\frac{1}{2}(B_{j_2j_2} \pm \sqrt{\Delta_{j_2j_2}}),\\
\{a_{j_2j_21}, a_{j_2j_22}\} &= \mathrm{sort}\{b_{j_2j_21}, b_{j_2j_22}\}, a_{j_2j_21} < a_{j_2j_22}.
\end{aligned}$$

(iii$_1$) In phase space, $a_{j_1j_21} \neq a_{j_2j_2s_1}$ $(s_1 = 1, 2)$

$$\frac{dx_{j_1}}{dx_{j_2}} = \frac{a_{j_1j_10}}{a_{j_2j_20}} \frac{(x_{j_2} - a_{j_1j_21})\left[(x_{j_1} - a_{j_1j_1})^2 + b_{j_1j_1}\right]}{(x_{j_1} - a_{j_2j_11})(x_{j_2} - a_{j_2j_21})(x_{j_2} - a_{j_2j_22})},$$

and the deformation of the foregoing equation is

$$\begin{aligned}
&\left[\frac{(x_{j_1} - a_{j_1j_1})}{(x_{j_1} - a_{j_1j_1})^2 + b_{j_1j_1}} + \frac{(a_{j_1j_1} - a_{j_2j_11})}{(x_{j_1} - a_{j_1j_1})^2 + b_{j_1j_1}}\right]dx_{j_1}\\
&= \frac{a_{j_1j_10}}{a_{j_2j_20}}\left[\frac{(a_{j_2j_21} - a_{j_1j_21})}{(a_{j_2j_21} - a_{j_2j_22})}\frac{1}{(x_{j_2} - a_{j_2j_21})} + \frac{(a_{j_2j_22} - a_{j_1j_21})}{(a_{j_2j_22} - a_{j_2j_21})}\frac{1}{(x_{j_2} - a_{j_2j_22})}\right]dx_{j_2}.
\end{aligned}$$

With an initial condition of (x_{j_10}, x_{j_20}) at $t = t_0$,, the integration of the above equation gives

$$\frac{1}{2}\ln\frac{|(x_{j_1}-a_{j_1j_1})^2+b_{j_1j_1}|}{|(x_{j_10}-a_{j_1j_1})^2+b_{j_1j_1}|}$$
$$+\frac{(a_{j_1j_1}-a_{j_2j_11})}{\sqrt{b_{j_1j_1}}}(\arctan\frac{x_{j_1}-a_{j_1j_1}}{\sqrt{b_{j_1j_1}}}-\arctan\frac{x_{j_10}-a_{j_1j_1}}{\sqrt{b_{j_1j_1}}})$$
$$=\frac{a_{j_1j_10}}{a_{j_2j_20}}\sum\nolimits_{s_1=1}^{2}\frac{(a_{j_2j_2s_1}-a_{j_1j_21})}{(a_{j_2j_2s_1}-a_{j_2j_2s_2})}\ln\frac{|x_{j_2}-a_{j_2j_2s_1}|}{|x_{j_2}-a_{j_2j_2s_1}|}.$$

($\mathrm{iii}_{1\mathrm{a}}$) Consider two cases (I) $x^*_{j_2}=a_{j_1j_21}$ and (II) $x^*_{j_1}=a_{j_2j_11}$.

(I) At $x^*_{j_2}=a_{j_1j_21}\neq a_{j_2j_2s_1},a_{j_2j_2s_2}$ (s_1, $s_2\in\{1,2\}$, $s_1\neq s_2$) with $\overline{x}_{j_1}\neq a_{j_2j_11}$, in phase space,

$$\frac{dx_{j_1}}{dx_{j_2}}\Big|_{x^*_{j_2}=a_{j_1j_21}}=\frac{a_{j_1j_10}}{a_{j_2j_20}}\frac{(x_{j_2}-a_{j_1j_21})\big[(\overline{x}_{j_1}-a_{j_1j_1})^2+b_{j_1j_1}\big]}{(\overline{x}_{j_1}-a_{j_2j_11})\prod_{s_1=1}^{2}(x_{j_2}-a_{j_2j_2s_1})}\Big|_{x^*_{j_2}=a_{j_1j_21}}=0.$$

If

$$\frac{d^2x_{j_1}}{dx^2_{j_2}}\Big|_{x^*_{j_2}=a_{j_1j_21}}=\frac{a_{j_1j_10}}{a_{j_2j_20}}\frac{(\overline{x}_{j_1}-a_{j_1j_1})^2+b_{j_1j_1}}{(\overline{x}_{j_1}-a_{j_2j_11})\prod_{s_1=1}^{2}(a_{j_1j_21}-a_{j_2j_2s_1})}>0,$$

there is an up-parabola flow. If

$$\frac{d^2x_{j_1}}{dx^2_{j_2}}\Big|_{x^*_{j_2}=a_{j_1j_21}}=\frac{a_{j_1j_10}}{a_{j_2j_20}}\frac{(\overline{x}_{j_1}-a_{j_1j_1})^2+b_{j_1j_1}}{(\overline{x}_{j_1}-a_{j_2j_11})\prod_{s_1=1}^{2}(a_{j_1j_21}-a_{j_2j_2s_1})}<0,$$

there is a down-parabola flow. Let

$$\dot{x}_{j_2}=a_{j_2j_20}(\overline{x}_{j_1}-a_{j_2j_11})\prod\nolimits_{s_1=1}^{2}(a_{j_1j_21}-a_{j_2j_2s_1})\neq 0.$$

The parabola flows at $x^*_{j_1}=a_{j_1j_11}$ in the x_{j_2}-direction are positive and negative, for $\dot{x}_{j_2}>0$ and $\dot{x}_{j_2}<0$, respectively.

$$(\overline{x}_{j_1}-a_{j_1j_1})^2+b_{j_1j_1}>0.$$

Therefore, the equilibrium of $x^*_{j_2}=a_{j_1j_21}\neq a_{j_2j_2s_1},a_{j_2j_2s_2}$ (s_1, $s_2\in\{1,2\}$, $s_1\neq s_2$) has the following properties:

- For $a_{j_1j_10}>0$ and $a_{j_2j_20}(\overline{x}_{j_1}-a_{j_2j_11})\prod_{s_1=1}^{2}(a_{j_1j_21}-a_{j_2j_2s_1})>0$,

$$(a_{j_1j_21},\dot{x}_{j_2})=\underbrace{(\text{UP,pF})}_{\text{up-parabola flow }(+)}.$$

- For $a_{j_1j_10}<0$ and $a_{j_2j_20}(\overline{x}_{j_1}-a_{j_2j_11})\prod_{s_1=1}^{2}(a_{j_1j_21}-a_{j_2j_2s_1})>0$,

$$(a_{j_1j_21},\dot{x}_{j_2})=\underbrace{(\text{DP,pF})}_{\text{down-parabola flow }(+)}.$$

- For $a_{j_1j_10}>0$ and $a_{j_2j_20}(\overline{x}_{j_1}-a_{j_2j_11})\prod_{s_1=1}^{2}(a_{j_1j_21}-a_{j_2j_2s_1})<0$,

$$(a_{j_1j_21},\dot{x}_{j_2})=\underbrace{(\text{DP,nF})}_{\text{down-parabola flow }(-)}.$$

- For $a_{j_1j_10}<0$ and $a_{j_2j_20}(\overline{x}_{j_1}-a_{j_2j_11})\prod_{s_1=1}^{2}(a_{j_1j_21}-a_{j_2j_2s_1})<0$,

$$(a_{j_1j_21},\dot{x}_{j_2})=\underbrace{(\text{UP,nF})}_{\text{up-parabola flow }(-)}.$$

(II) Similarly, at $x_{j_1}^*=a_{j_2j_11}$ with $\overline{x}_{j_2}=a_{j_1j_21}$, in phase space,

$$\frac{dx_{j_2}}{dx_{j_1}}\bigg|_{x_{j_1}^*=a_{j_2j_11}}=\frac{a_{j_2j_20}}{a_{j_1j_10}}\frac{(x_{j_1}-a_{j_2j_11})\prod_{s_1=1}^{2}(\overline{x}_{j_2}-a_{j_2j_2s_1})}{(\overline{x}_{j_2}-a_{j_1j_21})\left[(x_{j_1}-a_{j_1j_1})^2+b_{j_1j_1}\right]}\bigg|_{x_{j_1}^*=a_{j_2j_11}}=0.$$

If

$$\frac{d^2x_{j_2}}{dx_{j_1}^2}\bigg|_{x_{j_1}^*=a_{j_2j_11}}=\frac{a_{j_2j_20}}{a_{j_1j_10}}\frac{\prod_{s_1=1}^{2}(\overline{x}_{j_2}-a_{j_2j_2s_1})}{(\overline{x}_{j_2}-a_{j_1j_21})\left[(a_{j_2j_11}-a_{j_1j_1})^2+b_{j_1j_1}\right]}>0,$$

there is an up-parabola flow. If

$$\frac{d^2x_{j_2}}{dx_{j_1}^2}\bigg|_{x_{j_1}^*=a_{j_2j_11}}=\frac{a_{j_2j_20}}{a_{j_1j_10}}\frac{\prod_{s_1=1}^{2}(\overline{x}_{j_2}-a_{j_2j_2s_1})}{(\overline{x}_{j_2}-a_{j_1j_21})\left[(a_{j_2j_11}-a_{j_1j_1})^2+b_{j_1j_1}\right]}<0,$$

there is a down-parabola flow. Let

$$\dot{x}_{j_1} = a_{j_1j_10}(\overline{x}_{j_2} - a_{j_1j_21})\left[(a_{j_2j_11} - a_{j_1j_1})^2 + b_{j_1j_1}\right].$$

Because

$$(a_{j_2j_11} - a_{j_1j_1})^2 + b_{j_1j_1} > 0,$$

the parabola flows at $x_{j_1}^* = a_{j_1j_11}$ are positive and negative, for $a_{j_1j_10} > 0$ and $a_{j_1j_10} < 0$, respectively. In addition, for $\overline{x}_{j_2} \neq a_{j_2j_2s_1}$ $(s_1 = 1, 2)$,

$$\prod\nolimits_{s_1=1}^{2}(\overline{x}_{j_2} - a_{j_2j_2s_1}) \neq 0.$$

Thus, the equilibrium of $x_{j_1}^* = a_{j_2j_11}$ with $\overline{x}_{j_2} = a_{j_1j_21}$ has the following properties:

- For $a_{j_1j_10}(\overline{x}_{j_2} - a_{j_1j_21}) > 0$ and $a_{j_2j_20}\prod\nolimits_{s_1=1}^{2}(\overline{x}_{j_2} - a_{j_2j_2s_1}) > 0$,

$$(\dot{x}_{j_1}, a_{j_2j_11}) = \underbrace{(\text{pF,UP})}_{\text{up-parabola flow }(+)}.$$

- For $a_{j_1j_10}(\overline{x}_{j_2} - a_{j_1j_21}) < 0$ and $a_{j_2j_20}\prod\nolimits_{s_1=1}^{2}(\overline{x}_{j_2} - a_{j_2j_2s_1}) > 0$,

$$(\dot{x}_{j_1}, a_{j_2j_11}) = \underbrace{(\text{nF,DP})}_{\text{down-parabola flow }(-)}.$$

- For $a_{j_1j_10}(\overline{x}_{j_2} - a_{j_1j_21}) > 0$ and $a_{j_2j_20}\prod\nolimits_{s_1=1}^{2}(\overline{x}_{j_2} - a_{j_2j_2s_1}) < 0$,

$$(\dot{x}_{j_1}, a_{j_2j_11}) = \underbrace{(\text{pF,DP})}_{\text{down-parabola flow }(+)}.$$

- For $a_{j_1j_10}(\overline{x}_{j_2} - a_{j_1j_21}) < 0$ and $a_{j_2j_20}\prod\nolimits_{s_1=1}^{2}(\overline{x}_{j_2} - a_{j_2j_2s_1}) < 0$,

$$(\dot{x}_{j_1}, a_{j_2j_11}) = \underbrace{(\text{nF,UP})}_{\text{up-parabola flow }(-)}.$$

Therefore, from the analysis of cases (I) and (II), the equilibrium of $(x_{j_2}^*, x_{j_1}^*) = (a_{j_1j_21}, a_{j_2j_11})$ has the following properties as in Eqs. (1.33)–(1.36):

- *For* $a_{j_1j_10} > 0$ *and* $a_{j_2j_20}\prod\nolimits_{s_1=1}^{2}(\overline{x}_{j_2} - a_{j_2j_2s_1}) > 0$,

$$(a_{j_1j_21}, a_{j_2j_11}) = \underbrace{(\text{UP}_+,\text{UP}_+)}_{\text{positive saddle}}.$$

The equilibrium of $(x_{j_2}^*, x_{j_1}^*) = (a_{j_1j_21}, a_{j_2j_11})$ is a (UP_+, UP_+)-positive saddle.

- For $a_{j_1j_10} < 0$ and $a_{j_2j_20}\prod_{s_1=1}^{2}(\overline{x}_{j_2} - a_{j_2j_2s_1}) > 0$,

$$(a_{j_1j_21}, a_{j_2j_11}) = \underbrace{(\text{DP}_+,\text{DP}_-)}_{\text{CCW center}}.$$

The equilibrium of $(x_{j_2}^*, x_{j_1}^*) = (a_{j_1j_21}, a_{j_2j_11})$ is a (DP_+,DP_-)-counterclockwise center.

- For $a_{j_1j_10} > 0$ and $a_{j_2j_20}\prod_{s_1=1}^{2}(\overline{x}_{j_2} - a_{j_2j_2s_1}) < 0$,

$$(a_{j_1j_21}, a_{j_2j_11}) = \underbrace{(\text{DP}_-,\text{DP}_+)}_{\text{CW center}}.$$

The equilibrium of $(x_{j_2}^*, x_{j_1}^*) = (a_{j_1j_21}, a_{j_2j_11})$ is a (DP_-, DP_+)-clockwise center.

- For $a_{j_1j_10} < 0$ and $a_{j_2j_20}\prod_{s_1=1}^{2}(\overline{x}_{j_2} - a_{j_2j_2s_1}) < 0$,

$$(a_{j_1j_21}, a_{j_2j_11}) = \underbrace{(\text{UP}_-,\text{UP}_-)}_{\text{negative saddle}}.$$

The equilibrium of $(x_{j_2}^*, x_{j_1}^*) = (a_{j_1j_21}, a_{j_2j_11})$ is a (UP_-,UP_-)-negative saddle.

(iii_{1b}) At $x_{j_2}^* = a_{j_2j_2s_1} \neq a_{j_1j_21}$ (s_1, $s_2 \in \{1,2\}$, $s_1 \neq s_2$) with $\overline{x}_{j_1} \in (-\infty,\infty)$, the variational equation is

$$\begin{aligned}\Delta\dot{x}_{j_2} &= a_{j_2j_20}(\overline{x}_{j_1} - a_{j_2j_11})(a_{j_2j_2s_1} - a_{j_2j_2s_2})\Delta x_{j_2},\\ \dot{x}_{j_1} &= a_{j_1j_10}\left[(\overline{x}_{j_1} - a_{j_1j_1})^2 + b_{j_1j_1}\right](a_{j_2j_2s_1} - a_{j_1j_21})\end{aligned}$$

with

$$(\overline{x}_{j_1} - a_{j_1j_1})^2 + b_{j_1j_1} > 0.$$

The flows at $x_{j_2}^* = a_{j_2j_2s_1}$ are source and sink for $a_{j_2j_20}(\overline{x}_{j_1} - a_{j_2j_11})(a_{j_2j_2s_1} - a_{j_2j_2s_2}) > 0$ and $a_{j_2j_20}(\overline{x}_{j_1} - a_{j_2j_11})(a_{j_2j_2s_1} - a_{j_2j_2s_2}) < 0$ in the x_{j_2}-direction, respectively. The flows at $x_{j_2}^* = a_{j_2j_2s_1}$ in the x_{j_1}-direction are positive and negative for $a_{j_1j_10}(a_{j_2j_2s_1} - a_{j_1j_21}) > 0$ and $a_{j_1j_10}(a_{j_2j_2s_1} - a_{j_1j_21}) < 0$, respectively.

Thus, the equilibrium of $x_{j_2}^* = a_{j_2j_2s_1} \neq a_{j_1j_21}$ (s_1, $s_2 \in \{1,2\}$, $s_1 \neq s_2$) has the following properties:

- For $a_{j_1j_10}(a_{j_2j_2s_1} - a_{j_1j_21}) > 0$ and $a_{j_2j_20}(\overline{x}_{j_1} - a_{j_2j_11})(a_{j_2j_2s_1} - a_{j_2j_2s_2}) > 0$,

$$(\dot{x}_{j_1}, a_{j_2j_2s_1}) = \underbrace{(\text{pF,SO})}_{\text{source flow }(+)}.$$

- For $a_{j_1j_10}(a_{j_2j_2s_1} - a_{j_1j_21}) < 0$ and $a_{j_2j_20}(\overline{x}_{j_1} - a_{j_2j_11})(a_{j_2j_2s_1} - a_{j_2j_2s_2}) > 0$,

$$(\dot{x}_{j_1}, a_{j_2j_2s_1}) = \underbrace{(\text{nF,SO})}_{\text{source flow }(-)}.$$

- For $a_{j_1j_10}(a_{j_2j_2s_1} - a_{j_1j_21}) > 0$ and $a_{j_2j_20}(\overline{x}_{j_1} - a_{j_2j_11})(a_{j_2j_2s_1} - a_{j_2j_2s_2}) < 0$,

$$(\dot{x}_{j_1}, a_{j_2j_2s_1}) = \underbrace{(\text{pF,SI})}_{\text{sink flow }(+)}.$$

- For $a_{j_1j_10}(a_{j_2j_2s_1} - a_{j_1j_21}) < 0$ and $a_{j_2j_20}(\overline{x}_{j_1} - a_{j_2j_11})(a_{j_2j_2s_1} - a_{j_2j_2s_2}) < 0$,

$$(\dot{x}_{j_1}, a_{j_2j_2s_1}) = \underbrace{(\text{nF,SI})}_{\text{sink flow }(-)}.$$

From the above analysis with $\overline{x}_{j_1} = a_{j_2j_11}$ and the case (II) of (iii$_{1b}$) with $\overline{x}_{j_2} = a_{j_2j_2s_1}$, therefore, the equilibrium of $(x^*_{j_1}, x^*_{j_2}) = (a_{j_2j_11}, a_{j_2j_2s_1})$ (s_1, $s_2 \in \{1,2\}$, $s_1 \neq s_2$) with $a_{j_2j_2s_1} \neq a_{j_1j_21}$ has the following properties as in Eqs. (1.37)–(1.40):

- For $a_{j_1j_10}(a_{j_2j_2s_1} - a_{j_1j_21}) > 0$ and $a_{j_2j_20}(a_{j_2j_2s_1} - a_{j_2j_2s_2}) > 0$,

$$(a_{j_2j_11}, a_{j_2j_2s_1}) = \underbrace{(\text{pF,DP:UP})}_{\text{hyperbolic flow }(+)}.$$

  The equilibrium of $(x^*_{j_1}, x^*_{j_2}) = (a_{j_2j_11}, a_{j_2j_2s_1})$ is a (pF,DP:UP)-positive hyperbolic flow.

- For $a_{j_1j_10}(a_{j_2j_2s_1} - a_{j_1j_21}) < 0$ and $a_{j_2j_20}(a_{j_2j_2s_1} - a_{j_2j_2s_2}) > 0$,

$$(a_{j_2j_11}, a_{j_2j_2s_1}) = \underbrace{(\text{nF,UP:DP})}_{\text{hyperbolic-secant flow }(-)}.$$

  The equilibrium of $(x^*_{j_1}, x^*_{j_2}) = (a_{j_2j_11}, a_{j_2j_2s_1})$ is an (nF,UP:DP)-negative hyperbolic-secant flow.

- For $a_{j_1j_10}(a_{j_2j_2s_1} - a_{j_1j_21}) > 0$ and $a_{j_2j_20}(a_{j_2j_2s_1} - a_{j_2j_2s_2}) < 0$,

$$(a_{j_2j_1 1}, a_{j_2j_2s_1}) = \underbrace{(\text{pF,UP:DP})}_{\text{hyperbolic-secant flow } (+)}.$$

The equilibrium of $(x_{j_1}^*, x_{j_2}^*) = (a_{j_2j_1 1}, a_{j_2j_2s_1})$ is a (pF,UP:DP)-positive hyperbolic-secant flow.

- For $a_{j_1j_1 0}(a_{j_2j_2s_1} - a_{j_1j_2 1}) < 0$ and $a_{j_2j_2 0}(a_{j_2j_2s_1} - a_{j_2j_2s_2}) < 0$,

$$(a_{j_2j_1 1}, a_{j_2j_2s_1}) = \underbrace{(\text{nF,DP:UP})}_{\text{hyperbolic flow } (-)}.$$

The equilibrium of $(x_{j_1}^*, x_{j_2}^*) = (a_{j_2j_1 1}, a_{j_2j_2s_1})$ is an (nF,DP:UP)-negative hyperbolic flow.

(iii$_2$) At $a_{j_1j_2 1} = a_{j_2j_2s_1}$ $(s_1, s_2 \in \{1,2\}, s_1 \neq s_2)$, in phase space,

$$\frac{dx_{j_1}}{dx_{j_2}} = \frac{a_{j_1j_1 0}}{a_{j_2j_2 0}} \frac{[(x_{j_1} - a_{j_1j_1})^2 + b_{j_1j_1}]}{(x_{j_1} - a_{j_2j_1 1})(x_{j_2} - a_{j_2j_2s_2})},$$

and the deformation of the foregoing equation is

$$\left[\frac{(x_{j_1} - a_{j_1j_1})}{(x_{j_1} - a_{j_1j_1})^2 + b_{j_1j_1}} + \frac{(a_{j_1j_1} - a_{j_2j_1 1})}{(x_{j_1} - a_{j_1j_1})^2 + b_{j_1j_1}}\right] dx_{j_1} = \frac{a_{j_1j_1 0}}{a_{j_2j_2 0}} \frac{dx_{j_2}}{(x_{j_2} - a_{j_2j_2s_2})}.$$

With an initial condition of $(x_{j_1 0}, x_{j_2 0})$ at $t = t_0$, the integration of the above equation gives the first integral manifold for $a_{j_1j_2 1} = a_{j_2j_2s_1}$ $(s_1, s_2 \in \{1,2\}, s_1 \neq s_2)$ as

$$\begin{aligned}
&\frac{1}{2} \ln \frac{|(x_{j_1} - a_{j_1j_1})^2 + b_{j_1j_1}|}{|(x_{j_1 0} - a_{j_1j_1})^2 + b_{j_1j_1}|} \\
&\quad + \frac{(a_{j_1j_1} - a_{j_2j_1 1})}{\sqrt{b_{j_1j_1}}} \left(\arctan \frac{x_{j_1} - a_{j_1j_1}}{\sqrt{b_{j_1j_1}}} - \arctan \frac{x_{j_1 0} - a_{j_1j_1}}{\sqrt{b_{j_1j_1}}} \right) \\
&= \frac{a_{j_1j_1 0}}{a_{j_2j_2 0}} \ln \frac{|x_{j_2} - a_{j_2j_2s_2}|}{|x_{j_2} - a_{j_2j_2s_2}|}.
\end{aligned}$$

(iii$_{2a}$) At $x_{j_2}^* = a_{j_1j_2 1} = a_{j_2j_2s_1}$ $(s_1, s_2 \in \{1,2\}, s_1 \neq s_2)$ with $\bar{x}_{j_1} \neq a_{j_2j_1 1}$, in phase space,

$$\left.\frac{dx_{j_1}}{dx_{j_2}}\right|_{x_{j_2}^* = a_{j_1j_2 1}} = \frac{a_{j_1j_1 0}}{a_{j_2j_2 0}} \frac{[(\bar{x}_{j_1} - a_{j_1j_1})^2 + b_{j_1j_1}]}{(\bar{x}_{j_1} - a_{j_2j_1 1})(a_{j_1j_2s_1} - a_{j_2j_2s_2})} > 0,$$

there is an increasing inflection flow. If

$$\frac{dx_{j_1}}{dx_{j_2}}\bigg|_{x^*_{j_2}=a_{j_1j_21}} = \frac{a_{j_1j_10}}{a_{j_2j_20}} \frac{[(\overline{x}_{j_1}-a_{j_1j_1})^2+b_{j_1j_1}]}{(\overline{x}_{j_1}-a_{j_2j_11})(a_{j_1j_2s_1}-a_{j_2j_2s_2})} < 0,$$

there is a decreasing inflection flow. The corresponding variational equation is

$$\Delta\dot{x}_{j_2} = a_{j_2j_20}(\overline{x}_{j_1}-a_{j_2j_11})(a_{j_2j_2s_1}-a_{j_2j_2s_2})\Delta x_{j_2}.$$

The inflection flows at $x^*_{j_2}=a_{j_1j_21}=a_{j_2j_11}$ for $a_{j_2j_20}(\overline{x}_{j_1}-a_{j_2j_11})(a_{j_2j_2s_1}-a_{j_2j_2s_2})>0$ and $a_{j_2j_20}(\overline{x}_{j_1}-a_{j_2j_11})(a_{j_2j_2s_1}-a_{j_2j_2s_2})<0$ are sink and source in the x_{j_2} -direction, respectively.

Therefore, the infinite equilibrium of $x^*_{j_2}=a_{j_1j_21}=a_{j_2j_2s_1}$ has the following properties as in Eqs. (1.42)–(1.45):

- For $a_{j_1j_10}>0$ and $a_{j_2j_20}(\overline{x}_{j_1}-a_{j_2j_11})(a_{j_2j_2s_1}-a_{j_2j_2s_2})>0$,

$$(\overline{x}_{j_1}, a_{j_2j_2s_1}) = \underbrace{(\text{II,SO})}_{\text{increasing-inflection source}}.$$

  The infinite-equilibrium of $x^*_{j_2}=a_{j_1j_21}=a_{j_2j_2s_1}$ is an (II,SO)-increasing-inflection source.
- For $a_{j_1j_10}<0$ and $a_{j_2j_20}(\overline{x}_{j_1}-a_{j_2j_11})(a_{j_2j_2s_1}-a_{j_2j_2s_2})>0$,

$$(\overline{x}_{j_1}, a_{j_2j_2s_1}) = \underbrace{(\text{DI,SO})}_{\text{decreasing-inflection source}}.$$

  The infinite-equilibrium of $x^*_{j_2}=a_{j_1j_21}=a_{j_2j_2s_1}$ is a (DI,SO)-decreasing-inflection source.
- For $a_{j_1j_10}>0$ and $a_{j_2j_20}(\overline{x}_{j_1}-a_{j_2j_11})(a_{j_2j_2s_1}-a_{j_2j_2s_2})<0$,

$$(\overline{x}_{j_1}, a_{j_2j_2s_1}) = \underbrace{(\text{DI,SI})}_{\text{decreasing-inflection sink}}.$$

  The infinite-equilibrium of $x^*_{j_2}=a_{j_1j_21}=a_{j_2j_2s_1}$ is a (DI,SI)-decreasing-inflection sink.
- For $a_{j_1j_10}<0$ and $a_{j_2j_20}(\overline{x}_{j_1}-a_{j_2j_11})(a_{j_2j_2s_1}-a_{j_2j_2s_2})<0$,

$$(\overline{x}_{j_1}, a_{j_2j_2s_1}) = \underbrace{(\text{II,SI})}_{\text{increasing-inflection sink}}.$$

  The infinite-equilibrium of $x^*_{j_2}=a_{j_1j_21}=a_{j_2j_2s_1}$ is an (II,SI)-increasing-infection sink.

(iii$_{2b}$) From (iii$_{1a}$) with $\overline{x}_{j_1} = a_{j_2 j_1 1}$ and the case (II) of case (iii$_{1a}$) with $\overline{x}_{j_2} = a_{j_1 j_2 1}$, thus, the equilibrium of $(x^*_{j_2}, x^*_{j_1}) = (a_{j_1 j_2 1}, a_{j_2 j_1 1})$ with $a_{j_1 j_2 1} = a_{j_2 j_2 s_1}$ has the following properties as in Eqs. (1.46)–(1.49):

- For $a_{j_1 j_1 0} > 0$ and $a_{j_2 j_2 0}(a_{j_2 j_2 s_1} - a_{j_2 j_2 s_2}) > 0$,

$$(a_{j_1 j_2 1}, a_{j_2 j_1 1}) = \underbrace{(_{\text{SI:SO}}\text{US},_{\text{DI:II}}\text{UP})}_{\text{up-parabola upper-saddle}}.$$

  The equilibrium of $(x^*_{j_1}, x^*_{j_2}) = (a_{j_2 j_1 1}, a_{j_2 j_2 s_1})$ is an $(_{\text{SI:SO}}\text{US},_{\text{DI:II}}\text{DP})$-up-parabola upper-saddle.
- For $a_{j_1 j_1 0} < 0$ and $a_{j_2 j_2 0}(a_{j_2 j_2 s_1} - a_{j_2 j_2 s_2}) > 0$,

$$(a_{j_1 j_2 1}, a_{j_2 j_1 1}) = \underbrace{(_{\text{SI:SO}}\text{US},_{\text{II:DI}}\text{DP})}_{\text{down-parabola upper-saddle}}.$$

  The equilibrium of $(x^*_{j_1}, x^*_{j_2}) = (a_{j_2 j_1 1}, a_{j_2 j_2 s_1})$ is an $(_{\text{SI:SO}}\text{US},_{\text{II:DI}}\text{DP})$-down-parabola upper-saddle.
- For $a_{j_1 j_1 0} > 0$ and $a_{j_2 j_2 0}(a_{j_2 j_2 s_1} - a_{j_2 j_2 s_2}) < 0$,

$$(a_{j_1 j_2 1}, a_{j_2 j_1 1}) = \underbrace{(_{\text{SO:SI}}\text{LS},_{\text{II:DI}}\text{DP})}_{\text{down-parabola lower-saddle}}.$$

  The equilibrium of $(x^*_{j_1}, x^*_{j_2}) = (a_{j_2 j_1 1}, a_{j_2 j_2 s_1})$ is an $(_{\text{SO:SI}}\text{LS},_{\text{II:DI}}\text{DP})$-down-parabola lower-saddle.
- For $a_{j_1 j_1 0} < 0$ and $a_{j_2 j_2 0}(a_{j_2 j_2 s_1} - a_{j_2 j_2 s_2}) < 0$,

$$(a_{j_1 j_2 1}, a_{j_2 j_1 1}) = \underbrace{(_{\text{SO:SI}}\text{LS},_{\text{DI:II}}\text{UP})}_{\text{up-parabola lower-saddle}}.$$

  The equilibrium of $(x^*_{j_1}, x^*_{j_2}) = (a_{j_2 j_1 1}, a_{j_2 j_2 s_1})$ is an $(_{\text{SO:SI}}\text{LS},_{\text{DI:II}}\text{UP})$-up-parabola lower-saddle.

(iv) For $\Delta_{j_1 j_2} = 0$ and $\Delta_{j_2 j_2} < 0$, the standard form is

$$\begin{aligned}
\dot{x}_{j_2} &= a_{j_2 j_2 0}(x_{j_1} - a_{j_2 j_1 1})\left[(x_{j_2} - a_{j_2 j_2})^2 + b_{j_2 j_2}\right], \\
\dot{x}_{j_1} &= a_{j_1 j_1 0}(x_{j_1} - a_{j_1 j_1 1})^2 (x_{j_2} - a_{j_1 j_2 1})
\end{aligned}$$

where

$$
\begin{aligned}
a_{j_1j_1 1} &= -\frac{1}{2}B_{j_1j_1},\\
a_{j_2j_2} &= -\frac{1}{2}B_{j_2j_2}, b_{j_2j_2} = \frac{1}{4}(-\Delta_{j_2j_2}).
\end{aligned}
$$

(iv_1) In phase space,

$$
\frac{dx_{j_1}}{dx_{j_2}} = \frac{a_{j_1j_1 0}}{a_{j_2j_2 0}} \frac{(x_{j_1} - a_{j_1j_1 1})^2 (x_{j_2} - a_{j_1j_2 1})}{(x_{j_1} - a_{j_2j_1 1})\left[(x_{j_2} - a_{j_2j_2})^2 + b_{j_2j_2}\right]},
$$

and the deformation of the foregoing equation is

$$
\begin{aligned}
&\left[\frac{1}{(x_{j_1} - a_{j_1j_1 1})} + \frac{(a_{j_1j_1 1} - a_{j_2j_1 1})}{(x_{j_1} - a_{j_1j_1 1})^2}\right] dx_{j_1}\\
&= \frac{a_{j_1j_1 0}}{a_{j_2j_2 0}} \left[\frac{x_{j_2} - a_{j_2j_2}}{(x_{j_2} - a_{j_2j_2})^2 + b_{j_2j_2}} + \frac{a_{j_2j_2} - a_{j_1j_2 1}}{(x_{j_2} - a_{j_2j_2})^2 + b_{j_2j_2}}\right] dx_{j_2}.
\end{aligned}
$$

With an initial condition of $(x_{j_1 0}, x_{j_2 0})$ at $t = t_0$, the integration of the above equation gives

$$
\begin{aligned}
&\ln \frac{|x_{j_1} - a_{j_1j_1 1}|}{|x_{j_1 0} - a_{j_1j_1 1}|} - (a_{j_1j_1 1} - a_{j_2j_1 1})\left(\frac{1}{x_{j_1} - a_{j_1j_1 1}} - \frac{1}{x_{j_1 0} - a_{j_1j_1 1}}\right)\\
&= \frac{a_{j_1j_1 0}}{a_{j_2j_2 0}} \left[\frac{1}{2} \ln \frac{|(x_{j_2} - a_{j_2j_2})^2 + b_{j_2j_2}|}{|(x_{j_2 0} - a_{j_2j_2})^2 + b_{j_2j_2}|}\right.\\
&\left. + \frac{(a_{j_2j_2} - a_{j_1j_2 1})}{\sqrt{b_{j_2j_2}}} \left(\arctan \frac{x_{j_2} - a_{j_2j_2}}{\sqrt{b_{j_2j_2}}} - \arctan \frac{x_{j_2 0} - a_{j_2j_2}}{\sqrt{b_{j_2j_2}}}\right)\right].
\end{aligned}
$$

($\text{iv}_{1\text{a}}$) Consider two cases (I) $x_{j_2}^* = a_{j_1j_2 1}$ and (II) $x_{j_1}^* = a_{j_2j_1 1}$.

(I) In phase space, at $x_{j_2}^* = a_{j_1j_2 1}$ with $\bar{x}_{j_1} \neq a_{j_1j_1 1}$,

$$
\left.\frac{dx_{j_1}}{dx_{j_2}}\right|_{x_{j_2}^* = a_{j_1j_2 1}} = \left.\frac{a_{j_1j_1 0}}{a_{j_2j_2 0}} \frac{(\bar{x}_{j_1} - a_{j_1j_1 1})^2 (x_{j_2} - a_{j_1j_2 1})}{(\bar{x}_{j_1} - a_{j_2j_1 1})\left[(x_{j_2} - a_{j_2j_2})^2 + b_{j_2j_2}\right]}\right|_{x_{j_2}^* = a_{j_1j_2 1}} = 0.
$$

If

$$
\left.\frac{d^2 x_{j_1}}{dx_{j_2}^2}\right|_{x_{j_2}^* = a_{j_1j_2 1}} = \frac{a_{j_1j_1 0}}{a_{j_2j_2 0}} \frac{(\bar{x}_{j_1} - a_{j_1j_1 1})^2}{(\bar{x}_{j_1} - a_{j_2j_1 1})\left[(a_{j_1j_2 1} - a_{j_2j_2})^2 + b_{j_2j_2}\right]} > 0,
$$

there is an up-parabola flow at $x_{j_2}^* = a_{j_1j_2 1}$ with $\bar{x}_{j_1} \neq a_{j_1j_1 1}$. If

$$\frac{d^2x_{j_1}}{dx_{j_2}^2}\bigg|_{x_{j_2}^*=a_{j_1j_2 1}} = \frac{a_{j_1j_1 0}}{a_{j_2j_2 0}}\frac{(\overline{x}_{j_1}-a_{j_1j_1 1})^2}{(\overline{x}_{j_1}-a_{j_2j_1 1})\left[(a_{j_1j_2 1}-a_{j_2j_2})^2+b_{j_2j_2}\right]}<0,$$

there is a down-parabola flow at $x_{j_2}^*=a_{j_1j_2 1}$ with $\overline{x}_{j_1}\neq a_{j_1j_1 1}$. Let

$$\dot{x}_{j_2}=a_{j_2j_2 0}(\overline{x}_{j_1}-a_{j_2j_1 1})\left[(a_{j_1j_2 1}-a_{j_2j_2})^2+b_{j_2j_2}\right],$$

with

$$(a_{j_1j_2 1}-a_{j_2j_2})^2+b_{j_2j_2}>0.$$

The parabola flows at $x_{j_2}^*=a_{j_1j_2 1}$ with $\overline{x}_{j_1}\neq a_{j_1j_1 1}$ are positive and negative for $a_{j_2j_2 0}(\overline{x}_{j_1}-a_{j_2j_1 1})>0$ and $a_{j_2j_2 0}(\overline{x}_{j_1}-a_{j_2j_1 1})<0$, respectively.

Thus, the equilibrium of $x_{j_2}^*=a_{j_1j_2 1}$ with $\overline{x}_{j_1}\neq a_{j_1j_1 1}$ has the following properties:

- For $a_{j_1j_1 0}(\overline{x}_{j_1}-a_{j_1j_1 1})^2>0$ and $a_{j_2j_2 0}(\overline{x}_{j_1}-a_{j_2j_1 1})>0$,

$$(a_{j_1j_2 1},\dot{x}_{j_2})=\underbrace{(\text{UP,pF})}_{\text{up-parabola flow }(+)}.$$

- For $a_{j_1j_1 0}(\overline{x}_{j_1}-a_{j_1j_1 1})^2<0$ and $a_{j_2j_2 0}(\overline{x}_{j_1}-a_{j_2j_1 1})>0$,

$$(a_{j_1j_2 1},\dot{x}_{j_2})=\underbrace{(\text{DP,pF})}_{\text{down-parabola flow }(+)}.$$

- For $a_{j_1j_1 0}(\overline{x}_{j_1}-a_{j_1j_1 1})^2>0$ and $a_{j_2j_2 0}(\overline{x}_{j_1}-a_{j_2j_1 1})<0$,

$$(a_{j_1j_2 1},\dot{x}_{j_2})=\underbrace{(\text{DP,nF})}_{\text{dwon-parabola flow }(-)}.$$

- For $a_{j_1j_1 0}(\overline{x}_{j_1}-a_{j_1j_1 1})^2<0$ and $a_{j_2j_2 0}(\overline{x}_{j_1}-a_{j_2j_1 1})<0$,

$$(a_{j_1j_2 1},\dot{x}_{j_2})=\underbrace{(\text{UP,nF})}_{\text{up-parabola flow }(-)}.$$

(II) Similarly, at $x_{j_1}^*=a_{j_2j_1 1}$ with $\overline{x}_{j_2}=a_{j_1j_2 1}$, in phase space,

$$\frac{dx_{j_2}}{dx_{j_1}}\bigg|_{x_{j_1}^*=a_{j_2j_1 1}} = \frac{a_{j_2j_2 0}}{a_{j_1j_1 0}}\frac{(x_{j_1}-a_{j_2j_1 1})\left[(\overline{x}_{j_2}-a_{j_2j_2})^2+b_{j_2j_2}\right]}{(\overline{x}_{j_2}-a_{j_1j_2 1})(x_{j_1}-a_{j_1j_1 1})^2}\bigg|_{x_{j_1}^*=a_{j_2j_1 1}}=0.$$

If

$$\frac{d^2x_{j_2}}{dx_{j_1}^2}\bigg|_{x_{j_1}^*=a_{j_2j_1 1}} = \frac{a_{j_2j_2 0}}{a_{j_1j_1 0}} \frac{(\overline{x}_{j_2}-a_{j_2j_2})^2+b_{j_2j_2}}{(\overline{x}_{j_2}-a_{j_1j_2 1})(a_{j_2j_1 1}-a_{j_1j_1 1})^2} > 0,$$

there is an up-parabola flow. If

$$\frac{d^2x_{j_2}}{dx_{j_1}^2}\bigg|_{x_{j_1}^*=a_{j_2j_1 1}} = \frac{a_{j_2j_2 0}}{a_{j_1j_1 0}} \frac{(\overline{x}_{j_2}-a_{j_2j_2})^2+b_{j_2j_2}}{(\overline{x}_{j_2}-a_{j_1j_2 1})(a_{j_2j_1 1}-a_{j_1j_1 1})^2} < 0,$$

there is a down-parabola flow. For

$$\dot{x}_{j_1} = a_{j_1j_1 0}(\overline{x}_{j_2}-a_{j_1j_2 1})(a_{j_2j_1 1}-a_{j_1j_1 1})^2,$$

the parabola flows at $x_{j_1}^* = a_{j_1j_1 1}$ are positive and negative for $\dot{x}_{j_1} > 0$ and $\dot{x}_{j_1} < 0$, respectively. In addition,

$$(\overline{x}_{j_2}-a_{j_2j_2})^2+b_{j_2j_2} > 0.$$

Thus, the equilibrium of $x_{j_1}^* = a_{j_2j_1 1}$ with $\overline{x}_{j_2} = a_{j_1j_2 1}$ has the following properties:

- For $a_{j_1j_1 0}(\overline{x}_{j_2}-a_{j_1j_2 1})(a_{j_2j_1 1}-a_{j_1j_1 1})^2 > 0$ and $a_{j_2j_2 0} > 0$,

$$(\dot{x}_{j_1}, a_{j_2j_1 1}) = \underbrace{(\text{pF,UP})}_{\text{up-parabola flow } (+)}.$$

- For $a_{j_1j_1 0}(\overline{x}_{j_2}-a_{j_1j_2 1})(a_{j_2j_1 1}-a_{j_1j_1 1})^2 < 0$ and $a_{j_2j_2 0} > 0$,

$$(\dot{x}_{j_1}, a_{j_2j_1 1}) = \underbrace{(\text{nF,DP})}_{\text{down-parabola flow } (-)}.$$

- For $a_{j_1j_1 0}(\overline{x}_{j_2}-a_{j_1j_2 1})(a_{j_2j_1 1}-a_{j_1j_1 1})^2 > 0$ and $a_{j_2j_2 0} < 0$,

$$(\dot{x}_{j_1}, a_{j_2j_1 1}) = \underbrace{(\text{pF,DP})}_{\text{down-parabola flow } (+)}.$$

- For $a_{j_1j_1 0}(\overline{x}_{j_2}-a_{j_1j_2 1})(a_{j_2j_1 1}-a_{j_1j_1 1})^2 < 0$ and $a_{j_2j_2 0} < 0$,

$$(\dot{x}_{j_1}, a_{j_2 j_1 1}) = \underbrace{(\text{nF,UP})}_{\text{up-parabola flow }(-)}.$$

Therefore, from the analysis, the equilibrium of $(x_{j_2}^*, x_{j_1}^*) = (a_{j_1 j_2 1}, a_{j_2 j_1 1})$ has the following properties as in Eqs. (1.53)–(1.56):

- For $a_{j_1 j_1 0}(a_{j_2 j_1 1} - a_{j_1 j_1 1})^2 > 0$ and $a_{j_2 j_2 0} > 0$,

$$(a_{j_1 j_2 1}, a_{j_2 j_1 1}) = \underbrace{(\text{UP}_+,\text{UP}_+)}_{\text{positive saddle}}.$$

The equilibrium of $(x_{j_2}^*, x_{j_1}^*) = (a_{j_1 j_2 1}, a_{j_2 j_1 1})$ is a (UP$_+$, UP$_+$)-positive saddle.

- For $a_{j_1 j_1 0}(a_{j_2 j_1 1} - a_{j_1 j_1 1})^2 < 0$ and $a_{j_2 j_2 0} > 0$,

$$(a_{j_1 j_2 1}, a_{j_2 j_1 1}) = \underbrace{(\text{DP}_+,\text{DP}_-)}_{\text{CCW center}}.$$

The equilibrium of $(x_{j_2}^*, x_{j_1}^*) = (a_{j_1 j_2 1}, a_{j_2 j_1 1})$ is a (DP$_+$,DP$_-$)-counterclockwise center.

- For $a_{j_1 j_1 0}(a_{j_2 j_1 1} - a_{j_1 j_1 1})^2 > 0$ and $a_{j_2 j_2 0} < 0$,

$$(a_{j_1 j_2 1}, a_{j_2 j_1 1}) = \underbrace{(\text{DP}_-,\text{DP}_+)}_{\text{CW center}}.$$

The equilibrium of $(x_{j_2}^*, x_{j_1}^*) = (a_{j_1 j_2 1}, a_{j_2 j_1 1})$ is a (DP$_-$, DP$_+$)-clockwise center.

- For $a_{j_1 j_1 0}(a_{j_2 j_1 1} - a_{j_1 j_1 1})^2 < 0$ and $a_{j_2 j_2 0} < 0$,

$$(a_{j_1 j_2 1}, a_{j_2 j_1 1}) = \underbrace{(\text{UP}_-,\text{UP}_-)}_{\text{negative saddle}}.$$

The equilibrium of $(x_{j_2}^*, x_{j_1}^*) = (a_{j_1 j_2 1}, a_{j_2 j_1 1})$ is a (UP$_-$,UP$_-$)-negative saddle.

(iv$_{1b}$) The variational equation at $x_{j_1}^* = a_{j_1 j_1 1}$ with $\bar{x}_{j_2} \neq a_{j_1 j_2 1}$ is

$$\Delta \dot{x}_{j_1} = a_{j_1 j_1 0}(\bar{x}_{j_2} - a_{j_1 j_2 1})(\Delta x_{j_1})^2$$

and

$$\dot{x}_{j_2} = a_{j_2 j_2 0}(a_{j_1 j_1 1} - a_{j_2 j_1 1})\left[(\bar{x}_{j_2} - a_{j_2 j_2})^2 + b_{j_2 j_2}\right].$$

Thus, the equilibrium of $x_{j_2}^* = a_{j_1 j_1 1}$ with $\bar{x}_{j_2} \neq a_{j_1 j_2 1}$ has the following properties:

- For $a_{j_1j_1 0}(\overline{x}_{j_2} - a_{j_1j_2 1}) > 0$ and $a_{j_2j_2 0}(a_{j_1j_1 1} - a_{j_2j_1 1}) > 0$,

$$(a_{j_1j_1 1}, \dot{x}_{j_2}) = \underbrace{\text{(US,pF)}}_{\text{upper-saddle flow } (+)}.$$

- For $a_{j_1j_1 0}(\overline{x}_{j_2} - a_{j_1j_2 1}) < 0$ and $a_{j_2j_2 0}(a_{j_1j_1 1} - a_{j_2j_1 1}) > 0$,

$$(a_{j_1j_1 1}, \dot{x}_{j_2}) = \underbrace{\text{(LS,pF)}}_{\text{lower-saddle flow } (+)}.$$

- For $a_{j_1j_1 0}(\overline{x}_{j_2} - a_{j_1j_2 1}) > 0$ and $a_{j_2j_2 0}(a_{j_1j_1 1} - a_{j_2j_1 1}) < 0$,

$$(a_{j_1j_1 1}, \dot{x}_{j_2}) = \underbrace{\text{(US,nF)}}_{\text{upper-saddle flow } (-)}.$$

- For $a_{j_1j_1 0}(\overline{x}_{j_2} - a_{j_1j_2 1}) < 0$ and $a_{j_2j_2 0}(a_{j_1j_1 1} - a_{j_2j_1 1}) < 0$,

$$(a_{j_1j_1 1}, \dot{x}_{j_2}) = \underbrace{\text{(LS,nF)}}_{\text{lower-saddle flow } (-)}.$$

From the above analysis with $\overline{x}_{j_2} = a_{j_1j_2 1}$ and the case (I) of (iv_{1a}) with $\overline{x}_{j_1} = a_{j_1j_1 1}$, therefore, the equilibrium of $(x^*_{j_1}, x^*_{j_2}) = (a_{j_1j_1 1}, a_{j_1j_2 1})$ has the following properties as in Eqs. (1.57)–(1.60):

- For $a_{j_1j_1 0} > 0$ and $a_{j_2j_2 0}(a_{j_1j_1 1} - a_{j_2j_1 1}) > 0$,

$$(a_{j_1j_1 1}, a_{j_1j_2 1}) = \underbrace{\text{(UP:UP,pF)}}_{\text{hyperbolic-secant to hyperbolic flow } (+)}.$$

  The equilibrium of $(x^*_{j_1}, x^*_{j_2}) = (a_{j_1j_1 1}, a_{j_1j_2 1})$ is a (UP:UP,pF)-positive hyperbolic-secant-to-hyperbolic flow.

- For $a_{j_1j_1 0} < 0$ and $a_{j_2j_2 0}(a_{j_1j_1 1} - a_{j_2j_1 1}) > 0$,

$$(a_{j_1j_1 1}, a_{j_1j_2 1}) = \underbrace{\text{(DP:DP,pF)}}_{\text{hyperbolic to hyperbolic-secant flow } (+)}.$$

  The equilibrium of $(x^*_{j_1}, x^*_{j_2}) = (a_{j_1j_1 1}, a_{j_1j_2 1})$ is a (DP:DP,pF)-positive hyperbolic-to-hyperbolic-secant flow.

- For $a_{j_1j_1 0} > 0$ and $a_{j_2j_2 0}(a_{j_1j_1 1} - a_{j_2j_1 1}) < 0$,

$$(a_{j_1j_1 1}, a_{j_1j_2 1}) = \underbrace{\text{(DP:DP,nF)}}_{\text{hyperbolic to hyperbolic-secant flow } (-)}.$$

The equilibrium of $(x^*_{j_1}, x^*_{j_2}) = (a_{j_1j_1 1}, a_{j_1j_2 1})$ is a (DP:DP, nF)-negative hyperbolic-to-hyperbolic-secant flow.

- For $a_{j_1j_1 0} < 0$ and $a_{j_2j_2 0}(a_{j_1j_1 1} - a_{j_2j_1 1}) < 0$,

$$(a_{j_1j_1 1}, a_{j_1j_2 1}) = \underbrace{\text{(UP:UP,nF)}}_{\text{hyperbolic-secant to hyperbolic flow } (-)}.$$

The equilibrium of $(x^*_{j_1}, x^*_{j_2}) = (a_{j_1j_1 1}, a_{j_1j_2 1})$ is a (UP:UP,nF)-negative hyperbolic-secant-to-hyperbolic flow.

(iv$_2$) For $a_{j_2j_1 1} = a_{j_1j_1 1}$, in phase space,

$$\frac{dx_{j_1}}{dx_{j_2}} = \frac{a_{j_1j_1 0}}{a_{j_2j_2 0}} \frac{(x_{j_1} - a_{j_1j_1 1})(x_{j_2} - a_{j_1j_2 1})}{\left[(x_{j_2} - a_{j_2j_2})^2 + b_{j_2j_2}\right]},$$

and the deformation of the foregoing equation is

$$\frac{1}{(x_{j_1} - a_{j_1j_1 1})} dx_{j_1} = \frac{a_{j_1j_1 0}}{a_{j_2j_2 0}} \left[\frac{x_{j_2} - a_{j_2j_2}}{(x_{j_2} - a_{j_2j_2})^2 + b_{j_2j_2}} + \frac{a_{j_2j_2} - a_{j_1j_2 1}}{(x_{j_2} - a_{j_2j_2})^2 + b_{j_2j_2}}\right] dx_{j_2}.$$

With an initial condition of $(x_{j_1 0}, x_{j_2 0})$ at $t = t_0$,, the integration of the above equation gives

$$\begin{aligned}
\ln \frac{|x_{j_1} - a_{j_1j_1 1}|}{|x_{j_1 0} - a_{j_1j_1 1}|} = \frac{a_{j_1j_1 0}}{a_{j_2j_2 0}} \Bigg[& \frac{1}{2} \ln \frac{|(x_{j_2} - a_{j_2j_2})^2 + b_{j_2j_2}|}{|(x_{j_2 0} - a_{j_2j_2})^2 + b_{j_2j_2}|} \\
& + \frac{(a_{j_2j_2} - a_{j_1j_2 1})}{\sqrt{b_{j_2j_2}}} \left(\arctan \frac{x_{j_2} - a_{j_2j_2}}{\sqrt{b_{j_2j_2}}} - \arctan \frac{x_{j_2 0} - a_{j_2j_2}}{\sqrt{b_{j_2j_2}}} \right) \Bigg].
\end{aligned}$$

(iv$_{2a}$) For $x^*_{j_1} = a_{j_2j_1 1} = a_{j_1j_1 1}$ with $\bar{x}_{j_2} \neq a_{j_1j_2 1}$,

$$\left.\frac{dx_{j_2}}{dx_{j_1}}\right|_{x^*_{j_1} = a_{j_1j_1 1} \pm \varepsilon} = \pm \frac{a_{j_2j_2 0}}{a_{j_1j_1 0}} \frac{\left[(\bar{x}_{j_2} - a_{j_2j_2})^2 + b_{j_2j_2}\right]}{(\bar{x}_{j_2} - a_{j_1j_2 1})\varepsilon}.$$

If

$$\frac{a_{j_2j_2 0}}{a_{j_1j_1 0}} \frac{\left[(\bar{x}_{j_2} - a_{j_2j_2})^2 + b_{j_2j_2}\right]}{(\bar{x}_{j_2} - a_{j_1j_2 1})} > 0,$$

there is a down-up asymptotical flow at $x^*_{j_1} = a_{j_2j_1 1} = a_{j_1j_1 1}$. If

$$\frac{a_{j_2j_20}}{a_{j_1j_10}}\frac{\left[(\overline{x}_{j_2}-a_{j_2j_2})^2+b_{j_2j_2}\right]}{(\overline{x}_{j_2}-a_{j_1j_21})}<0,$$

there is an up-down asymptotic flow at $x^*_{j_1}=a_{j_2j_11}=a_{j_1j_11}$. At $x^*_{j_1}=a_{j_2j_11}=a_{j_1j_11}$ *with* $\overline{x}_{j_2}\neq a_{j_1j_21}$, the variational equation is

$$\Delta\dot{x}_{j_1}=a_{j_1j_10}(\overline{x}_{j_2}-a_{j_1j_21})(\Delta x_{j_1})^2.$$

The up-down and down-up asymptotic flows are upper saddle and lower saddle for $a_{j_1j_10}(\overline{x}_{j_2}-a_{j_1j_21})>0$ and $a_{j_1j_10}(\overline{x}_{j_2}-a_{j_1j_21})<0$, respectively.

Therefore, the infinite equilibrium of $x^*_{j_1}=a_{j_2j_11}=a_{j_1j_11}$ with $\overline{x}_{j_2}\neq a_{j_1j_21}$ has the following properties as in Eqs. (1.62)–(1.65):

- For $a_{j_1j_10}(\overline{x}_{j_2}-a_{j_1j_21})>0$ and $a_{j_2j_20}>0$,

$$(a_{j_1j_11},\overline{x}_{j_2})=\underbrace{(\text{US,DU})}_{\text{down-up upper-saddle}}.$$

  The infinite equilibrium of $x^*_{j_1}=a_{j_2j_11}=a_{j_1j_11}$ is a (US,DU)-down-up upper-saddle.
- For $a_{j_1j_10}(\overline{x}_{j_2}-a_{j_1j_21})<0$ and $a_{j_2j_20}>0$,

$$(a_{j_1j_11},\overline{x}_{j_2})=\underbrace{(\text{LS,UD})}_{\text{up-down lower-saddle}}.$$

  The infinite equilibrium of $x^*_{j_1}=a_{j_2j_11}=a_{j_1j_11}$ is an (LS,UD)-up-down lower-saddle.
- For $a_{j_1j_10}(\overline{x}_{j_2}-a_{j_1j_21})>0$ and $a_{j_2j_20}<0$,

$$(a_{j_1j_11},\overline{x}_{j_2})=\underbrace{(\text{US,UD})}_{\text{up-down upper-saddle}}.$$

  The infinite equilibrium of $x^*_{j_1}=a_{j_2j_11}=a_{j_1j_11}$ is a (US,UD)-up-down upper-saddle.
- For $a_{j_1j_10}(\overline{x}_{j_2}-a_{j_1j_21})<0$ and $a_{j_2j_20}<0$,

$$(a_{j_1j_11},\overline{x}_{j_2})=\underbrace{(\text{LS,DU})}_{\text{down-up lower-saddle}}.$$

  The infinite equilibrium of $x^*_{j_1}=a_{j_2j_11}=a_{j_1j_11}$ is an (LS,DU)-down-up lower-saddle.

(iv$_{2b}$) From (iv$_{2a}$) with $\bar{x}_{j_2} = a_{j_1 j_2 1}$ and the case (I) of (iv$_{1a}$) with $\bar{x}_{j_1} = a_{j_1 j_1 1} = a_{j_2 j_1 1}$, the equilibrium of $(x^*_{j_1}, x^*_{j_2}) = (a_{j_1 j_1 1}, a_{j_1 j_2 1})$ has the following properties as in Eqs. (1.66)–(1.69):

- For $a_{j_1 j_1 0} > 0$ and $a_{j_2 j_2 0} > 0$,

$$(a_{j_1 j_1 1}, a_{j_1 j_2 1}) = \underbrace{(\mathrm{DP}_-:\mathrm{UP}_+, {}_{\mathrm{UD}}\mathrm{LS}:{}_{\mathrm{DU}}\mathrm{US})}_{\text{hyperbolic lower-to-upper saddle}}.$$

The equilibrium of $(x^*_{j_1}, x^*_{j_2}) = (a_{j_1 j_1 1}, a_{j_1 j_2 1})$ is a (DP$_-$:UP$_+$,$_{\mathrm{UD}}$LS:$_{\mathrm{DU}}$US)-hyperbolic lower-to-upper saddle.

- For $a_{j_1 j_1 0} < 0$ and $a_{j_2 j_2 0} > 0$,

$$(a_{j_1 j_1 1}, a_{j_1 j_2 1}) = \underbrace{(\mathrm{UP}_-:\mathrm{DP}_+, {}_{\mathrm{DU}}\mathrm{US}:{}_{\mathrm{UD}}\mathrm{LS})}_{\text{hyperbolic-secant upper-to-lower saddle}}.$$

The equilibrium of $(x^*_{j_1}, x^*_{j_2}) = (a_{j_1 j_1 1}, a_{j_1 j_2 1})$ is a (UP$_-$:DP$_+$,$_{\mathrm{DU}}$US:$_{\mathrm{UD}}$LS)-hyperbolic secant upper-to-lower saddle.

- For $a_{j_1 j_1 0} > 0$ and $a_{j_2 j_2 0} < 0$,

$$(a_{j_1 j_1 1}, a_{j_1 j_2 1}) = \underbrace{(\mathrm{UP}_+:\mathrm{DP}_-, {}_{\mathrm{DU}}\mathrm{LS}:{}_{\mathrm{UD}}\mathrm{US})}_{\text{hyperbolic-secant lower-to-upper saddle}}.$$

The equilibrium of $(x^*_{j_1}, x^*_{j_2}) = (a_{j_1 j_1 1}, a_{j_1 j_2 1})$ is a (UP$_+$:DP$_-$, $_{\mathrm{DU}}$LS:$_{\mathrm{UD}}$US)-hyperbolic secant lower-to-upper saddle.

- For $a_{j_1 j_1 0} < 0$ and $a_{j_2 j_2 0} < 0$,

$$(a_{j_1 j_1 1}, a_{j_1 j_2 1}) = \underbrace{(\mathrm{DP}_+:\mathrm{UP}_-, {}_{\mathrm{UD}}\mathrm{US}:{}_{\mathrm{DU}}\mathrm{LS})}_{\text{hyperbolic upper-to-lower saddle}}.$$

The equilibrium of $(x^*_{j_2}, x^*_{j_1}) = (a_{j_1 j_2 1}, a_{j_2 j_1 1})$ is a (DP$_+$:UP$_-$, $_{\mathrm{UD}}$US:$_{\mathrm{DU}}$LS)-hyperbolic upper-to-lower saddle.

(v) For $\Delta_{j_1 j_2} = 0$ and $\Delta_{j_2 j_2} = 0$, the standard form is

$$\begin{aligned}\dot{x}_{j_2} &= a_{j_2 j_2 0}(x_{j_1} - a_{j_2 j_1 1})(x_{j_2} - a_{j_2 j_2 1})^2,\\ \dot{x}_{j_1} &= a_{j_1 j_1 0}(x_{j_1} - a_{j_1 j_1 1})^2(x_{j_2} - a_{j_1 j_2 1})\end{aligned}$$

where

$$a_{j_1 j_1 1} = -\frac{1}{2}B_{j_1 j_1}, a_{j_2 j_2 1} = -\frac{1}{2}B_{j_2 j_2}.$$

(v$_1$) For $a_{j_2 j_2 1} \neq a_{j_1 j_2 1}$ and $a_{j_2 j_1 1} \neq a_{j_1 j_1 1}$, in phase space,

$$\frac{dx_{j_1}}{dx_{j_2}} = \frac{a_{j_1j_10}}{a_{j_2j_20}} \frac{(x_{j_1} - a_{j_1j_11})^2 (x_{j_2} - a_{j_1j_21})}{(x_{j_1} - a_{j_2j_11})(x_{j_2} - a_{j_2j_21})^2},$$

and the deformation of the foregoing equation is

$$\left[\frac{1}{(x_{j_1} - a_{j_1j_11})} + \frac{(a_{j_1j_11} - a_{j_2j_11})}{(x_{j_1} - a_{j_1j_11})^2}\right] dx_{j_1} = \frac{a_{j_1j_10}}{a_{j_2j_20}} \left[\frac{1}{(x_{j_2} - a_{j_2j_21})} + \frac{a_{j_2j_21} - a_{j_1j_21}}{(x_{j_2} - a_{j_2j_21})^2}\right] dx_{j_2}.$$

With an initial condition of (x_{j_10}, x_{j_20}) at $t = t_0$, the integration of the above equation gives

$$\begin{aligned} &\ln \frac{|x_{j_1} - a_{j_1j_11}|}{|x_{j_10} - a_{j_1j_11}|} - (a_{j_1j_11} - a_{j_2j_11})\left(\frac{1}{x_{j_1} - a_{j_1j_11}} - \frac{1}{x_{j_10} - a_{j_1j_11}}\right) \\ &= \frac{a_{j_1j_10}}{a_{j_2j_20}} \left[\ln \frac{|x_{j_2} - a_{j_2j_21}|}{|x_{j_20} - a_{j_2j_21}|} - (a_{j_2j_21} - a_{j_1j_21})\left(\frac{1}{x_{j_2} - a_{j_2j_21}} - \frac{1}{x_{j_20} - a_{j_2j_21}}\right)\right]. \end{aligned}$$

(v_{1a}) Similar to (iv_{1a}), this case can be proved. Consider two cases (I) $x_{j_2}^* = a_{j_1j_21}$ and (II) $x_{j_1}^* = a_{j_2j_11}$.

(I) In phase space, at $x_{j_2}^* = a_{j_1j_21}$ with $\bar{x}_{j_1} \neq a_{j_1j_11}, a_{j_2j_11}$,

$$\frac{dx_{j_1}}{dx_{j_2}}\bigg|_{x_{j_2}^* = a_{j_1j_21}} = \frac{a_{j_1j_10}}{a_{j_2j_20}} \frac{(\bar{x}_{j_1} - a_{j_1j_11})^2 (x_{j_2} - a_{j_1j_21})}{(\bar{x}_{j_1} - a_{j_2j_11})(x_{j_2} - a_{j_2j_21})^2}\bigg|_{x_{j_2}^* = a_{j_1j_21}} = 0.$$

If

$$\frac{d^2 x_{j_1}}{dx_{j_2}^2}\bigg|_{x_{j_2}^* = a_{j_1j_21}} = \frac{a_{j_1j_10}}{a_{j_2j_20}} \frac{(\bar{x}_{j_1} - a_{j_1j_11})^2}{(\bar{x}_{j_1} - a_{j_2j_11})(a_{j_1j_21} - a_{j_2j_21})^2} > 0,$$

there is an up-parabola flow at $x_{j_2}^* = a_{j_1j_21} \neq a_{j_2j_21}$ with $\bar{x}_{j_1} \neq a_{j_1j_11}, a_{j_2j_11}$. If

$$\frac{d^2x_{j_1}}{dx_{j_2}^2}\bigg|_{x_{j_2}^*=a_{j_1j_21}} = \frac{a_{j_1j_10}}{a_{j_2j_20}} \frac{(\overline{x}_{j_1}-a_{j_1j_11})^2}{(\overline{x}_{j_1}-a_{j_2j_11})(a_{j_1j_21}-a_{j_2j_21})^2} < 0,$$

there is a down-parabola flow at $x_{j_2}^* = a_{j_1j_21} \neq a_{j_2j_21}$ with $\overline{x}_{j_1} \neq a_{j_1j_11}, a_{j_2j_11}$. Let

$$\dot{x}_{j_2} = a_{j_2j_20}(\overline{x}_{j_1}-a_{j_2j_11})(a_{j_1j_21}-a_{j_2j_21})^2,$$

and for $a_{j_1j_21} \neq a_{j_2j_21}$,

$$(a_{j_1j_21}-a_{j_2j_21})^2 > 0.$$

The parabola flows at $x_{j_2}^* = a_{j_1j_21} \neq a_{j_2j_21}$ with $x_{j_2}^* = a_{j_1j_21} \neq a_{j_2j_21}$ are positive and negative for $\dot{x}_{j_2} > 0$ and $\dot{x}_{j_2} < 0$, respectively.

Thus, the equilibrium of $x_{j_2}^* = a_{j_1j_21} \neq a_{j_2j_21}$ with $x_{j_2}^* = a_{j_1j_21} \neq a_{j_2j_21}$ has the following properties:

- For $a_{j_1j_10}(\overline{x}_{j_1}-a_{j_1j_11})^2 > 0$ and $a_{j_2j_20}(\overline{x}_{j_1}-a_{j_2j_11})(a_{j_1j_21}-a_{j_2j_21})^2 > 0$,

$$(a_{j_1j_21}, \dot{x}_{j_2}) = \underbrace{(\text{UP,pF})}_{\text{up-parabola flow } (+)}.$$

- For $a_{j_1j_10}(\overline{x}_{j_1}-a_{j_1j_11})^2 < 0$ and $a_{j_2j_20}(\overline{x}_{j_1}-a_{j_2j_11})(a_{j_1j_21}-a_{j_2j_21})^2 > 0$,

$$(a_{j_1j_21}, \dot{x}_{j_2}) = \underbrace{(\text{DP,pF})}_{\text{down-parabola flow } (+)}.$$

- For $a_{j_1j_10}(\overline{x}_{j_1}-a_{j_1j_11})^2 > 0$ and $a_{j_2j_20}(\overline{x}_{j_1}-a_{j_2j_11})(a_{j_1j_21}-a_{j_2j_21})^2 < 0$,

$$(a_{j_1j_21}, \dot{x}_{j_2}) = \underbrace{(\text{DP,nF})}_{\text{dwon-parabola flow } (-)}.$$

- For $a_{j_1j_10}(\overline{x}_{j_1}-a_{j_1j_11})^2 < 0$ and $a_{j_2j_20}(\overline{x}_{j_1}-a_{j_2j_11})(a_{j_1j_21}-a_{j_2j_21})^2 < 0$,

$$(a_{j_1j_21}, \dot{x}_{j_2}) = \underbrace{(\text{UP,nF})}_{\text{up-parabola flow } (-)}.$$

(II) Similarly, at $x_{j_1}^* = a_{j_2j_11} \neq a_{j_1j_11}$ with $\overline{x}_{j_2} \neq a_{j_1j_21}, a_{j_2j_21}$, in phase space,

$$\frac{dx_{j_2}}{dx_{j_1}}\Big|_{x^*_{j_1}=a_{j_2j_1 1}} = \frac{a_{j_2j_2 0}}{a_{j_1j_1 0}}\frac{(x_{j_1}-a_{j_2j_1 1})(\overline{x}_{j_2}-a_{j_2j_2 1})^2}{(\overline{x}_{j_2}-a_{j_1j_2 1})(x_{j_1}-a_{j_1j_1 1})^2}\Big|_{x^*_{j_1}=a_{j_2j_1 1}} = 0.$$

If

$$\frac{d^2x_{j_2}}{dx^2_{j_1}}\Big|_{x^*_{j_1}=a_{j_2j_1 1}} = \frac{a_{j_2j_2 0}}{a_{j_1j_1 0}}\frac{(\overline{x}_{j_2}-a_{j_2j_2 1})^2}{(\overline{x}_{j_2}-a_{j_1j_2 1})(a_{j_2j_1 1}-a_{j_1j_1 1})^2} > 0,$$

there is an up-parabola flow. If

$$\frac{d^2x_{j_2}}{dx^2_{j_1}}\Big|_{x^*_{j_1}=a_{j_2j_1 1}} = \frac{a_{j_2j_2 0}}{a_{j_1j_1 0}}\frac{(\overline{x}_{j_2}-a_{j_2j_2 1})^2}{(\overline{x}_{j_2}-a_{j_1j_2 1})(a_{j_2j_1 1}-a_{j_1j_1 1})^2} < 0,$$

there is a down-parabola flow. For

$$\dot{x}_{j_1} = a_{j_1j_1 0}(\overline{x}_{j_2}-a_{j_1j_2 1})(a_{j_2j_1 1}-a_{j_1j_1 1})^2,$$

the parabola flows at $x^*_{j_1}=a_{j_2j_1 1}\neq a_{j_1j_1 1}$ are positive and negative for $\dot{x}_{j_1}>0$ and $\dot{x}_{j_1}<0$, respectively. Thus, the equilibrium of $x^*_{j_1}=a_{j_2j_1 1}\neq a_{j_1j_1 1}$ with $\overline{x}_{j_2}\neq a_{j_1j_2 1}, a_{j_2j_2 1}$ has the following properties:

- For $a_{j_1j_1 0}(\overline{x}_{j_2}-a_{j_1j_2 1})(a_{j_2j_1 1}-a_{j_1j_1 1})^2>0$ and $a_{j_2j_2 0}(\overline{x}_{j_2}-a_{j_2j_2 1})^2>0$,

$$(\dot{x}_{j_1}, a_{j_2j_1 1}) = \underbrace{(\text{pF,UP})}_{\text{up-parabola flow }(+)}.$$

- For $a_{j_1j_1 0}(\overline{x}_{j_2}-a_{j_1j_2 1})(a_{j_2j_1 1}-a_{j_1j_1 1})^2<0$ and $a_{j_2j_2 0}(\overline{x}_{j_2}-a_{j_2j_2 1})^2>0$,

$$(\dot{x}_{j_1}, a_{j_2j_1 1}) = \underbrace{(\text{nF,DP})}_{\text{down-parabola flow }(-)}.$$

- For $a_{j_1j_1 0}(\overline{x}_{j_2}-a_{j_1j_2 1})(a_{j_2j_1 1}-a_{j_1j_1 1})^2>0$ and $a_{j_2j_2 0}(\overline{x}_{j_2}-a_{j_2j_2 1})^2<0$,

$$(\dot{x}_{j_1}, a_{j_2j_1 1}) = \underbrace{(\text{pF,DP})}_{\text{down-parabola flow }(+)}.$$

- For $a_{j_1j_1 0}(\overline{x}_{j_2}-a_{j_1j_2 1})(a_{j_2j_1 1}-a_{j_1j_1 1})^2<0$ and $a_{j_2j_2 0}(\overline{x}_{j_2}-a_{j_2j_2 1})^2<0$,

$$(\dot{x}_{j_1}, a_{j_2j_1 1}) = \underbrace{(\text{nF,UP})}_{\text{up-parabola flow }(-)}.$$

Therefore, from the above analysis of cases (I) and (II), the equilibrium of $(x^*_{j_2}, x^*_{j_1}) = (a_{j_1j_2 1}, a_{j_2j_1 1})$ has the following properties as in Eqs. (1.73)–(1.76):

- For $a_{j_1j_1 0}(a_{j_2j_1 1}-a_{j_1j_1 1})^2>0$ and $a_{j_2j_2 0}(a_{j_1j_2 1}-a_{j_2j_2 1})^2>0$,

$$(a_{j_1j_2 1},a_{j_2j_1 1})=\underbrace{(\mathrm{UP}_+,\mathrm{UP}_+)}_{\text{positive saddle}}.$$

The equilibrium of $(x^*_{j_2},x^*_{j_1})=(a_{j_1j_2 1},a_{j_2j_1 1})$ is a (UP$_+$, UP$_+$)-positive saddle.

- For $a_{j_1j_1 0}(a_{j_2j_1 1}-a_{j_1j_1 1})^2<0$ and $a_{j_2j_2 0}(a_{j_1j_2 1}-a_{j_2j_2 1})^2>0$,

$$(a_{j_1j_2 1},a_{j_2j_1 1})=\underbrace{(\mathrm{DP}_+,\mathrm{DP}_-)}_{\text{CCW center}}.$$

The equilibrium of $(x^*_{j_2},x^*_{j_1})=(a_{j_1j_2 1},a_{j_2j_1 1})$ is a (DP$_+$,DP$_-$)-counterclockwise center.

- For $a_{j_1j_1 0}(a_{j_2j_1 1}-a_{j_1j_1 1})^2>0$ and $a_{j_2j_2 0}(a_{j_1j_2 1}-a_{j_2j_2 1})^2<0$,

$$(a_{j_1j_2 1},a_{j_2j_1 1})=\underbrace{(\mathrm{DP}_-,\mathrm{DP}_+)}_{\text{CW center}}.$$

The equilibrium of $(x^*_{j_2},x^*_{j_1})=(a_{j_1j_2 1},a_{j_2j_1 1})$ is a (DP$_-$, DP$_+$)-clockwise center.

- For $a_{j_1j_1 0}(a_{j_2j_1 1}-a_{j_1j_1 1})^2<0$ and $a_{j_2j_2 0}(a_{j_1j_2 1}-a_{j_2j_2 1})^2<0$,

$$(a_{j_1j_2 1},a_{j_2j_1 1})=\underbrace{(\mathrm{UP}_-,\mathrm{UP}_-)}_{\text{negative saddle}}.$$

The equilibrium of $(x^*_{j_2},x^*_{j_1})=(a_{j_1j_2 1},a_{j_2j_1 1})$ is a (UP$_-$,UP$_-$)-negative saddle.

($\mathrm{v}_{1\mathrm{b}}$) For $(x^*_{j_1},x^*_{j_2})=(a_{j_1j_1 1},a_{j_2j_2 1})$, the variational equations are

$$\begin{aligned}\Delta\dot{x}_{j_2}&=a_{j_2j_2 0}(a_{j_1j_1 1}-a_{j_2j_1 1})(\Delta x_{j_2})^2,\\ \Delta\dot{x}_{j_1}&=a_{j_1j_1 0}(a_{j_2j_2 1}-a_{j_1j_2 1})(\Delta x_{j_1})^2.\end{aligned}$$

the flows at $(x^*_{j_1},x^*_{j_2})=(a_{j_1j_1 1},a_{j_2j_2 1})$ in the x_{j_1}-direction are upper-saddle and lower-saddle for $a_{j_1j_1 0}(a_{j_2j_2 1}-a_{j_1j_2 1})>0$ and $a_{j_1j_1 0}(a_{j_2j_2 1}-a_{j_1j_2 1})<0$, respectively. The flows at $(x^*_{j_1},x^*_{j_2})=(a_{j_1j_1 1},a_{j_2j_2 1})$ in the x_{j_2}-direction are upper-saddle and lower-saddle for $a_{j_2j_2 0}(a_{j_1j_1 1}-a_{j_2j_1 1})>0$ and $a_{j_2j_2 0}(a_{j_1j_1 1}-a_{j_2j_1 1})<0$, respectively.

Therefore, the equilibrium of $(x^*_{j_1},x^*_{j_2})=(a_{j_1j_1 1},a_{j_2j_2 1})$ with $a_{j_2j_2 1}\neq a_{j_1j_2 1}$ has the following properties as in Eqs. (1.77)–(1.80):

- For $a_{j_1j_1 0}(a_{j_2j_2 1}-a_{j_1j_2 1})>0$ and $a_{j_2j_2 0}(a_{j_1j_1 1}-a_{j_2j_1 1})>0$,

$$(a_{j_1j_1 1}, a_{j_2j_2 1}) = \underbrace{(\text{US,US})}_{\text{double-saddle}}.$$

The equilibrium of $(x^*_{j_1}, x^*_{j_2}) = (a_{j_1j_1 1}, a_{j_2j_2 1})$ is a (US,US)-double-saddle.

- For $a_{j_1j_1 0}(a_{j_2j_2 1} - a_{j_1j_2 1}) < 0$ and $a_{j_2j_2 0}(a_{j_1j_1 1} - a_{j_2j_1 1}) > 0$,

$$(a_{j_1j_1 1}, a_{j_2j_2 1}) = \underbrace{(\text{LS,US})}_{\text{double-saddle}}.$$

The equilibrium of $(x^*_{j_1}, x^*_{j_2}) = (a_{j_1j_1 1}, a_{j_2j_2 1})$ is an (LS,US)-double-saddle.

- For $a_{j_1j_1 0}(a_{j_2j_2 1} - a_{j_1j_2 1}) > 0$ and $a_{j_2j_2 0}(a_{j_1j_1 1} - a_{j_2j_1 1}) < 0$,

$$(a_{j_1j_1 1}, a_{j_2j_2 1}) = \underbrace{(\text{US,LS})}_{\text{double-saddle}}.$$

The equilibrium of $(x^*_{j_1}, x^*_{j_2}) = (a_{j_1j_1 1}, a_{j_2j_2 1})$ is a (US,LS)-double-saddle.

- For $a_{j_1j_1 0}(a_{j_2j_2 1} - a_{j_1j_2 1}) < 0$ and $a_{j_2j_2 0}(a_{j_1j_1 1} - a_{j_2j_1 1}) < 0$,

$$(a_{j_1j_1 1}, a_{j_2j_2 1}) = \underbrace{(\text{LS,LS})}_{\text{double-saddle}}.$$

The equilibrium of $(x^*_{j_1}, x^*_{j_2}) = (a_{j_1j_1 1}, a_{j_2j_2 1})$ is an (LS,LS)-double-saddle.

– The double-saddles are the appearing bifurcations of saddle source and saddle sink, and such double saddles are also the appearing bifurcations of the network of saddles, source, and sink.

(v_{1c}) For $(x^*_{j_1}, x^*_{j_2}) = (a_{j_1j_1 1}, a_{j_2j_2 1})$, consider the variational equation at $x^*_{j_1} = a_{j_1j_1 1}$ with $\overline{x}_{j_2} \neq a_{j_1j_2 1}, a_{j_2j_2 1}$ as

$$\Delta\dot{x}_{j_1} = a_{j_1j_1 0}(\overline{x}_{j_2} - a_{j_1j_2 1})(\Delta x_{j_1})^2$$

and

$$\dot{x}_{j_2} = a_{j_2j_2 0}(a_{j_1j_1 1} - a_{j_2j_1 1})(\overline{x}_{j_2} - a_{j_2j_2 1})^2 \neq 0.$$

Thus, the equilibrium of $x^*_{j_1} = a_{j_1j_1 1}$ with $\overline{x}_{j_2} \neq a_{j_1j_2 1}, a_{j_2j_2 1}$ has the following properties:

- For $a_{j_1j_10}(\bar{x}_{j_2}-a_{j_1j_21})>0$ and $a_{j_2j_20}(a_{j_1j_11}-a_{j_2j_11})(\bar{x}_{j_2}-a_{j_2j_21})^2>0$,

$$(a_{j_1j_11},\dot{x}_{j_2})=\underbrace{(\text{US,pF})}_{\text{upper-saddle flow }(+)}.$$

- For $a_{j_1j_10}(\bar{x}_{j_2}-a_{j_1j_21})<0$ and $a_{j_2j_20}(a_{j_1j_11}-a_{j_2j_11})(\bar{x}_{j_2}-a_{j_2j_21})^2>0$,

$$(a_{j_1j_11},\dot{x}_{j_2})=\underbrace{(\text{LS,pF})}_{\text{lower-saddle flow }(+)}.$$

- For $a_{j_1j_10}(\bar{x}_{j_2}-a_{j_1j_21})>0$ and $a_{j_2j_20}(a_{j_1j_11}-a_{j_2j_11})(\bar{x}_{j_2}-a_{j_2j_21})^2<0$,

$$(a_{j_1j_11},\dot{x}_{j_2})=\underbrace{(\text{US,nF})}_{\text{upper-saddle flow }(-)}.$$

- For $a_{j_1j_10}(\bar{x}_{j_2}-a_{j_1j_21})<0$ and $a_{j_2j_20}(a_{j_1j_11}-a_{j_2j_11})(\bar{x}_{j_2}-a_{j_2j_21})^2<0$,

$$(a_{j_1j_11},\dot{x}_{j_2})=\underbrace{(\text{LS,nF})}_{\text{lower-saddle flow }(-)}.$$

From the above analysis with $\bar{x}_{j_2}=a_{j_1j_21}$ and the case (I) of $(\text{v}_{1\text{a}})$ with $\bar{x}_{j_1}=a_{j_1j_11}$, therefore, the equilibrium of $(x^*_{j_1},x^*_{j_2})=(a_{j_1j_11},a_{j_1j_21})$ with $a_{j_2j_21}\neq a_{j_1j_21}$ has the following properties as in Eqs. (1.81)–(1.84):

- For $a_{j_1j_10}>0$ and $a_{j_2j_20}(a_{j_1j_11}-a_{j_2j_11})(a_{j_1j_21}-a_{j_2j_21})^2>0$,

$$(a_{j_1j_11},a_{j_1j_21})=\underbrace{(\text{UP:UP,pF})}_{\text{hyperbolic-secant to hyperbolic flow }(+)}.$$

The equilibrium of $(x^*_{j_1},x^*_{j_2})=(a_{j_1j_11},a_{j_1j_21})$ is a (UP:UP, pF)-positive hyperbolic-secant-to-hyperbolic flow.

- For $a_{j_1j_10}<0$ and $a_{j_2j_20}(a_{j_1j_11}-a_{j_2j_11})(a_{j_1j_21}-a_{j_2j_21})^2>0$,

$$(a_{j_1j_11},a_{j_1j_21})=\underbrace{(\text{DP:DP,pF})}_{\text{hyperbolic to hyperbolic-secant flow }(+)}.$$

The equilibrium of $(x^*_{j_1},x^*_{j_2})=(a_{j_1j_11},a_{j_1j_21})$ is a (DP:DP, pF)-positive hyperbolic-to-hyperbolic-secant flow.

- For $a_{j_1j_10}>0$ and $a_{j_2j_20}(a_{j_1j_11}-a_{j_2j_11})(a_{j_1j_21}-a_{j_2j_21})^2<0$,

$$(a_{j_1j_11}, a_{j_1j_21}) = \underbrace{(\text{DP:DP,nF})}_{\text{hyperbolic to hyperbolic-secant flow }(-)}.$$

The equilibrium of $(x_{j_1}^*, x_{j_2}^*) = (a_{j_1j_11}, a_{j_1j_21})$ is an (DP:DP,nF)-negative hyperbolic-to-hyperbolic-secant flow.

- For $a_{j_1j_10} < 0$ and $a_{j_2j_20}(a_{j_1j_11} - a_{j_2j_11})(a_{j_1j_21} - a_{j_2j_21})^2 < 0$,

$$(a_{j_1j_11}, a_{j_1j_21}) = \underbrace{(\text{UP:UP,nF})}_{\text{hyperbolic-secant to hyperbolic flow }(-)}.$$

The equilibrium of $(x_{j_1}^*, x_{j_2}^*) = (a_{j_1j_11}, a_{j_1j_21})$ is a (UP:UP, nF)-negative hyperbolic-secant-to-hyperbolic flow.

($\mathrm{v}_{1\mathrm{d}}$) For $(x_{j_1}^*, x_{j_2}^*) = (a_{j_2j_11}, a_{j_2j_21})$, consider the variational equation at $x_{j_2}^* = a_{j_2j_21} \neq a_{j_1j_21}$ with $\overline{x}_{j_1} \neq a_{j_1j_11}, a_{j_2j_11}$ as

$$\Delta\dot{x}_{j_2} = a_{j_2j_20}(\overline{x}_{j_1} - a_{j_2j_11})(\Delta x_{j_2})^2$$

and

$$\dot{x}_{j_1} = a_{j_1j_10}(a_{j_2j_21} - a_{j_1j_21})(\overline{x}_{j_1} - a_{j_1j_11})^2 \neq 0.$$

Thus, the equilibrium of $x_{j_2}^* = a_{j_2j_21} \neq a_{j_1j_21}$ with $\overline{x}_{j_1} \neq a_{j_1j_11}, a_{j_2j_11}$ has the following properties:

- For $a_{j_1j_10}(a_{j_2j_21} - a_{j_1j_21})(\overline{x}_{j_1} - a_{j_1j_11})^2 > 0$ and $a_{j_2j_20}(\overline{x}_{j_1} - a_{j_2j_11}) > 0$,

$$(\dot{x}_{j_1}, a_{j_2j_21}) = \underbrace{(\text{pF,US})}_{\text{upper-saddle flow }(+)}.$$

- For $a_{j_1j_10}(a_{j_2j_21} - a_{j_1j_21})(\overline{x}_{j_1} - a_{j_1j_11})^2 < 0$ and $a_{j_2j_20}(\overline{x}_{j_1} - a_{j_2j_11}) > 0$,

$$(\dot{x}_{j_1}, a_{j_2j_21}) = \underbrace{(\text{nF,US})}_{\text{upper-saddle flow }(-)}.$$

- For $a_{j_1j_10}(a_{j_2j_21} - a_{j_1j_21})(\overline{x}_{j_1} - a_{j_1j_11})^2 > 0$ and $a_{j_2j_20}(\overline{x}_{j_1} - a_{j_2j_11}) < 0$,

$$(\dot{x}_{j_1}, a_{j_2j_21}) = \underbrace{(\text{pF,LS})}_{\text{lower-saddle flow }(+)}.$$

- For $a_{j_1j_10}(a_{j_2j_21} - a_{j_1j_21})(\overline{x}_{j_1} - a_{j_1j_11})^2 < 0$ and $a_{j_2j_20}(\overline{x}_{j_1} - a_{j_2j_11}) < 0$,

$$(\dot{x}_{j_1}, a_{j_2j_2 1}) = \underbrace{(\text{nF,LS})}_{\text{lower-saddle flow }(-)}.$$

From the above analysis with $\bar{x}_{j_1} = a_{j_2j_1 1}$ and the case (II) of ($\mathrm{v}_{1\mathrm{a}}$) with $\bar{x}_{j_2} = a_{j_2j_1 1}$, therefore, the equilibrium of $(x^*_{j_1}, x^*_{j_2}) = (a_{j_2j_1 1}, a_{j_2j_2 1})$ with $a_{j_2j_2 1} \neq a_{j_1j_2 1}$ has the following properties as in Eqs. (1.85)–(1.88):

- For $a_{j_1j_1 0}(a_{j_2j_2 1} - a_{j_1j_2 1})(a_{j_2j_1 1} - a_{j_1j_1 1})^2 > 0$ and $a_{j_2j_2 0} > 0$,

$$(a_{j_2j_1 1}, a_{j_2j_2 1}) = \underbrace{(\text{pF,UP:UP})}_{\text{hyperbolic-secant to hyperbolic flow }(+)}.$$

  The equilibrium of $(x^*_{j_1}, x^*_{j_2}) = (a_{j_2j_1 1}, a_{j_2j_2 1})$ is a (pF, UP:UP)-positive hyperbolic-secant-to-hyperbolic flow.
- For $a_{j_1j_1 0}(a_{j_2j_2 1} - a_{j_1j_2 1})(a_{j_2j_1 1} - a_{j_1j_1 1})^2 < 0$ and $a_{j_2j_2 0} > 0$,

$$(a_{j_2j_1 1}, a_{j_2j_2 1}) = \underbrace{(\text{nF,DP:DP})}_{\text{hyperbolic to hyperbolic-secant flow }(-)}.$$

  The equilibrium of $(x^*_{j_1}, x^*_{j_2}) = (a_{j_2j_1 1}, a_{j_2j_2 1})$ is an (nF, DP:DP)-negative hyperbolic-to-hyperbolic-secant flow.
- For $a_{j_1j_1 0}(a_{j_2j_2 1} - a_{j_1j_2 1})(a_{j_2j_1 1} - a_{j_1j_1 1})^2 > 0$ and $a_{j_2j_2 0} < 0$,

$$(a_{j_2j_1 1}, a_{j_2j_2 1}) = \underbrace{(\text{pF,DP:DP})}_{\text{hyperbolic to hyperbolic-secant flow }(+)}.$$

  The equilibrium of $(x^*_{j_1}, x^*_{j_2}) = (a_{j_2j_1 1}, a_{j_2j_2 1})$ is a (pF, DP:DP)-positive hyperbolic-to-hyperbolic-secant flow.
- For $a_{j_1j_1 0}(a_{j_2j_2 1} - a_{j_1j_2 1})(a_{j_2j_1 1} - a_{j_1j_1 1})^2 < 0$ and $a_{j_2j_2 0} < 0$,

$$(a_{j_2j_1 1}, a_{j_2j_2 1}) = \underbrace{(\text{nF,UP:UP})}_{\text{hyperbolic-secant to hyperbolic flow }(-)}.$$

  The equilibrium of $(x^*_{j_1}, x^*_{j_2}) = (a_{j_2j_1 1}, a_{j_2j_2 1})$ is an (nF, UP:UP)-negative hyperbolic-secant-to-hyperbolic flow.

(v_2) For $a_{j_2j_2 1} = a_{j_1j_2 1}$ and $a_{j_1j_1 1} \neq a_{j_2j_1 1}$, in phase space,

$$\frac{dx_{j_1}}{dx_{j_2}} = \frac{a_{j_1j_10}}{a_{j_2j_20}} \frac{(x_{j_1} - a_{j_1j_11})^2 (x_{j_2} - a_{j_1j_21})}{(x_{j_1} - a_{j_2j_11})(x_{j_2} - a_{j_2j_21})^2}$$
$$= \frac{a_{j_1j_10}}{a_{j_2j_20}} \frac{(x_{j_1} - a_{j_1j_11})^2}{(x_{j_1} - a_{j_2j_11})(x_{j_2} - a_{j_2j_21})},$$

and the deformation of the foregoing equation is

$$\left[\frac{1}{(x_{j_1} - a_{j_1j_11})} + \frac{(a_{j_1j_11} - a_{j_2j_11})}{(x_{j_1} - a_{j_1j_11})^2}\right] dx_{j_1} = \frac{a_{j_1j_10}}{a_{j_2j_20}} \frac{dx_{j_2}}{(x_{j_2} - a_{j_2j_21})}.$$

With an initial condition of (x_{j_10}, x_{j_20}) at $t = t_0$, the integration of the above equation gives the first integral manifold for $a_{j_2j_21} = a_{j_1j_21}$ and $a_{j_1j_11} \neq a_{j_2j_11}$ as

$$\ln \frac{|x_{j_1} - a_{j_1j_11}|}{|x_{j_10} - a_{j_1j_11}|} - (a_{j_1j_11} - a_{j_2j_11})\left(\frac{1}{x_{j_1} - a_{j_1j_11}} - \frac{1}{x_{j_10} - a_{j_1j_11}}\right)$$
$$= \frac{a_{j_1j_10}}{a_{j_2j_20}} \ln \frac{|x_{j_2} - a_{j_2j_21}|}{|x_{j_20} - a_{j_2j_21}|}.$$

(v_{2a}) At $x^*_{j_2} = a_{j_1j_21} = a_{j_2j_21}$ with $\bar{x}_{j_1} \neq a_{j_1j_11}$, in phase space, for a small $\varepsilon > 0$,

$$\left.\frac{dx_{j_1}}{dx_{j_2}}\right|_{x^*_{j_2} = a_{j_2j_21} \pm \varepsilon} = \pm \frac{a_{j_1j_10}}{a_{j_2j_20}} \frac{(\bar{x}_{j_1} - a_{j_1j_11})^2}{(\bar{x}_{j_1} - a_{j_2j_11})\varepsilon}.$$

If

$$\frac{a_{j_1j_10}}{a_{j_2j_20}} \frac{(\bar{x}_{j_1} - a_{j_1j_11})^2}{(\bar{x}_{j_1} - a_{j_2j_11})} > 0,$$

there is a down-up flow. If

$$\frac{a_{j_1j_10}}{a_{j_2j_20}} \frac{(\bar{x}_{j_1} - a_{j_1j_11})^2}{(\bar{x}_{j_1} - a_{j_2j_11})} < 0,$$

there is an up-down flow. The variational equation is

$$\Delta\dot{x}_{j_2} = a_{j_2j_20}(\bar{x}_{j_1} - a_{j_2j_11})(\Delta x_{j_2})^2.$$

The down-up and up-down flows in the x_{j_2}-direction are upper-saddle and lower-saddle for $a_{j_2j_20}(\bar{x}_{j_1} - a_{j_2j_11}) > 0$ and $a_{j_2j_20}(\bar{x}_{j_1} - a_{j_2j_11}) < 0$, respectively.

Therefore, the infinite-equilibrium of $x^*_{j_2} = a_{j_1j_21} = a_{j_2j_21}$ with $\bar{x}_{j_1} \neq a_{j_1j_11}$ has the following properties as in Eqs. (1.90)–(1.93):

- For $a_{j_1j_10}(\overline{x}_{j_1} - a_{j_1j_11})^2 > 0$ and $a_{j_2j_20}(\overline{x}_{j_1} - a_{j_2j_11}) > 0$,

$$(\overline{x}_{j_1}, a_{j_1j_21}) = \underbrace{(\mathrm{DU,US})}_{\text{down-up upper-saddle}}.$$

The infinite-equilibrium of $x_{j_2}^* = a_{j_1j_21} = a_{j_2j_21}$ is an (DU,US)-down-up upper saddle.

- For $a_{j_1j_10}(\overline{x}_{j_1} - a_{j_1j_11})^2 < 0$ and $a_{j_2j_20}(\overline{x}_{j_1} - a_{j_2j_11}) > 0$,

$$(\overline{x}_{j_1}, a_{j_1j_21}) = \underbrace{(\mathrm{UD,US})}_{\text{up-down upper-saddle}}.$$

The infinite-equilibrium of $x_{j_2}^* = a_{j_1j_21} = a_{j_2j_21}$ is a (UD,US)-up-down upper saddle.

- For $a_{j_1j_10}(\overline{x}_{j_1} - a_{j_1j_11})^2 > 0$ and $a_{j_2j_20}(\overline{x}_{j_1} - a_{j_2j_11}) < 0$,

$$(\overline{x}_{j_1}, a_{j_1j_21}) = \underbrace{(\mathrm{UD,LS})}_{\text{up-down lower-saddle}}.$$

The infinite-equilibrium of $x_{j_2}^* = a_{j_1j_21} = a_{j_2j_21}$ is a (UD,LS)-up-down lower saddle.

- For $a_{j_1j_10}(\overline{x}_{j_1} - a_{j_1j_11})^2 < 0$ and $a_{j_2j_20}(\overline{x}_{j_1} - a_{j_2j_11}) < 0$,

$$(\overline{x}_{j_1}, a_{j_1j_21}) = \underbrace{(\mathrm{DU,LS})}_{\text{down-up lower-saddle}}.$$

The infinite-equilibrium of $x_{j_2}^* = a_{j_1j_21} = a_{j_2j_21}$ is a (DU,LS)-down-up lower saddle.

(v_{2b}) From case (v_{2a}) with $\overline{x}_{j_1} = a_{j_1j_11}$, the equilibrium of $(x_{j_1}^*, x_{j_2}^*) = (a_{j_1j_11}, a_{j_1j_21})$ with $a_{j_2j_21} = a_{j_1j_21}$ has the following properties as in Eqs. (1.94)–(1.97):

- For $a_{j_1j_10} > 0$ and $a_{j_2j_20}(a_{j_1j_11} - a_{j_2j_11}) > 0$,

$$(a_{j_1j_11}, a_{j_2j_21}) = \underbrace{(\mathrm{DU{:}DU,US})}_{\text{down-up upper-saddle}}.$$

The equilibrium of $(x_{j_1}^*, x_{j_2}^*) = (a_{j_1j_11}, a_{j_2j_21})$ is a (DU:DU,US)-double down-up upper-saddle.

- For $a_{j_1j_10} < 0$ and $a_{j_2j_20}(a_{j_1j_11} - a_{j_2j_11}) > 0$,

$$(a_{j_1j_11}, a_{j_2j_21}) = \underbrace{(\text{UD:UD,US})}_{\text{up-down upper-saddle}}.$$

The equilibrium of $(x_{j_1}^*, x_{j_2}^*) = (a_{j_1j_11}, a_{j_2j_21})$ is a (UD:UD,US)-double up-down upper-saddle.

- For $a_{j_1j_10} > 0$ and $a_{j_2j_20}(a_{j_1j_11} - a_{j_2j_11}) < 0$,

$$(a_{j_1j_11}, a_{j_2j_21}) = \underbrace{(\text{UD:UD,LS})}_{\text{up-down lower-saddle}}.$$

The equilibrium of $(x_{j_1}^*, x_{j_2}^*) = (a_{j_1j_11}, a_{j_2j_21})$ is a (UD:UD,LS)-double up-down lower-saddle.

- For $a_{j_1j_10} < 0$ and $a_{j_2j_20}(a_{j_1j_11} - a_{j_2j_11}) < 0$,

$$(a_{j_1j_11}, a_{j_2j_21}) = \underbrace{(\text{DU:DU,LS})}_{\text{down-up upper-saddle}}.$$

The equilibrium of $(x_{j_1}^*, x_{j_2}^*) = (a_{j_1j_11}, a_{j_2j_21})$ is a (DU:DU,LS)-double down-up lower-saddle.

(v_{2c}) From case (v_{2a}) with $\bar{x}_{j_1} = a_{j_2j_11}$ and the case (II) of (v_{1a}) with $\bar{x}_{j_2} = a_{j_2j_21}$, the equilibrium of $(x_{j_1}^*, x_{j_2}^*) = (a_{j_2j_11}, a_{j_2j_21})$ with $a_{j_2j_21} = a_{j_1j_21}$ and $a_{j_1j_11} \neq a_{j_2j_11}$ has the following properties as in Eqs. (1.98)–(1.101):

- For $a_{j_1j_10}(a_{j_2j_11} - a_{j_1j_11})^2 > 0$ and $a_{j_2j_20} > 0$,

$$(a_{j_2j_11}, a_{j_2j_21}) = \underbrace{({}_{\text{DU}}\text{LS}:{}_{\text{UD}}\text{US,DP}_{-}:\text{UP}_{+})}_{\text{hyperbolic lower-to-upper saddle}}.$$

The equilibrium of $(x_{j_1}^*, x_{j_2}^*) = (a_{j_2j_11}, a_{j_2j_21})$ is a (${}_{\text{DU}}$LS:${}_{\text{UD}}$US, DP$_-$:UP$_+$)-hyperbolic lower-to-upper saddle.

- For $a_{j_1j_10}(a_{j_2j_11} - a_{j_1j_11})^2 < 0$ and $a_{j_2j_20} > 0$,

$$(a_{j_2j_11}, a_{j_2j_21}) = \underbrace{({}_{\text{UD}}\text{US}:{}_{\text{DU}}\text{LS,DP}_{+}:\text{UP}_{-})}_{\text{hyperbolic-secant upper-to-lower saddle}}.$$

The equilibrium of $(x_{j_1}^*, x_{j_2}^*) = (a_{j_2j_11}, a_{j_2j_21})$ is a (${}_{\text{UD}}$US:${}_{\text{DU}}$LS, UP$_+$:DP$_-$)-hyperbolic secant upper-to-lower saddle.

- For $a_{j_1j_10}(a_{j_2j_11} - a_{j_1j_11})^2 > 0$ and $a_{j_2j_20} < 0$,

$$(a_{j_2 j_1 1}, a_{j_2 j_2 1}) = \underbrace{({}_{\mathrm{UD}}\mathrm{LS}{:}{}_{\mathrm{DU}}\mathrm{US}, \mathrm{UP}_{-}{:}\mathrm{DP}_{+})}_{\text{hyperbolic-secant lower-to-upper saddle}}.$$

The equilibrium of $(x^*_{j_1}, x^*_{j_2}) = (a_{j_2 j_1 1}, a_{j_2 j_2 1})$ is a $({}_{\mathrm{UD}}\mathrm{LS}{:}{}_{\mathrm{DU}}\mathrm{US}, \mathrm{UP}_{-}{:}\mathrm{DP}_{+})$-hyperbolic secant lower-to-upper saddle.

- For $a_{j_1 j_1 0}(a_{j_2 j_1 1} - a_{j_1 j_1 1})^2 < 0$ and $a_{j_2 j_2 0} < 0$,

$$(a_{j_2 j_1 1}, a_{j_2 j_2 1}) = \underbrace{({}_{\mathrm{DU}}\mathrm{US}{:}{}_{\mathrm{UD}}\mathrm{LS}, \mathrm{DP}_{+}{:}\mathrm{UP}_{-})}_{\text{hyperbolic upper-to-lower saddle}}.$$

The equilibrium of $(x^*_{j_1}, x^*_{j_2}) = (a_{j_2 j_1 1}, a_{j_2 j_2 1})$ is a $({}_{\mathrm{DU}}\mathrm{LS}{:}{}_{\mathrm{UD}}\mathrm{US}, \mathrm{DP}_{+}{:}\mathrm{UP}_{-})$-hyperbolic upper-to-lower saddle.

(v_3) For $a_{j_1 j_1 1} = a_{j_2 j_1 1}$ and $a_{j_2 j_2 1} \neq a_{j_1 j_2 1}$, in phase space,

$$\begin{aligned}\frac{dx_{j_1}}{dx_{j_2}} &= \frac{a_{j_1 j_1 0}\,(x_{j_1} - a_{j_1 j_1 1})^2 (x_{j_2} - a_{j_1 j_2 1})}{a_{j_2 j_2 0}\,(x_{j_1} - a_{j_2 j_1 1})(x_{j_2} - a_{j_2 j_2 1})^2}\\ &= \frac{a_{j_1 j_1 0}}{a_{j_2 j_2 0}} \frac{(x_{j_1} - a_{j_1 j_1 1})(x_{j_2} - a_{j_1 j_2 1})}{(x_{j_2} - a_{j_2 j_2 1})^2},\end{aligned}$$

and the deformation of the foregoing equation is

$$\frac{dx_{j_1}}{(x_{j_1} - a_{j_1 j_1 1})} = \frac{a_{j_1 j_1 0}}{a_{j_2 j_2 0}} \left[\frac{1}{(x_{j_2} - a_{j_2 j_2 1})} + \frac{(a_{j_1 j_2 1} - a_{j_2 j_2 1})}{(x_{j_2} - a_{j_2 j_2 1})^2}\right] dx_{j_2}.$$

With an initial condition of $(x_{j_1 0}, x_{j_2 0})$ at $t = t_0$, the integration of the above equation gives the first integral manifold $a_{j_1 j_1 1} = a_{j_2 j_1 1}$ and $a_{j_2 j_2 1} \neq a_{j_1 j_2 1}$ *as*

$$\begin{aligned}\ln \frac{|x_{j_1} - a_{j_1 j_1 1}|}{|x_{j_1 0} - a_{j_1 j_1 1}|} = \frac{a_{j_1 j_1 0}}{a_{j_2 j_2 0}} \Bigg[&\ln \frac{|x_{j_2} - a_{j_2 j_2 1}|}{|x_{j_2 0} - a_{j_2 j_2 1}|}\\ &- (a_{j_2 j_2 1} - a_{j_1 j_2 1})\Big(\frac{1}{x_{j_2} - a_{j_2 j_2 1}} - \frac{1}{x_{j_2 0} - a_{j_2 j_2 1}}\Big)\Bigg].\end{aligned}$$

$(\mathrm{v}_{3\mathrm{a}})$ For $x^*_{j_1} = a_{j_1 j_1 1} = a_{j_2 j_1 1}$ and $\overline{x}_{j_2} \neq a_{j_2 j_2 1}, a_{j_1 j_2 1}$ with $a_{j_2 j_2 1} \neq a_{j_1 j_2 1}$, in phase space,

$$\left.\frac{dx_{j_2}}{dx_{j_1}}\right|_{x^*_{j_1} = a_{j_1 j_1 1} \pm \varepsilon} = \pm \frac{a_{j_2 j_2 0}}{a_{j_1 j_1 0}} \frac{(\overline{x}_{j_2} - a_{j_2 j_2 1})^2}{\varepsilon(\overline{x}_{j_2} - a_{j_1 j_2 1})}.$$

If

$$\frac{a_{j_2j_20}}{a_{j_1j_10}} \frac{(\overline{x}_{j_2} - a_{j_2j_21})^2}{(\overline{x}_{j_2} - a_{j_1j_21})} > 0,$$

there is a down-up asymptotic flow. If

$$\frac{a_{j_2j_20}}{a_{j_1j_10}} \frac{(\overline{x}_{j_2} - a_{j_2j_21})^2}{(\overline{x}_{j_2} - a_{j_1j_21})} < 0,$$

there is an up-down asymptotic flow. The variational equation at $x_{j_1}^* = a_{j_1j_11} = a_{j_2j_11}$ is

$$\Delta\dot{x}_{j_1} = a_{j_1j_10}(\overline{x}_{j_2} - a_{j_1j_21})(\Delta x_{j_1})^2.$$

The up-down and down-up asymptotic flows are upper-saddle and lower-saddle for $a_{j_1j_10}(\overline{x}_{j_2} - a_{j_1j_21}) > 0$ and $a_{j_1j_10}(\overline{x}_{j_2} - a_{j_1j_21}) < 0$, respectively.

Therefore, the infinite equilibrium of $x_{j_1}^* = a_{j_1j_11} = a_{j_2j_11}$ and $\overline{x}_{j_2} \neq a_{j_2j_21}, a_{j_1j_21}$ with $a_{j_2j_21} \neq a_{j_1j_21}$ has the following properties as in Eqs. (1.103)–(1.106):

- For $a_{j_1j_10}(\overline{x}_{j_2} - a_{j_1j_21}) > 0$ and $a_{j_2j_20}(\overline{x}_{j_2} - a_{j_2j_21})^2 > 0$,

$$(a_{j_1j_11}, \overline{x}_{j_2}) = \underbrace{(\text{US,DU})}_{\text{down-up upper-saddle}}.$$

  The infinite-equilibrium of $x_{j_1}^* = a_{j_1j_11} = a_{j_2j_11}$ is a (US,DU)-down-up upper-saddle.

- For $a_{j_1j_10}(\overline{x}_{j_2} - a_{j_1j_21}) < 0$ and $a_{j_2j_20}(\overline{x}_{j_2} - a_{j_2j_21})^2 > 0$,

$$(a_{j_1j_11}, \overline{x}_{j_2}) = \underbrace{(\text{LS,UD})}_{\text{up-down lower-saddle}}.$$

  The infinite-equilibrium of $x_{j_1}^* = a_{j_1j_11} = a_{j_2j_11}$ is an (LS,UD)-up-down lower-saddle.

- For $a_{j_1j_10}(\overline{x}_{j_2} - a_{j_1j_21}) > 0$ and $a_{j_2j_20}(\overline{x}_{j_2} - a_{j_2j_21})^2 < 0$,

$$(a_{j_1j_11}, \overline{x}_{j_2}) = \underbrace{(\text{US,UD})}_{\text{up-down upper-saddle}}.$$

  The infinite-equilibrium of $x_{j_1}^* = a_{j_1j_11} = a_{j_2j_11}$ is a (US.UD)-up-down upper-saddle.

- For $a_{j_1j_10}(\overline{x}_{j_2} - a_{j_1j_21}) < 0$ and $a_{j_2j_20}(\overline{x}_{j_2} - a_{j_2j_21})^2 < 0$,

$$(a_{j_1j_11}, \bar{x}_{j_2}) = \underbrace{(\text{LS,DU})}_{\text{down-up lower-saddle}}.$$

The infinite-equilibrium of $x^*_{j_1} = a_{j_1j_11} = a_{j_2j_11}$ is an (LS,DU)-down-up lower-saddle.

(v_{3b}) From (v_{3a}) with $\bar{x}_{j_2} = a_{j_2j_21}$, thus, the equilibrium of $(x^*_{j_1}, x^*_{j_2}) = (a_{j_1j_11}, a_{j_2j_21})$ with $a_{j_1j_11} = a_{j_2j_11}$ and $a_{j_2j_21} \neq a_{j_1j_21}$ has the following properties as in Eqs. (1.107)–(1.110):

- For $a_{j_1j_10}(a_{j_2j_21} - a_{j_1j_21}) > 0$ and $a_{j_2j_20} > 0$

$$(a_{j_1j_11}, a_{j_2j_21}) = \underbrace{(\text{US,DU:DU})}_{\text{down-up upper-saddle}}.$$

  The equilibrium of $(x^*_{j_1}, x^*_{j_2}) = (a_{j_1j_11}, a_{j_2j_21})$ is a (US,DU:DU)-double down-up upper-saddle.
- For $a_{j_1j_10}(a_{j_2j_21} - a_{j_1j_21}) < 0$ and $a_{j_2j_20} > 0$

$$(a_{j_1j_11}, a_{j_2j_21}) = \underbrace{(\text{LS,UD:UD})}_{\text{up-down lower-saddle}}.$$

  The equilibrium of $(x^*_{j_1}, x^*_{j_2}) = (a_{j_1j_11}, a_{j_2j_21})$ is an (LS,UD:UD)-double up-down lower-saddle.
- For $a_{j_1j_10}(a_{j_2j_21} - a_{j_1j_21}) > 0$ and $a_{j_2j_20} < 0$

$$(a_{j_1j_11}, a_{j_2j_21}) = \underbrace{(\text{US,UD:UD})}_{\text{up-down upper-saddle}}.$$

  The equilibrium of $(x^*_{j_1}, x^*_{j_2}) = (a_{j_1j_11}, a_{j_2j_21})$ is a (US,UD:UD)-double up-down upper-saddle.
- For $a_{j_1j_10}(a_{j_2j_21} - a_{j_1j_21}) < 0$ and $a_{j_2j_20} < 0$

$$(a_{j_1j_11}, a_{j_2j_21}) = \underbrace{(\text{LS,DU:DU})}_{\text{down-up lower-saddle}}.$$

  The equilibrium of $(x^*_{j_1}, x^*_{j_2}) = (a_{j_1j_11}, a_{j_2j_21})$ is an (LS,DU:DU)-double down-up lower-saddle.

(v_{3c}) From (v_{3a}) with $\bar{x}_{j_2} = a_{j_1j_21}$ and the case (I) of (v_{1a}) with $\bar{x}_{j_1} = a_{j_1j_11}$, therefore, the equilibrium of $(x^*_{j_1}, x^*_{j_2}) = (a_{j_1j_11}, a_{j_1j_21})$ has the following properties as in Eqs. (1.111)–(1.114):

- For $a_{j_1j_10} > 0$ and $a_{j_2j_20}(a_{j_1j_21} - a_{j_2j_21})^2 > 0$,

$$(a_{j_1j_11}, a_{j_1j_21}) = \underbrace{(\mathrm{DP}_{-}{:}\mathrm{UP}_{+}, {}_{\mathrm{DU}}\mathrm{LS}{:}_{\mathrm{UD}}\mathrm{US})}_{\text{hyperbolic lower-to-upper saddle}}.$$

The equilibrium of $(x^*_{j_1}, x^*_{j_2}) = (a_{j_1j_11}, a_{j_1j_21})$ is a $(\mathrm{DP}_{-}{:}\mathrm{UP}_{+}, {}_{\mathrm{DU}}\mathrm{LS}{:}_{\mathrm{UD}}\mathrm{US})$-hyperbolic lower-to-upper saddle.

- For $a_{j_1j_10} < 0$ and $a_{j_2j_20}(a_{j_1j_21} - a_{j_2j_21})^2 > 0$,

$$(a_{j_1j_11}, a_{j_1j_21}) = \underbrace{(\mathrm{UP}_{+}{:}\mathrm{DP}_{-}, {}_{\mathrm{UD}}\mathrm{US}{:}_{\mathrm{DU}}\mathrm{LS})}_{\text{hyperbolic-secant upper-to-lower saddle}}.$$

The equilibrium of $(x^*_{j_1}, x^*_{j_2}) = (a_{j_1j_11}, a_{j_1j_21})$ is a $(\mathrm{UP}_{+}{:}\mathrm{DP}_{-}, {}_{\mathrm{UD}}\mathrm{US}{:}_{\mathrm{DU}}\mathrm{LS})$-hyper-secant upper-to-lower saddle.

- For $a_{j_1j_10} > 0$ and $a_{j_2j_20}(a_{j_1j_21} - a_{j_2j_21})^2 < 0$,

$$(a_{j_1j_11}, a_{j_1j_21}) = \underbrace{(\mathrm{UP}_{+}{:}\mathrm{DP}_{-}, {}_{\mathrm{UD}}\mathrm{LS}{:}_{\mathrm{DU}}\mathrm{US})}_{\text{hyperbolic-secant lower-to-upper saddle}}.$$

The equilibrium of $(x^*_{j_1}, x^*_{j_2}) = (a_{j_1j_11}, a_{j_1j_21})$ is an $(\mathrm{UP}_{+}{:}\mathrm{DP}_{-}, {}_{\mathrm{UD}}\mathrm{LS}{:}_{\mathrm{DU}}\mathrm{US})$-hyper-secant lower-to-upper saddle.

- For $a_{j_1j_10} < 0$ and $a_{j_2j_20}(a_{j_1j_21} - a_{j_2j_21})^2 < 0$,

$$(a_{j_1j_11}, a_{j_1j_21}) = \underbrace{(\mathrm{DP}_{+}{:}\mathrm{UP}_{-}, {}_{\mathrm{DU}}\mathrm{US}{:}_{\mathrm{UD}}\mathrm{LS})}_{\text{hyperbolic upper-to-lower saddle}}.$$

The equilibrium of $(x^*_{j_1}, x^*_{j_2}) = (a_{j_1j_11}, a_{j_1j_21})$ is a $(\mathrm{DP}_{+}{:}\mathrm{UP}_{-}, {}_{\mathrm{DU}}\mathrm{US}{:}_{\mathrm{UD}}\mathrm{LS})$-hyperbolic upper-to-lower saddle.

(v_4) For $a_{j_1j_11} = a_{j_2j_11}$ and $a_{j_2j_21} = a_{j_1j_21}$, in phase space,

$$\begin{aligned}\frac{dx_{j_1}}{dx_{j_2}} &= \frac{a_{j_1j_10}\,(x_{j_1} - a_{j_1j_11})^2(x_{j_2} - a_{j_1j_21})}{a_{j_2j_20}\,(x_{j_1} - a_{j_2j_11})(x_{j_2} - a_{j_2j_21})^2}\\ &= \frac{a_{j_1j_10}\,(x_{j_1} - a_{j_1j_11})}{a_{j_2j_20}\,(x_{j_2} - a_{j_2j_21})},\end{aligned}$$

and the deformation of the foregoing equation is

$$\frac{dx_{j_1}}{(x_{j_1} - a_{j_1j_11})} = \frac{a_{j_1j_10}}{a_{j_2j_20}}\frac{dx_{j_2}}{(x_{j_2} - a_{j_2j_21})}.$$

With an initial condition of (x_{j_10}, x_{j_20}) at $t = t_0$, the integration of the above equation gives the first integral manifold $a_{j_1j_11} = a_{j_2j_11}$ and $a_{j_2j_21} = a_{j_1j_21}$ *as*

$$\ln \frac{|x_{j_1} - a_{j_1 j_1 1}|}{|x_{j_1 0} - a_{j_1 j_1 1}|} = \frac{a_{j_1 j_1 0}}{a_{j_2 j_2 0}} \ln \frac{|x_{j_2} - a_{j_2 j_2 1}|}{|x_{j_2 0} - a_{j_2 j_2 1}|}.$$

Similar to the case (v_{2a}), for $x^*_{j_2} = a_{j_2 j_2 1} = a_{j_1 j_2 1}$ with $\bar{x}_{j_1} \neq a_{j_1 j_1 1} = a_{j_2 j_1 1}$,

$$\left.\frac{dx_{j_1}}{dx_{j_2}}\right|_{x_{j_2} = a_{j_2 j_2 1} \pm \varepsilon} = \pm \frac{a_{j_1 j_1 0}}{a_{j_2 j_2 0}} \frac{(\bar{x}_{j_1} - a_{j_1 j_1 1})}{\varepsilon}.$$

If

$$\frac{a_{j_1 j_1 0}}{a_{j_2 j_2 0}} (\bar{x}_{j_1} - a_{j_1 j_1 1}) > 0,$$

there is a down-up asymptotic flow. If

$$\frac{a_{j_1 j_1 0}}{a_{j_2 j_2 0}} (\bar{x}_{j_1} - a_{j_1 j_1 1}) < 0,$$

there is an up-down asymptotic flow. The corresponding variational equation is

$$\Delta \dot{x}_{j_2} = a_{j_2 j_2 0} (\bar{x}_{j_1} - a_{j_2 j_1 1})(\Delta x_{j_2})^2.$$

The up-down and down-up flows for $a_{j_2 j_2 0}(\bar{x}_{j_1} - a_{j_2 j_1 1}) > 0$ and $a_{j_2 j_2 0}(\bar{x}_{j_1} - a_{j_2 j_1 1}) < 0$ are upper-saddle and lower-saddle, respectively.

Similar to the case (v_{3a}), for $x^*_{j_1} = a_{j_1 j_1 1} = a_{j_2 j_1 1}$ with $\bar{x}_{j_2} \neq a_{j_2 j_2 1} = a_{j_1 j_2 1}$,

$$\left.\frac{dx_{j_2}}{dx_{j_1}}\right|_{x_{j_1} = a_{j_1 j_1 1} \pm \varepsilon} = \pm \frac{a_{j_2 j_2 0}}{a_{j_1 j_1 0}} \frac{(\bar{x}_{j_2} - a_{j_2 j_2 1})}{\varepsilon}.$$

If

$$\frac{a_{j_2 j_2 0}}{a_{j_1 j_1 0}} (\bar{x}_{j_2} - a_{j_2 j_2 1}) > 0,$$

there is a down-up asymptotic flow. If

$$\frac{a_{j_2 j_2 0}}{a_{j_1 j_1 0}} (\bar{x}_{j_2} - a_{j_2 j_2 1}) < 0,$$

there is an up-down asymptotic flow. The corresponding variational equation is

$$\Delta \dot{x}_{j_1} = a_{j_1 j_1 0} (\bar{x}_{j_2} - a_{j_1 j_2 1})(\Delta x_{j_1})^2.$$

The up-down and down-up flows for $a_{j_1j_10}(\overline{x}_{j_2}-a_{j_1j_21})>0$ and $a_{j_1j_10}(\overline{x}_{j_2}-a_{j_1j_21})<0$ are upper saddle and lower saddle, respectively.

Therefore, the equilibrium of $(x^*_{j_2},x^*_{j_1})=(a_{j_1j_21},a_{j_2j_11})$ with $a_{j_1j_11}=a_{j_2j_11}$ and $a_{j_2j_21}=a_{j_1j_21}$ has the following properties as in Eqs. (1.116)–(1.119):

- For $a_{j_1j_10}>0$ and $a_{j_2j_20}>0$,

$$(a_{j_1j_21},a_{j_2j_11})=\underbrace{(\mathrm{DU}_+,\mathrm{DU}_+)}_{\text{positive down-up saddle}}.$$

  The equilibrium of $(x^*_{j_2},x^*_{j_1})=(a_{j_1j_21},a_{j_2j_11})$ is an $(\mathrm{DU}_+,\mathrm{DU}_+)$-positive down-up saddle.

- For $a_{j_1j_10}<0$ and $a_{j_2j_20}>0$,

$$(a_{j_1j_21},a_{j_2j_11})=\underbrace{(\mathrm{UD}_+,\mathrm{UD}_-)}_{\text{CCW up-down saddle}}.$$

  The equilibrium of $(x^*_{j_2},x^*_{j_1})=(a_{j_1j_21},a_{j_2j_11})$ is a $(\mathrm{UD}_+,\mathrm{UD}_-)$-counterclockwise up-down saddle.

- For $a_{j_1j_10}>0$ and $a_{j_2j_20}<0$,

$$(a_{j_1j_21},a_{j_2j_11})=\underbrace{(\mathrm{UD}_-,\mathrm{UD}_+)}_{\text{CW up-down saddle}},$$

  The equilibrium of $(x^*_{j_2},x^*_{j_1})=(a_{j_1j_21},a_{j_2j_11})$ is a $(\mathrm{UD}_-,\mathrm{UD}_+)$-clockwise up-down saddle.

- For $a_{j_1j_10}<0$ and $a_{j_2j_20}<0$,

$$(a_{j_1j_21},a_{j_2j_11})=\underbrace{(\mathrm{DU}_-,\mathrm{DU}_-)}_{\text{negative down-up saddle}}.$$

  The equilibrium of $(x^*_{j_2},x^*_{j_1})=(a_{j_1j_21},a_{j_2j_11})$ is a $(\mathrm{DU}_-,\mathrm{DU}_-)$ negative down-up saddle.

(vi) For $\Delta_{j_1j_1}=0$ and $\Delta_{j_2j_2}>0$, the standard form is

$$\begin{aligned}\dot{x}_{j_2}&=a_{j_2j_20}(x_{j_1}-a_{j_2j_11})(x_{j_2}-a_{j_2j_21})(x_{j_2}-a_{j_2j_22}),\\ \dot{x}_{j_1}&=a_{j_1j_10}(x_{j_1}-a_{j_1j_11})^2(x_{j_2}-a_{j_1j_21})\end{aligned}$$

where

$$
\begin{aligned}
& a_{j_1 j_1 1} = -\frac{1}{2} B_{j_1 j_1}, \\
& b_{j_2 j_2 1}, b_{j_2 j_2 2} = -\frac{1}{2}(B_{j_2 j_2} \pm \sqrt{\Delta_{j_2 j_2}}), \\
& \{a_{j_2 j_2 1}, a_{j_2 j_2 2}\} = \text{sort}\{b_{j_2 j_2 1}, b_{j_2 j_2 2}\}, a_{j_2 j_2 1} < a_{j_2 j_2 2}.
\end{aligned}
$$

(vi_1) In phase space, $a_{j_2 j_2 s_1} \neq a_{j_1 j_2 1}$ and $a_{j_1 j_1 1} \neq a_{j_2 j_1 1}$, we have

$$
\frac{dx_{j_1}}{dx_{j_2}} = \frac{a_{j_1 j_1 0}}{a_{j_2 j_2 0}} \frac{(x_{j_1} - a_{j_1 j_1 1})^2 (x_{j_2} - a_{j_1 j_2 1})}{(x_{j_1} - a_{j_2 j_1 1})(x_{j_2} - a_{j_2 j_2 1})(x_{j_2} - a_{j_2 j_2 2})},
$$

and the deformation of the foregoing equation is

$$
\begin{aligned}
\left[\frac{1}{(x_{j_1} - a_{j_1 j_1 1})} + \frac{(a_{j_1 j_1 1} - a_{j_2 j_1 1})}{(x_{j_1} - a_{j_1 j_1 1})^2}\right] dx_{j_1} = & \frac{a_{j_1 j_1 0}}{a_{j_2 j_2 0}} \left[\frac{a_{j_2 j_2 1} - a_{j_1 j_2 1}}{(a_{j_2 j_2 1} - a_{j_2 j_2 2})(x_{j_2} - a_{j_2 j_2 1})}\right. \\
& \left. + \frac{a_{j_2 j_2 2} - a_{j_1 j_2 1}}{(a_{j_2 j_2 2} - a_{j_2 j_2 1})(x_{j_2} - a_{j_2 j_2 2})}\right] dx_{j_2}.
\end{aligned}
$$

With an initial condition of $(x_{j_1 0}, x_{j_2 0})$ at $t = t_0$, the integration of the above equation gives the first integral manifold ($a_{j_2 j_2 s_1} \neq a_{j_1 j_2 1}$ and $a_{j_1 j_1 1} \neq a_{j_2 j_1 1}$) as

$$
\begin{aligned}
& \ln \frac{|x_{j_1} - a_{j_1 j_1 1}|}{|x_{j_1 0} - a_{j_1 j_1 1}|} - (a_{j_1 j_1 1} - a_{j_2 j_1 1})\left(\frac{1}{x_{j_1} - a_{j_1 j_1 1}} - \frac{1}{x_{j_1 0} - a_{j_1 j_1 1}}\right) \\
& = \frac{a_{j_1 j_1 0}}{a_{j_2 j_2 0}} \sum\nolimits_{s_1 = 1}^{2} \frac{(a_{j_2 j_2 s_1} - a_{j_1 j_2 1})}{(a_{j_2 j_2 s_1} - a_{j_2 j_2 s_2})} \ln \frac{|x_{j_2} - a_{j_2 j_2 s_1}|}{|x_{j_2 0} - a_{j_2 j_2 s_1}|}.
\end{aligned}
$$

($\text{vi}_{1\text{a}}$) Similar to the case ($\text{v}_{1\text{a}}$), this case can be proved. Consider two cases (I) $x^*_{j_2} = a_{j_1 j_2 1}$ and (II)$x^*_{j_1} = a_{j_2 j_1 1}$.

(I) In phase space, at $x^*_{j_2} = a_{j_1 j_2 1} \neq a_{j_2 j_2 s_1}$ ($s_1 = 1, 2$) with $\bar{x}_{j_1} \neq a_{j_1 j_1 1}, a_{j_2 j_1 1}$,

$$
\left.\frac{dx_{j_1}}{dx_{j_2}}\right|_{x^*_{j_2} = a_{j_1 j_2 1}} = \frac{a_{j_1 j_1 0}}{a_{j_2 j_2 0}} \left.\frac{(\bar{x}_{j_1} - a_{j_1 j_1 1})^2 (x_{j_2} - a_{j_1 j_2 1})}{(\bar{x}_{j_1} - a_{j_2 j_1 1}) \prod_{s_1 = 1}^{2} (x_{j_2} - a_{j_2 j_2 s_1})}\right|_{x^*_{j_2} = a_{j_1 j_2 1}} = 0.
$$

If

$$
\left.\frac{d^2 x_{j_1}}{dx^2_{j_2}}\right|_{x^*_{j_2} = a_{j_1 j_2 1}} = \frac{a_{j_1 j_1 0}}{a_{j_2 j_2 0}} \frac{(\bar{x}_{j_1} - a_{j_1 j_1 1})^2}{(\bar{x}_{j_1} - a_{j_2 j_1 1}) \prod_{s_1 = 1}^{2} (a_{j_1 j_2 1} - a_{j_2 j_2 s_1})} > 0,
$$

there is an up-parabola flow at $x^*_{j_2} = a_{j_1 j_2 1} \neq a_{j_2 j_2 s_1}$ ($s_1 = 1, 2$) with $\bar{x}_{j_1} \neq a_{j_1 j_1 1}, a_{j_2 j_1 1}$. If

$$\left.\frac{d^2x_{j_1}}{dx_{j_2}^2}\right|_{x_{j_2}^*=a_{j_1j_21}} = \frac{a_{j_1j_10}}{a_{j_2j_20}}\frac{(\overline{x}_{j_1}-a_{j_1j_11})^2}{(\overline{x}_{j_1}-a_{j_2j_11})\prod_{s_1=1}^{2}(a_{j_1j_21}-a_{j_2j_2s_1})} < 0,$$

there is a down-parabola flow at $x_{j_2}^* = a_{j_1j_21} \neq a_{j_2j_2s_1}$ $(s_1 = 1, 2)$ with $\overline{x}_{j_1} \neq a_{j_1j_11}, a_{j_2j_11}$. Let

$$\dot{x}_{j_2} = a_{j_2j_20}(\overline{x}_{j_1} - a_{j_2j_11})\prod_{s_1=1}^{2}(a_{j_1j_21} - a_{j_2j_2s_1}).$$

The parabola flows at $x_{j_2}^* = a_{j_1j_21} \neq a_{j_2j_21}$ with $x_{j_2}^* = a_{j_1j_21} \neq a_{j_2j_21}$ are positive and negative for $\dot{x}_{j_2} > 0$ and $\dot{x}_{j_2} < 0$, respectively.

Thus, the equilibrium of $x_{j_2}^* = a_{j_1j_21} \neq a_{j_2j_2s_1}$ $(s_1 = 1, 2)$ with $x_{j_2}^* = a_{j_1j_21} \neq a_{j_2j_21}$ has the following properties:

- For $a_{j_1j_10}(\overline{x}_{j_1} - a_{j_1j_11})^2 > 0$ and $a_{j_2j_20}(\overline{x}_{j_1} - a_{j_2j_11})\prod_{s_1=1}^{2}(a_{j_1j_21} - a_{j_2j_2s_1}) > 0$,

$$(a_{j_1j_21}, \dot{x}_{j_2}) = \underbrace{(\text{UP,pF})}_{\text{up-parabola flow }(+)}.$$

- For $a_{j_1j_10}(\overline{x}_{j_1} - a_{j_1j_11})^2 < 0$ and $a_{j_2j_20}(\overline{x}_{j_1} - a_{j_2j_11})\prod_{s_1=1}^{2}(a_{j_1j_21} - a_{j_2j_2s_1}) > 0$,

$$(a_{j_1j_21}, \dot{x}_{j_2}) = \underbrace{(\text{DP,pF})}_{\text{down-parabola flow }(+)}.$$

- For $a_{j_1j_10}(\overline{x}_{j_1} - a_{j_1j_11})^2 > 0$ and $a_{j_2j_20}(\overline{x}_{j_1} - a_{j_2j_11})\prod_{s_1=1}^{2}(a_{j_1j_21} - a_{j_2j_2s_1}) < 0$,

$$(a_{j_1j_21}, \dot{x}_{j_2}) = \underbrace{(\text{DP,nF})}_{\text{dwon-parabola flow }(-)}.$$

- For $a_{j_1j_10}(\overline{x}_{j_1} - a_{j_1j_11})^2 < 0$ and $a_{j_2j_20}(\overline{x}_{j_1} - a_{j_2j_11})\prod_{s_1=1}^{2}(a_{j_1j_21} - a_{j_2j_2s_1}) < 0$,

$$(a_{j_1j_21}, \dot{x}_{j_2}) = \underbrace{(\text{UP,nF})}_{\text{up-parabola flow }(-)}.$$

(II) Similarly, at $x_{j_1}^* = a_{j_2j_11} \neq a_{j_1j_11}$ with $\overline{x}_{j_2} \neq a_{j_1j_21}, a_{j_2j_21}$, in phase space,

$$\frac{dx_{j_2}}{dx_{j_1}}\Big|_{x^*_{j_1}=a_{j_2j_11}}=\frac{a_{j_2j_20}}{a_{j_1j_10}}\frac{(x_{j_1}-a_{j_2j_11})\prod_{s_1=1}^{2}(\overline{x}_{j_2}-a_{j_2j_2s_1})}{(\overline{x}_{j_2}-a_{j_1j_21})(x_{j_1}-a_{j_1j_11})^2}\Big|_{x^*_{j_1}=a_{j_2j_11}}=0.$$

If

$$\frac{d^2x_{j_2}}{dx^2_{j_1}}\Big|_{x^*_{j_1}=a_{j_2j_11}}=\frac{a_{j_2j_20}}{a_{j_1j_10}}\frac{\prod_{s_1=1}^{2}(\overline{x}_{j_2}-a_{j_2j_2s_1})}{(\overline{x}_{j_2}-a_{j_1j_21})(a_{j_2j_11}-a_{j_1j_11})^2}>0,$$

there is an up-parabola flow. If

$$\frac{d^2x_{j_2}}{dx^2_{j_1}}\Big|_{x^*_{j_1}=a_{j_2j_11}}=\frac{a_{j_2j_20}}{a_{j_1j_10}}\frac{\prod_{s_1=1}^{2}(\overline{x}_{j_2}-a_{j_2j_2s_1})}{(\overline{x}_{j_2}-a_{j_1j_21})(a_{j_2j_11}-a_{j_1j_11})^2}<0,$$

there is a down-parabola flow. For

$$\dot{x}_{j_1}=a_{j_1j_10}(\overline{x}_{j_2}-a_{j_1j_21})(a_{j_2j_11}-a_{j_1j_11})^2,$$

the parabola flows at $x^*_{j_1}=a_{j_2j_11}\neq a_{j_1j_11}$ are positive and negative for $\dot{x}_{j_1}>0$ and $\dot{x}_{j_1}<0$, respectively. Thus, the equilibrium of $x^*_{j_1}=a_{j_2j_11}\neq a_{j_1j_11}$ with $\overline{x}_{j_2}\neq a_{j_1j_21},a_{j_2j_21}$ has the following properties:

- For $a_{j_1j_10}(\overline{x}_{j_2}-a_{j_1j_21})(a_{j_2j_11}-a_{j_1j_11})^2>0$ and $a_{j_2j_20}\prod_{s_1=1}^{2}(\overline{x}_{j_2}-a_{j_2j_2s_1})>0$,

$$(\dot{x}_{j_1},a_{j_2j_11})=\underbrace{(\text{pF},\text{UP})}_{\text{up-parabola flow }(+)}.$$

- For $a_{j_1j_10}(\overline{x}_{j_2}-a_{j_1j_21})(a_{j_2j_11}-a_{j_1j_11})^2<0$ and $a_{j_2j_20}\prod_{s_1=1}^{2}(\overline{x}_{j_2}-a_{j_2j_2s_1})>0$,

$$(\dot{x}_{j_1},a_{j_2j_11})=\underbrace{(\text{nF},\text{DP})}_{\text{down-parabola flow }(-)}.$$

- For $a_{j_1j_10}(\overline{x}_{j_2}-a_{j_1j_21})(a_{j_2j_11}-a_{j_1j_11})^2>0$ and $a_{j_2j_20}\prod_{s_1=1}^{2}(\overline{x}_{j_2}-a_{j_2j_2s_1})<0$,

$$(\dot{x}_{j_1},a_{j_2j_11})=\underbrace{(\text{pF},\text{DP})}_{\text{down-parabola flow }(+)}.$$

- For $a_{j_1j_10}(\bar{x}_{j_2} - a_{j_1j_21})(a_{j_2j_11} - a_{j_1j_11})^2 < 0$ and $a_{j_2j_20}\prod_{s_1=1}^{2}(\bar{x}_{j_2} - a_{j_2j_2s_1}) < 0$,

$$(\dot{x}_{j_1}, a_{j_2j_11}) = \underbrace{\text{(nF,UP)}}_{\text{up-parabola flow }(-)}.$$

Therefore, from the above analysis of cases (I) and (II), *the equilibrium of* $(x_{j_2}^*, x_{j_1}^*) = (a_{j_1j_21}, a_{j_2j_11})$ has the following properties as in Eqs. (1.123)–(1.126):

- For $a_{j_1j_10}(a_{j_2j_11} - a_{j_1j_11})^2 > 0$ and $a_{j_2j_20}\prod_{s_1=1}^{2}(a_{j_1j_21} - a_{j_2j_2s_1}) > 0$,

$$(a_{j_1j_21}, a_{j_2j_11}) = \underbrace{(\text{UP}_+,\text{UP}_+)}_{\text{positive saddle}}.$$

  The equilibrium of $(x_{j_2}^*, x_{j_1}^*) = (a_{j_1j_21}, a_{j_2j_11})$ is a (UP$_+$, UP$_+$)-positive saddle.

- For $a_{j_1j_10}(a_{j_2j_11} - a_{j_1j_11})^2 < 0$ and $a_{j_2j_20}\prod_{s_1=1}^{2}(a_{j_1j_21} - a_{j_2j_2s_1}) > 0$,

$$(a_{j_1j_21}, a_{j_2j_11}) = \underbrace{(\text{DP}_+,\text{DP}_-)}_{\text{CCW center}}.$$

  The equilibrium of $(x_{j_2}^*, x_{j_1}^*) = (a_{j_1j_21}, a_{j_2j_11})$ is a (DP$_+$,DP$_-$)-counterclockwise center.

- For $a_{j_1j_10}(a_{j_2j_11} - a_{j_1j_11})^2 > 0$ and $a_{j_2j_20}\prod_{s_1=1}^{2}(a_{j_1j_21} - a_{j_2j_2s_1}) < 0$,

$$(a_{j_1j_21}, a_{j_2j_11}) = \underbrace{(\text{DP}_-,\text{DP}_+)}_{\text{CW center}}.$$

  The equilibrium of $(x_{j_2}^*, x_{j_1}^*) = (a_{j_1j_21}, a_{j_2j_11})$ is a (DP$_-$, DP$_+$)-clockwise center.

- For $a_{j_1j_10}(a_{j_2j_11} - a_{j_1j_11})^2 < 0$ and $a_{j_2j_20}\prod_{s_1=1}^{2}(a_{j_1j_21} - a_{j_2j_2s_1}) < 0$,

$$(a_{j_1j_21}, a_{j_2j_11}) = \underbrace{(\text{UP}_-,\text{UP}_-)}_{\text{negative saddle}}.$$

  The equilibrium of $(x_{j_2}^*, x_{j_1}^*) = (a_{j_1j_21}, a_{j_2j_11})$ is a (UP$_-$,UP$_-$)-negative saddle.

(vi$_{1b}$) At $(x_{j_1}^*, x_{j_2}^*) = (a_{j_1j_11}, a_{j_2j_2s_1})$, the variational equations are

$$\begin{aligned}
\Delta\dot{x}_{j_2} &= a_{j_2j_20}(a_{j_1j_11} - a_{j_2j_11})(a_{j_2j_2s_1} - a_{j_2j_2s_2})\Delta x_{j_2},\\
\Delta\dot{x}_{j_1} &= a_{j_1j_10}(a_{j_2j_2s_1} - a_{j_1j_21})(\Delta x_{j_1})^2.
\end{aligned}$$

The flows $(x_{j_1}^*, x_{j_2}^*) = (a_{j_1j_11}, a_{j_2j_2s_1})$ in the x_{j_1}-direction are upper saddle and lower saddle for $a_{j_1j_10}(a_{j_2j_2s_1} - a_{j_1j_21}) > 0$ and $a_{j_1j_10}(a_{j_2j_2s_1} - a_{j_1j_21}) < 0$, respectively. The flows $(x_{j_1}^*, x_{j_2}^*) = (a_{j_1j_1 1}, a_{j_2j_2s_1})$ in the x_{j_2} -direction are source and sink for $a_{j_2j_2 0}(a_{j_1j_1 1} - a_{j_2j_1 1})(a_{j_2j_2s_1} - a_{j_2j_2s_2}) > 0$ and $a_{j_2j_2 0}(a_{j_1j_1 1} - a_{j_2j_1 1})(a_{j_2j_2s_1} - a_{j_2j_2s_2}) < 0$, respectively.

Therefore, the equilibrium of $(x_{j_1}^*, x_{j_2}^*) = (a_{j_1j_1 1}, a_{j_2j_2s_1})$ ($s_1,\ s_2 \in \{1,\ 2\}, s_1 \neq s_2$) has the following properties as in Eqs. (1.127)–(1.130):

- For $a_{j_1j_1 0}(a_{j_2j_2s_1} - a_{j_1j_2 1}) > 0$ and $a_{j_2j_2 0}(a_{j_1j_1 1} - a_{j_2j_1 1})(a_{j_2j_2s_1} - a_{j_2j_2s_2}) > 0$,

$$(a_{j_1j_1 1}, a_{j_2j_2s_1}) = \underbrace{(\text{US,SO})}_{\text{upper-saddle source}}.$$

The equilibrium of $(x_{j_1}^*, x_{j_2}^*) = (a_{j_1j_1 1}, a_{j_2j_2s_1})$ is a (US,SO)-upper-saddle source.

- For $a_{j_1j_1 0}(a_{j_2j_2s_1} - a_{j_1j_2 1}) < 0$ and $a_{j_2j_2 0}(a_{j_1j_1 1} - a_{j_2j_1 1})(a_{j_2j_2s_1} - a_{j_2j_2s_2}) > 0$,

$$(a_{j_1j_1 1}, a_{j_2j_2s_1}) = \underbrace{(\text{LS,SO})}_{\text{lower-saddle source}}.$$

The equilibrium of $(x_{j_1}^*, x_{j_2}^*) = (a_{j_1j_1 1}, a_{j_2j_2s_1})$ is an (LS,SO)-lower-saddle source.

- For $a_{j_1j_1 0}(a_{j_2j_2s_1} - a_{j_1j_2 1}) > 0$ and $a_{j_2j_2 0}(a_{j_1j_1 1} - a_{j_2j_1 1})(a_{j_2j_2s_1} - a_{j_2j_2s_2}) < 0$,

$$(a_{j_1j_1 1}, a_{j_2j_2s_1}) = \underbrace{(\text{US,SI})}_{\text{upper-saddle sink}}.$$

The equilibrium of $(x_{j_1}^*, x_{j_2}^*) = (a_{j_1j_1 1}, a_{j_2j_2s_1})$ is a (US,SI)-upper-saddle sink.

- For $a_{j_1j_1 0}(a_{j_2j_2s_1} - a_{j_1j_2 1}) < 0$ and $a_{j_2j_2 0}(a_{j_1j_1 1} - a_{j_2j_1 1})(a_{j_2j_2s_1} - a_{j_2j_2s_2}) < 0$,

$$(a_{j_1j_1 1}, a_{j_2j_2s_1}) = \underbrace{(\text{LS,SI})}_{\text{lower-saddle sink}}.$$

The equilibrium of $(x_{j_1}^*, x_{j_2}^*) = (a_{j_1j_1 1}, a_{j_2j_2s_1})$ is an (LS,SI)-lower-saddle sink.

(vi$_{1c}$) The variational equation at $x_{j_1}^* = a_{j_1j_1 1}$ with $\bar{x}_{j_2} \neq a_{j_1j_2 1}$ is

$$\Delta\dot{x}_{j_1} = a_{j_1j_1 0}(\bar{x}_{j_2} - a_{j_1j_2 1})(\Delta x_{j_1})^2$$

The flows at $x_{j_1}^* = a_{j_1j_1 1}$ with $\bar{x}_{j_2} \neq a_{j_1j_2 1}$ are upper-saddle and lower-saddle for $a_{j_1j_1 0}(\bar{x}_{j_2} - a_{j_1j_2 1}) > 0$ and $a_{j_1j_1 0}(\bar{x}_{j_2} - a_{j_1j_2 1}) < 0$, respectively.

$$\dot{x}_{j_2} = a_{j_2j_2 0}(a_{j_1j_1 1} - a_{j_2j_1 1})\prod\nolimits_{s_1=1}^{2}(\bar{x}_{j_2} - a_{j_2j_2s_1}).$$

The upper-saddle and lower-saddle are positive and negative for $\dot{x}_{j_2} > 0$ and $\dot{x}_{j_2} < 0$, respectively. Thus, the equilibrium of $x_{j_2}^* = a_{j_1j_2 1}$ with $x_{j_1}^* = a_{j_1j_1 1}$ has the following properties:

- For $a_{j_1j_10}(\overline{x}_{j_2}-a_{j_1j_21})>0$ and $a_{j_2j_20}(a_{j_1j_11}-a_{j_2j_11})\prod_{s_1=1}^{2}(\overline{x}_{j_2}-a_{j_2j_2s_1})>0$,

$$(a_{j_1j_11},\dot{x}_{j_2})=\underbrace{(\text{US,pF})}_{\text{upper-saddle flow }(+)}.$$

- For $a_{j_1j_10}(\overline{x}_{j_2}-a_{j_1j_21})<0$ and $a_{j_2j_20}(a_{j_1j_11}-a_{j_2j_11})\prod_{s_1=1}^{2}(\overline{x}_{j_2}-a_{j_2j_2s_1})>0$,

$$(a_{j_1j_11},\dot{x}_{j_2})=\underbrace{(\text{LS,pF})}_{\text{lower-saddle flow }(+)}.$$

- For $a_{j_1j_10}(\overline{x}_{j_2}-a_{j_1j_21})>0$ and $a_{j_2j_20}(a_{j_1j_11}-a_{j_2j_11})\prod_{s_1=1}^{2}(\overline{x}_{j_2}-a_{j_2j_2s_1})<0$,

$$(a_{j_1j_11},\dot{x}_{j_2})=\underbrace{(\text{US,nF})}_{\text{upper-saddle flow }(-)}.$$

- For $a_{j_1j_10}(\overline{x}_{j_2}-a_{j_1j_21})<0$ and $a_{j_2j_20}(a_{j_1j_11}-a_{j_2j_11})\prod_{s_1=1}^{2}(\overline{x}_{j_2}-a_{j_2j_2s_1})<0$,

$$(a_{j_1j_11},\dot{x}_{j_2})=\underbrace{(\text{LS,nF})}_{\text{lower-saddle flow }(-)}.$$

From the above analysis with $\overline{x}_{j_2}=a_{j_1j_21}$ and the case (I) of (vi_{1a}), the equilibrium of $(x_{j_1}^*,x_{j_2}^*)=(a_{j_1j_11},a_{j_1j_21})$ with $a_{j_1j_21}\neq a_{j_2j_21},a_{j_2j_22}$ has the following properties as in Eqs. (1.131)–(1.134):

- For $a_{j_1j_10}>0$ and $a_{j_2j_20}(a_{j_1j_11}-a_{j_2j_11})\prod_{s_1=1}^{2}(a_{j_1j_21}-a_{j_2j_2s_1})>0$,

$$(a_{j_1j_11},a_{j_1j_21})=\underbrace{(\text{UP:UP,pF})}_{\text{hyperbolic-secant to hyperbolic flow }(+)}.$$

  The equilibrium of $(x_{j_1}^*,x_{j_2}^*)=(a_{j_1j_11},a_{j_1j_21})$ is a (UP:UP, pF)-positive hyperbolic-secant-to-hyperbolic flow.
- For $a_{j_1j_10}<0$ and $a_{j_2j_20}(a_{j_1j_11}-a_{j_2j_11})\prod_{s_1=1}^{2}(a_{j_1j_21}-a_{j_2j_2s_1})>0$,

$$(a_{j_1j_11},a_{j_1j_21})=\underbrace{(\text{DP:DP,pF})}_{\text{hyperbolic to hyperbolic-secant flow }(+)}.$$

  The equilibrium of $(x_{j_1}^*,x_{j_2}^*)=(a_{j_1j_11},a_{j_1j_21})$ is a (DP:DP, pF)-positive hyperbolic-to-hyperbolic-secant flow.
- For $a_{j_1j_10}>0$ and $a_{j_2j_20}(a_{j_1j_11}-a_{j_2j_11})\prod_{s_1=1}^{2}(a_{j_1j_21}-a_{j_2j_2s_1})<0$,

$$(a_{j_1j_11}, a_{j_1j_21}) = \underbrace{\text{(DP:DP,nF)}}_{\text{hyperbolic to hyperbolic-secant flow }(-)}.$$

The equilibrium of $(x^*_{j_1}, x^*_{j_2}) = (a_{j_1j_11}, a_{j_1j_21})$ is a (DP:DP, nF)-negative hyperbolic-to-hyperbolic-secant flow.

- For $a_{j_1j_10} < 0$ and $a_{j_2j_20}(a_{j_1j_11} - a_{j_2j_11})\prod_{s_1=1}^{2}(a_{j_1j_21} - a_{j_2j_2s_1}) < 0$,

$$(a_{j_1j_11}, a_{j_1j_21}) = \underbrace{\text{(UP:UP,nF)}}_{\text{hyperbolic-secant to hyperbolic flow }(-)}.$$

The equilibrium of $(x^*_{j_1}, x^*_{j_2}) = (a_{j_1j_11}, a_{j_1j_21})$ is a (UP:UP, nF)-negative hyperbolic-secant-to-hyperbolic flow.

($\text{vi}_{1\text{d}}$) From the case (II) of ($\text{vi}_{1\text{a}}$) with $\bar{x}_{j_2} = a_{j_2j_2s_1}$, the equilibrium of $(x^*_{j_1}, x^*_{j_2}) = (a_{j_2j_11}, a_{j_2j_2s_1})$ ($s_1, s_2 \in \{1,2\}$, $s_1 \neq s_2$) with $a_{j_2j_2s_1} \neq a_{j_1j_21}$ has the following properties as in Eqs. (1.135)–(1.138):

- For $a_{j_1j_10}(a_{j_2j_2s_1} - a_{j_1j_21})(a_{j_2j_11} - a_{j_1j_11})^2 > 0$ and $a_{j_2j_20}(a_{j_2j_2s_1} - a_{j_2j_2s_2}) > 0$,

$$(a_{j_2j_11}, a_{j_2j_2s_1}) = \underbrace{\text{(pF,DP:UP)}}_{\text{hyperbolic flow }(+)}.$$

The equilibrium of $(x^*_{j_1}, x^*_{j_2}) = (a_{j_2j_11}, a_{j_2j_2s_1})$ is a (pF,DP:UP)-positive hyperbolic flow.

- For $a_{j_1j_10}(a_{j_2j_2s_1} - a_{j_1j_21})(a_{j_2j_11} - a_{j_1j_11})^2 < 0$ and $a_{j_2j_20}(a_{j_2j_2s_1} - a_{j_2j_2s_2}) > 0$,

$$(a_{j_2j_11}, a_{j_2j_2s_1}) = \underbrace{\text{(nF,UP:DP)}}_{\text{hyperbolic-secant flow }(-)}.$$

The equilibrium of $(x^*_{j_1}, x^*_{j_2}) = (a_{j_2j_11}, a_{j_2j_2s_1})$ is an (nF,UP:DP)-negative hyperbolic-secant flow.

- For $a_{j_1j_10}(a_{j_2j_2s_1} - a_{j_1j_21})(a_{j_2j_11} - a_{j_1j_11})^2 > 0$ and $a_{j_2j_20}(a_{j_2j_2s_1} - a_{j_2j_2s_2}) < 0$,

$$(a_{j_2j_11}, a_{j_2j_2s_1}) = \underbrace{\text{(pF,UP:DP)}}_{\text{hyperbolic-secant flow }(+)}.$$

The equilibrium of $(x^*_{j_1}, x^*_{j_2}) = (a_{j_2j_11}, a_{j_2j_2s_1})$ is a (pF,UP:DP)-positive hyperbolic-secant flow.

- For $a_{j_1j_10}(a_{j_2j_2s_1} - a_{j_1j_21})(a_{j_2j_11} - a_{j_1j_11})^2 < 0$ and $a_{j_2j_20}(a_{j_2j_2s_1} - a_{j_2j_2s_2}) < 0$,

$$(a_{j_2j_11}, a_{j_2j_2s_1}) = \underbrace{(\text{nF,DP:UP})}_{\text{hyperbolic flow }(-)}.$$

The equilibrium of $(x_{j_1}^*, x_{j_2}^*) = (a_{j_2j_11}, a_{j_2j_2s_1})$ is an (nF,DP:UP)-negative hyperbolic flow.

(vi$_2$) For $a_{j_2j_2s_1} = a_{j_1j_21}$ and $a_{j_1j_11} \neq a_{j_2j_11}$, we have

$$\frac{dx_{j_1}}{dx_{j_2}} = \frac{a_{j_1j_10}}{a_{j_2j_20}} \frac{(x_{j_1} - a_{j_1j_11})^2}{(x_{j_1} - a_{j_2j_11})(x_{j_2} - a_{j_2j_2s_2})},$$

and the deformation of the foregoing equation is

$$\left[\frac{1}{(x_{j_1} - a_{j_1j_11})} + \frac{(a_{j_1j_11} - a_{j_2j_11})}{(x_{j_1} - a_{j_1j_11})^2}\right] dx_{j_1} = \frac{a_{j_1j_10}}{a_{j_2j_20}} \frac{dx_{j_2}}{(x_{j_2} - a_{j_2j_2s_2})}.$$

With an initial condition of (x_{j_10}, x_{j_20}) at $t = t_0$, the integration of the above equation gives the first integral manifold ($a_{j_2j_2s_1} = a_{j_1j_21}$ and $a_{j_1j_11} \neq a_{j_2j_11}$,) as

$$\ln \frac{|x_{j_1} - a_{j_1j_11}|}{|x_{j_10} - a_{j_1j_11}|} - (a_{j_1j_11} - a_{j_2j_11})(\frac{1}{x_{j_1} - a_{j_1j_11}} - \frac{1}{x_{j_11} - a_{j_1j_11}})$$
$$= \frac{a_{j_1j_10}}{a_{j_2j_20}} \ln \frac{|x_{j_2} - a_{j_2j_2s_2}|}{|x_{j_2} - a_{j_2j_2s_2}|}.$$

(vi$_{2a}$) At $x_{j_2}^* = a_{j_1j_21} = a_{j_2j_2s_1}$ with $\bar{x}_{j_1} \neq a_{j_1j_11}, a_{j_2j_11}$, if

$$\left.\frac{dx_{j_1}}{dx_{j_2}}\right|_{x_{j_2}^* = a_{j_1j_21}} = \frac{a_{j_1j_10}}{a_{j_2j_20}} \frac{(\bar{x}_{j_1} - a_{j_1j_11})^2}{(\bar{x}_{j_1} - a_{j_2j_11})(a_{j_2j_2s_1} - a_{j_2j_2s_2})} > 0,$$

there is an increasing inflection flow, and if

$$\left.\frac{dx_{j_1}}{dx_{j_2}}\right|_{x_{j_2}^* = a_{j_1j_21}} = \frac{a_{j_1j_10}}{a_{j_2j_20}} \frac{(\bar{x}_{j_1} - a_{j_1j_11})^2}{(\bar{x}_{j_1} - a_{j_2j_11})(a_{j_2j_2s_1} - a_{j_2j_2s_2})} < 0,$$

there is a decreasing inflection flow. The corresponding variational equation is

$$\Delta\dot{x}_{j_2} = a_{j_2j_20}(\bar{x}_{j_1} - a_{j_2j_11})(a_{j_2j_2s_1} - a_{j_2j_2s_2})\Delta x_{j_2}.$$

The parabola flow at $x_{j_2}^* = a_{j_1j_21} = a_{j_2j_2s_1}$ with $\bar{x}_{j_1} \neq a_{j_1j_11}$ are source and sink flows for $a_{j_2j_20}(\bar{x}_{j_1} - a_{j_2j_11})(a_{j_2j_2s_1} - a_{j_2j_2s_2}) > 0$ and $a_{j_2j_20}(\bar{x}_{j_1} - a_{j_2j_11})$ $(a_{j_2j_2s_1} - a_{j_2j_2s_2}) < 0$, respectively.

Therefore, the infinite equilibrium of $x_{j_2}^* = a_{j_1 j_2 1} = a_{j_2 j_2 s_1}$ with $\bar{x}_{j_1} \neq a_{j_1 j_1 1}, a_{j_2 j_1 1}$ has the following properties as in Eqs. (1.140)–(1.143):

- For $a_{j_1 j_1 0}(\bar{x}_{j_1} - a_{j_1 j_1 1})^2 > 0$ and $a_{j_2 j_2 0}(\bar{x}_{j_1} - a_{j_2 j_1 1})(a_{j_2 j_2 s_1} - a_{j_2 j_2 s_2}) > 0$,

$$(\bar{x}_{j_1}, a_{j_2 j_2 s_1}) = \underbrace{(\text{II,SO})}_{\text{increasing-inflection source}}.$$

  The infinite-equilibrium of $x_{j_2}^* = a_{j_1 j_2 1} = a_{j_2 j_2 s_1}$ is an (II,SO)-increasing-inflection source.
- For $a_{j_1 j_1 0}(\bar{x}_{j_1} - a_{j_1 j_1 1}) < 0$ and $a_{j_2 j_2 0}(\bar{x}_{j_1} - a_{j_2 j_1 1})(a_{j_2 j_2 s_1} - a_{j_2 j_2 s_2}) > 0$,

$$(\bar{x}_{j_1}, a_{j_2 j_2 s_1}) = \underbrace{(\text{DI,SO})}_{\text{decreasing-inflection source}}.$$

  The infinite-equilibrium of $x_{j_2}^* = a_{j_1 j_2 1} = a_{j_2 j_2 s_1}$ is a (DI,SO)-decreasing-inflection source.
- For $a_{j_1 j_1 0}(\bar{x}_{j_1} - a_{j_1 j_1 1}) > 0$ and $a_{j_2 j_2 0}(\bar{x}_{j_1} - a_{j_2 j_1 1})(a_{j_2 j_2 s_1} - a_{j_2 j_2 s_2}) < 0$,

$$(\bar{x}_{j_1}, a_{j_2 j_2 s_1}) = \underbrace{(\text{DI,SI})}_{\text{decreasing-inflection sink}}.$$

  The infinite-equilibrium of $x_{j_2}^* = a_{j_1 j_2 1} = a_{j_2 j_2 s_1}$ is a (DI,SI)-decreasing-inflection sink.
- For $a_{j_1 j_1 0}(\bar{x}_{j_1} - a_{j_1 j_1 1}) < 0$ and $a_{j_2 j_2 0}(\bar{x}_{j_1} - a_{j_2 j_1 1})(a_{j_2 j_2 s_1} - a_{j_2 j_2 s_2}) < 0$,

$$(\bar{x}_{j_1}, a_{j_2 j_2 s_1}) = \underbrace{(\text{II,SI})}_{\text{increasing-inflection sink}}.$$

  The infinite-equilibrium of $x_{j_2}^* = a_{j_1 j_2 1} = a_{j_2 j_2 s_1}$ is an (II,SI)-increasing inflection sink.

(vi_{2b}) From case (vi_{2a}), for $\bar{x}_{j_1} = a_{j_1 j_1 1}$, the equilibrium of $(x_{j_1}^*, x_{j_2}^*) = (a_{j_1 j_1 1}, a_{j_1 j_2 s_2})$ with $a_{j_2 j_2 s_1} = a_{j_1 j_2 1}$ has the following properties as in Eqs. (1.144)–(1.147):

- For $a_{j_1 j_1 0} > 0$ and $a_{j_2 j_2 0}(a_{j_1 j_1 1} - a_{j_2 j_1 1})(a_{j_2 j_2 s_1} - a_{j_2 j_2 s_2}) > 0$,

$$(a_{j_1 j_1 1}, a_{j_2 j_2 s_1}) = \underbrace{(\text{II:II,SO})}_{\text{increasing-inflection source}}.$$

  The equilibrium of $(x_{j_1}^*, x_{j_2}^*) = (a_{j_1 j_1 1}, a_{j_2 j_2 s_1})$ is an (II:II,SO)-increasing-inflection source of the second kind.
- For $a_{j_1 j_1 0} < 0$ and $a_{j_2 j_2 0}(a_{j_1 j_1 1} - a_{j_2 j_1 1})(a_{j_2 j_2 s_1} - a_{j_2 j_2 s_2}) > 0$,

$$(a_{j_1j_11}, a_{j_2j_2s_1}) = \underbrace{(\text{DI:DI,SO})}_{\text{decreasing-inflection source}}.$$

The equilibrium of $(x^*_{j_1}, x^*_{j_2}) = (a_{j_1j_11}, a_{j_2j_2s_1})$ is a (DI:DI,SO)-decreasing-inflection source of the second kind.

- For $a_{j_1j_10} > 0$ and $a_{j_2j_20}(a_{j_1j_11} - a_{j_2j_11})(a_{j_2j_2s_1} - a_{j_2j_2s_2}) < 0$,

$$(a_{j_1j_11}, a_{j_2j_2s_1}) = \underbrace{(\text{DI:DI,SI})}_{\text{decreasing-inflection sink}}.$$

The equilibrium of $(x^*_{j_1}, x^*_{j_2}) = (a_{j_1j_11}, a_{j_2j_2s_1})$ is a (DI:DI,SI)-decreasing-inflection sink of the second kind.

- For $a_{j_1j_10} < 0$ and $a_{j_2j_20}(a_{j_1j_11} - a_{j_2j_11})(a_{j_2j_2s_1} - a_{j_2j_2s_2}) < 0$,

$$(a_{j_1j_11}, a_{j_2j_2s_1}) = \underbrace{(\text{II:II,SI})}_{\text{increasing-inflection sink}}.$$

The equilibrium of $(x^*_{j_1}, x^*_{j_2}) = (a_{j_1j_11}, a_{j_2j_2s_1})$ is an (II:II,SI)-increasing-inflection sink of the second kind.

(vi_{2c}) From (vi_{2a}) with $\bar{x}_{j_1} = a_{j_2j_11}$, the equilibrium of $(x^*_{j_1}, x^*_{j_2}) = (a_{j_1j_11}, a_{j_1j_2s_2})$ with $a_{j_2j_2s_1} = a_{j_1j_21}$ has the following properties as in Eqs. (1.148)–(1.151):

- For $a_{j_1j_10}(a_{j_2j_11} - a_{j_1j_11})^2 > 0$ and $a_{j_2j_20}(a_{j_2j_2s_1} - a_{j_2j_2s_2}) > 0$,

$$(a_{j_1j_21}, a_{j_2j_11}) = \underbrace{({}_{\text{SI:SO}}\text{US}, {}_{\text{DI:II}}\text{UP})}_{\text{up-parabola upper-saddle}}.$$

The equilibrium of $(x^*_{j_2}, x^*_{j_1}) = (a_{j_1j_21}, a_{j_2j_11})$ is a (${}_{\text{SI:SO}}\text{US}, {}_{\text{DI:II}}\text{UP}$)-up-parabola upper-saddle.

- For $a_{j_1j_10}(a_{j_2j_11} - a_{j_1j_11})^2 < 0$ and $a_{j_2j_20}(a_{j_2j_2s_1} - a_{j_2j_2s_2}) > 0$,

$$(a_{j_1j_21}, a_{j_2j_11}) = \underbrace{({}_{\text{SI:SO}}\text{US}, {}_{\text{II:DI}}\text{DP})}_{\text{down-parabola upper-saddle}}.$$

The equilibrium of $(x^*_{j_2}, x^*_{j_1}) = (a_{j_1j_21}, a_{j_2j_11})$ is an (${}_{\text{SI:SO}}\text{US}, {}_{\text{II:DI}}\text{DP}$)-down-parabola upper-saddle.

- For $a_{j_1j_10}(a_{j_2j_11} - a_{j_1j_11})^2 > 0$ and $a_{j_2j_20}(a_{j_2j_2s_1} - a_{j_2j_2s_2}) < 0$,

$$(a_{j_1j_21}, a_{j_2j_11}) = \underbrace{({}_{\text{SO:SI}}\text{LS}, {}_{\text{II:DI}}\text{DP})}_{\text{down-parabola lower-saddle}}.$$

The equilibrium of $(x^*_{j_2}, x^*_{j_1}) = (a_{j_1j_21}, a_{j_2j_11})$ is an $({}_{\text{SO:SI}}\text{LS}, {}_{\text{II:DI}}\text{DP})$-down-parabola lower-saddle.

- For $a_{j_1j_10}(a_{j_2j_11} - a_{j_1j_11})^2 < 0$ and $a_{j_2j_20}(a_{j_2j_2s_1} - a_{j_2j_2s_2}) < 0$,

$$(a_{j_1j_21}, a_{j_2j_11}) = \underbrace{({}_{\text{SO:SI}}\text{LS}, {}_{\text{DI:II}}\text{UP})}_{\text{up-parabola lower-saddle}}.$$

The equilibrium of $(x^*_{j_2}, x^*_{j_1}) = (a_{j_1j_21}, a_{j_2j_11})$ is an $({}_{\text{SO:SI}}\text{LS}, {}_{\text{DI:II}}\text{UP})$-up-parabola lower-saddle.

(vi_3) For $a_{j_1j_11} = a_{j_2j_11}$ and $a_{j_2j_2s_1} \neq a_{j_1j_21}$, we have

$$\frac{dx_{j_1}}{dx_{j_2}} = \frac{a_{j_1j_10}}{a_{j_2j_20}} \frac{(x_{j_2} - a_{j_1j_21})(x_{j_1} - a_{j_1j_11})}{(x_{j_2} - a_{j_2j_21})(x_{j_2} - a_{j_2j_22})},$$

and the deformation of the foregoing equation is

$$\frac{dx_{j_1}}{(x_{j_1} - a_{j_1j_11})} = \frac{a_{j_1j_10}}{a_{j_2j_20}} \frac{(x_{j_2} - a_{j_1j_21})dx_{j_2}}{(x_{j_2} - a_{j_2j_2s_1})(x_{j_2} - a_{j_2j_2s_2})}.$$

With an initial condition of (x_{j_10}, x_{j_20}) at $t = t_0$, the integration of the above equation gives the first integral manifold ($a_{j_1j_11} = a_{j_2j_11}$ and $a_{j_1j_21} \neq a_{j_2j_2s_1}, a_{j_2j_2s_2}$) as

$$\ln \frac{|x_{j_1} - a_{j_1j_11}|}{|x_{j_10} - a_{j_1j_11}|} = \frac{a_{j_1j_10}}{a_{j_2j_20}} \sum\nolimits_{s_1=1, s_2 \neq s_1}^{2} \frac{(a_{j_2j_2s_1} - a_{j_1j_21})}{(a_{j_2j_2s_1} - a_{j_2j_2s_2})} \ln \frac{|x_{j_2} - a_{j_2j_2s_1}|}{|x_{j_2} - a_{j_2j_2s_1}|}.$$

($\text{vi}_{2\text{a}}$) At $x^*_{j_1} = a_{j_1j_11} = a_{j_2j_11}$ with $\overline{x}_{j_2} \neq a_{j_1j_21}, a_{j_2j_21}, a_{j_2j_22}$, we have

$$\left.\frac{dx_{j_2}}{dx_{j_1}}\right|_{x^*_{j_1} = a_{j_1j_11} \pm \varepsilon} = \frac{a_{j_2j_20}}{a_{j_1j_10}} \frac{\prod_{s_1=1}^{2} (\overline{x}_{j_2} - a_{j_2j_2s_1})}{(\overline{x}_{j_2} - a_{j_1j_21})\varepsilon}.$$

If

$$\frac{a_{j_2j_20}}{a_{j_1j_10}} \frac{\prod_{s_1=1}^{2} (\overline{x}_{j_2} - a_{j_2j_2s_1})}{(\overline{x}_{j_2} - a_{j_1j_21})} > 0,$$

there is a down-up asymptotic flow. If

$$\frac{a_{j_2j_20}}{a_{j_1j_10}} \frac{\prod_{s_1=1}^{2}(\overline{x}_{j_2} - a_{j_2j_2s_1})}{(\overline{x}_{j_2} - a_{j_1j_21})} < 0,$$

there is an up-down asymptotic flow. The corresponding variational equation is

$$\Delta\dot{x}_{j_1} = a_{j_1j_10}(\overline{x}_{j_2} - a_{j_1j_21})(\Delta x_{j_1})^2.$$

The up-down and down-up flow at $x_{j_1}^* = a_{j_1j_11} = a_{j_2j_11}$ with $\overline{x}_{j_2} \neq a_{j_1j_21}, a_{j_2j_21}, a_{j_2j_22}$ are upper saddle and lower saddle for $a_{j_1j_10}(\overline{x}_{j_2} - a_{j_1j_21}) > 0$ and $a_{j_1j_10}(\overline{x}_{j_2} - a_{j_1j_21}) < 0$, respectively.

Therefore, the infinite-equilibrium of $x_{j_1}^* = a_{j_1j_11} = a_{j_2j_11}$ and $\overline{x}_{j_2} \neq a_{j_2j_21}, a_{j_1j_21}$ with $a_{j_2j_21} \neq a_{j_1j_21}$ has the following properties as in Eqs. (1.153)–(1.156):

- For $a_{j_1j_10}(\overline{x}_{j_2} - a_{j_1j_21}) > 0$ and $a_{j_2j_20}\prod_{s_1=1}^{2}(\overline{x}_{j_2} - a_{j_2j_2s_1}) > 0$,

$$(a_{j_1j_11}, \overline{x}_{j_2}) = \underbrace{(\text{US,DU})}_{\text{down-up upper-saddle}}.$$

  The infinite-equilibrium of $x_{j_1}^* = a_{j_1j_11} = a_{j_2j_11}$ is a (US,DU)-down-up upper-saddle.
- For $a_{j_1j_10}(\overline{x}_{j_2} - a_{j_1j_21}) < 0$ and $a_{j_2j_20}\prod_{s_1=1}^{2}(\overline{x}_{j_2} - a_{j_2j_2s_1}) > 0$,

$$(a_{j_1j_11}, \overline{x}_{j_2}) = \underbrace{(\text{LS,UD})}_{\text{up-down lower-saddle}}.$$

  The infinite-equilibrium of $x_{j_1}^* = a_{j_1j_11} = a_{j_2j_11}$ is an (LS,UD)-up-down lower-saddle.
- For $a_{j_1j_10}(\overline{x}_{j_2} - a_{j_1j_21}) > 0$ and $a_{j_2j_20}\prod_{s_1=1}^{2}(\overline{x}_{j_2} - a_{j_2j_2s_1}) < 0$,

$$(a_{j_1j_11}, \overline{x}_{j_2}) = \underbrace{(\text{US,UD})}_{\text{up-down upper-saddle}}.$$

  The infinite-equilibrium of $x_{j_1}^* = a_{j_1j_11} = a_{j_2j_11}$ is a (US,UD)-up-down upper-saddle.
- For $a_{j_1j_10}(\overline{x}_{j_2} - a_{j_1j_21}) < 0$ and $a_{j_2j_20}\prod_{s_1=1}^{2}(\overline{x}_{j_2} - a_{j_2j_2s_1}) < 0$,

$$(a_{j_1j_11}, \overline{x}_{j_2}) = \underbrace{(\text{LS,DU})}_{\text{down-up lower-saddle}}.$$

The infinite-equilibrium of $x_{j_1}^* = a_{j_1 j_1 1} = a_{j_2 j_1 1}$ is an (LS,DU)-down-up lower-saddle.

(vi$_{3b}$) From (vi$_{3a}$) with $\bar{x}_{j_2} = a_{j_2 j_2 1}$, thus, the equilibrium of $(x_{j_1}^*, x_{j_2}^*) = (a_{j_1 j_1 1}, a_{j_2 j_2 s_1})$ with $a_{j_1 j_1 1} = a_{j_2 j_1 1}$ and $a_{j_2 j_2 s_1} \neq a_{j_1 j_2 1}$ has the following properties as in Eqs. (1.157)–(1.160):

- For $a_{j_1 j_1 0}(a_{j_2 j_2 s_1} - a_{j_1 j_2 1}) > 0$ and $a_{j_2 j_2 0}(a_{j_2 j_2 s_1} - a_{j_2 j_2 s_2}) > 0$

$$(a_{j_1 j_1 1}, a_{j_2 j_2 s_1}) = \underbrace{(\text{US,UD:DU})}_{\text{up-down-to-down-up upper-saddle}}.$$

  The equilibrium of $(x_{j_1}^*, x_{j_2}^*) = (a_{j_1 j_1 1}, a_{j_2 j_2 s_1})$ is a (US,UD:DU)-up-down-to-down-up upper-saddle.
- For $a_{j_1 j_1 0}(a_{j_2 j_2 s_1} - a_{j_1 j_2 1}) < 0$ and $a_{j_2 j_2 0}(a_{j_2 j_2 s_1} - a_{j_2 j_2 s_2}) > 0$

$$(a_{j_1 j_1 1}, a_{j_2 j_2 s_1}) = \underbrace{(\text{LS,DU:UD})}_{\text{down-up-to-up-down lower-saddle}}.$$

  The equilibrium of $(x_{j_1}^*, x_{j_2}^*) = (a_{j_1 j_1 1}, a_{j_2 j_2 s_1})$ is an (LS,DU:UD)-down-up- to-up-down lower-saddle.
- For $a_{j_1 j_1 0}(a_{j_2 j_2 s_1} - a_{j_1 j_2 1}) > 0$ and $a_{j_2 j_2 0}(a_{j_2 j_2 s_1} - a_{j_2 j_2 s_2}) < 0$

$$(a_{j_1 j_1 1}, a_{j_2 j_2 s_1}) = \underbrace{(\text{US,DU:UD})}_{\text{down-up-to-up-down lower-saddle}}.$$

  The equilibrium of $(x_{j_1}^*, x_{j_2}^*) = (a_{j_1 j_1 1}, a_{j_2 j_2 s_1})$ is a (US,DU:UD)-down-up-to-up-down upper-saddle.
- For $a_{j_1 j_1 0}(a_{j_2 j_2 s_1} - a_{j_1 j_2 1}) < 0$ and $a_{j_2 j_2 0}(a_{j_2 j_2 s_1} - a_{j_2 j_2 s_2}) < 0$

$$(a_{j_1 j_1 1}, a_{j_2 j_2 s_1}) = \underbrace{(\text{LS,UD:DU})}_{\text{up-down-to-down-up lower-saddle}}.$$

  The equilibrium of $(x_{j_1}^*, x_{j_2}^*) = (a_{j_1 j_1 1}, a_{j_2 j_2 s_1})$ is an (LS, UD:DU)-up-down-to-down-up lower-saddle.

(vi$_{3c}$) From (vi$_{3a}$) with $\bar{x}_{j_2} = a_{j_1 j_2 1}$ and the case (I) of (vi$_{1a}$) with $\bar{x}_{j_1} = a_{j_1 j_1 1}$, therefore, the equilibrium of $(x_{j_1}^*, x_{j_2}^*) = (a_{j_1 j_1 1}, a_{j_1 j_2 1})$ with $a_{j_1 j_1 1} = a_{j_2 j_1 1}$ has the following properties as in Eqs. (1.161)–(1.164):

- For $a_{j_1 j_1 0} > 0$ and $a_{j_2 j_2 0} \prod_{s_1=1}^{2} (a_{j_1 j_2 1} - a_{j_2 j_2 s_1}) > 0$,

$$(a_{j_1j_1 1}, a_{j_1j_2 1}) = \underbrace{(\mathrm{DP}_-:\mathrm{UP}_+, {}_{\mathrm{UD}}\mathrm{LS}:{}_{\mathrm{DU}}\mathrm{US})}_{\text{hyperbolic lower-to-upper saddle}}.$$

The equilibrium of $(x^*_{j_1}, x^*_{j_2}) = (a_{j_1j_1 1}, a_{j_1j_2 1})$ is a (DP$_-$:UP$_+$, $_{\mathrm{UD}}$LS:$_{\mathrm{DU}}$US)-hyperbolic lower-to-upper saddle.

- For $a_{j_1j_1 0} < 0$ and $a_{j_2j_2 0}\prod_{s_1=1}^{2}(a_{j_1j_2 1} - a_{j_2j_2 s_1}) > 0$,

$$(a_{j_1j_1 1}, a_{j_1j_2 1}) = \underbrace{(\mathrm{UP}_-:\mathrm{DP}_+, {}_{\mathrm{DU}}\mathrm{US}:{}_{\mathrm{UD}}\mathrm{LS})}_{\text{hyperbolic-secant upper-to-lower saddle}}.$$

The equilibrium of $(x^*_{j_1}, x^*_{j_2}) = (a_{j_1j_1 1}, a_{j_1j_2 1})$ is a (UP$_-$:DP$_+$,$_{\mathrm{DU}}$US:$_{\mathrm{UD}}$LS)-hyperbolic-secant upper-to-lower saddle.

- For $a_{j_1j_1 0} > 0$ and $a_{j_2j_2 0}\prod_{s_1=1}^{2}(a_{j_1j_2 1} - a_{j_2j_2 s_1}) < 0$,

$$(a_{j_1j_1 1}, a_{j_1j_2 1}) = \underbrace{(\mathrm{UP}_+:\mathrm{DP}_-, {}_{\mathrm{DU}}\mathrm{LS}:{}_{\mathrm{UD}}\mathrm{US})}_{\text{hyperbolic-secant lower-to-upper saddle}}.$$

The equilibrium of $(x^*_{j_1}, x^*_{j_2}) = (a_{j_1j_1 1}, a_{j_1j_2 1})$ is an (UP$_+$:DP$_-$, $_{\mathrm{DU}}$LS:$_{\mathrm{UD}}$US)-hyperbolic-secant lower-to-upper saddle.

- For $a_{j_1j_1 0} < 0$ and $a_{j_2j_2 0}\prod_{s_1=1}^{2}(a_{j_1j_2 1} - a_{j_2j_2 s_1}) < 0$,

$$(a_{j_1j_1 1}, a_{j_1j_2 1}) = \underbrace{(\mathrm{DP}_+:\mathrm{UP}_-, {}_{\mathrm{DU}}\mathrm{US}:{}_{\mathrm{UD}}\mathrm{LS})}_{\text{hyperbolic upper-to-lower saddle}}.$$

The equilibrium of $(x^*_{j_1}, x^*_{j_2}) = (a_{j_1j_1 1}, a_{j_1j_2 1})$ is a (DP$_+$:UP$_-$, $_{\mathrm{DU}}$US:$_{\mathrm{UD}}$LS)-hyperbolic upper-to-lower saddle.

(vi$_4$) For $a_{j_1j_1 1} = a_{j_2j_1 1}$ and $a_{j_2j_2 s_1} = a_{j_1j_2 1}$, in phase space,

$$\begin{aligned}\frac{dx_{j_1}}{dx_{j_2}} &= \frac{a_{j_1j_1 0}}{a_{j_2j_2 0}}\frac{(x_{j_1} - a_{j_1j_1 1})^2(x_{j_2} - a_{j_1j_2 1})}{(x_{j_1} - a_{j_2j_1 1})(x_{j_2} - a_{j_2j_2 s_1})(x_{j_2} - a_{j_2j_2 s_2})}\\ &= \frac{a_{j_1j_1 0}}{a_{j_2j_2 0}}\frac{(x_{j_1} - a_{j_1j_1 1})}{(x_{j_2} - a_{j_2j_2 s_2})},\end{aligned}$$

and the deformation of the foregoing equation is

$$\frac{dx_{j_1}}{(x_{j_1} - a_{j_1j_1 1})} = \frac{a_{j_1j_1 0}}{a_{j_2j_2 0}}\frac{dx_{j_2}}{(x_{j_2} - a_{j_2j_2 s_2})}.$$

With an initial condition of $(x_{j_1 0}, x_{j_2 0})$ at $t = t_0$, the integration of the above equation gives the first integral manifold $a_{j_1j_1 1} = a_{j_2j_1 1}$ and $a_{j_2j_2 1} = a_{j_1j_2 1}$ *as*

$$\ln \frac{|x_{j_1} - a_{j_1 j_1 1}|}{|x_{j_1 0} - a_{j_1 j_1 1}|} = \frac{a_{j_1 j_1 0}}{a_{j_2 j_2 0}} \ln \frac{|x_{j_2} - a_{j_2 j_2 s_2}|}{|x_{j_2 0} - a_{j_2 j_2 s_2}|}.$$

For $x_{j_2}^* = a_{j_2 j_2 s_1} = a_{j_1 j_2 1}$ with $\bar{x}_{j_1} \neq a_{j_1 j_1 1} = a_{j_2 j_1 1}$, if

$$\left.\frac{dx_{j_1}}{dx_{j_2}}\right|_{x_{j_2} = a_{j_2 j_2 s_1}} = \frac{a_{j_1 j_1 0}}{a_{j_2 j_2 0}} \frac{(\bar{x}_{j_1} - a_{j_1 j_1 1})}{(a_{j_2 j_2 s_1} - a_{j_2 j_2 s_2})} > 0,$$

there is an increasing inflection flow. If

$$\left.\frac{dx_{j_1}}{dx_{j_2}}\right|_{x_{j_2} = a_{j_2 j_2 s_1}} = \frac{a_{j_1 j_1 0}}{a_{j_2 j_2 0}} \frac{(\bar{x}_{j_1} - a_{j_1 j_1 1})}{(a_{j_2 j_2 s_1} - a_{j_2 j_2 s_2})} < 0,$$

there is a decreasing inflection flow. The corresponding variational equation is

$$\Delta \dot{x}_{j_2} = a_{j_2 j_2 0}(\bar{x}_{j_1} - a_{j_2 j_1 1})(a_{j_2 j_2 s_1} - a_{j_2 j_2 s_2})\Delta x_{j_2}.$$

The inflection flows are source and sink for $a_{j_2 j_2 0}(\bar{x}_{j_1} - a_{j_2 j_1 1})(a_{j_2 j_2 s_1} - a_{j_2 j_2 s_2}) > 0$ and $a_{j_2 j_2 0}(\bar{x}_{j_1} - a_{j_2 j_1 1})(a_{j_2 j_2 s_1} - a_{j_2 j_2 s_2}) < 0$, respectively.

Similar to the case ($\mathrm{v}_{3\mathrm{a}}$), for $x_{j_1}^* = a_{j_1 j_1 1} = a_{j_2 j_1 1}$ with $\bar{x}_{j_2} \neq a_{j_2 j_2 1} = a_{j_1 j_2 1}$,

$$\left.\frac{dx_{j_2}}{dx_{j_1}}\right|_{x_{j_1} = a_{j_1 j_1 1} \pm \varepsilon} = \pm \frac{a_{j_2 j_2 0}}{a_{j_1 j_1 0}} \frac{(\bar{x}_{j_2} - a_{j_2 j_2 1})}{\varepsilon}.$$

If

$$\frac{a_{j_2 j_2 0}}{a_{j_1 j_1 0}}(\bar{x}_{j_2} - a_{j_2 j_2 1}) > 0,$$

there is a down-up asymptotic flow. If

$$\frac{a_{j_2 j_2 0}}{a_{j_1 j_1 0}}(\bar{x}_{j_2} - a_{j_2 j_2 1}) < 0,$$

there is an up-down asymptotic flow. The corresponding variational equation is

$$\Delta \dot{x}_{j_1} = a_{j_1 j_1 0}(\bar{x}_{j_2} - a_{j_1 j_2 1})(\Delta x_{j_1})^2.$$

The up-down and down-up flows for $a_{j_1 j_1 0}(\bar{x}_{j_2} - a_{j_1 j_2 1}) > 0$ and $a_{j_1 j_1 0}(\bar{x}_{j_2} - a_{j_1 j_2 1}) < 0$ are upper-saddle and lower-saddle, respectively.

Therefore, the equilibrium of $(x_{j_2}^*, x_{j_1}^*) = (a_{j_1 j_2 1}, a_{j_2 j_1 1})$ with $a_{j_1 j_1 1} = a_{j_2 j_1 1}$ and $a_{j_2 j_2 s_1} = a_{j_1 j_2 1}$ has the following properties as in Eqs. (1.166)–(1.169):

- For $a_{j_1 j_1 0} > 0$ and $a_{j_2 j_2 0}(a_{j_2 j_2 s_1} - a_{j_2 j_2 s_2}) > 0$

$$(a_{j_1j_21}, a_{j_2j_11}) = \underbrace{({}_{\text{LS:US}}\text{DU}, {}_{\text{DI}}\text{SI:}{}_{\text{II}}\text{SO})}_{\text{down-up inflection sink-to-source}}.$$

The equilibrium of $(x^*_{j_2}, x^*_{j_1}) = (a_{j_1j_21}, a_{j_2j_11})$ is an $({}_{\text{LS:US}}\text{DU},{}_{\text{DI}}\text{SI:}{}_{\text{II}}\text{SO})$- down-up inflection sink-to-source.

- For $a_{j_1j_10} < 0$ and $a_{j_2j_20}(a_{j_2j_2s_1} - a_{j_2j_2s_2}) > 0$

$$(a_{j_1j_21}, a_{j_2j_11}) = \underbrace{({}_{\text{US:LS}}\text{UD}, {}_{\text{II}}\text{SI:}{}_{\text{DI}}\text{SO})}_{\text{up-down inflection sink-to-source}}.$$

The equilibrium of $(x^*_{j_2}, x^*_{j_1}) = (a_{j_1j_21}, a_{j_2j_11})$ is a $({}_{\text{US:LS}}\text{UD},{}_{\text{II}}\text{SI:}{}_{\text{DI}}\text{SO})$-up-down inflection sink-to-source.

- For $a_{j_1j_10} > 0$ and $a_{j_2j_20}(a_{j_2j_2s_1} - a_{j_2j_2s_2}) < 0$

$$(a_{j_1j_21}, a_{j_2j_11}) = \underbrace{({}_{\text{LS:US}}\text{UD}, {}_{\text{II}}\text{SO:}{}_{\text{DI}}\text{SI})}_{\text{up-down inflection source-to-sink}}.$$

The equilibrium of $(x^*_{j_2}, x^*_{j_1}) = (a_{j_1j_21}, a_{j_2j_11})$ is an $({}_{\text{LS:US}}\text{UD},{}_{\text{II}}\text{SO:}{}_{\text{DI}}\text{SI})$-up-down inflection source-to-sink.

- For $a_{j_1j_10} < 0$ and $a_{j_2j_20}(a_{j_2j_2s_1} - a_{j_2j_2s_2}) < 0$

$$(a_{j_1j_21}, a_{j_2j_11}) = \underbrace{({}_{\text{US:LS}}\text{DU}, {}_{\text{DI}}\text{SO:}{}_{\text{II}}\text{SI})}_{\text{down-up inflection source-to-sink}}.$$

The equilibrium of $(x^*_{j_2}, x^*_{j_1}) = (a_{j_1j_21}, a_{j_2j_11})$ is a $({}_{\text{US:LS}}\text{DU},{}_{\text{DI}}\text{SO:}{}_{\text{II}}\text{SI})$-down-up inflection source-to-sink.

(vii) For $\Delta_{j_1j_1} > 0$ and $\Delta_{j_2j_2} < 0$, the standard form is

$$\begin{aligned}
\dot{x}_{j_2} &= a_{j_2j_20}(x_{j_1} - a_{j_2j_11})\left[(x_{j_2} - a_{j_2j_2})^2 + b_{j_2j_2}\right],\\
\dot{x}_{j_1} &= a_{j_1j_10}(x_{j_1} - a_{j_1j_11})(x_{j_1} - a_{j_1j_12})(x_{j_2} - a_{j_1j_21})
\end{aligned}$$

where

$$\begin{aligned}
b_{j_1j_11}, b_{j_1j_12} &= -\frac{1}{2}(B_{j_1j_1} \pm \sqrt{\Delta_{j_1j_1}}),\\
\{a_{j_1j_11}, a_{j_1j_12}\} &= \text{sort}\{b_{j_1j_11}, b_{j_1j_12}\}, a_{j_1j_11} < a_{j_1j_12};\\
a_{j_2j_2} &= -\frac{1}{2}B_{j_2j_2}, b_{j_2j_2} = \frac{1}{4}(-\Delta_{j_2j_2}).
\end{aligned}$$

(vii$_1$) In phase space,

$$\frac{dx_{j_1}}{dx_{j_2}} = \frac{a_{j_1j_10}}{a_{j_2j_20}} \frac{(x_{j_1} - a_{j_1j_11})(x_{j_1} - a_{j_1j_12})(x_{j_2} - a_{j_1j_21})}{(x_{j_1} - a_{j_2j_11})\left[(x_{j_2} - a_{j_2j_2})^2 + b_{j_2j_2}\right]},$$

and the deformation of the foregoing equation is

$$\frac{(x_{j_1} - a_{j_2j_11})}{(x_{j_1} - a_{j_1j_11})(x_{j_1} - a_{j_1j_12})} dx_{j_1} = \frac{a_{j_1j_10}}{a_{j_2j_20}} \left[\frac{(x_{j_2} - a_{j_2j_2})}{(x_{j_2} - a_{j_2j_2})^2 + b_{j_2j_2}} + \frac{(a_{j_2j_2} - a_{j_1j_21})}{(x_{j_2} - a_{j_2j_2})^2 + b_{j_2j_2}} \right] dx_{j_2}.$$

With an initial condition of (x_{j_10}, x_{j_20}) at $t = t_0$, the integration of the above equation gives the first integral manifold as

$$\begin{aligned} &\sum\nolimits_{s_1=1}^{2} \frac{a_{j_1j_1s_1} - a_{j_2j_11}}{a_{j_1j_1s\cdot} - a_{j_1j_11}} \ln \frac{|x_{j_1} - a_{j_1j_1s_1}|}{|x_{j_10} - a_{j_1j_1s_1}|} \\ &= \frac{a_{j_1j_10}}{a_{j_2j_20}} \left[\frac{1}{2} \ln \frac{|(x_{j_2} - a_{j_2j_2})^2 + b_{j_2j_2}|}{|(x_{j_20} - a_{j_2j_2})^2 + b_{j_2j_2}|} \right. \\ &\left. + \frac{a_{j_2j_2} - a_{j_1j_21}}{\sqrt{b_{j_2j_2}}} \left(\arctan \frac{x_{j_2} - a_{j_2j_2}}{\sqrt{b_{j_2j_2}}} - \arctan \frac{x_{j_20} - a_{j_2j_2}}{\sqrt{b_{j_2j_2}}} \right) \right]. \end{aligned}$$

(vii$_{1a}$) Consider two cases (I) $x_{j_2}^* = a_{j_1j_21}$ and (II) $x_{j_1}^* = a_{j_2j_11}$.

(I) In phase space, at $x_{j_2}^* = a_{j_1j_21}$ with $\bar{x}_{j_1} \neq a_{j_1j_11}$,

$$\left.\frac{dx_{j_1}}{dx_{j_2}}\right|_{x_{j_2}^* = a_{j_1j_21}} = \frac{a_{j_1j_10}}{a_{j_2j_20}} \left.\frac{\prod_{s_1=1}^{2}(\bar{x}_{j_1} - a_{j_1j_1s_1})(x_{j_2} - a_{j_1j_21})}{(\bar{x}_{j_1} - a_{j_2j_11})\left[(x_{j_2} - a_{j_2j_2})^2 + b_{j_2j_2}\right]}\right|_{x_{j_2}^* = a_{j_1j_21}} = 0.$$

If

$$\left.\frac{d^2x_{j_1}}{dx_{j_2}^2}\right|_{x_{j_2}^* = a_{j_1j_21}} = \frac{a_{j_1j_10}}{a_{j_2j_20}} \frac{\prod_{s_1=1}^{2}(\bar{x}_{j_1} - a_{j_1j_1s_1})}{(\bar{x}_{j_1} - a_{j_2j_11})\left[(a_{j_1j_21} - a_{j_2j_2})^2 + b_{j_2j_2}\right]} > 0,$$

there is an up-parabola flow at $x_{j_2}^* = a_{j_1j_21}$ with $\bar{x}_{j_1} \neq a_{j_1j_11}$. If

$$\left.\frac{d^2x_{j_1}}{dx_{j_2}^2}\right|_{x_{j_2}^* = a_{j_1j_21}} = \frac{a_{j_1j_10}}{a_{j_2j_20}} \frac{\prod_{s_1=1}^{2}(\bar{x}_{j_1} - a_{j_1j_1s_1})}{(\bar{x}_{j_1} - a_{j_2j_11})\left[(a_{j_1j_21} - a_{j_2j_2})^2 + b_{j_2j_2}\right]} < 0,$$

there is a down-parabola flow at $x_{j_2}^* = a_{j_1j_21}$ with $\bar{x}_{j_1} \neq a_{j_1j_11}$. Let

$$\dot{x}_{j_2} = a_{j_2 j_2 0}(\overline{x}_{j_1} - a_{j_2 j_1 1})\left[(a_{j_1 j_2 1} - a_{j_2 j_2})^2 + b_{j_2 j_2}\right],$$

with

$$(a_{j_1 j_2 1} - a_{j_2 j_2})^2 + b_{j_2 j_2} > 0.$$

The parabola flows at $x_{j_2}^* = a_{j_1 j_2 1}$ with $\overline{x}_{j_1} \neq a_{j_1 j_1 1}$ are positive and negative for $a_{j_2 j_2 0}(\overline{x}_{j_1} - a_{j_2 j_1 1}) > 0$ and $a_{j_2 j_2 0}(\overline{x}_{j_1} - a_{j_2 j_1 1}) < 0$, respectively.

Thus, the equilibrium of $x_{j_2}^* = a_{j_1 j_2 1}$ with $\overline{x}_{j_1} \neq a_{j_1 j_1 1}$ has the following properties:

- For $a_{j_1 j_1 0}\prod_{s_1=1}^{2}(\overline{x}_{j_1} - a_{j_1 j_1 s_1}) > 0$ and $a_{j_2 j_2 0}(\overline{x}_{j_1} - a_{j_2 j_1 1}) > 0$,

$$(a_{j_1 j_2 1}, \dot{x}_{j_2}) = \underbrace{(\mathrm{UP,pF})}_{\text{up-parabola flow }(+)}.$$

- For $a_{j_1 j_1 0}\prod_{s_1=1}^{2}(\overline{x}_{j_1} - a_{j_1 j_1 s_1}) < 0$ and $a_{j_2 j_2 0}(\overline{x}_{j_1} - a_{j_2 j_1 1}) > 0$,

$$(a_{j_1 j_2 1}, \dot{x}_{j_2}) = \underbrace{(\mathrm{DP,pF})}_{\text{down-parabola flow }(+)}.$$

- For $a_{j_1 j_1 0}\prod_{s_1=1}^{2}(\overline{x}_{j_1} - a_{j_1 j_1 s_1}) > 0$ and $a_{j_2 j_2 0}(\overline{x}_{j_1} - a_{j_2 j_1 1}) < 0$,

$$(a_{j_1 j_2 1}, \dot{x}_{j_2}) = \underbrace{(\mathrm{DP,nF})}_{\text{dwon-parabola flow }(-)}.$$

- For $a_{j_1 j_1 0}\prod_{s_1=1}^{2}(\overline{x}_{j_1} - a_{j_1 j_1 s_1}) < 0$ and $a_{j_2 j_2 0}(\overline{x}_{j_1} - a_{j_2 j_1 1}) < 0$,

$$(a_{j_1 j_2 1}, \dot{x}_{j_2}) = \underbrace{(\mathrm{UP,nF})}_{\text{up-parabola flow }(-)}.$$

(II) Similarly, at $x_{j_1}^* = a_{j_2 j_1 1}$ with $\overline{x}_{j_2} = a_{j_1 j_2 1}$, in phase space,

$$\frac{dx_{j_2}}{dx_{j_1}}\bigg|_{x_{j_1}^* = a_{j_2 j_1 1}} = \frac{a_{j_2 j_2 0}}{a_{j_1 j_1 0}} \frac{(x_{j_1} - a_{j_2 j_1 1})\left[(\overline{x}_{j_2} - a_{j_2 j_2})^2 + b_{j_2 j_2}\right]}{(\overline{x}_{j_2} - a_{j_1 j_2 1})\prod_{s_1=1}^{2}(x_{j_1} - a_{j_1 j_1 s_1})}\bigg|_{x_{j_1}^* = a_{j_2 j_1 1}} = 0.$$

If

$$\frac{d^2x_{j_2}}{dx_{j_1}^2}\bigg|_{x_{j_1}^*=a_{j_2j_11}} = \frac{a_{j_2j_20}}{a_{j_1j_10}} \frac{(\overline{x}_{j_2}-a_{j_2j_2})^2+b_{j_2j_2}}{(\overline{x}_{j_2}-a_{j_1j_21})\prod_{s_1=1}^{2}(a_{j_2j_11}-a_{j_1j_1s_1})} > 0,$$

there is an up-parabola flow. If

$$\frac{d^2x_{j_2}}{dx_{j_1}^2}\bigg|_{x_{j_1}^*=a_{j_2j_11}} = \frac{a_{j_2j_20}}{a_{j_1j_10}} \frac{(\overline{x}_{j_2}-a_{j_2j_2})^2+b_{j_2j_2}}{(\overline{x}_{j_2}-a_{j_1j_21})\prod_{s_1=1}^{2}(a_{j_2j_11}-a_{j_1j_1s_1})} < 0,$$

there is a down-parabola flow. For

$$\dot{x}_{j_1} = a_{j_1j_10}(\overline{x}_{j_2}-a_{j_1j_21})\prod\nolimits_{s_1=1}^{2}(a_{j_2j_11}-a_{j_1j_1s_1}),$$

the parabola flows at $x_{j_1}^* = a_{j_1j_11}$ are positive and negative for $\dot{x}_{j_1} > 0$ and $\dot{x}_{j_1} < 0$, respectively. In addition,

$$(\overline{x}_{j_2}-a_{j_2j_2})^2+b_{j_2j_2} > 0.$$

Thus, the equilibrium of $x_{j_1}^* = a_{j_2j_11}$ with $\overline{x}_{j_2} = a_{j_1j_21}$ has the following properties:

- For $a_{j_1j_10}(\overline{x}_{j_2}-a_{j_1j_21})\prod_{s_1=1}^{2}(a_{j_2j_11}-a_{j_1j_1s_1}) > 0$ and $a_{j_2j_20} > 0$,

$$(\dot{x}_{j_1}, a_{j_2j_11}) = \underbrace{(\text{pF,UP})}_{\text{up-parabola flow }(+)}.$$

- For $a_{j_1j_10}(\overline{x}_{j_2}-a_{j_1j_21})\prod_{s_1=1}^{2}(a_{j_2j_11}-a_{j_1j_1s_1}) < 0$ and $a_{j_2j_20} > 0$,

$$(\dot{x}_{j_1}, a_{j_2j_11}) = \underbrace{(\text{nF,DP})}_{\text{down-parabola flow }(-)}.$$

- For $a_{j_1j_10}(\overline{x}_{j_2}-a_{j_1j_21})\prod_{s_1=1}^{2}(a_{j_2j_11}-a_{j_1j_1s_1}) > 0$ and $a_{j_2j_20} < 0$,

$$(\dot{x}_{j_1}, a_{j_2j_11}) = \underbrace{(\text{pF,DP})}_{\text{down-parabola flow }(+)}.$$

- For $a_{j_1j_10}(\overline{x}_{j_2}-a_{j_1j_21})\prod_{s_1=1}^{2}(a_{j_2j_11}-a_{j_1j_1s_1}) < 0$ and $a_{j_2j_20} < 0$,

$$(\dot{x}_{j_1}, a_{j_2j_11}) = \underbrace{(\text{nF,UP})}_{\text{up-parabola flow }(-)}.$$

Therefore, from the analysis, the equilibrium of $(x_{j_2}^*, x_{j_1}^*) = (a_{j_1j_21}, a_{j_2j_11})$ has the following properties as in Eqs. (1.173)–(1.176):

- For $a_{j_1 j_1 0}\prod_{s_1=1}^{2}(a_{j_2 j_1 1}-a_{j_1 j_1 s_1})>0$ and $a_{j_2 j_2 0}>0$,

$$(a_{j_1 j_2 1}, a_{j_2 j_1 1}) = \underbrace{(\mathrm{UP}_+, \mathrm{UP}_+)}_{\text{positive saddle}}.$$

The equilibrium of $(x_{j_2}^*, x_{j_1}^*) = (a_{j_1 j_2 1}, a_{j_2 j_1 1})$ is a $(\mathrm{UP}_+, \mathrm{UP}_+)$-positive saddle.

- For $a_{j_1 j_1 0}\prod_{s_1=1}^{2}(a_{j_2 j_1 1}-a_{j_1 j_1 s_1})<0$ and $a_{j_2 j_2 0}>0$,

$$(a_{j_1 j_2 1}, a_{j_2 j_1 1}) = \underbrace{(\mathrm{DP}_+, \mathrm{DP}_-)}_{\text{CCW center}}.$$

The equilibrium of $(x_{j_2}^*, x_{j_1}^*) = (a_{j_1 j_2 1}, a_{j_2 j_1 1})$ is a $(\mathrm{DP}_+, \mathrm{DP}_-)$-counterclockwise center.

- For $a_{j_1 j_1 0}\prod_{s_1=1}^{2}(a_{j_2 j_1 1}-a_{j_1 j_1 s_1})>0$ and $a_{j_2 j_2 0}<0$,

$$(a_{j_1 j_2 1}, a_{j_2 j_1 1}) = \underbrace{(\mathrm{DP}_-, \mathrm{DP}_+)}_{\text{CW center}}.$$

The equilibrium of $(x_{j_2}^*, x_{j_1}^*) = (a_{j_1 j_2 1}, a_{j_2 j_1 1})$ is a $(\mathrm{DP}_-, \mathrm{DP}_+)$-clockwise center.

- For $a_{j_1 j_1 0}\prod_{s_1=1}^{2}(a_{j_2 j_1 1}-a_{j_1 j_1 s_1})<0$ and $a_{j_2 j_2 0}<0$,

$$(a_{j_1 j_2 1}, a_{j_2 j_1 1}) = \underbrace{(\mathrm{UP}_-, \mathrm{UP}_-)}_{\text{negative saddle}}.$$

The equilibrium of $(x_{j_2}^*, x_{j_1}^*) = (a_{j_1 j_2 1}, a_{j_2 j_1 1})$ is a $(\mathrm{UP}_-, \mathrm{UP}_-)$-negative saddle.

($\mathrm{vii}_{1\mathrm{b}}$) From the case (I) of ($\mathrm{vii}_{1\mathrm{a}}$) with $\bar{x}_{j_1} = a_{j_1 j_1 s_1}$ ($s_1, s_2 \in \{1,2\}, s_1 \neq s_2$), thus, the equilibrium of $(x_{j_1}^*, x_{j_2}^*) = (a_{j_1 j_1 s_1}, a_{j_1 j_2 1})$ (s_1, $s_2 \in \{1,2\}$, $s_1 \neq s_2$) has the following properties as in Eqs. (1.177)–(1.180):

- For $a_{j_1 j_1 0}(a_{j_1 j_1 s_1}-a_{j_1 j_1 s_2})>0$ and $a_{j_2 j_2 0}(a_{j_1 j_1 s_1}-a_{j_2 j_1 1})>0$,

$$(a_{j_1 j_1 s_1}, a_{j_1 j_2 1}) = \underbrace{(\mathrm{DP{:}UP,pF})}_{\text{hyperbolic flow }(+)}.$$

The equilibrium of $(x_{j_1}^*, x_{j_2}^*) = (a_{j_1 j_1 s_1}, a_{j_1 j_2 1})$ is a (DP:UP,pF)-positive hyperbolic flow.

- For $a_{j_1 j_1 0}(a_{j_1 j_1 s_1}-a_{j_1 j_1 s_2})<0$ and $a_{j_2 j_2 0}(a_{j_1 j_1 s_1}-a_{j_2 j_1 1})>0$,

$$(a_{j_1j_1s_1}, a_{j_1j_21}) = \underbrace{\text{(UP:DP,pF)}}_{\text{hyperbolic-secant flow }(+)}.$$

The equilibrium of $(x^*_{j_1}, x^*_{j_2}) = (a_{j_1j_1s_1}, a_{j_1j_21})$ is a (UP:DP,pF)-positive hyperbolic-secant flow.

- For $a_{j_1j_10}(a_{j_1j_1s_1} - a_{j_1j_1s_2}) > 0$ and $a_{j_2j_20}(a_{j_1j_1s_1} - a_{j_2j_11}) < 0$,

$$(a_{j_1j_1s_1}, a_{j_1j_21}) = \underbrace{\text{(UP:DP,nF)}}_{\text{hyperbolic-secant flow }(-)}.$$

The equilibrium of $(x^*_{j_1}, x^*_{j_2}) = (a_{j_1j_1s_1}, a_{j_1j_21})$ is a (UP:DP,nF)-negative hyperbolic-secant flow.

- For $a_{j_1j_10}(a_{j_1j_1s_1} - a_{j_1j_1s_2}) < 0$ and $a_{j_2j_20}(a_{j_1j_1s_1} - a_{j_2j_11}) < 0$,

$$(a_{j_1j_1s_1}, a_{j_1j_21}) = \underbrace{\text{(DP:UP,nF)}}_{\text{hyperbolic flow }(-)}.$$

The equilibrium of $(x^*_{j_1}, x^*_{j_2}) = (a_{j_1j_1s_1}, a_{j_1j_21})$ is a (DP:UP,nF)-negative hyperbolic flow.

(vii$_2$) For $a_{j_2j_11} = a_{j_1j_1s_1}$ (s_1, $s_2 \in \{1,2\}$, $s_1 \neq s_2$), in phase space,

$$\frac{dx_{j_1}}{dx_{j_2}} = \frac{a_{j_1j_10}}{a_{j_2j_20}} \frac{(x_{j_1} - a_{j_1j_1s_2})(x_{j_2} - a_{j_1j_21})}{\left[(x_{j_2} - a_{j_2j_2})^2 + b_{j_2j_2}\right]},$$

and the deformation of the foregoing equation is

$$\frac{1}{(x_{j_1} - a_{j_1j_1s_2})} dx_{j_1} = \frac{a_{j_1j_10}}{a_{j_2j_20}} \left[\frac{x_{j_2} - a_{j_2j_2}}{(x_{j_2} - a_{j_2j_2})^2 + b_{j_2j_2}} + \frac{a_{j_2j_2} - a_{j_1j_21}}{(x_{j_2} - a_{j_2j_2})^2 + b_{j_2j_2}}\right] dx_{j_2}.$$

With an initial condition of (x_{j_10}, x_{j_20}) at $t = t_0$, the integration of the above equation gives

$$\begin{aligned}\ln \frac{|x_{j_1} - a_{j_1j_1s_2}|}{|x_{j_10} - a_{j_1j_1s_2}|} = \frac{a_{j_1j_10}}{a_{j_2j_20}} &\left[\frac{1}{2} \ln \frac{|(x_{j_2} - a_{j_2j_2})^2 + b_{j_2j_2}|}{|(x_{j_20} - a_{j_2j_2})^2 + b_{j_2j_2}|}\right. \\ &\left. + \frac{(a_{j_2j_2} - a_{j_1j_21})}{\sqrt{b_{j_2j_2}}} \left(\arctan \frac{x_{j_2} - a_{j_2j_2}}{\sqrt{b_{j_2j_2}}} - \arctan \frac{x_{j_20} - a_{j_2j_2}}{\sqrt{b_{j_2j_2}}}\right)\right].\end{aligned}$$

(vii$_{2a}$) For $x^*_{j_1} = a_{j_2j_11} = a_{j_1j_1s_1}$ (s_1, $s_2 \in \{1,2\}$, $s_1 \neq s_2$), with $\bar{x}_{j_2} \neq a_{j_1j_21}$, if

$$\frac{dx_{j_2}}{dx_{j_1}}\bigg|_{x_{j_1}^*=a_{j_1j_1s_1}} = \frac{a_{j_2j_20}}{a_{j_1j_10}} \frac{[(\overline{x}_{j_2}-a_{j_2j_2})^2+b_{j_2j_2}]}{(\overline{x}_{j_2}-a_{j_1j_21})(a_{j_1j_2s_1}-a_{j_1j_2s_2})} > 0,$$

there is an increasing-inflection flow at $x_{j_1}^* = a_{j_2j_11} = a_{j_1j_1s_1}$, and if

$$\frac{dx_{j_2}}{dx_{j_1}}\bigg|_{x_{j_1}^*=a_{j_1j_1s_1}} = \frac{a_{j_2j_20}}{a_{j_1j_10}} \frac{[(\overline{x}_{j_2}-a_{j_2j_2})^2+b_{j_2j_2}]}{(\overline{x}_{j_2}-a_{j_1j_21})(a_{j_1j_2s_1}-a_{j_1j_2s_2})} < 0,$$

there is a decreasing-inflection flow at $x_{j_1}^* = a_{j_2j_11} = a_{j_1j_1s_1}$. At $x_{j_1}^* = a_{j_2j_11} = a_{j_1j_1s_1}$ with $\overline{x}_{j_2} \neq a_{j_1j_21}$, the variational equation is

$$\Delta\dot{x}_{j_1} = a_{j_1j_10}(\overline{x}_{j_2}-a_{j_1j_21})(a_{j_1j_2s_1}-a_{j_1j_2s_2})\Delta x_{j_1}.$$

The inflection flows are source and sink for $a_{j_1j_10}(\overline{x}_{j_2}-a_{j_1j_21})(a_{j_1j_2s_1}-a_{j_1j_2s_2}) > 0$ and $a_{j_1j_10}(\overline{x}_{j_2}-a_{j_1j_21})(a_{j_1j_2s_1}-a_{j_1j_2s_2}) < 0$, respectively.

Therefore, the infinite equilibrium of $x_{j_1}^* = a_{j_2j_11} = a_{j_1j_1s_1}$ with $\overline{x}_{j_2} \neq a_{j_1j_21}$ has the following properties as in Eqs. (1.182)–(1.185):

- For $a_{j_1j_10}(\overline{x}_{j_2}-a_{j_1j_21})(a_{j_1j_2s_1}-a_{j_1j_2s_2}) > 0$ and $a_{j_2j_20} > 0$,

$$(a_{j_1j_1s_1}, \overline{x}_{j_2}) = \underbrace{(\text{SO,II})}_{\text{increasing-inflection source}}.$$

  The infinite-equilibrium of $x_{j_1}^* = a_{j_2j_11} = a_{j_1j_1s_1}$ is an (SO,II)-increasing-inflection source.
- For $a_{j_1j_10}(\overline{x}_{j_2}-a_{j_1j_21})(a_{j_1j_2s_1}-a_{j_1j_2s_2}) < 0$ and $a_{j_2j_20} > 0$,

$$(a_{j_1j_1s_1}, \overline{x}_{j_2}) = \underbrace{(\text{SI,DI})}_{\text{decreasing-inflection sink}}.$$

  The infinite-equilibrium of $x_{j_1}^* = a_{j_2j_11} = a_{j_1j_1s_1}$ is an (SI,DI)-decreasing-inflection sink.
- For $a_{j_1j_10}(\overline{x}_{j_2}-a_{j_1j_21})(a_{j_1j_2s_1}-a_{j_1j_2s_2}) > 0$ and $a_{j_2j_20} < 0$,

$$(a_{j_1j_1s_1}, \overline{x}_{j_2}) = \underbrace{(\text{SO,DI})}_{\text{decreasing-inflection source}}.$$

  The infinite-equilibrium of $x_{j_1}^* = a_{j_2j_11} = a_{j_1j_1s_1}$ is an (SO,DI)-decreasing-inflection source.

  For $a_{j_1j_10}(\overline{x}_{j_2}-a_{j_1j_21})(a_{j_1j_2s_1}-a_{j_1j_2s_2}) < 0$ and $a_{j_2j_20} < 0$,

$$(a_{j_1j_1s_1}, \overline{x}_{j_2}) = \underbrace{(\text{SI,II})}_{\text{increasing-inflection sink}}.$$

The infinite equilibrium of $x_{j_1}^* = a_{j_2j_11} = a_{j_1j_1s_1}$ is an (SO,II)-increasing-inflection sink.

(vii$_{2b}$) From (vii$_{2a}$) with $\bar{x}_{j_2} \neq a_{j_1j_21}$ and the case (I) of (vii$_{1a}$) with $\bar{x}_{j_1} = a_{j_1j_1s_1} = a_{j_2j_11}$. The equilibrium of $(x_{j_1}^*, x_{j_2}^*) = (a_{j_1j_11}, a_{j_1j_21})$ has the following properties as in Eqs. (1.186)–(1.189):

- For $a_{j_1j_10}(a_{j_2j_11} - a_{j_1j_1s_2}) > 0$ and $a_{j_2j_20} > 0$,

$$(a_{j_1j_21}, a_{j_2j_11}) = \underbrace{({}_{\text{DI:II}}\text{UP}, {}_{\text{SI:SO}}\text{US})}_{\text{inflection up-parabola upper-saddle}}.$$

  The equilibrium of $(x_{j_2}^*, x_{j_1}^*) = (a_{j_1j_21}, a_{j_2j_11})$ is a (${}_{\text{DI:II}}$UP, ${}_{\text{SI:SO}}$US)-inflection up-parabola upper-saddle.
- For $a_{j_1j_10}(a_{j_2j_11} - a_{j_1j_1s_2}) < 0$ and $a_{j_2j_20} > 0$,

$$(a_{j_1j_21}, a_{j_2j_11}) = \underbrace{({}_{\text{II:DI}}\text{DP}, {}_{\text{SO:SI}}\text{LS})}_{\text{inflection down-parabola lower-saddle}}.$$

  The equilibrium of $(x_{j_2}^*, x_{j_1}^*) = (a_{j_1j_21}, a_{j_2j_11})$ is an (${}_{\text{II:DI}}$DP: ${}_{\text{SO:SI}}$LS)-inflection down-parabola lower-saddle.
- For $a_{j_1j_10}(a_{j_2j_11} - a_{j_1j_1s_2}) > 0$ and $a_{j_2j_20} < 0$,

$$(a_{j_1j_21}, a_{j_2j_11}) = \underbrace{({}_{\text{II:DI}}\text{DP}, {}_{\text{SI:SO}}\text{US})}_{\text{inflection down-parabola upper-saddle}}.$$

  The equilibrium of $(x_{j_2}^*, x_{j_1}^*) = (a_{j_1j_21}, a_{j_2j_11})$ is an (${}_{\text{II:DI}}$DP, ${}_{\text{SI:SO}}$US)-inflection down-parabola upper-saddle.
- For $a_{j_1j_10}(a_{j_2j_11} - a_{j_1j_1s_2}) < 0$ and $a_{j_2j_20} < 0$,

$$(a_{j_1j_21}, a_{j_2j_11}) = \underbrace{({}_{\text{DI:II}}\text{UP}, {}_{\text{SO:SI}}\text{LS})}_{\text{inflection up-parabola lower-saddle}}.$$

  The equilibrium of $(x_{j_2}^*, x_{j_1}^*) = (a_{j_1j_21}, a_{j_2j_11})$ is a (${}_{\text{DI:II}}$UP, ${}_{\text{SO:SI}}$LS)-inflection up-parabola lower-saddle.

(viii) For $\Delta_{j_1j_1} > 0$ and $\Delta_{j_2j_2} = 0$, the standard form is

$$\begin{aligned}\dot{x}_{j_2} &= a_{j_2j_20}(x_{j_1} - a_{j_2j_11})(x_{j_2} - a_{j_2j_21})^2,\\ \dot{x}_{j_1} &= a_{j_1j_10}(x_{j_1} - a_{j_1j_11})(x_{j_1} - a_{j_1j_12})(x_{j_2} - a_{j_1j_21})\end{aligned}$$

where

$$b_{j_1j_11}, b_{j_1j_12} = -\frac{1}{2}(B_{j_1j_1} \pm \sqrt{\Delta_{j_1j_1}}),$$
$$\{a_{j_1j_11}, a_{j_1j_12}\} = \text{sort}\{b_{j_1j_11}, b_{j_1j_12}\}, a_{j_1j_11} < a_{j_1j_12};$$
$$a_{j_2j_21} = -\frac{1}{2}B_{j_2j_2}.$$

(viii$_1$) For $a_{j_2j_21} \neq a_{j_1j_21}, a_{j_1j_22}$, in phase space,

$$\frac{dx_{j_1}}{dx_{j_2}} = \frac{a_{j_1j_10}}{a_{j_2j_20}} \frac{(x_{j_1} - a_{j_1j_11})(x_{j_1} - a_{j_1j_12})(x_{j_2} - a_{j_1j_21})}{(x_{j_1} - a_{j_2j_11})(x_{j_2} - a_{j_2j_21})^2},$$

and the deformation of the foregoing equation is

$$\frac{(x_{j_1} - a_{j_2j_11})}{(x_{j_1} - a_{j_1j_11})(x_{j_1} - a_{j_1j_12})} dx_{j_1} = \frac{a_{j_1j_10}}{a_{j_2j_20}} \left[\frac{1}{(x_{j_2} - a_{j_2j_21})} + \frac{(a_{j_2j_21} - a_{j_1j_21})}{(x_{j_2} - a_{j_2j_21})^2}\right] dx_{j_2}.$$

With an initial condition of (x_{j_10}, x_{j_20}) at $t = t_0$, the integration of the above equation gives the first integral manifold for $a_{j_2j_21} \neq a_{j_1j_21}, a_{j_1j_22}$ as

$$\begin{aligned} &\sum\nolimits_{s_1=1}^{2} \frac{a_{j_1j_1s_1} - a_{j_2j_11}}{a_{j_1j_1s_1} - a_{j_1j_11}} \ln \frac{|x_{j_1} - a_{j_1j_1s_1}|}{|x_{j_10} - a_{j_1j_1s_1}|} \\ &= \frac{a_{j_1j_10}}{a_{j_2j_20}} \left[\ln \frac{|x_{j_2} - a_{j_2j_21}|}{|x_{j_20} - a_{j_2j_21}|} - (a_{j_2j_21} - a_{j_1j_21})(\frac{1}{x_{j_2} - a_{j_2j_21}} - \frac{1}{x_{j_20} - a_{j_2j_21}})\right]. \end{aligned}$$

(viii$_{1a}$) Consider two cases (I) $x_{j_2}^* = a_{j_1j_21}$ and (II) $x_{j_1}^* = a_{j_2j_11}$.

(I) In phase space, at $x_{j_2}^* = a_{j_1j_21}$ with $\bar{x}_{j_1} \neq a_{j_1j_11}$,

$$\left.\frac{dx_{j_1}}{dx_{j_2}}\right|_{x_{j_2}^* = a_{j_1j_21}} = \left.\frac{a_{j_1j_10}}{a_{j_2j_20}} \frac{\prod_{s_1=1}^{2}(\bar{x}_{j_1} - a_{j_1j_1s_1})(x_{j_2} - a_{j_1j_21})}{(\bar{x}_{j_1} - a_{j_2j_11})(x_{j_2} - a_{j_2j_21})^2}\right|_{x_{j_2}^* = a_{j_1j_21}} = 0.$$

If

$$\left.\frac{d^2x_{j_1}}{dx_{j_2}^2}\right|_{x_{j_2}^* = a_{j_1j_21}} = \frac{a_{j_1j_10}}{a_{j_2j_20}} \frac{\prod_{s_1=1}^{2}(\bar{x}_{j_1} - a_{j_1j_1s_1})}{(\bar{x}_{j_1} - a_{j_2j_11})(a_{j_1j_21} - a_{j_2j_21})^2} > 0,$$

there is an up-parabola flow at $x_{j_2}^* = a_{j_1j_21}$ with $\bar{x}_{j_1} \neq a_{j_1j_11}$. If

$$\left.\frac{d^2x_{j_1}}{dx_{j_2}^2}\right|_{x_{j_2}^* = a_{j_1j_21}} = \frac{a_{j_1j_10}}{a_{j_2j_20}} \frac{\prod_{s_1=1}^{2}(\bar{x}_{j_1} - a_{j_1j_1s_1})}{(\bar{x}_{j_1} - a_{j_2j_11})(a_{j_1j_21} - a_{j_2j_21})^2} < 0,$$

there is a down-parabola flow at $x_{j_2}^* = a_{j_1j_2 1}$ with $\overline{x}_{j_1} \neq a_{j_1j_1 1}$. Let

$$\dot{x}_{j_2} = a_{j_2j_2 0}(\overline{x}_{j_1} - a_{j_2j_1 1})(a_{j_1j_2 1} - a_{j_2j_2 1})^2,$$

with for $a_{j_1j_2 1} \neq a_{j_2j_2 1}$,

$$(a_{j_1j_2 1} - a_{j_2j_2})^2 > 0.$$

The parabola flows at $x_{j_2}^* = a_{j_1j_2 1}$ with $\overline{x}_{j_1} \neq a_{j_1j_1 1}$ are positive and negative for $\dot{x}_{j_2} > 0$ and $\dot{x}_{j_2} < 0$, respectively.

Thus, the equilibrium of $x_{j_2}^* = a_{j_1j_2 1}$ with $\overline{x}_{j_1} \neq a_{j_1j_1 1}$ has the following properties:

- For $a_{j_1j_1 0}\prod_{s_1=1}^2(\overline{x}_{j_1} - a_{j_1j_1 s_1}) > 0$ and $a_{j_2j_2 0}(\overline{x}_{j_1} - a_{j_2j_1 1})(a_{j_1j_2 1} - a_{j_2j_2 1})^2 > 0$,

$$(a_{j_1j_2 1}, \dot{x}_{j_2}) = \underbrace{(\text{UP,pF})}_{\text{up-parabola flow } (+)}.$$

- For $a_{j_1j_1 0}\prod_{s_1=1}^2(\overline{x}_{j_1} - a_{j_1j_1 s_1}) < 0$ and $a_{j_2j_2 0}(\overline{x}_{j_1} - a_{j_2j_1 1})(a_{j_1j_2 1} - a_{j_2j_2 1})^2 > 0$,

$$(a_{j_1j_2 1}, \dot{x}_{j_2}) = \underbrace{(\text{DP,pF})}_{\text{down-parabola flow } (+)}.$$

- For $a_{j_1j_1 0}\prod_{s_1=1}^2(\overline{x}_{j_1} - a_{j_1j_1 s_1}) > 0$ and $a_{j_2j_2 0}(\overline{x}_{j_1} - a_{j_2j_1 1})(a_{j_1j_2 1} - a_{j_2j_2 1})^2 < 0$,

$$(a_{j_1j_2 1}, \dot{x}_{j_2}) = \underbrace{(\text{DP,nF})}_{\text{dwon-parabola flow } (-)}.$$

- For $a_{j_1j_1 0}\prod_{s_1=1}^2(\overline{x}_{j_1} - a_{j_1j_1 s_1}) < 0$ and $a_{j_2j_2 0}(\overline{x}_{j_1} - a_{j_2j_1 1})(a_{j_1j_2 1} - a_{j_2j_2 1})^2 < 0$,

$$(a_{j_1j_2 1}, \dot{x}_{j_2}) = \underbrace{(\text{UP,nF})}_{\text{up-parabola flow } (-)}.$$

(II) Similarly, at $x_{j_1}^* = a_{j_2j_1 1}$ with $\overline{x}_{j_2} \neq a_{j_1j_2 1}$, in phase space,

$$\frac{dx_{j_2}}{dx_{j_1}}\bigg|_{x_{j_1}^* = a_{j_2j_1 1}} = \frac{a_{j_2j_2 0}}{a_{j_1j_1 0}} \frac{(x_{j_1} - a_{j_2j_1 1})(\overline{x}_{j_2} - a_{j_2j_2 1})^2}{(\overline{x}_{j_2} - a_{j_1j_2 1})\prod_{s_1=1}^2(x_{j_1} - a_{j_1j_1 s_1})}\bigg|_{x_{j_1}^* = a_{j_2j_1 1}} = 0.$$

If

$$\left.\frac{d^2x_{j_2}}{dx_{j_1}^2}\right|_{x_{j_1}^*=a_{j_2j_11}}=\frac{a_{j_2j_20}}{a_{j_1j_10}}\frac{(\bar{x}_{j_2}-a_{j_2j_21})^2}{(\bar{x}_{j_2}-a_{j_1j_21})\prod_{s_1=1}^{2}(a_{j_2j_11}-a_{j_1j_1s_1})}>0,$$

there is an up-parabola flow. If

$$\left.\frac{d^2x_{j_2}}{dx_{j_1}^2}\right|_{x_{j_1}^*=a_{j_2j_11}}=\frac{a_{j_2j_20}}{a_{j_1j_10}}\frac{(\bar{x}_{j_2}-a_{j_2j_21})^2}{(\bar{x}_{j_2}-a_{j_1j_21})\prod_{s_1=1}^{2}(a_{j_2j_11}-a_{j_1j_1s_1})}<0,$$

there is a down-parabola flow. For

$$\dot{x}_{j_1}=a_{j_1j_10}(\bar{x}_{j_2}-a_{j_1j_21})\prod\nolimits_{s_1=1}^{2}(a_{j_2j_11}-a_{j_1j_1s_1}),$$

the parabola flows at $x_{j_1}^*=a_{j_1j_11}$ are positive and negative for $\dot{x}_{j_1}>0$ and $\dot{x}_{j_1}<0$, respectively. Thus, the equilibrium of $x_{j_1}^*=a_{j_2j_11}$ with $\bar{x}_{j_2}=a_{j_1j_21}$ has the following properties:

- For $a_{j_1j_10}(\bar{x}_{j_2}-a_{j_1j_21})\prod_{s_1=1}^{2}(a_{j_2j_11}-a_{j_1j_1s_1})>0$ and $a_{j_2j_20}(\bar{x}_{j_2}-a_{j_2j_21})^2>0$,

$$(\dot{x}_{j_1},a_{j_2j_11})=\underbrace{(\text{pF,UP})}_{\text{up-parabola flow }(+)}.$$

- For $a_{j_1j_10}(\bar{x}_{j_2}-a_{j_1j_21})\prod_{s_1=1}^{2}(a_{j_2j_11}-a_{j_1j_1s_1})<0$ and $a_{j_2j_20}(\bar{x}_{j_2}-a_{j_2j_21})^2>0$,

$$(\dot{x}_{j_1},a_{j_2j_11})=\underbrace{(\text{nF,DP})}_{\text{down-parabola flow }(-)}.$$

- For $a_{j_1j_10}(\bar{x}_{j_2}-a_{j_1j_21})\prod_{s_1=1}^{2}(a_{j_2j_11}-a_{j_1j_1s_1})>0$ and $a_{j_2j_20}(\bar{x}_{j_2}-a_{j_2j_21})^2<0$,

$$(\dot{x}_{j_1},a_{j_2j_11})=\underbrace{(\text{pF,DP})}_{\text{down-parabola flow }(+)}.$$

- For $a_{j_1j_10}(\bar{x}_{j_2}-a_{j_1j_21})\prod_{s_1=1}^{2}(a_{j_2j_11}-a_{j_1j_1s_1})<0$ and $a_{j_2j_20}(\bar{x}_{j_2}-a_{j_2j_21})^2<0$,

$$(\dot{x}_{j_1},a_{j_2j_11})=\underbrace{(\text{nF,UP})}_{\text{up-parabola flow }(-)}.$$

Therefore, from the analysis, the equilibrium of $(x_{j_2}^*,x_{j_1}^*)=(a_{j_1j_21},a_{j_2j_11})$ has the following properties as in Eqs. (1.193)–(1.196):

- For $a_{j_1j_1 0}\prod_{s_1=1}^{2}(a_{j_2j_1 1}-a_{j_1j_1 s_1})>0$ and $a_{j_2j_2 0}(a_{j_1j_2 1}-a_{j_2j_2 1})^2>0$,

$$(a_{j_1j_2 1},a_{j_2j_1 1})=\underbrace{(\mathrm{UP}_+,\mathrm{UP}_+)}_{\text{positive saddle}}.$$

The equilibrium of $(x_{j_2}^*,x_{j_1}^*)=(a_{j_1j_2 1},a_{j_2j_1 1})$ is a (UP_+, UP_+)-positive saddle.

- For $a_{j_1j_1 0}\prod_{s_1=1}^{2}(a_{j_2j_1 1}-a_{j_1j_1 s_1})<0$ and $a_{j_2j_2 0}(a_{j_1j_2 1}-a_{j_2j_2 1})^2>0$,

$$(a_{j_1j_2 1},a_{j_2j_1 1})=\underbrace{(\mathrm{DP}_+,\mathrm{DP}_-)}_{\text{CCW center}}.$$

The equilibrium of $(x_{j_2}^*,x_{j_1}^*)=(a_{j_1j_2 1},a_{j_2j_1 1})$ is a (DP_+,DP_-)-counterclockwise center.

- For $a_{j_1j_1 0}\prod_{s_1=1}^{2}(a_{j_2j_1 1}-a_{j_1j_1 s_1})>0$ and $a_{j_2j_2 0}(a_{j_1j_2 1}-a_{j_2j_2 1})^2<0$,

$$(a_{j_1j_2 1},a_{j_2j_1 1})=\underbrace{(\mathrm{DP}_-,\mathrm{DP}_+)}_{\text{CW center}}.$$

The equilibrium of $(x_{j_2}^*,x_{j_1}^*)=(a_{j_1j_2 1},a_{j_2j_1 1})$ is a (DP_-, DP_+)-clockwise center.

- For $a_{j_1j_1 0}\prod_{s_1=1}^{2}(a_{j_2j_1 1}-a_{j_1j_1 s_1})<0$ and $a_{j_2j_2 0}(a_{j_1j_2 1}-a_{j_2j_2 1})^2<0$,

$$(a_{j_1j_2 1},a_{j_2j_1 1})=\underbrace{(\mathrm{UP}_-,\mathrm{UP}_-)}_{\text{negative saddle}}.$$

The equilibrium of $(x_{j_2}^*,x_{j_1}^*)=(a_{j_1j_2 1},a_{j_2j_1 1})$ is a (UP_-,UP_-)-negative saddle.

($\mathrm{viii}_{1\mathrm{b}}$) At $(x_{j_1}^*,x_{j_2}^*)=(a_{j_1j_1 s_1},a_{j_2j_2 1})$, the variational equations are

$$\begin{aligned}\Delta\dot{x}_{j_2}&=a_{j_2j_2 0}(a_{j_1j_1 s_1}-a_{j_2j_1 1})(\Delta x_{j_2})^2,\\ \Delta\dot{x}_{j_1}&=a_{j_1j_1 0}(a_{j_2j_2 1}-a_{j_1j_2 1})(a_{j_1j_1 s_1}-a_{j_1j_1 s_2})\Delta x_{j_1}.\end{aligned}$$

The flows at $(x_{j_1}^*,x_{j_2}^*)=(a_{j_1j_1 s_1},a_{j_2j_2 1})$ in the x_{j_1}-direction are source and sink for $a_{j_1j_1 0}(a_{j_2j_2 1}-a_{j_1j_2 1})(a_{j_1j_1 s_1}-a_{j_1j_1 s_2})>0$ and $a_{j_1j_1 0}(a_{j_2j_2 1}-a_{j_1j_2 1})$ $(a_{j_1j_1 s_1}-a_{j_1j_1 s_2})<0$, respectively. The flows at $(x_{j_1}^*,x_{j_2}^*)=(a_{j_1j_1 s_1},a_{j_2j_2 1})$ in the x_{j_2}-direction are upper saddle and lower saddle for $a_{j_2j_2 0}(a_{j_1j_1 s_1}-a_{j_2j_1 1})>0$ and $a_{j_2j_2 0}(a_{j_1j_1 s_1}-a_{j_2j_1 1})<0$, respectively.

Therefore, the equilibrium of $(x_{j_1}^*,x_{j_2}^*)=(a_{j_1j_1 s_1},a_{j_2j_2 1})$ has the following properties as in Eqs. (1.197)–(1.200):

- For $a_{j_1j_1 0}(a_{j_1j_1 s_1}-a_{j_1j_1 s_2})(a_{j_2j_2 1}-a_{j_1j_2 1})>0$ and $a_{j_2j_2 0}(a_{j_1j_1 s_1}-a_{j_2j_1 1})>0$,

$$(a_{j_1j_1s_1}, a_{j_2j_2 1}) = \underbrace{(\text{SO,US})}_{\text{upper-saddle source}}.$$

The equilibrium of $(x^*_{j_1}, x^*_{j_2}) = (a_{j_1j_1s_1}, a_{j_2j_2 1})$ is an (SO,US)-upper-saddle source.

- For $a_{j_1j_1 0}(a_{j_1j_1s_1} - a_{j_1j_1s_2})(a_{j_2j_2 1} - a_{j_1j_2 1}) < 0$ and $a_{j_2j_2 0}(a_{j_1j_1s_1} - a_{j_2j_1 1}) > 0$,

$$(a_{j_1j_1s_1}, a_{j_2j_2 1}) = \underbrace{(\text{SI,US})}_{\text{upper-saddle sink}}.$$

The equilibrium of $(x^*_{j_1}, x^*_{j_2}) = (a_{j_1j_1s_1}, a_{j_2j_2 1})$ is an (SI,US)-upper-saddle sink.

- For $a_{j_1j_1 0}(a_{j_1j_1s_1} - a_{j_1j_1s_2})(a_{j_2j_2 1} - a_{j_1j_2 1}) > 0$ and $a_{j_2j_2 0}(a_{j_1j_1s_1} - a_{j_2j_1 1}) < 0$,

$$(a_{j_1j_1s_1}, a_{j_2j_2 1}) = \underbrace{(\text{SO,LS})}_{\text{lower-saddle source}}.$$

The equilibrium of $(x^*_{j_1}, x^*_{j_2}) = (a_{j_1j_1s_1}, a_{j_2j_2 1})$ is an (SO,LS)-lower-saddle source.

- For $a_{j_1j_1 0}(a_{j_1j_1s_1} - a_{j_1j_1s_2})(a_{j_2j_2 1} - a_{j_1j_2 1}) < 0$ and $a_{j_2j_2 0}(a_{j_1j_1s_1} - a_{j_2j_1 1}) < 0$,

$$(a_{j_1j_1s_1}, a_{j_2j_2 1}) = \underbrace{(\text{SI,LS})}_{\text{lower-saddle sink}}.$$

The equilibrium of $(x^*_{j_1}, x^*_{j_2}) = (a_{j_1j_1s_1}, a_{j_2j_2 1})$ is an (SI,LS)-lower-saddle sink.

– The saddle-source and saddle-sink are the appearing bifurcations of saddle-source and saddle-sink.

(viii$_{1c}$) For $x^*_{j_2} = a_{j_1j_2 1} \neq a_{j_2j_2 1}$ and $\bar{x}_{j_1} \neq a_{j_1j_1s_1}$ ($s_1, s_2 \in \{1, 2\}, s_1 \neq s_2$), from the above analysis in the case (I) of (viii$_{1a}$) with $\bar{x}_{j_1} = a_{j_1j_1s_1}$ ($s_1, s_2 \in \{1, 2\}, s_1 \neq s_2$), the equilibrium of $(x^*_{j_1}, x^*_{j_2}) = (a_{j_1j_1s_1}, a_{j_1j_2 1})$ with $a_{j_1j_2 1} \neq a_{j_2j_2 1}$ ($s_1, s_2 \in \{1, 2\}, s_1 \neq s_2$) has the following properties as in Eqs. (1.201)–(1.204):

- For $a_{j_1j_1 0}(a_{j_1j_1s_1} - a_{j_1j_1s_2}) > 0$ and $a_{j_2j_2 0}(a_{j_1j_1s_1} - a_{j_1j_1s_2})(a_{j_1j_2 1} - a_{j_2j_2 1})^2 > 0$,

$$(a_{j_1j_1s_1}, a_{j_1j_2 1}) = \underbrace{(\text{DP:UP,pF})}_{\text{hyperbolic flow } (+)}.$$

The equilibrium of $(x^*_{j_1}, x^*_{j_2}) = (a_{j_1j_1s_1}, a_{j_1j_2 1})$ is a (DP:UP, pF)-positive hyperbolic flow.

- For $a_{j_1j_1 0}(a_{j_1j_1s_1} - a_{j_1j_1s_2}) < 0$ and $a_{j_2j_2 0}(a_{j_1j_1s_1} - a_{j_1j_1s_2})(a_{j_1j_2 1} - a_{j_2j_2 1})^2 > 0$,

$$(a_{j_1j_1s_1}, a_{j_1j_2 1}) = \underbrace{(\text{UP:DP,pF})}_{\text{hyperbolic-secant flow } (+)}.$$

The equilibrium of $(x^*_{j_1}, x^*_{j_2}) = (a_{j_1j_1s_1}, a_{j_1j_2 1})$ is a (UP:DP, pF)-positive hyperbolic-secant flow.

- For $a_{j_1j_1 0}(a_{j_1j_1s_1} - a_{j_1j_1s_2}) > 0$ and $a_{j_2j_2 0}(a_{j_1j_1s_1} - a_{j_1j_1s_2})(a_{j_1j_2 1} - a_{j_2j_2 1})^2 < 0$,

$$(a_{j_1j_1s_1}, a_{j_1j_2 1}) = \underbrace{(\text{UP:DP,nF})}_{\text{hyperbolic-secant flow } (-)}.$$

The equilibrium of $(x^*_{j_1}, x^*_{j_2}) = (a_{j_1j_1s_1}, a_{j_1j_2 1})$ is a (UP:DP,nF)-negative hyperbolic-secant flow

- For $a_{j_1j_1 0}(a_{j_1j_1s_1} - a_{j_1j_1s_2}) < 0$ and $a_{j_2j_2 0}(a_{j_1j_1s_1} - a_{j_1j_1s_2})(a_{j_1j_2 1} - a_{j_2j_2 1})^2 < 0$,

$$(a_{j_1j_1s_1}, a_{j_1j_2 1}) = \underbrace{(\text{DP:UP,nF})}_{\text{hyperbolic flow } (-)}.$$

The equilibrium of $(x^*_{j_1}, x^*_{j_2}) = (a_{j_1j_1s_1}, a_{j_1j_2 1})$ is a (DP:UP, nF)-negative hyperbolic flow.

(viii$_{1d}$) For $(x^*_{j_1}, x^*_{j_2}) = (a_{j_2j_1}1, a_{j_2j_2}1)$ with $a_{j_1j_2}1 \neq a_{j_2j_2}1$ and $a_{j_1j_1}1 \neq a_{j_2j_1}1$, from the case (II) of (viii$_{1a}$), the equilibrium of $(x^*_{j_1}, x^*_{j_2}) = (a_{j_2j_1}1, a_{j_2j_2}1)$ with $a_{j_1j_2}1 \neq a_{j_2j_2}1$ and $a_{j_1j_1}1 \neq a_{j_2j_1}1$ has the following properties as in Eqs. (1.205)–(1.208):

- For $a_{j_1j_1 0}(a_{j_2j_2}1 - a_{j_1j_2}1)\prod_{s_1=1}^{2}(a_{j_2j_1}1 - a_{j_1j_1s_1}) > 0$ and $a_{j_2j_2 0} > 0$,

$$(a_{j_2j_1}1, a_{j_2j_2}1) = \underbrace{(\text{pF,UP:UP})}_{\text{hyperbolic-secant to hyperbolic flow } (+)}.$$

The equilibrium of $(x^*_{j_1}, x^*_{j_2}) = (a_{j_2j_1}1, a_{j_2j_2}1)$ is a (pF,UP:UP)-positive hyperbolic-secant-to-hyperbolic flow.

- For $a_{j_1j_1 0}(a_{j_2j_2}1 - a_{j_1j_2}1)\prod_{s_1=1}^{2}(a_{j_2j_1}1 - a_{j_1j_1s_1}) < 0$ and $a_{j_2j_2 0} > 0$,

$$(a_{j_2j_1}1, a_{j_2j_2}1) = \underbrace{(\text{nF,DP:DP})}_{\text{hyperbolic to hyperbolic-secant flow } (-)}.$$

The equilibrium of $(x^*_{j_1}, x^*_{j_2}) = (a_{j_2j_1}1, a_{j_2j_2}1)$ is an (nF, DP:DP)-negative hyperbolic-to-hyperbolic-secant flow.

- For $a_{j_1j_1 0}(a_{j_2j_2}1 - a_{j_1j_2}1)\prod_{s_1=1}^{2}(a_{j_2j_1}1 - a_{j_1j_1s_1}) > 0$ and $a_{j_2j_2 0} < 0$,

$$(a_{j_2j_11}, a_{j_2j_21}) = \underbrace{\text{(pF,DP:DP)}}_{\text{hyperbolic to hyperbolic-secant flow } (+)}.$$

The equilibrium of $(x^*_{j_1}, x^*_{j_2}) = (a_{j_2j_11}, a_{j_2j_21})$ is a (DP:DP,nF)-negative hyperbolic-to-hyperbolic-secant flow.

- For $a_{j_1j_10}(a_{j_2j_21} - a_{j_1j_21})\prod_{s_1=1}^{2}(a_{j_2j_11} - a_{j_1j_1s_1}) < 0$ and $a_{j_2j_20} < 0$,

$$(a_{j_2j_11}, a_{j_2j_21}) = \underbrace{\text{(nF,UP:UP)}}_{\text{hyperbolic-secant to hyperbolic flow } (-)}.$$

The equilibrium of $(x^*_{j_1}, x^*_{j_2}) = (a_{j_2j_11}, a_{j_2j_21})$ is an (nF,UP:UP)-negative hyperbolic-secant-to-hyperbolic flow.

(viii$_2$) For $a_{j_2j_21} = a_{j_1j_21}$ and $a_{j_1j_1s_1} \neq a_{j_2j_11}$, we have

$$\begin{aligned}\frac{dx_{j_1}}{dx_{j_2}} &= \frac{a_{j_1j_10}}{a_{j_2j_20}} \frac{(x_{j_1} - a_{j_1j_11})(x_{j_1} - a_{j_1j_12})(x_{j_2} - a_{j_1j_21})}{(x_{j_1} - a_{j_2j_11})(x_{j_2} - a_{j_2j_21})^2} \\ &= \frac{a_{j_1j_10}\,(x_{j_1} - a_{j_1j_11})(x_{j_1} - a_{j_1j_12})}{a_{j_2j_20}\,(x_{j_1} - a_{j_2j_11})(x_{j_2} - a_{j_2j_21})},\end{aligned}$$

and the deformation of the foregoing equation is

$$\sum_{s_1=1, s_2 \neq s_1}^{2} \frac{a_{j_1j_1s_1} - a_{j_2j_11}}{a_{j_1j_1s_1} - a_{j_1j_1s_2}} \frac{1}{x_{j_1} - a_{j_1j_1s_1}} dx_{j_1} = \frac{a_{j_1j_10}}{a_{j_2j_20}} \frac{dx_{j_2}}{x_{j_2} - a_{j_2j_21}}.$$

With an initial condition of (x_{j_10}, x_{j_20}) at $t = t_0$,, the integration of the above equation gives the first integral manifold for $a_{j_2j_21} = a_{j_1j_21}$ as

$$\sum_{s_1=1, s_2 \neq s_1}^{2} \frac{a_{j_1j_1s_1} - a_{j_2j_11}}{a_{j_1j_1s_1} - a_{j_1j_1s_2}} \ln \frac{|x_{j_1} - a_{j_1j_1s_1}|}{|x_{j_10} - a_{j_1j_1s_1}|} = \frac{a_{j_1j_10}}{a_{j_2j_20}} \ln \frac{|x_{j_2} - a_{j_2j_21}|}{|x_{j_20} - a_{j_2j_21}|}.$$

(viii$_{2a}$) At $x^*_{j_2} = a_{j_1j_21} = a_{j_2j_21}$ with $\bar{x}_{j_1} \neq a_{j_1j_11}, a_{j_1j_12}$, for a small $\varepsilon > 0$,

$$\left.\frac{dx_{j_1}}{dx_{j_2}}\right|_{x^*_{j_2} = a_{j_2j_21} \pm \varepsilon} = \pm \frac{a_{j_1j_10}}{a_{j_2j_20}} \frac{(\bar{x}_{j_1} - a_{j_1j_11})(\bar{x}_{j_1} - a_{j_1j_12})}{(\bar{x}_{j_1} - a_{j_2j_11})\varepsilon}.$$

From the first integral manifold, the flow is an asymptotic flow. If

$$\frac{a_{j_1j_10}}{a_{j_2j_20}}\frac{\prod_{s_1=1}^{2}(\overline{x}_{j_1}-a_{j_1j_1s_1})}{(\overline{x}_{j_1}-a_{j_2j_11})}>0,$$

the flow at $x_{j_2}^*=a_{j_1j_21}=a_{j_2j_21}$ is a down-up flow. If

$$\frac{a_{j_1j_10}}{a_{j_2j_20}}\frac{\prod_{s_1=1}^{2}(\overline{x}_{j_1}-a_{j_1j_1s_1})}{(\overline{x}_{j_1}-a_{j_2j_11})}<0,$$

the flow at $x_{j_2}^*=a_{j_1j_21}=a_{j_2j_21}$ is an up-down flow. The variational equation at $x_{j_2}^*=a_{j_1j_2s_1}=a_{j_2j_21}$ is

$$\Delta\dot{x}_{j_2}=a_{j_2j_20}(\overline{x}_{j_1}-a_{j_2j_11})(\Delta x_{j_2})^2.$$

The asymptotic flows in the x_{j_2}-direction are upper saddle and lower saddle for $a_{j_2j_20}(\overline{x}_{j_1}-a_{j_2j_11})>0$ and $a_{j_2j_20}(\overline{x}_{j_1}-a_{j_2j_11})<0$, respectively.

Therefore, the infinite equilibrium of $x_{j_2}^*=a_{j_1j_21}=a_{j_2j_21}$ with $\overline{x}_{j_1}\neq a_{j_1j_11}$ has the following properties as in Eqs. (1.210)–(1.213):

- For $a_{j_1j_10}\prod_{s_1=1}^{2}(\overline{x}_{j_1}-a_{j_1j_1s_1})>0$ and $a_{j_2j_20}(\overline{x}_{j_1}-a_{j_2j_11})>0$,

$$(\overline{x}_{j_1},a_{j_1j_21})=\underbrace{(\text{DU,US})}_{\text{down-up upper-saddle}}.$$

  The infinite equilibrium of $x_{j_2}^*=a_{j_1j_21}=a_{j_2j_21}$ is an (DU,US)-down-up upper-saddle.

- For $a_{j_1j_10}\prod_{s_1=1}^{2}(\overline{x}_{j_1}-a_{j_1j_1s_1})<0$ and $a_{j_2j_20}(\overline{x}_{j_1}-a_{j_2j_11})>0$,

$$(\overline{x}_{j_1},a_{j_1j_21})=\underbrace{(\text{UD,US})}_{\text{up-down upper-saddle}}.$$

  The infinite equilibrium of $x_{j_2}^*=a_{j_1j_21}=a_{j_2j_21}$ is a (UD,US)-up-down upper-saddle.

- For $a_{j_1j_10}\prod_{s_1=1}^{2}(\overline{x}_{j_1}-a_{j_1j_1s_1})>0$ and $a_{j_2j_20}(\overline{x}_{j_1}-a_{j_2j_11})<0$,

$$(\overline{x}_{j_1},a_{j_1j_21})=\underbrace{(\text{UD,LS})}_{\text{up-down lower-saddle}}.$$

  The infinite equilibrium of $x_{j_2}^*=a_{j_1j_21}=a_{j_2j_21}$ is a (UD,LS)-up-down lower-saddle.

- For $a_{j_1j_10}\prod_{s_1=1}^{2}(\overline{x}_{j_1}-a_{j_1j_1s_1})<0$ and $a_{j_2j_20}(\overline{x}_{j_1}-a_{j_2j_11})<0$,

$$(\overline{x}_{j_1}, a_{j_1j_2 1}) = \underbrace{\text{(DU,LS)}}_{\text{down-up lower-saddle}}.$$

The infinite equilibrium of $x^*_{j_2} = a_{j_1j_2 1} = a_{j_2j_2 1}$ is a (DU,LS)-down-up lower-saddle.

(viii$_{2b}$) From case (viii$_{2a}$), the equilibrium of $(x^*_{j_1}, x^*_{j_2}) = (a_{j_1j_1s_1}, a_{j_2j_2 1})$ with $a_{j_1j_2 1} = a_{j_2j_2 1}$ has the following properties as in Eqs. (1.214)–(1.217):

- For $a_{j_1j_1 0}(a_{j_1j_1s_1} - a_{j_1j_1s_2}) > 0$ and $a_{j_2j_2 0}(a_{j_1j_1s_1} - a_{j_2j_1 1}) > 0$,

$$(a_{j_1j_1s_1}, a_{j_2j_2 1}) = \underbrace{\text{(UD:DU,US)}}_{\text{up-down-to-down-up upper-saddle}}.$$

  The equilibrium of $(x^*_{j_1}, x^*_{j_2}) = (a_{j_1j_1s_1}, a_{j_2j_2 1})$ is a (UD:DU,US)-up-down-to-down-up upper-saddle.
- For $a_{j_1j_1 0}(a_{j_1j_1s_1} - a_{j_1j_1s_2}) < 0$ and $a_{j_2j_2 0}(a_{j_1j_1s_1} - a_{j_2j_1 1}) > 0$,

$$(a_{j_1j_1s_1}, a_{j_2j_2 1}) = \underbrace{\text{(DU:UD,US)}}_{\text{down-up-to-up-down upper-saddle}}.$$

  The equilibrium of $(x^*_{j_1}, x^*_{j_2}) = (a_{j_1j_1s_1}, a_{j_2j_2 1})$ is a (DU:UD,US)-down-up-to-up-down upper-saddle.
- For $a_{j_1j_1 0}(a_{j_1j_1s_1} - a_{j_1j_1s_2}) > 0$ and $a_{j_2j_2 0}(a_{j_1j_1s_1} - a_{j_2j_1 1}) < 0$,

$$(a_{j_1j_1s_1}, a_{j_2j_2 1}) = \underbrace{\text{(DU:UD,LS)}}_{\text{down-up-to-up-down lower-saddle}}.$$

  The equilibrium of $(x^*_{j_1}, x^*_{j_2}) = (a_{j_1j_1s_1}, a_{j_2j_2 1})$ is a (DU:UD,LS)-down-up-to-up-down lower-saddle.
- For $a_{j_1j_1 0}(a_{j_1j_1s_1} - a_{j_1j_1s_2}) < 0$ and $a_{j_2j_2 0}(a_{j_1j_1s_1} - a_{j_2j_1 1}) < 0$,

$$(a_{j_1j_1s_1}, a_{j_2j_2 1}) = \underbrace{\text{(UD:DU,LS)}}_{\text{up-down-to-down-up lower-saddle}}.$$

  The equilibrium of $(x^*_{j_1}, x^*_{j_2}) = (a_{j_1j_1s_1}, a_{j_2j_2 1})$ is a (UD:DU,LS)-up-down-to-down-up lower-saddle.

(viii$_{2a}$) with $\overline{x}_{j_1} = a_{j_2j_1 1}$ and the case (II) of (viii$_{1a}$) with $\overline{x}_{j_2} = a_{j_2j_2 1}$, the equilibrium of $(x^*_{j_1}, x^*_{j_2}) = (a_{j_2j_1 1}, a_{j_2j_2 1})$ with $a_{j_2j_2 1} = a_{j_1j_2 1}$ and $a_{j_1j_1s_1} \neq a_{j_2j_1 1}$ has the following properties as in Eqs. (1.218)–(1.221):

- For $a_{j_1j_1 0}(a_{j_1j_1s_1} - a_{j_1j_1s_2}) > 0$ and $a_{j_2j_2 0}(a_{j_1j_1s_1} - a_{j_2j_1 1}) > 0$,

$$(a_{j_2j_11}, a_{j_2j_21}) = \underbrace{({}_{\text{UD}}\text{LS:}{}_{\text{DU}}\text{US,DP}_{-}\text{:UP}_{+})}_{\text{hyperbolic lower-to-upper saddle}}.$$

The equilibrium of $(x_{j_1}^*, x_{j_2}^*) = (a_{j_2j_11}, a_{j_2j_21})$ is a (${}_{\text{UD}}$LS:${}_{\text{DU}}$US,DP$_-$:DP$_+$)-hyperbolic-lower-to-upper saddle.

- For $a_{j_1j_10}(a_{j_1j_1s_1} - a_{j_1j_1s_2}) < 0$ and $a_{j_2j_20}(a_{j_1j_1s_1} - a_{j_2j_11}) > 0$,

$$(a_{j_2j_11}, a_{j_2j_21}) = \underbrace{({}_{\text{DU}}\text{LS:}{}_{\text{UD}}\text{US,UP}_{+}\text{:DP}_{-})}_{\text{hyperbolic-secant lower-to-upper saddle}}.$$

The equilibrium of $(x_{j_1}^*, x_{j_2}^*) = (a_{j_2j_11}, a_{j_2j_21})$ is a (${}_{\text{DU}}$LS:${}_{\text{UD}}$US, UP$_+$:DP$_-$)-hyperbolic-secant lower-to-upper saddle.

- For $a_{j_1j_10}(a_{j_1j_1s_1} - a_{j_1j_1s_2}) > 0$ and $a_{j_2j_20}(a_{j_1j_1s_1} - a_{j_2j_11}) < 0$,

$$(a_{j_2j_11}, a_{j_2j_21}) = \underbrace{({}_{\text{DU}}\text{US:}{}_{\text{UD}}\text{LS,UP}_{-}\text{:DP}_{+})}_{\text{hyperbolic-secant upper-to-lower saddle}}.$$

The equilibrium of $(x_{j_1}^*, x_{j_2}^*) = (a_{j_2j_11}, a_{j_2j_21})$ is a (${}_{\text{DU}}$US:${}_{\text{UD}}$LS, UP$_-$:DP$_+$)-hyperbolic-secant upper-to-lower saddle.

- For $a_{j_1j_10}(a_{j_1j_1s_1} - a_{j_1j_1s_2}) < 0$ and $a_{j_2j_20}(a_{j_1j_1s_1} - a_{j_2j_11}) < 0$,

$$(a_{j_2j_11}, a_{j_2j_21}) = \underbrace{({}_{\text{UD}}\text{US:}{}_{\text{DU}}\text{LS,DP}_{+}\text{:UP}_{-})}_{\text{hyperbolic upper-to-lower saddle}}.$$

The equilibrium of $(x_{j_1}^*, x_{j_2}^*) = (a_{j_2j_11}, a_{j_2j_21})$ is a (${}_{\text{UD}}$US:${}_{\text{DU}}$LS, DP$_+$:UP$_-$)-hyperbolic upper-to-lower saddle.

(viii$_3$) For $a_{j_1j_1s_1} = a_{j_2j_11}$ and $a_{j_1j_21} \neq a_{j_2j_21}$, in phase space,

$$\begin{aligned}\frac{dx_{j_1}}{dx_{j_2}} &= \frac{a_{j_1j_10}}{a_{j_2j_20}} \frac{(x_{j_1} - a_{j_1j_1s_1})(x_{j_1} - a_{j_1j_1s_2})(x_{j_2} - a_{j_1j_21})}{(x_{j_1} - a_{j_2j_11})(x_{j_2} - a_{j_2j_21})^2} \\ &= \frac{a_{j_1j_10}}{a_{j_2j_20}} \frac{(x_{j_1} - a_{j_1j_1s_2})(x_{j_2} - a_{j_1j_21})}{(x_{j_2} - a_{j_2j_21})^2},\end{aligned}$$

and the deformation of the foregoing equation is

$$\frac{dx_{j_1}}{(x_{j_1} - a_{j_1j_1s_2})} = \frac{a_{j_1j_10}}{a_{j_2j_20}} \left[\frac{1}{(x_{j_2} - a_{j_2j_21})} + \frac{(a_{j_2j_21} - a_{j_1j_21})}{(x_{j_2} - a_{j_2j_21})^2} \right] dx_{j_2}.$$

With an initial condition of (x_{j_10}, x_{j_20}) at $t = t_0$, the integration of the above equation gives the first integral manifold for $a_{j_1j_1s_1} = a_{j_2j_11}$ and $a_{j_1j_21} \neq a_{j_2j_21}$ as

$$\ln \frac{|x_{j_1} - a_{j_1 j_1 s_2}|}{|x_{j_1 0} - a_{j_1 j_1 s_2}|} = \frac{a_{j_1 j_1 0}}{a_{j_2 j_2 0}} \Big[\ln \frac{|x_{j_2} - a_{j_2 j_2 1}|}{|x_{j_2 0} - a_{j_2 j_2 1}|} - (a_{j_2 j_2 1} - a_{j_1 j_2 1})(\frac{1}{x_{j_2} - a_{j_2 j_2 1}} - \frac{1}{x_{j_2 0} - a_{j_2 j_2 1}}) \Big].$$

(viii$_{3a}$) For $x_{j_1}^* = a_{j_1 j_1 s_1} = a_{j_2 j_1 1}$ ($s_1 = 1, 2$) with $\overline{x}_{j_2} \neq a_{j_2 j_2 1}, a_{j_1 j_2 1},$, in phase space, if

$$\frac{dx_{j_2}}{dx_{j_1}}\Big|_{x_{j_1}^* = a_{j_2 j_1 1}} = \frac{a_{j_2 j_2 0}}{a_{j_1 j_1 0}} \frac{(\overline{x}_{j_2} - a_{j_2 j_2 1})^2}{(a_{j_1 j_1 s_1} - a_{j_1 j_1 s_2})(\overline{x}_{j_2} - a_{j_1 j_2 1})} > 0,$$

there is an increasing-inflection flow, and if

$$\frac{dx_{j_2}}{dx_{j_1}}\Big|_{x_{j_1}^* = a_{j_2 j_1 1}} = \frac{a_{j_2 j_2 0}}{a_{j_1 j_1 0}} \frac{(\overline{x}_{j_2} - a_{j_2 j_2 1})^2}{(a_{j_1 j_1 s_1} - a_{j_1 j_1 s_2})(\overline{x}_{j_2} - a_{j_1 j_2 1})} < 0,$$

there is a decreasing-inflection flow. The variational equation is

$$\Delta \dot{x}_{j_1} = (a_{j_1 j_1 s_1} - a_{j_1 j_1 s_2})(\overline{x}_{j_2} - a_{j_1 j_2 1})\Delta x_{j_1}.$$

The inflection flows are source and sink for $(a_{j_1 j_1 s_1} - a_{j_1 j_1 s_2})(\overline{x}_{j_2} - a_{j_1 j_2 1}) > 0$ and $(a_{j_1 j_1 s_1} - a_{j_1 j_1 s_2})(\overline{x}_{j_2} - a_{j_1 j_2 1}) < 0$, respectively.

From the above analysis, therefore, the infinite equilibrium of $x_{j_1}^* = a_{j_1 j_1 s_1} = a_{j_2 j_1 1}$ ($s_1 = 1, 2$) with $\overline{x}_{j_2} \neq a_{j_2 j_2 1}, a_{j_1 j_2 1}$ has the following properties as in Eqs. (1.223)–(1.226):

- For $a_{j_1 j_1 0}(\overline{x}_{j_2} - a_{j_1 j_2 1})(a_{j_1 j_1 s_1} - a_{j_1 j_1 s_2}) > 0$ and $a_{j_2 j_2 0}(\overline{x}_{j_2} - a_{j_2 j_2 1})^2 > 0$,

$$(a_{j_1 j_1 s_1}, \overline{x}_{j_2}) = \underbrace{(\text{SO,II})}_{\text{increasing-inflection source}}.$$

  The infinite-equilibrium of $x_{j_1}^* = a_{j_2 j_1 1} = a_{j_1 j_1 s_1}$ is an (SO, II)-increasing-inflection source.
- For $a_{j_1 j_1 0}(\overline{x}_{j_2} - a_{j_1 j_2 1})(a_{j_1 j_1 s_1} - a_{j_1 j_1 s_2}) < 0$ and $a_{j_2 j_2 0}(\overline{x}_{j_2} - a_{j_2 j_2 1})^2 > 0$,

$$(a_{j_1 j_1 s_1}, \overline{x}_{j_2}) = \underbrace{(\text{SI,DI})}_{\text{decreasing-inflection sink}}.$$

  The infinite-equilibrium of $x_{j_1}^* = a_{j_2 j_1 1} = a_{j_1 j_1 s_1}$ is an (SI,DI)-decreasing-inflection sink.
- For $a_{j_1 j_1 0}(\overline{x}_{j_2} - a_{j_1 j_2 1})(a_{j_1 j_1 s_1} - a_{j_1 j_1 s_2}) > 0$ and $a_{j_2 j_2 0}(\overline{x}_{j_2} - a_{j_2 j_2 1})^2 < 0$,

$$(a_{j_1j_1s_1},\bar{x}_{j_2})=\underbrace{(\text{SO,DI})}_{\text{decreasing-inflection source}}.$$

The infinite-equilibrium of $x^*_{j_1}=a_{j_2j_1 1}=a_{j_1j_1s_1}$ is an (SO,DI)-decreasing-inflection source.

- For $a_{j_1j_1 0}(\bar{x}_{j_2}-a_{j_1j_2 1})(a_{j_1j_1s_1}-a_{j_1j_1s_2})<0$ and $a_{j_2j_2 0}(\bar{x}_{j_2}-a_{j_2j_2 1})^2<0$,

$$(a_{j_1j_1s_1},\bar{x}_{j_2})=\underbrace{(\text{SI,II})}_{\text{increasing-inflection sink}}.$$

The infinite-equilibrium of $x^*_{j_1}=a_{j_2j_1 1}=a_{j_1j_1s_1}$ is an (SI, II)-increasing-inflection sink.

(viii$_{3b}$) From (viii$_{3b}$) with $\bar{x}_{j_2}=a_{j_1j_2 1}\neq a_{j_2j_2 1}$, therefore, the equilibrium of $(x^*_{j_2},x^*_{j_1})=(a_{j_1j_2 1},a_{j_2j_1 1})$ for $a_{j_1j_1s_1}=a_{j_2j_1 1}$ and $a_{j_1j_2 1}\neq a_{j_2j_2 1}$ has the following properties as in Eqs. (1.227)–(1.230):

- For $a_{j_1j_1 0}(a_{j_2j_1 1}-a_{j_1j_1s_2})>0$ and $a_{j_2j_2 0}(a_{j_1j_2 1}-a_{j_2j_2 1})^2>0$,

$$(a_{j_1j_2 1},a_{j_2j_1 1})=\underbrace{({}_{\text{DI:II}}\text{UP},{}_{\text{SI:SO}}\text{US})}_{\text{inflection up-parabola upper-saddle}}.$$

The equilibrium of $(x^*_{j_2},x^*_{j_1})=(a_{j_1j_2 1},a_{j_2j_1 1})$ is a $({}_{\text{DI:II}}\text{UP},\ {}_{\text{SI:SO}}\text{US})$-inflection up-parabola upper-saddle.

- For $a_{j_1j_1 0}(a_{j_2j_1 1}-a_{j_1j_1s_2})<0$ and $a_{j_2j_2 0}(a_{j_1j_2 1}-a_{j_2j_2 1})^2>0$,

$$(a_{j_1j_2 1},a_{j_2j_1 1})=\underbrace{({}_{\text{II:DI}}\text{DP},{}_{\text{SO:SI}}\text{LS})}_{\text{inflection down-parabola lower-saddle}}.$$

The equilibrium of $(x^*_{j_2},x^*_{j_1})=(a_{j_1j_2 1},a_{j_2j_1 1})$ is an $({}_{\text{II:DI}}\text{DP},\ {}_{\text{SO:SI}}\text{LS})$-inflection down-parabola lower-saddle.

- For $a_{j_1j_1 0}(a_{j_2j_1 1}-a_{j_1j_1s_2})>0$ and $a_{j_2j_2 0}(a_{j_1j_2 1}-a_{j_2j_2 1})^2<0$,

$$(a_{j_1j_2 1},a_{j_2j_1 1})=\underbrace{({}_{\text{II:DI}}\text{DP},{}_{\text{SO:SI}}\text{US})}_{\text{inflection down-parabola upper-saddle}}.$$

The equilibrium of $(x^*_{j_2},x^*_{j_1})=(a_{j_1j_2 1},a_{j_2j_1 1})$ is an $({}_{\text{II:DI}}\text{DP},{}_{\text{SO:SI}}\text{US})$-inflection down-parabola lower-saddle.

- For $a_{j_1j_1 0}(a_{j_2j_1 1}-a_{j_1j_1s_2})<0$ and $a_{j_2j_2 0}(a_{j_1j_2 1}-a_{j_2j_2 1})^2<0$,

$$(a_{j_1j_2 1}, a_{j_2j_1 1}) = \underbrace{({}_{\mathrm{DI:II}}\mathrm{UP}, {}_{\mathrm{SO:SI}}\mathrm{LS})}_{\text{inflection up-parabola lower-saddle}}.$$

The equilibrium of $(x^*_{j_2}, x^*_{j_1}) = (a_{j_1j_2 1}, a_{j_2j_1 1})$ is a $({}_{\mathrm{DI:II}}\mathrm{UP}, {}_{\mathrm{SO:SI}}\mathrm{LS})$-inflection up-parabola lower-saddle.

(viii$_{3c}$) From (viii$_{3a}$) with $\bar{x}_{j_2} = a_{j_2j_2 1} \neq a_{j_1j_2 1}$, therefore, the equilibrium of $(x^*_{j_2}, x^*_{j_1}) = (a_{j_1j_1s_1}, a_{j_2j_2 1})$ for $a_{j_1j_1s_1} = a_{j_2j_1 1}$ and $a_{j_1j_2 1} \neq a_{j_2j_2 1}$ has the following properties as in Eqs. (1.231)–(1.234):

- For $a_{j_1j_1 0}(a_{j_2j_2 1} - a_{j_1j_2 1})(a_{j_1j_1s_1} - a_{j_1j_1s_2}) > 0$ and $a_{j_2j_2 0} > 0$,

$$(a_{j_1j_1s_1}, a_{j_2j_2 1}) = \underbrace{(\mathrm{SO,II:II})}_{\text{increasing-inflection source}}.$$

  The equilibrium of $(x^*_{j_2}, x^*_{j_1}) = (a_{j_1j_1s_1}, a_{j_2j_2 1})$ is an (SO,II:II)-increasing-inflection source of the second kind.
- For $a_{j_1j_1 0}(a_{j_2j_2 1} - a_{j_1j_2 1})(a_{j_1j_1s_1} - a_{j_1j_1s_2}) < 0$ and $a_{j_2j_2 0} > 0$,

$$(a_{j_1j_1s_1}, a_{j_2j_2 1}) = \underbrace{(\mathrm{SI,DI:DI})}_{\text{decreasing-inflection sink}}.$$

  The equilibrium of $(x^*_{j_2}, x^*_{j_1}) = (a_{j_1j_1s_1}, a_{j_2j_2 1})$ is an (SI:DI:DI)-decreasing-inflection sink of the second kind.
- For $a_{j_1j_1 0}(a_{j_2j_2 1} - a_{j_1j_2 1})(a_{j_1j_1s_1} - a_{j_1j_1s_2}) > 0$ and $a_{j_2j_2 0} < 0$,

$$(a_{j_1j_1s_1}, a_{j_2j_2 1}) = \underbrace{(\mathrm{SO,DI:DI})}_{\text{decreasing-inflection source}}.$$

  The equilibrium of $(x^*_{j_2}, x^*_{j_1}) = (a_{j_1j_1s_1}, a_{j_2j_2 1})$ is an (SO,DI:DI)-decreasing-inflection source of the second kind.
- For $a_{j_1j_1 0}(a_{j_2j_2 1} - a_{j_1j_2 1})(a_{j_1j_1s_1} - a_{j_1j_1s_2}) < 0$ and $a_{j_2j_2 0} < 0$,

$$(a_{j_1j_1s_1}, a_{j_2j_2 1}) = \underbrace{(\mathrm{SI,II:II})}_{\text{increasing-inflection source}}.$$

  The equilibrium of $(x^*_{j_2}, x^*_{j_1}) = (a_{j_1j_1s_1}, a_{j_2j_2 1})$ is an (SI, II:II)-increasing-inflection sink of the second kind.

(viii$_4$) For $a_{j_1j_1s_1} = a_{j_2j_1 1}$ and $a_{j_1j_2 1} = a_{j_2j_2 1}$, in phase space,

$$\begin{aligned}\frac{dx_{j_1}}{dx_{j_2}} &= \frac{a_{j_1j_10}}{a_{j_2j_20}}\frac{(x_{j_1}-a_{j_1j_1s_1})(x_{j_1}-a_{j_1j_1s_2})(x_{j_2}-a_{j_1j_21})}{(x_{j_1}-a_{j_2j_11})(x_{j_2}-a_{j_2j_21})^2}\\ &= \frac{a_{j_1j_10}}{a_{j_2j_20}}\frac{(x_{j_1}-a_{j_1j_1s_2})}{(x_{j_2}-a_{j_2j_21})},\end{aligned}$$

and the deformation of the foregoing equation is

$$\frac{dx_{j_1}}{(x_{j_1}-a_{j_1j_1s_2})} = \frac{a_{j_1j_10}}{a_{j_2j_20}}\frac{dx_{j_2}}{(x_{j_2}-a_{j_2j_21})}.$$

With an initial condition of (x_{j_10}, x_{j_20}) at $t = t_0$, the integration of the above equation gives the first integral manifold for $a_{j_1j_1s_1} = a_{j_2j_11}$ and $a_{j_1j_21} = a_{j_2j_21}$ as

$$\ln\frac{|x_{j_1}-a_{j_1j_1s_2}|}{|x_{j_10}-a_{j_1j_1s_2}|} = \frac{a_{j_1j_10}}{a_{j_2j_20}}\ln\frac{|x_{j_2}-a_{j_2j_21}|}{|x_{j_20}-a_{j_2j_21}|}.$$

From (viii$_{2a}$) with $\bar{x}_{j_1} = a_{j_1j_1s_1} = a_{j_2j_11}$ and (viii$_{3a}$) with $\bar{x}_{j_2} = a_{j_2j_21} = a_{j_1j_21}$, therefore, the equilibrium of $(x^*_{j_2}, x^*_{j_1}) = (a_{j_1j_1s_1}, a_{j_2j_21})$ has the following properties as in Eqs. (1.236)–(1.239):

- For $a_{j_1j_10}(a_{j_1j_1s_1} - a_{j_1j_1s_2}) > 0$ and $a_{j_2j_20} > 0$,

$$(a_{j_1j_21}, a_{j_2j_11}) = \underbrace{({}_{\mathrm{DI}}\mathrm{SI}{:}{}_{\mathrm{II}}\mathrm{SO}, {}_{\mathrm{LS:US}}\mathrm{DU})}_{\text{down-up inflection sink-to-source}}.$$

  The equilibrium of $(x^*_{j_2}, x^*_{j_1}) = (a_{j_1j_1s_1}, a_{j_2j_21})$ is a (${}_{\mathrm{DI}}$SI:${}_{\mathrm{II}}$SO, ${}_{\mathrm{LS:US}}$DU)- down-up inflection sink to source.
- For $a_{j_1j_10}(a_{j_2j_11} - a_{j_1j_1s_2}) < 0$ and $a_{j_2j_20} > 0$,

$$(a_{j_1j_21}, a_{j_2j_11}) = \underbrace{({}_{\mathrm{II}}\mathrm{SO}{:}{}_{\mathrm{DI}}\mathrm{SI}, {}_{\mathrm{LS:US}}\mathrm{UD})}_{\text{up-down inflection source-to-sink}}.$$

  The equilibrium of $(x^*_{j_2}, x^*_{j_1}) = (a_{j_1j_1s_1}, a_{j_2j_21})$ is an (${}_{\mathrm{II}}$SO:${}_{\mathrm{DI}}$SI, ${}_{\mathrm{LS:US}}$UD)-up-down inflection source to sink.
- For $a_{j_1j_10}(a_{j_2j_11} - a_{j_1j_1s_2}) > 0$ and $a_{j_2j_20} < 0$,

$$(a_{j_1j_21}, a_{j_2j_11}) = \underbrace{({}_{\mathrm{II}}\mathrm{SI}{:}{}_{\mathrm{DI}}\mathrm{SO}, {}_{\mathrm{LS:US}}\mathrm{UD})}_{\text{up-down inflection sink-to-source}}.$$

  The equilibrium of $(x^*_{j_2}, x^*_{j_1}) = (a_{j_1j_1s_1}, a_{j_2j_21})$ is a (${}_{\mathrm{II}}$SI:${}_{\mathrm{DI}}$SO, ${}_{\mathrm{US:LS}}$UD)-up-down inflection sink to source.

- For $a_{j_1j_10}(a_{j_2j_11}-a_{j_1j_1s_2})<0$ and $a_{j_2j_20}<0$,

$$(a_{j_1j_21},a_{j_2j_11})=\underbrace{({}_{\rm DI}{\rm SO}{:}_{\rm II}{\rm SI},{}_{\rm US:LS}{\rm DU})}_{\text{down-up inflection source-to-sink}}.$$

The equilibrium of $(x^*_{j_2},x^*_{j_1})=(a_{j_1j_1s_1},a_{j_2j_21})$ is a $({}_{\rm DI}{\rm SO}{:}_{\rm II}{\rm SI},\ {}_{\rm US:LS}{\rm DU})$- down-up inflection upper-to-lower saddle.

(ix) For $\Delta_{j_1j_1}>0$ and $\Delta_{j_2j_2}>0$, the standard form is

$$\begin{aligned}\dot{x}_{j_2}&=a_{j_2j_20}(x_{j_1}-a_{j_2j_11})(x_{j_2}-a_{j_2j_21})(x_{j_2}-a_{j_2j_22}),\\ \dot{x}_{j_1}&=a_{j_1j_10}(x_{j_1}-a_{j_1j_11})(x_{j_1}-a_{j_1j_12})(x_{j_2}-a_{j_1j_21})\end{aligned}$$

where

$$\begin{aligned}&b_{j_1j_11},b_{j_1j_12}=-\frac{1}{2}(B_{j_1j_1}\pm\sqrt{\Delta_{j_1j_1}}),\\ &\{a_{j_1j_11},a_{j_1j_12}\}=\mathrm{sort}\{b_{j_1j_11},b_{j_1j_12}\},a_{j_1j_11}<a_{j_1j_12};\\ &b_{j_2j_21},b_{j_2j_22}=-\frac{1}{2}(B_{j_2j_2}\pm\sqrt{\Delta_{j_2j_2}}),\\ &\{a_{j_2j_21},a_{j_2j_22}\}=\mathrm{sort}\{b_{j_2j_21},b_{j_2j_22}\},a_{j_2j_21}<a_{j_2j_22}.\end{aligned}$$

(ix_1) For $a_{j_1j_21}\neq a_{j_2j_21},a_{j_2j_22}$ $(s_1=1,\ 2)$, in phase space,

$$\frac{dx_{j_1}}{dx_{j_2}}=\frac{a_{j_1j_10}}{a_{j_2j_20}}\frac{(x_{j_1}-a_{j_1j_11})(x_{j_1}-a_{j_1j_12})(x_{j_2}-a_{j_1j_21})}{(x_{j_1}-a_{j_2j_11})(x_{j_2}-a_{j_2j_21})(x_{j_2}-a_{j_2j_22})},$$

and the deformation of the foregoing equation is

$$\begin{aligned}&\sum\nolimits_{s_1=1,s_2\neq s_1}^{2}\frac{(a_{j_1j_1s_1}-a_{j_2j_11})}{(a_{j_1j_1s_1}-a_{j_1j_1s_2})}\frac{dx_{j_1}}{(x_{j_1}-a_{j_1j_1s_1})}\\ &=\frac{a_{j_1j_10}}{a_{j_2j_20}}\sum\nolimits_{l_1=1,l_2\neq l_1}^{2}\frac{(a_{j_2j_2l_1}-a_{j_1j_21})}{(a_{j_2j_2l_1}-a_{j_2j_2l_2})}\frac{dx_{j_2}}{(x_{j_2}-a_{j_2j_2l_1})}.\end{aligned}$$

With an initial condition of (x_{j_10},x_{j_20}) at $t=t_0$, the integration of the above equation gives the first integral manifold as

$$\begin{aligned}&\sum\nolimits_{s_1=1,s_1\neq s_1}^{2}\frac{(a_{j_1j_1s_1}-a_{j_2j_11})}{(a_{j_1j_1s_1}-a_{j_1j_1s_2})}\ln\frac{|x_{j_1}-a_{j_1j_1s_1}|}{|x_{j_10}-a_{j_1j_1s_1}|}\\ &=\frac{a_{j_1j_10}}{a_{j_2j_20}}\sum\nolimits_{l_1=1,l_2\neq l_1}^{2}\frac{(a_{j_2j_2l_1}-a_{j_1j_21})}{(a_{j_2j_2l_1}-a_{j_2j_2l_2})}\ln\frac{|x_{j_2}-a_{j_2j_2l_1}|}{|x_{j_20}-a_{j_2j_2l_1}|}.\end{aligned}$$

($\mathrm{ix}_{1\mathrm{a}}$) Consider two cases (I) $x^*_{j_2}=a_{j_1j_21}$ and (II) $x^*_{j_1}=a_{j_2j_11}$.

(I) In phase space, at $x^*_{j_2} = a_{j_1j_21} \neq a_{j_1j_11}, a_{j_1j_12}$ with $\bar{x}_{j_1} \neq a_{j_1j_11}$,

$$\frac{dx_{j_1}}{dx_{j_2}}\bigg|_{x^*_{j_2}=a_{j_1j_21}} = \frac{a_{j_1j_10}}{a_{j_2j_20}} \frac{\prod_{s_1=1}^{2}(\bar{x}_{j_1} - a_{j_1j_1s_1})(x_{j_2} - a_{j_1j_21})}{(\bar{x}_{j_1} - a_{j_2j_11})\prod_{l_1=1}^{2}(x_{j_2} - a_{j_2j_2l_1})}\bigg|_{x^*_{j_2}=a_{j_1j_21}} = 0.$$

If

$$\frac{d^2x_{j_1}}{dx^2_{j_2}}\bigg|_{x^*_{j_2}=a_{j_1j_21}} = \frac{a_{j_1j_10}}{a_{j_2j_20}} \frac{\prod_{s_1=1}^{2}(\bar{x}_{j_1} - a_{j_1j_1s_1})}{(\bar{x}_{j_1} - a_{j_2j_11})\prod_{l_1=1}^{2}(a_{j_1j_21} - a_{j_2j_2l_1})} > 0,$$

there is an up-parabola flow at $x^*_{j_2} = a_{j_1j_21}$ with $\bar{x}_{j_1} \neq a_{j_1j_11}$. If

$$\frac{d^2x_{j_1}}{dx^2_{j_2}}\bigg|_{x^*_{j_2}=a_{j_1j_21}} = \frac{a_{j_1j_10}}{a_{j_2j_20}} \frac{\prod_{s_1=1}^{2}(\bar{x}_{j_1} - a_{j_1j_1s_1})}{(\bar{x}_{j_1} - a_{j_2j_11})\prod_{l_1=1}^{2}(a_{j_1j_21} - a_{j_2j_2l_1})} < 0,$$

there is a down-parabola flow at $x^*_{j_2} = a_{j_1j_21}$ with $\bar{x}_{j_1} \neq a_{j_1j_11}$. Let

$$\dot{x}_{j_2} = a_{j_2j_20}(\bar{x}_{j_1} - a_{j_2j_11})\prod_{l_1=1}^{2}(a_{j_1j_21} - a_{j_2j_2l_1}).$$

The parabola flows at $x^*_{j_2} = a_{j_1j_21}$ with $\bar{x}_{j_1} \neq a_{j_1j_11}$ are positive and negative for $\dot{x}_{j_2} > 0$ and $\dot{x}_{j_2} < 0$, respectively. Thus, the equilibrium of $x^*_{j_2} = a_{j_1j_21}$ with $\bar{x}_{j_1} \neq a_{j_1j_11}$ has the following properties:

- For $a_{j_1j_10}\prod_{s_1=1}^{2}(\bar{x}_{j_1} - a_{j_1j_1s_1}) > 0$ and $a_{j_2j_20}(\bar{x}_{j_1} - a_{j_2j_11})\prod_{l_1=1}^{2}(a_{j_1j_21} - a_{j_2j_2l_1}) > 0$,

$$(a_{j_1j_21}, \dot{x}_{j_2}) = \underbrace{(\text{UP,pF})}_{\text{up-parabola flow }(+)}.$$

- For $a_{j_1j_10}\prod_{s_1=1}^{2}(\bar{x}_{j_1} - a_{j_1j_1s_1}) < 0$ and $a_{j_2j_20}(\bar{x}_{j_1} - a_{j_2j_11})\prod_{l_1=1}^{2}(a_{j_1j_21} - a_{j_2j_2l_1}) > 0$,

$$(a_{j_1j_21}, \dot{x}_{j_2}) = \underbrace{(\text{DP,pF})}_{\text{down-parabola flow }(+)}.$$

- For $a_{j_1j_10}\prod_{s_1=1}^{2}(\bar{x}_{j_1} - a_{j_1j_1s_1}) > 0$ and $a_{j_2j_20}(\bar{x}_{j_1} - a_{j_2j_11})\prod_{l_1=1}^{2}(a_{j_1j_21} - a_{j_2j_2l_1}) < 0$,

$$(a_{j_1j_21}, \dot{x}_{j_2}) = \underbrace{(\text{DP,nF})}_{\text{dwon-parabola flow }(-)}.$$

- For $a_{j_1j_10}\prod_{s_1=1}^{2}(\bar{x}_{j_1} - a_{j_1j_1s_1}) < 0$ and $a_{j_2j_20}(\bar{x}_{j_1} - a_{j_2j_11})\prod_{l_1=1}^{2}(a_{j_1j_21} - a_{j_2j_2l_1}) < 0$,

$$(a_{j_1j_2 1}, \dot{x}_{j_2}) = \underbrace{\text{(UP,nF)}}_{\text{up-parabola flow }(-)}.$$

(II) Similarly, at $x^*_{j_1} = a_{j_2j_1 1} \neq a_{j_2j_2 1}, a_{j_2j_2 2}$ with $\overline{x}_{j_2} \neq a_{j_1j_2 1}$, in phase space,

$$\frac{dx_{j_2}}{dx_{j_1}}\Big|_{x^*_{j_1}=a_{j_2j_1 1}} = \frac{a_{j_2j_2 0}}{a_{j_1j_1 0}} \frac{(x_{j_1} - a_{j_2j_1 1})\prod_{l_1=1}^{2}(\overline{x}_{j_2} - a_{j_2j_2 l_1})}{(\overline{x}_{j_2} - a_{j_1j_2 1})\prod_{s_1=1}^{2}(x_{j_1} - a_{j_1j_1 s_1})}\Big|_{x^*_{j_1}=a_{j_2j_1 1}} = 0.$$

If

$$\frac{d^2x_{j_2}}{dx^2_{j_1}}\Big|_{x^*_{j_1}=a_{j_2j_1 1}} = \frac{a_{j_2j_2 0}}{a_{j_1j_1 0}} \frac{\prod_{l_1=1}^{2}(\overline{x}_{j_2} - a_{j_2j_2 l_1})}{(\overline{x}_{j_2} - a_{j_1j_2 1})\prod_{s_1=1}^{2}(a_{j_2j_1 1} - a_{j_1j_1 s_1})} > 0,$$

there is an up-parabola flow. If

$$\frac{d^2x_{j_2}}{dx^2_{j_1}}\Big|_{x^*_{j_1}=a_{j_2j_1 1}} = \frac{a_{j_2j_2 0}}{a_{j_1j_1 0}} \frac{\prod_{l_1=1}^{2}(\overline{x}_{j_2} - a_{j_2j_2 l_1})}{(\overline{x}_{j_2} - a_{j_1j_2 1})\prod_{s_1=1}^{2}(a_{j_2j_1 1} - a_{j_1j_1 s_1})} < 0,$$

there is a down-parabola flow. For

$$\dot{x}_{j_1} = a_{j_1j_1 0}(\overline{x}_{j_2} - a_{j_1j_2 1})\prod_{s_1=1}^{2}(a_{j_2j_1 1} - a_{j_1j_1 s_1}),$$

the parabola flows at $x^*_{j_1} = a_{j_1j_1 1}$ are positive and negative for $\dot{x}_{j_1} > 0$ and $\dot{x}_{j_1} < 0$, respectively. Thus, the equilibrium of $x^*_{j_1} = a_{j_2j_1 1}$ with $\overline{x}_{j_2} = a_{j_1j_2 1}$ has the following properties:

- For $a_{j_1j_1 0}(\overline{x}_{j_2} - a_{j_1j_2 1})\prod_{s_1=1}^{2}(a_{j_2j_1 1} - a_{j_1j_1 s_1}) > 0$ and $a_{j_2j_2 0}\prod_{l_1=1}^{2}(\overline{x}_{j_2} - a_{j_2j_2 l_1}) > 0$,

$$(\dot{x}_{j_1}, a_{j_2j_1 1}) = \underbrace{\text{(pF,UP)}}_{\text{up-parabola flow }(+)}.$$

- For $a_{j_1j_1 0}(\overline{x}_{j_2} - a_{j_1j_2 1})\prod_{s_1=1}^{2}(a_{j_2j_1 1} - a_{j_1j_1 s_1}) < 0$ and $a_{j_2j_2 0}\prod_{l_1=1}^{2}(\overline{x}_{j_2} - a_{j_2j_2 l_1}) > 0$,

$$(\dot{x}_{j_1}, a_{j_2j_1 1}) = \underbrace{\text{(nF,DP)}}_{\text{down-parabola flow }(-)}.$$

- For $a_{j_1j_1 0}(\overline{x}_{j_2} - a_{j_1j_2 1})\prod_{s_1=1}^{2}(a_{j_2j_1 1} - a_{j_1j_1 s_1}) > 0$ and $a_{j_2j_2 0}\prod_{l_1=1}^{2}(\overline{x}_{j_2} - a_{j_2j_2 l_1}) < 0$,

$$(\dot{x}_{j_1}, a_{j_2j_1 1}) = \underbrace{(\text{pF,DP})}_{\text{down-parabola flow }(+)}.$$

- For $a_{j_1j_1 0}(\bar{x}_{j_2} - a_{j_1j_2 1})\prod_{s_1=1}^{2}(a_{j_2j_1 1} - a_{j_1j_1 s_1}) < 0$ and $a_{j_2j_2 0}\prod_{l_1=1}^{2}(\bar{x}_{j_2} - a_{j_2j_2 l_1}) < 0$,

$$(\dot{x}_{j_1}, a_{j_2j_1 1}) = \underbrace{(\text{nF,UP})}_{\text{up-parabola flow }(-)}.$$

Therefore, from the analysis, the equilibrium of $(x_{j_2}^*, x_{j_1}^*) = (a_{j_1j_2 1}, a_{j_2j_1 1})$ has the following properties as in Eqs. (1.243)–(1.246):

- For $a_{j_1j_1 0}\prod_{s_1=1}^{2}(a_{j_2j_1 1} - a_{j_1j_1 s_1}) > 0$ and $a_{j_2j_2 0}\prod_{l_1=1}^{2}(\bar{x}_{j_2} - a_{j_2j_2 l_1}) > 0$,

$$(a_{j_1j_2 1}, a_{j_2j_1 1}) = \underbrace{(\text{UP}_+,\text{UP}_+)}_{\text{positive saddle}}.$$

  The equilibrium of $(x_{j_1}^*, x_{j_2}^*) = (a_{j_1j_2 1}, a_{j_2j_1 1})$ is a (UP_+, UP_+)-positive saddle.

- For $a_{j_1j_1 0}\prod_{s_1=1}^{2}(a_{j_2j_1 1} - a_{j_1j_1 s_1}) < 0$ and $a_{j_2j_2 0}\prod_{l_1=1}^{2}(\bar{x}_{j_2} - a_{j_2j_2 l_1}) > 0$,

$$(a_{j_1j_2 1}, a_{j_2j_1 1}) = \underbrace{(\text{DP}_+,\text{DP}_-)}_{\text{CCW center}}.$$

  The equilibrium of $(x_{j_2}^*, x_{j_1}^*) = (a_{j_1j_2 1}, a_{j_2j_1 1})$ is a (DP_+,DP_-)-counterclockwise center.

- For $a_{j_1j_1 0}\prod_{s_1=1}^{2}(a_{j_2j_1 1} - a_{j_1j_1 s_1}) > 0$ and $a_{j_2j_2 0}\prod_{l_1=1}^{2}(\bar{x}_{j_2} - a_{j_2j_2 l_1}) < 0$,

$$(a_{j_1j_2 1}, a_{j_2j_1 1}) = \underbrace{(\text{DP}_-,\text{DP}_+)}_{\text{CW center}}.$$

  The equilibrium of $(x_{j_2}^*, x_{j_1}^*) = (a_{j_1j_2 1}, a_{j_2j_1 1})$ is a (DP_-, DP_+)-clockwise center.

- For $a_{j_1j_1 0}\prod_{s_1=1}^{2}(a_{j_2j_1 1} - a_{j_1j_1 s_1}) < 0$ and $a_{j_2j_2 0}\prod_{l_1=1}^{2}(\bar{x}_{j_2} - a_{j_2j_2 l_1}) < 0$,

$$(a_{j_1j_2 1}, a_{j_2j_1 1}) = \underbrace{(\text{UP}_-,\text{UP}_-)}_{\text{negative saddle}}.$$

  The equilibrium of $(x_{j_2}^*, x_{j_1}^*) = (a_{j_1j_2 1}, a_{j_2j_1 1})$ is a (UP_-,UP_-)-negative saddle.

(ix_{1b}) At $(x_{j_1}^*, x_{j_2}^*) = (a_{j_1j_1 s_1}, a_{j_1j_2 l_1})$ (s_1, s_2, l_1, l_2 $\in \{1, 2\}$, $s_1 \neq s_2$, $l_1 \neq l_2$), the variational equations are

$$\Delta\dot{x}_{j_2} = a_{j_2j_20}(a_{j_1j_1s_1} - a_{j_2j_11})(a_{j_2j_2l_1} - a_{j_2j_2l_2})\Delta x_{j_2},$$
$$\Delta\dot{x}_{j_1} = a_{j_1j_10}(a_{j_1j_1s_1} - a_{j_1j_1s_2})(a_{j_2j_2l_1} - a_{j_1j_21})\Delta x_{j_1}.$$

The flows in the x_{j_1}-direction for $a_{j_1j_10}(a_{j_1j_1s_1} - a_{j_1j_1s_2})(a_{j_2j_2l_1} - a_{j_1j_21}) > 0$ and $a_{j_1j_10}(a_{j_1j_1s_1} - a_{j_1j_1s_2})(a_{j_2j_2l_1} - a_{j_1j_21}) < 0$ are source and sink, respectively. The flows in the x_{j_2}-direction are source and sink for $a_{j_2j_20}(a_{j_1j_1s_1} - a_{j_2j_11})(a_{j_2j_2l_1} - a_{j_2j_2l_2}) > 0$ and $a_{j_2j_20}(a_{j_1j_1s_1} - a_{j_2j_11})(a_{j_2j_2l_1} - a_{j_2j_2l_2}) < 0$, respectively.

Therefore, the equilibrium of $(x^*_{j_1}, x^*_{j_2}) = (a_{j_1j_1s_1}, a_{j_2j_2l_1})$ (s_1, s_2, l_1, l_2 $\in \{1, 2\}$, $s_1 \neq s_2$, $l_1 \neq l_2$) has the following properties as in Eqs. (1.247)–(1.250):

- For $a_{j_1j_10}(a_{j_2j_2l_1} - a_{j_1j_21})(a_{j_1j_1s_1} - a_{j_1j_1s_2}) > 0$ and $a_{j_2j_20}(a_{j_1j_1s_1} - a_{j_2j_11})(a_{j_2j_2l_1} - a_{j_2j_2l_2}) > 0$,
$$(a_{j_1j_1s_1}, a_{j_2j_2l_1}) = \underbrace{(\text{SO,SO})}_{\text{source}}.$$
The equilibrium of $(x^*_{j_1}, x^*_{j_2}) = (a_{j_1j_1s_1}, a_{j_2j_2l_1})$ is an (SO,SO) source.
- For $a_{j_1j_10}(a_{j_2j_2l_1} - a_{j_1j_21})(a_{j_1j_1s_1} - a_{j_1j_1s_2}) < 0$ and $a_{j_2j_20}(a_{j_1j_1s_1} - a_{j_2j_11})(a_{j_2j_2l_1} - a_{j_2j_2l_2}) > 0$,
$$(a_{j_1j_11}, a_{j_2j_2s_1}) = \underbrace{(\text{SI,SO})}_{\text{saddle}}.$$
The equilibrium of $(x^*_{j_1}, x^*_{j_2}) = (a_{j_1j_1s_1}, a_{j_2j_2l_1})$ is an (SI,SO) saddle.
- For $a_{j_1j_10}(a_{j_2j_2l_1} - a_{j_1j_21})(a_{j_1j_1s_1} - a_{j_1j_1s_2}) > 0$ and $a_{j_2j_20}(a_{j_1j_1s_1} - a_{j_2j_11})(a_{j_2j_2l_1} - a_{j_2j_2l_2}) < 0$,
$$(a_{j_1j_11}, a_{j_2j_2s_1}) = \underbrace{(\text{SO,SI})}_{\text{saddle}}.$$
The equilibrium of $(x^*_{j_1}, x^*_{j_2}) = (a_{j_1j_1s_1}, a_{j_2j_2l_1})$ is an (SO,SI) saddle.
- For $a_{j_1j_10}(a_{j_2j_2l_1} - a_{j_1j_21})(a_{j_1j_1s_1} - a_{j_1j_1s_2}) < 0$ and $a_{j_2j_20}(a_{j_1j_1s_1} - a_{j_2j_11})(a_{j_2j_2l_1} - a_{j_2j_2l_2}) < 0$,
$$(a_{j_1j_11}, a_{j_2j_2s_1}) = \underbrace{(\text{SI,SI})}_{\text{sink}}.$$
The equilibrium of $(x^*_{j_1}, x^*_{j_2}) = (a_{j_1j_1s_1}, a_{j_2j_2l_1})$ is an (SI,SI) sink.

($\mathrm{ix}_{1\mathrm{c}}$) From the case (I) of ($\mathrm{ix}_{1\mathrm{a}}$) with $\bar{x}_{j_1} = a_{j_1 j_1 s_1}$ ($s_1 = 1, 2$), the equilibrium of $(x^*_{j_1}, x^*_{j_2}) = (a_{j_1 j_1 s_1}, a_{j_1 j_2 1})$ ($s_1, s_2 \in \{1, 2\}, s_1 \neq s_2$) with $a_{j_1 j_2 1} \neq a_{j_2 j_2 1}, a_{j_2 j_2 2}$ has the following properties as in Eqs. (1.251)–(1.254):

- For $a_{j_1 j_1 0}(a_{j_1 j_1 s_1} - a_{j_1 j_1 s_2}) > 0$ and $a_{j_2 j_2 0}(a_{j_1 j_1 s_1} - a_{j_2 j_1 1}) \prod_{l_1=1}^{2} (a_{j_1 j_2 1} - a_{j_2 j_2 l_1}) > 0$,

$$(a_{j_1 j_1 s_1}, a_{j_1 j_2 1}) = \underbrace{(\text{DP:UP,pF})}_{\text{hyperbolic flow }(+)}.$$

  The equilibrium of $(x^*_{j_1}, x^*_{j_2}) = (a_{j_1 j_1 s_1}, a_{j_1 j_2 1})$ is a (DP:UP, pF)-positive hyperbolic flow.

- For $a_{j_1 j_1 0}(a_{j_1 j_1 s_1} - a_{j_1 j_1 s_2}) < 0$ and $a_{j_2 j_2 0}(a_{j_1 j_1 s_1} - a_{j_2 j_1 1}) \prod_{l_1=1}^{2} (a_{j_1 j_2 1} - a_{j_2 j_2 l_1}) > 0$,

$$(a_{j_1 j_1 s_1}, a_{j_1 j_2 1}) = \underbrace{(\text{UP:DP,pF})}_{\text{hyperbolic-secant flow }(+)}.$$

  The equilibrium of $(x^*_{j_1}, x^*_{j_2}) = (a_{j_1 j_1 s_1}, a_{j_1 j_2 1})$ is a (UP:DP, pF)-positive hyperbolic-secant flow.

- For $a_{j_1 j_1 0}(a_{j_1 j_1 s_1} - a_{j_1 j_1 s_2}) > 0$ and $a_{j_2 j_2 0}(a_{j_1 j_1 s_1} - a_{j_2 j_1 1}) \prod_{l_1=1}^{2} (a_{j_1 j_2 1} - a_{j_2 j_2 l_1}) < 0$,

$$(a_{j_1 j_1 s_1}, a_{j_1 j_2 1}) = \underbrace{(\text{UP:DP,nF})}_{\text{hyperbolic-secant flow }(-)}.$$

  The equilibrium of $(x^*_{j_1}, x^*_{j_2}) = (a_{j_1 j_1 s_1}, a_{j_1 j_2 1})$ is a (UP:DP, nF)-negative hyperbolic-secant flow.

- For $a_{j_1 j_1 0}(a_{j_1 j_1 s_1} - a_{j_1 j_1 s_2}) < 0$ and $a_{j_2 j_2 0}(a_{j_1 j_1 s_1} - a_{j_2 j_1 1}) \prod_{l_1=1}^{2} (a_{j_1 j_2 1} - a_{j_2 j_2 l_1}) < 0$,

$$(a_{j_1 j_1 s_1}, a_{j_1 j_2 1}) = \underbrace{(\text{DP:UP,nF})}_{\text{hyperbolic flow }(-)}.$$

  The equilibrium of $(x^*_{j_1}, x^*_{j_2}) = (a_{j_1 j_1 s_1}, a_{j_1 j_2 1})$ is a (DP:UP, nF)-negative hyperbolic flow.

($\mathrm{ix}_{1\mathrm{d}}$) From the case (II) of ($\mathrm{ix}_{1\mathrm{a}}$) with $\bar{x}_{j_2} = a_{j_2 j_2 l_1}$, the equilibrium of $(x^*_{j_1}, x^*_{j_2}) = (a_{j_2 j_1 1}, a_{j_2 j_2 l_1})$ ($l_1, l_2 \in \{1, 2\}, l_1 \neq l_2$) with $a_{j_2 j_1 1} \neq a_{j_1 j_1 1}, a_{j_1 j_1 2}$ has the following properties as in Eqs. (1.255)–(1.258):

- For $a_{j_1 j_1 0}(a_{j_2 j_2 l_1} - a_{j_1 j_2 1}) \prod_{s_1=1}^{2} (a_{j_2 j_1 1} - a_{j_1 j_1 s_1}) > 0$ and $a_{j_2 j_2 0}(a_{j_1 j_2 l_1} - a_{j_2 j_2 l_2}) > 0$,

$$(a_{j_2j_1 1}, a_{j_2j_2l_1}) = \underbrace{(\text{pF,DP:UP})}_{\text{hyperbolic flow }(+)}.$$

The equilibrium of $(x_{j_1}^*, x_{j_2}^*) = (a_{j_2j_1 1}, a_{j_2j_2l_1})$is a (pF, DP:UP)-positive hyperbolic flow.

- For $a_{j_1j_1 0}(a_{j_2j_2l_1} - a_{j_1j_2 1})\prod_{s_1=1}^{2}(a_{j_2j_1 1} - a_{j_1j_1s_1}) < 0$ and $a_{j_2j_2 0}(a_{j_1j_2l_1} - a_{j_2j_2l_2}) > 0$,

$$(a_{j_1j_1s_1}, a_{j_1j_2 1}) = \underbrace{(\text{nF,UP:DP})}_{\text{hyperbolic-secant flow }(-)}.$$

The equilibrium of $(x_{j_1}^*, x_{j_2}^*) = (a_{j_1j_1s_1}, a_{j_1j_2 1})$ is an (nF, UP:DP)-negative hyperbolic-secant flow.

- For $a_{j_1j_1 0}(a_{j_2j_2l_1} - a_{j_1j_2 1})\prod_{s_1=1}^{2}(a_{j_2j_1 1} - a_{j_1j_1s_1}) > 0$ and $a_{j_2j_2 0}(a_{j_1j_2l_1} - a_{j_2j_2l_2}) < 0$,

$$(a_{j_1j_1s_1}, a_{j_1j_2 1}) = \underbrace{(\text{pF,UP:DP})}_{\text{hyperbolic-secant flow }(+)}.$$

The equilibrium of $(x_{j_1}^*, x_{j_2}^*) = (a_{j_1j_1s_1}, a_{j_1j_2 1})$ is a (pF, UP:DP)-positive hyperbolic-secant flow.

- For $a_{j_1j_1 0}(a_{j_2j_2l_1} - a_{j_1j_2 1})\prod_{s_1=1}^{2}(a_{j_2j_1 1} - a_{j_1j_1s_1}) < 0$ and $a_{j_2j_2 0}(a_{j_1j_2l_1} - a_{j_2j_2l_2}) < 0$,

$$(a_{j_1j_1s_1}, a_{j_1j_2 1}) = \underbrace{(\text{nF,DP:UP})}_{\text{hyperbolic flow }(-)}.$$

The equilibrium of $(x_{j_1}^*, x_{j_2}^*) = (a_{j_1j_1s_1}, a_{j_1j_2 1})$ is an (nF, DP:UP)-negative hyperbolic flow.

(ix$_2$) For $a_{j_1j_2 1} = a_{j_2j_2l_1}$ and $a_{j_2j_1 1} \neq a_{j_1j_1s_1}$ $(s_1, s_2, l_1, l_2 \in \{1,2\}, s_1 \neq s_2, l_1 \neq l_2)$, in phase space,

$$\begin{aligned}\frac{dx_{j_1}}{dx_{j_2}} &= \frac{a_{j_1j_1 0}}{a_{j_2j_2 0}} \frac{(x_{j_1} - a_{j_1j_1s_1})(x_{j_1} - a_{j_1j_1s_2})(x_{j_2} - a_{j_1j_2 1})}{(x_{j_1} - a_{j_2j_1 1})(x_{j_2} - a_{j_2j_2l_1})(x_{j_2} - a_{j_2j_2l_2})} \\ &= \frac{a_{j_1j_1 0}}{a_{j_2j_2 0}} \frac{(x_{j_1} - a_{j_1j_1s_1})(x_{j_1} - a_{j_1j_1s_2})}{(x_{j_1} - a_{j_2j_1 1})(x_{j_2} - a_{j_2j_2l_2})},\end{aligned}$$

and the deformation of the foregoing equation is

$$\frac{(x_{j_1}-a_{j_2j_1 1})dx_{j_1}}{(x_{j_1}-a_{j_1j_1s_1})(x_{j_1}-a_{j_1j_1s_2})}=\frac{a_{j_1j_1 0}}{a_{j_2j_2 0}}\frac{dx_{j_2}}{(x_{j_2}-a_{j_2j_2l_2})}.$$

With an initial condition of $(x_{j_1 0},x_{j_2 0})$ at $t=t_0$, the integration of the above equation gives the first integral manifold as

$$\sum\nolimits_{s_1=1,s_2\neq s_1}^{2}\frac{a_{j_1j_1s_1}-a_{j_2j_1 1}}{a_{j_1j_1s_1}-a_{j_1j_1 1}}\ln\frac{|x_{j_1}-a_{j_1j_1s_1}|}{|x_{j_1 0}-a_{j_1j_1s_1}|}=\frac{a_{j_1j_1 0}}{a_{j_2j_2 0}}\ln\frac{|x_{j_2}-a_{j_2j_2l_2}|}{|x_{j_2}-a_{j_2j_2l_2}|}.$$

($\text{ix}_{2\text{a}}$) For $x_{j_2}^*=a_{j_1j_2 1}=a_{j_2j_2l_1}$ with $\bar{x}_{j_1}\neq a_{j_1j_1s_1},a_{j_1j_1s_2}$, if

$$\left.\frac{dx_{j_1}}{dx_{j_2}}\right|_{x_{j_2}^*=a_{j_2j_2l_1}}=\frac{a_{j_1j_1 0}}{a_{j_2j_2 0}}\frac{\prod_{s_1=1}^{2}(\bar{x}_{j_1}-a_{j_1j_1s_1})}{(\bar{x}_{j_1}-a_{j_2j_1 1})(a_{j_2j_2l_1}-a_{j_2j_2l_2})}>0,$$

there is an increasing inflection flow, and if

$$\left.\frac{dx_{j_1}}{dx_{j_2}}\right|_{x_{j_2}^*=a_{j_2j_2l_1}}=\frac{a_{j_1j_1 0}}{a_{j_2j_2 0}}\frac{\prod_{s_1=1}^{2}(\bar{x}_{j_1}-a_{j_1j_1s_1})}{(\bar{x}_{j_1}-a_{j_2j_1 1})(a_{j_2j_2l_1}-a_{j_2j_2l_2})}<0,$$

there is a decreasing inflection flow. The variational equation at $x_{j_2}^*=a_{j_1j_2s_1}=a_{j_2j_2 1}$ is

$$\Delta\dot{x}_{j_2}=a_{j_2j_2 0}(\bar{x}_{j_1}-a_{j_2j_1 1})(a_{j_2j_2l_1}-a_{j_2j_2l_2})\Delta x_{j_2}.$$

The asymptotic flows are source and sink for $a_{j_2j_2 0}(\bar{x}_{j_1}-a_{j_2j_1 1})\ (a_{j_2j_2l_1}-a_{j_2j_2l_2})>0$ and $a_{j_2j_2 0}(\bar{x}_{j_1}-a_{j_2j_1 1})(a_{j_2j_2l_1}-a_{j_2j_2l_2})<0$in the x_{j_2} -direction, respectively.

Therefore, the infinite equilibrium of $x_{j_2}^*=a_{j_1j_2 1}=a_{j_2j_2l_1}$ with $\bar{x}_{j_1}\neq a_{j_1j_1s_1},a_{j_1j_1s_2}$ has the following properties as in Eqs. (1.260)–(1.263):

- For $a_{j_1j_1 0}\prod_{s_1=1}^{2}(\bar{x}_{j_1}-a_{j_1j_1s_1})>0$ and $a_{j_2j_2 0}(\bar{x}_{j_1}-a_{j_2j_1 1})(a_{j_2j_2l_1}-a_{j_2j_2l_2})>0$,

$$(\bar{x}_{j_1},a_{j_2j_2l_1})=\underbrace{(\text{II,SO})}_{\text{increasing-inflection source}}.$$

  The infinite-equilibrium of $x_{j_2}^*=a_{j_1j_2 1}=a_{j_2j_2l_1}$ is an (II,SO)-increasing-inflection source.

- For $a_{j_1j_1 0}\prod_{s_1=1}^{2}(\bar{x}_{j_1}-a_{j_1j_1s_1})<0$ and $a_{j_2j_2 0}(\bar{x}_{j_1}-a_{j_2j_1 1})(a_{j_2j_2l_1}-a_{j_2j_2l_2})>0$,

$$(\bar{x}_{j_1},a_{j_2j_2l_1})=\underbrace{(\text{DI,SO})}_{\text{decreasing-inflection source}}.$$

  The infinite-equilibrium of $x_{j_2}^*=a_{j_1j_2 1}=a_{j_2j_2l_1}$ is a (DI,SO)-decreasing-inflection source.

- For $a_{j_1j_1 0}\prod_{s_1=1}^{2}(\bar{x}_{j_1}-a_{j_1j_1s_1})>0$ and $a_{j_2j_2 0}(\bar{x}_{j_1}-a_{j_2j_1 1})(a_{j_2j_2l_1}-a_{j_2j_2l_2})<0$,

$$(\overline{x}_{j_1}, a_{j_2j_2l_1}) = \underbrace{\text{(DI,SI)}}_{\text{decreasing-inflection sink}}.$$

The infinite-equilibrium of $x^*_{j_2} = a_{j_1j_2 1} = a_{j_2j_2l_1}$ is a (DI,SI)-decreasing-inflection sink.

- For $a_{j_1j_1 0}\prod_{s_1=1}^{2}(\overline{x}_{j_1} - a_{j_1j_1s_1}) < 0$ and $a_{j_2j_2 0}(\overline{x}_{j_1} - a_{j_2j_1 1})(a_{j_2j_2l_1} - a_{j_2j_2l_2}) < 0$,

$$(\overline{x}_{j_1}, a_{j_2j_2l_1}) = \underbrace{\text{(II,SI)}}_{\text{increasing-inflection sink}}.$$

The infinite-equilibrium of $x^*_{j_2} = a_{j_1j_2 1} = a_{j_2j_2l_1}$ is an (II,SI)-increasing-inflection sink.

(ix$_{2b}$) From case (ix$_{2a}$), let $\overline{x}_{j_1} \neq a_{j_1j_1s_1}$.. Therefore, the equilibrium of $(x^*_{j_1}, x^*_{j_2}) = (a_{j_1j_1s_1}, a_{j_2j_2l_1})$ with $a_{j_2j_2s_1} = a_{j_1j_2l_1}$ has the following properties as in Eqs. (1.264)–(1.267):

- For $a_{j_1j_1 0}(a_{j_1j_1s_1} - a_{j_1j_1s_2}) > 0$ and $a_{j_2j_2 0}(a_{j_1j_1s_1} - a_{j_2j_1 1})(a_{j_2j_2l_1} - a_{j_2j_2l_2}) > 0$,

$$(a_{j_1j_1s_1}, a_{j_2j_2l_1}) = \underbrace{\text{(DI:II,SO)}}_{\text{inflection source}}.$$

The equilibrium of $(x^*_{j_1}, x^*_{j_2}) = (a_{j_1j_1s_1}, a_{j_2j_2l_1})$ is a (DI:II,SO)-inflection source.

- For $a_{j_1j_1 0}(a_{j_1j_1s_1} - a_{j_1j_1s_2}) < 0$ and $a_{j_2j_2 0}(a_{j_1j_1s_1} - a_{j_2j_1 1})(a_{j_2j_2l_1} - a_{j_2j_2l_2}) > 0$,

$$(a_{j_1j_1s_1}, a_{j_2j_2l_1}) = \underbrace{\text{(II:DI,SO)}}_{\text{inflection source}}.$$

The equilibrium of $(x^*_{j_1}, x^*_{j_2}) = (a_{j_1j_1s_1}, a_{j_2j_2l_1})$ is an (II:DI,SO)-inflection source.

- For $a_{j_1j_1 0}(a_{j_1j_1s_1} - a_{j_1j_1s_2}) > 0$ and $a_{j_2j_2 0}(a_{j_1j_1s_1} - a_{j_2j_1 1})(a_{j_2j_2l_1} - a_{j_2j_2l_2}) < 0$,

$$(a_{j_1j_1s_1}, a_{j_2j_2l_1}) = \underbrace{\text{(II:DI,SI)}}_{\text{inflection sink}}.$$

The equilibrium of $(x^*_{j_1}, x^*_{j_2}) = (a_{j_1j_1s_1}, a_{j_2j_2l_1})$ is an (II:DI,SI)-inflection sink.

- For $a_{j_1j_1 0}(a_{j_1j_1s_1} - a_{j_1j_1s_2}) < 0$ and $a_{j_2j_2 0}(a_{j_1j_1s_1} - a_{j_2j_1 1})(a_{j_2j_2l_1} - a_{j_2j_2l_2}) < 0$,

$$(a_{j_1j_1s_1}, a_{j_2j_2l_1}) = \underbrace{\text{(DI:II,SI)}}_{\text{inflection sink}}.$$

The equilibrium of $(x^*_{j_1}, x^*_{j_2}) = (a_{j_1j_1s_1}, a_{j_2j_2l_1})$ is a (DI:II,SI)-inflection sink.

(ix_{2c}) From case (ix_{1a}) with $\bar{x}_{j_1} = a_{j_2 j_1 1}$, the equilibrium of $(x_{j_2}^*, x_{j_1}^*) = (a_{j_1 j_2 1}, a_{j_2 j_1 1})$ with $a_{j_1 j_2 1} = a_{j_2 j_2 l_1}$ and $a_{j_2 j_1 1} \neq a_{j_1 j_1 s_1}$ has the following properties as in Eqs. (1.268)–(1.271):

- For $a_{j_1 j_1 0} \prod_{s_1=1}^{2} (a_{j_2 j_1 1} - a_{j_1 j_1 s_1}) > 0$ and $a_{j_2 j_2 0}(a_{j_2 j_2 l_1} - a_{j_2 j_2 l_2}) > 0$,

$$(a_{j_1 j_2 1}, a_{j_2 j_1 1}) = \underbrace{(_{\text{SI:SO}}\text{US}, _{\text{DI:II}}\text{UP})}_{\text{up-parabola upper-saddle}}.$$

  The equilibrium of $(x_{j_2}^*, x_{j_1}^*) = (a_{j_1 j_2 1}, a_{j_2 j_1 1})$ is an ($_{\text{SI:SO}}$US, $_{\text{DI:II}}$UP)-up-parabola upper-saddle.

- For $a_{j_1 j_1 0} \prod_{s_1=1}^{2} (a_{j_2 j_1 1} - a_{j_1 j_1 s_1}) < 0$ and $a_{j_2 j_2 0}(a_{j_2 j_2 l_1} - a_{j_2 j_2 l_2}) > 0$,

$$(a_{j_1 j_2 1}, a_{j_2 j_1 1}) = \underbrace{(_{\text{SI:SO}}\text{US}, _{\text{II:DI}}\text{DP})}_{\text{down-parabola upper-saddle}}.$$

  The equilibrium of $(x_{j_2}^*, x_{j_1}^*) = (a_{j_1 j_2 1}, a_{j_2 j_1 1})$ is an ($_{\text{SI:SO}}$US, $_{\text{II:DI}}$DP)-down-parabola upper-saddle.

- For $a_{j_1 j_1 0} \prod_{s_1=1}^{2} (a_{j_2 j_1 1} - a_{j_1 j_1 s_1}) > 0$ and $a_{j_2 j_2 0}(a_{j_2 j_2 l_1} - a_{j_2 j_2 l_2}) < 0$,

$$(a_{j_1 j_2 1}, a_{j_2 j_1 1}) = \underbrace{(_{\text{SO:SI}}\text{LS}, _{\text{II:DI}}\text{DP})}_{\text{down-parabola lower-saddle}}.$$

  The equilibrium of $(x_{j_2}^*, x_{j_1}^*) = (a_{j_1 j_2 1}, a_{j_2 j_1 1})$ is an ($_{\text{SO:SI}}$LS, $_{\text{II:DI}}$DP)-down-parabola lower-saddle.

- For $a_{j_1 j_1 0} \prod_{s_1=1}^{2} (a_{j_2 j_1 1} - a_{j_1 j_1 s_1}) < 0$ and $a_{j_2 j_2 0}(a_{j_2 j_2 l_1} - a_{j_2 j_2 l_2}) < 0$,

$$(a_{j_1 j_2 1}, a_{j_2 j_1 1}) = \underbrace{(_{\text{SO:SI}}\text{LS}, _{\text{DI:II}}\text{UP})}_{\text{up-parabola lower-saddle}}.$$

  The equilibrium of $(x_{j_2}^*, x_{j_1}^*) = (a_{j_1 j_2 1}, a_{j_2 j_1 1})$ is an ($_{\text{SO:SI}}$LS, $_{\text{DI:II}}$UP)-up-parabola lower-saddle.

(ix_3) For $a_{j_1 j_1 s_1} = a_{j_2 j_1 1}$ and $a_{j_1 j_2 1} \neq a_{j_2 j_2 l_1}$ $(s_1, s_2, l_1, l_2 \in \{1, 2\}, s_1 \neq s_2, l_1 \neq l_2)$, in phase space,

$$\begin{aligned}\frac{dx_{j_1}}{dx_{j_2}} &= \frac{a_{j_1 j_1 0}\,(x_{j_1} - a_{j_1 j_1 s_1})(x_{j_1} - a_{j_1 j_1 s_2})(x_{j_2} - a_{j_1 j_2 1})}{a_{j_2 j_2 0}\,(x_{j_1} - a_{j_2 j_1 1})(x_{j_2} - a_{j_2 j_2 l_1})(x_{j_2} - a_{j_2 j_2 l_2})}\\ &= \frac{a_{j_1 j_1 0}\,(x_{j_1} - a_{j_1 j_1 s_2})(x_{j_2} - a_{j_1 j_2 1})}{a_{j_2 j_2 0}\,(x_{j_2} - a_{j_2 j_2 l_1})(x_{j_2} - a_{j_2 j_2 l_2})},\end{aligned}$$

and the deformation of the foregoing equation is

$$\frac{dx_{j_1}}{(x_{j_1}-a_{j_1j_1s_2})}=\frac{a_{j_1j_10}}{a_{j_2j_20}}\frac{(x_{j_2}-a_{j_1j_21})dx_{j_2}}{(x_{j_2}-a_{j_2j_2l_1})(x_{j_2}-a_{j_2j_2l_2})},$$

With an initial condition of (x_{j_10},x_{j_20}) at $t=t_0$, the integration of the above equation gives the first integral manifold for $a_{j_1j_1s_1}=a_{j_2j_11}$ and $a_{j_1j_21}\neq a_{j_2j_2l_1}$ (s_1, s_2, l_1, l_2 $\in\{1,2\}$, $s_1\neq s_2$, $l_1\neq l_2$) as

$$\ln\frac{|x_{j_1}-a_{j_1j_1s_2}|}{|x_{j_10}-a_{j_1j_1s_2}|}=\frac{a_{j_1j_10}}{a_{j_2j_20}}\sum\nolimits_{l_1=1,l_2\neq l_1}^{2}\frac{a_{j_2j_2l_1}-a_{j_1j_21}}{a_{j_2j_2l_1}-a_{j_2j_2l_2}}\ln\frac{|x_{j_2}-a_{j_2j_2l_1}|}{|x_{j_20}-a_{j_2j_2l_1}|}.$$

(ix_{3a}) For $x_{j_1}^*=a_{j_2j_11}=a_{j_1j_1s_1}$ with $\bar{x}_{j_2}\neq a_{j_2j_2l_1},a_{j_2j_2l_2},a_{j_1j_21}$, if

$$\left.\frac{dx_{j_2}}{dx_{j_1}}\right|_{x_{j_1}^*=a_{j_1j_1s_1}}=\frac{a_{j_2j_20}}{a_{j_1j_10}}\frac{\prod_{s_1=1}^{2}(\bar{x}_{j_2}-a_{j_2j_2l_1})}{(x_{j_1}-a_{j_1j_1s_2})(\bar{x}_{j_2}-a_{j_1j_21})}>0,$$

there is an increasing inflection flow, and if

$$\left.\frac{dx_{j_2}}{dx_{j_1}}\right|_{x_{j_1}^*=a_{j_1j_1s_1}}=\frac{a_{j_2j_20}}{a_{j_1j_10}}\frac{\prod_{s_1=1}^{2}(\bar{x}_{j_2}-a_{j_2j_2l_1})}{(x_{j_1}-a_{j_1j_1s_2})(\bar{x}_{j_2}-a_{j_1j_21})}<0,$$

there is a decreasing inflection flow. The variational equation at $x_{j_1}^*=a_{j_1j_1s_1}=a_{j_2j_11}$ is

$$\Delta\dot{x}_{j_1}=a_{j_1j_10}(x_{j_1}-a_{j_1j_1s_2})(\bar{x}_{j_2}-a_{j_1j_21})\Delta x_{j_1}.$$

The asymptotic flows are source and sink for $a_{j_1j_10}(x_{j_1}-a_{j_1j_1s_2})(\bar{x}_{j_2}-a_{j_1j_21})>0$ and $a_{j_1j_10}(x_{j_1}-a_{j_1j_1s_2})(\bar{x}_{j_2}-a_{j_1j_21})<0$ in the x_{j_1}-direction, respectively.

Therefore, the infinite equilibrium of $x_{j_1}^*=a_{j_1j_1s_1}=a_{j_2j_11}$ (s_1, $s_2\in\{1,2\}$) with $\bar{x}_{j_2}\neq a_{j_2j_21},a_{j_2j_22},a_{j_1j_21}$ has the following properties as in Eqs. (1.273)–(1.276):

- For $a_{j_1j_10}(\bar{x}_{j_2}-a_{j_1j_21})(a_{j_1j_1s_1}-a_{j_1j_1s_2})>0$ and $a_{j_2j_20}\prod_{l_1=1}^{2}(\bar{x}_{j_2}-a_{j_2j_2l_1})>0$,

$$(a_{j_1j_1s_1},\bar{x}_{j_2})=\underbrace{(\mathrm{SO,II})}_{\text{increasing-inflection source}}.$$

  The infinite equilibrium of $x_{j_1}^*=a_{j_2j_11}=a_{j_1j_1s_1}$ is an (SO, II)-increasing-inflection source.
- For $a_{j_1j_10}(\bar{x}_{j_2}-a_{j_1j_21})(a_{j_1j_1s_1}-a_{j_1j_1s_2})<0$ and $a_{j_2j_20}\prod_{l_1=1}^{2}(\bar{x}_{j_2}-a_{j_2j_2l_1})>0$,

$$(a_{j_1j_1s_1}, \bar{x}_{j_2}) = \underbrace{(\mathrm{SI,DI})}_{\text{decreasing-inflection sink}}.$$

The infinite equilibrium of $x^*_{j_1} = a_{j_2j_1 1} = a_{j_1j_1s_1}$ is an (SI, DI)-decreasing-inflection sink.

- For $a_{j_1j_1 0}(\bar{x}_{j_2} - a_{j_1j_2 1})(a_{j_1j_1s_1} - a_{j_1j_1s_2}) > 0$ and $a_{j_2j_2 0}\prod_{l_1=1}^{2}(\bar{x}_{j_2} - a_{j_2j_2l_1}) < 0$,

$$(a_{j_1j_1s_1}, \bar{x}_{j_2}) = \underbrace{(\mathrm{SO,DI})}_{\text{decreasing-inflection source}}.$$

The infinite equilibrium of $x^*_{j_1} = a_{j_2j_1 1} = a_{j_1j_1s_1}$ is an (SO, DI)-decreasing-inflection source.

- For $a_{j_1j_1 0}(\bar{x}_{j_2} - a_{j_1j_2 1})(a_{j_1j_1s_1} - a_{j_1j_1s_2}) < 0$ and $a_{j_2j_2 0}\prod_{l_1=1}^{2}(\bar{x}_{j_2} - a_{j_2j_2l_1}) < 0$,

$$(a_{j_1j_1s_1}, \bar{x}_{j_2}) = \underbrace{(\mathrm{SI,II})}_{\text{increasing-inflection sink}}.$$

The infinite equilibrium of $x^*_{j_1} = a_{j_2j_1 1} = a_{j_1j_1s_1}$ is an (SI, II)-increasing-inflection sink.

(ix_{3b}) From (ix_{3a}) with $\bar{x}_{j_2} = a_{j_1j_2 1}$, the equilibrium of $(x^*_{j_2}, x^*_{j_1}) = (a_{j_1j_2 1}, a_{j_2j_1 1})$ for $a_{j_1j_1s_1} = a_{j_2j_1 1}$ and $a_{j_1j_2 1} \neq a_{j_2j_2l_1}$ has the following properties as in Eqs. (1.277)–(1.280):

- For $a_{j_1j_1 0}(a_{j_2j_1 1} - a_{j_1j_1s_2}) > 0$ and $a_{j_2j_2 0}\prod_{l_1=1}^{2}(a_{j_1j_2 1} - a_{j_2j_2l_1}) > 0$,

$$(a_{j_1j_2 1}, a_{j_2j_1 1}) = \underbrace{({}_{\mathrm{DII}}\mathrm{UP}, {}_{\mathrm{SI:SO}}\mathrm{US})}_{\text{up-parabola upper-saddle}}.$$

The equilibrium of $(x^*_{j_2}, x^*_{j_1}) = (a_{j_1j_2 1}, a_{j_2j_1 1})$ is a (${}_{\mathrm{DI:II}}\mathrm{UP}$, ${}_{\mathrm{SI:SO}}\mathrm{US}$)-up-parabola upper-saddle.

- For $a_{j_1j_1 0}(a_{j_2j_1 1} - a_{j_1j_1s_2}) < 0$ and $a_{j_2j_2 0}\prod_{l_1=1}^{2}(a_{j_1j_2 1} - a_{j_2j_2l_1}) > 0$,

$$(a_{j_1j_2 1}, a_{j_2j_1 1}) = \underbrace{({}_{\mathrm{II:DI}}\mathrm{DP}, {}_{\mathrm{SO:SI}}\mathrm{LS})}_{\text{down-parabola lower-saddle}}.$$

The equilibrium of $(x^*_{j_2}, x^*_{j_1}) = (a_{j_1j_2 1}, a_{j_2j_1 1})$ is a (${}_{\mathrm{II:DI}}\mathrm{DP}$:${}_{\mathrm{SO:SI}}\mathrm{LS}$)-down-parabola lower-saddle.

- For $a_{j_1j_10}(a_{j_2j_11}-a_{j_1j_1s_2})>0$ and $a_{j_2j_20}\prod_{l_1=1}^{2}(a_{j_1j_21}-a_{j_2j_2l_1})<0$,

$$(a_{j_1j_21},a_{j_2j_11})=\underbrace{({}_{\mathrm{II:DI}}\mathrm{DP},{}_{\mathrm{SI:SO}}\mathrm{US})}_{\text{inflection down-parabola upper-saddle}}.$$

The equilibrium of $(x_{j_2}^*,x_{j_1}^*)=(a_{j_1j_21},a_{j_2j_11})$ is a (${}_{\mathrm{II:DI}}$DP,${}_{\mathrm{SO:SI}}$US)-down-parabola lower-saddle.

- For $a_{j_1j_10}(a_{j_2j_11}-a_{j_1j_1s_2})<0$ and $a_{j_2j_20}\prod_{l_1=1}^{2}(a_{j_1j_21}-a_{j_2j_2l_1})<0$,

$$(a_{j_1j_21},a_{j_2j_11})=\underbrace{({}_{\mathrm{DI:II}}\mathrm{UP},{}_{\mathrm{SO:SI}}\mathrm{LS})}_{\text{up-parabola lower-saddle}}.$$

The equilibrium of $(x_{j_2}^*,x_{j_1}^*)=(a_{j_1j_21},a_{j_2j_11})$ is a (${}_{\mathrm{DI:II}}$UP, ${}_{\mathrm{SO:SI}}$LS)-up-parabola lower-saddle.

(ix$_{3c}$) From (ix$_{3a}$) with $\overline{x}_{j_2}=a_{j_2j_2l_1}$, The equilibrium of $(x_{j_1}^*,x_{j_2}^*)=(a_{j_1j_1s_1},a_{j_2j_2l_1})$ for $a_{j_1j_1s_1}=a_{j_2j_11}$ and $a_{j_1j_21}\neq a_{j_2j_21},a_{j_2j_22}$ has the following properties as in Eqs. (1.281)–(1.284):

- For $a_{j_1j_10}(a_{j_2j_2l_1}-a_{j_1j_21})(a_{j_1j_1s_1}-a_{j_1j_1s_2})>0$ and $a_{j_2j_20}(a_{j_2j_2l_1}-a_{j_2j_2l_2})>0$,

$$(a_{j_1j_1s_1},a_{j_2j_2l_1})=\underbrace{(\mathrm{SO,DI:II})}_{\text{inflection source}}.$$

The equilibrium of $(x_{j_1}^*,x_{j_2}^*)=(a_{j_1j_1s_1},a_{j_2j_2l_1})$ is an (SO,DI:II) inflection source.

- For $a_{j_1j_10}(a_{j_2j_2l_1}-a_{j_1j_21})(a_{j_1j_1s_1}-a_{j_1j_1s_2})<0$ and $a_{j_2j_20}(a_{j_2j_2l_1}-a_{j_2j_2l_2})>0$,

$$(a_{j_1j_1s_1},a_{j_2j_2l_1})=\underbrace{(\mathrm{SI,II:DI})}_{\text{inflection sink}}.$$

The equilibrium of $(x_{j_1}^*,x_{j_2}^*)=(a_{j_1j_1s_1},a_{j_2j_2l_1})$ is an (SI:II:DI) inflection sink.

- For $a_{j_1j_10}(a_{j_2j_2l_1}-a_{j_1j_11})(a_{j_1j_1s_1}-a_{j_1j_1s_2})>0$ and $a_{j_2j_20}(a_{j_2j_2l_1}-a_{j_2j_2l_2})<0$,

$$(a_{j_1j_1s_1},a_{j_2j_2l_1})=\underbrace{(\mathrm{SO,II:DI})}_{\text{inflection source}}.$$

The equilibrium of $(x_{j_1}^*,x_{j_2}^*)=(a_{j_1j_1s_1},a_{j_2j_2l_1})$ is an (SO,II:DI) inflection source.

- For $a_{j_1j_10}(a_{j_2j_2l_1}-a_{j_1j_21})(a_{j_1j_1s_1}-a_{j_1j_1s_2})<0$ and $a_{j_2j_20}(a_{j_2j_2l_1}-a_{j_2j_2l_2})<0$,

$$(a_{j_1j_1s_1}, a_{j_2j_2l_1}) = \underbrace{(\text{SI,DI:II})}_{\text{inflection sink}}.$$

The equilibrium of $(x_{j_1}^*, x_{j_2}^*) = (a_{j_1j_1s_1}, a_{j_2j_2l_1})$ is an (SI,DI:II) inflection sink.

(xi$_4$) For $a_{j_1j_1s_1} = a_{j_2j_11}$ and $a_{j_1j_21} = a_{j_2j_2l_1}$ $(s_1, s_2, l_1, l_2 \in \{1,2\}, s_1 \neq s_2, l_1 \neq l_2)$, in phase space,

$$\begin{aligned}\frac{dx_{j_1}}{dx_{j_2}} &= \frac{a_{j_1j_10}\,(x_{j_1} - a_{j_1j_1s_1})(x_{j_1} - a_{j_1j_1s_2})(x_{j_2} - a_{j_1j_21})}{a_{j_2j_20}\,(x_{j_1} - a_{j_2j_11})(x_{j_2} - a_{j_2j_2l_1})(x_{j_2} - a_{j_2j_2l_2})}\\ &= \frac{a_{j_1j_10}\,(x_{j_1} - a_{j_1j_1s_2})}{a_{j_2j_20}\,(x_{j_2} - a_{j_2j_2l_2})},\end{aligned}$$

and the deformation of the foregoing equation is

$$\frac{dx_{j_1}}{(x_{j_1} - a_{j_1j_1s_2})} = \frac{a_{j_1j_10}}{a_{j_2j_20}}\frac{dx_{j_2}}{(x_{j_2} - a_{j_2j_2l_2})}.$$

With an initial condition of (x_{j_10}, x_{j_20}) at $t = t_0$,, the integration of the above equation gives the *first integral manifold for* $a_{j_1j_1s_1} = a_{j_2j_11}$ and $a_{j_1j_21} = a_{j_2j_2l_1}$ $(s_1, s_2, l_1, l_2 \in \{1,2\}, s_1 \neq s_2, l_1 \neq l_2)$ *as*

$$\ln\frac{|x_{j_1} - a_{j_1j_1s_2}|}{|x_{j_10} - a_{j_1j_1s_2}|} = \frac{a_{j_1j_10}}{a_{j_2j_20}}\ln\frac{|x_{j_2} - a_{j_2j_2l_2}|}{|x_{j_20} - a_{j_2j_2l_2}|}.$$

Similar to (ix$_2$) and (ix$_3$), therefore, the equilibrium of $(x_{j_1}^*, x_{j_2}^*) = (a_{j_1j_1s_1}, a_{j_2j_2l_1})$ for $a_{j_1j_1s_1} = a_{j_2j_11}$ and $a_{j_1j_21} = a_{j_2j_2l_1}$ has the following properties as in Eqs. (1.286)–(1.289): The equilibrium of for $a_{j_1j_1s_1} = a_{j_2j_11}$ and $a_{j_1j_21} = a_{j_2j_2l_1}$ has the following properties.

- For $a_{j_1j_10}(a_{j_1j_1s_1} - a_{j_1j_1s_2}) > 0$ and $a_{j_2j_20}(a_{j_2j_2l_1} - a_{j_2j_2l_2}) > 0$,

$$(a_{j_1j_1s_1}, a_{j_2j_2l_1}) = \underbrace{({}_{\text{SI:SO}}\text{II}, {}_{\text{SI:SO}}\text{II})}_{\text{increasing-inflection diagonal saddle}}.$$

  The equilibrium of $(x_{j_1}^*, x_{j_2}^*) = (a_{j_1j_1s_1}, a_{j_2j_2l_1})$ is an (${}_{\text{SI:SO}}$II:${}_{\text{SI:SO}}$II)-increasing-inflection diagonal-saddle.
- For $a_{j_1j_10}(a_{j_1j_1s_1} - a_{j_1j_1s_2}) < 0$ and $a_{j_2j_20}(a_{j_2j_2l_1} - a_{j_2j_2l_2}) > 0$,

$$(a_{j_1j_1s_1}, a_{j_2j_2l_1}) = \underbrace{({}_{\text{SI:SO}}\text{DI}, {}_{\text{SO:SI}}\text{DI})}_{\text{decreasing-inflection diagonal saddle}}.$$

  The equilibrium of $(x_{j_1}^*, x_{j_2}^*) = (a_{j_1j_1s_1}, a_{j_2j_2l_1})$ is an (${}_{\text{SI:SO}}$DI:${}_{\text{SO:SI}}$DI)-decreasing-inflection diagonal-saddle.

- For $a_{j_1j_1 0}(a_{j_1j_1s_1} - a_{j_1j_1s_2}) > 0$ and $a_{j_2j_2 0}(a_{j_2j_2l_1} - a_{j_2j_2l_2}) < 0$,

$$(a_{j_1j_1s_1}, a_{j_2j_2l_1}) = \underbrace{({}_{\text{SO:SI}}\text{DI}, {}_{\text{SI:SO}}\text{DI})}_{\text{decreasing-inflection diagonal saddle}}.$$

  The equilibrium of $(x^*_{j_1}, x^*_{j_2}) = (a_{j_1j_1s_1}, a_{j_2j_2l_1})$ is an $({}_{\text{SO:SI}}\text{DI}:{}_{\text{SI:SO}}\text{DI})$-decreasing-inflection diagonal saddle.
- For $a_{j_1j_1 0}(a_{j_1j_1s_1} - a_{j_1j_1s_2}) < 0$ and $a_{j_2j_2 0}(a_{j_2j_2l_1} - a_{j_2j_2l_2}) < 0$,

$$(a_{j_1j_1s_1}, a_{j_2j_2l_1}) = \underbrace{({}_{\text{SO:SI}}\text{II}, {}_{\text{SO:SI}}\text{II})}_{\text{increasing-inflection diagonal saddle}}.$$

  The equilibrium of $(x^*_{j_1}, x^*_{j_2}) = (a_{j_1j_1s_1}, a_{j_2j_2l_1})$ is an $({}_{\text{SO:SI}}\text{II}, {}_{\text{SO:SI}}\text{II})$-increasing-inflection diagonal saddle.

In the end, this theorem is proved. ■

Chapter 2
Double Saddles and Switching Dynamics

In this chapter, double-saddles and hyperbolic flow singularity in cubic systems with two crossing-linear and self-quadratic products vector fields are discussed. The double-saddle equilibriums are the appearing bifurcations of the saddle source and saddle-sink, and such double-saddle equilibriums are also the appearing bifurcations of the network of saddles, sink, and source. The hyperbolic flow singularity generates the appearing bifurcations of hyperbolic and hyperbolic-secant flows. The switching bifurcations are presented through the infinite-equilibriums. The up-down and down-up infinite-equilibriums are the switching bifurcation. The hyperbolic lower-to-upper saddle is for the switching of positive (negative) saddle with paralleled hyperbolic-secant-to-hyperbolic flow. The hyperbolic-secant lower-to-upper saddle is for the switching of a center with paralleled hyperbolic-to-hyperbolic-secant flow. The up-down (down-up) saddles are for the switching bifurcation of double-saddles with a connected hyperbolic-to-hyperbolic flow.

2.1 Double Saddles with Hyperbolic Flow Singularity

In this section, double-saddles and corresponding bifurcation dynamics will be discussed. The hyperbolic flow singularity will be presented for appearing hyperbolic and hyperbolic-secant flows.

Consider a dynamical system as

$$
\begin{aligned}
\dot{x}_1 &= a_{110}\,(x_1 - a_{111})^2 (x_2 - a_{121}),\\
\dot{x}_2 &= a_{220}\,(x_1 - a_{211})\,(x_2 - a_{221})^2,
\end{aligned}
\tag{2.1}
$$

and the corresponding first integral manifold is

A. C. J. Luo, *Two-Dimensional Two-Product Cubic Systems, Vol. II*,
https://doi.org/10.1007/978-3-031-48491-9_2

$$\ln\frac{|x_1-a_{111}|}{|x_{10}-a_{111}|}-(a_{111}-a_{211})(\frac{1}{x_1-a_{111}}-\frac{1}{x_{10}-a_{111}})$$
$$=\frac{a_{110}}{a_{220}}\left[\ln\frac{|x_2-a_{221}|}{|x_{20}-a_{221}|}-(a_{221}-a_{121})(\frac{1}{x_2-a_{221}}-\frac{1}{x_{20}-a_{221}})\right]. \tag{2.2}$$

(i) The equilibriums of $x_1^*=a_{111},a_{211}$ and $x_2^*=a_{121},a_{221}$ for $a_{121}<a_{221}$ are

$$\begin{Bmatrix}(a_{211},a_{221}) & (a_{111},a_{221})\\ (a_{211},a_{121}) & (a_{111},a_{121})\end{Bmatrix}=\begin{Bmatrix}\underbrace{\text{(pF,UP:UP)}}_{\text{HS to HB flow }(+)} & \underbrace{\text{(US,US)}}_{\text{double-saddle}}\\ \underbrace{(\text{UP}_+,\text{UP}_+)}_{\text{positive saddle}} & \underbrace{\text{(UP:UP,pF)}}_{\text{HS to HB flow }(+)}\end{Bmatrix} \tag{2.3}$$

for $a_{110}>0$ and $a_{220}>0$;

$$\begin{Bmatrix}(a_{211},a_{221}) & (a_{111},a_{221})\\ (a_{211},a_{121}) & (a_{111},a_{121})\end{Bmatrix}=\begin{Bmatrix}\underbrace{\text{(nF,DP:DP)}}_{\text{HB to HS flow }(-)} & \underbrace{\text{(LS,US)}}_{\text{double-saddle}}\\ \underbrace{(\text{DP}_+,\text{DP}_-,)}_{\text{CCW center}} & \underbrace{\text{(DP:DP,pF)}}_{\text{HB to HS flow }(+)}\end{Bmatrix} \tag{2.4}$$

for $a_{110}<0$ and $a_{220}>0$;

$$\begin{Bmatrix}(a_{211},a_{221}) & (a_{111},a_{221})\\ (a_{211},a_{121}) & (a_{111},a_{121})\end{Bmatrix}=\begin{Bmatrix}\underbrace{\text{(pF,DP:DP)}}_{\text{HB to HS flow }(+)} & \underbrace{\text{(US,LS)}}_{\text{double-saddle}}\\ \underbrace{(\text{DP}_-,\text{DP}_+)}_{\text{CW center}} & \underbrace{\text{(DP:DP,nF)}}_{\text{HB to HS flow }(-)}\end{Bmatrix} \tag{2.5}$$

for $a_{110}>0$ and $a_{220}<0$;

$$\begin{Bmatrix}(a_{211},a_{221}) & (a_{111},a_{221})\\ (a_{211},a_{121}) & (a_{111},a_{121})\end{Bmatrix}=\begin{Bmatrix}\underbrace{\text{(nF,UP:UP)}}_{\text{HS to HB flow }(-)} & \underbrace{\text{(LS,LS)}}_{\text{double-saddle}}\\ \underbrace{(\text{UP}_-,\text{UP}_-)}_{\text{negative saddle}} & \underbrace{\text{(UP:UP,nF)}}_{\text{HS to HB flow }(-)}\end{Bmatrix} \tag{2.6}$$

for $a_{110}<0$ and $a_{220}<0$;

as shown in Fig. 2.1. "HS to HB" and "HB to HS" are for the hyperbolic-secant to hyperbolic asymptotic flow and for the hyperbolic-to-hyperbolic-secant asymptotic flow, respectively. The double-saddle bifurcations of the upper saddle and lower saddle are presented for saddle sink and saddle source states and also for the appearing bifurcation of saddles, sink, and source. The hyperbolic-secant-to-hyperbolic flow bifurcations are the appearing bifurcations of the hyperbolic and hyperbolic-secant flows, which are also presented.

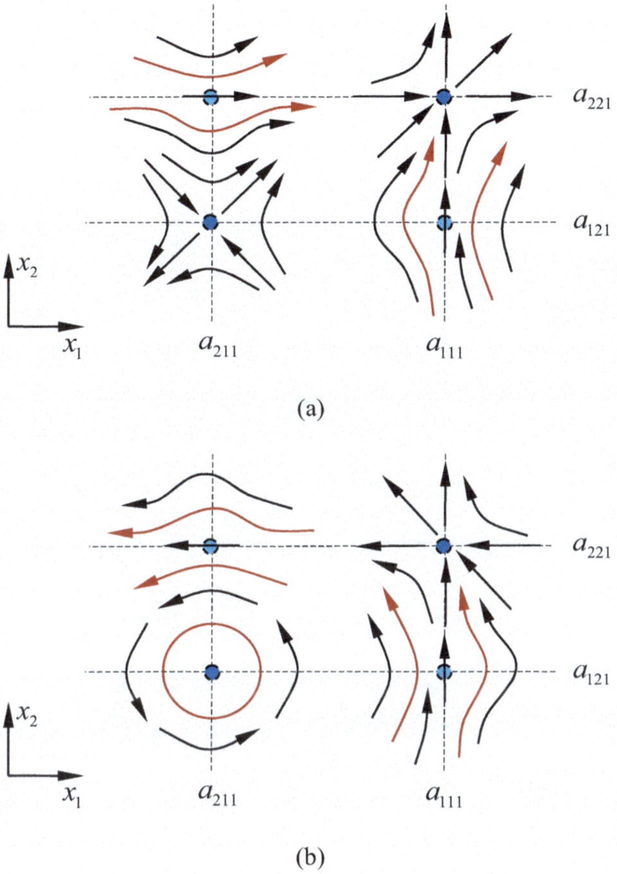

Fig. 2.1 Phase portraits ($a_{121} < a_{221}$ and $a_{211} < a_{111}$) for two-dimensional systems on the x_1-direction with $x_1^* = a_{111}, a_{211}$ and on the x_2-direction with $x_2^* = a_{121}, a_{221}$. **(a)** ($a_{110} > 0$, $a_{220} > 0$), **(b)** ($a_{110} < 0$, $a_{220} > 0$), **(c)** ($a_{110} > 0$, $a_{220} < 0$), and **(d)** ($a_{110} < 0$, $a_{220} < 0$)

(ii) The equilibriums of $x_1^* = a_{111}$ and $x_2^* = a_{121}, a_{221}$ for $a_{221} < a_{121}$ are

$$\begin{Bmatrix} (a_{211}, a_{121}) & (a_{111}, a_{121}) \\ (a_{211}, a_{221}) & (a_{111}, a_{221}) \end{Bmatrix} = \begin{Bmatrix} \underbrace{(\mathrm{UP}_+, \mathrm{UP}_+)}_{\text{positive saddle}} & \underbrace{(\mathrm{UP{:}UP,pF})}_{\text{HS to HB flow } (+)} \\ \underbrace{(\mathrm{nF, DP{:}DP})}_{\text{HS to HB flow } (-)} & \underbrace{(\mathrm{LS, US})}_{\text{double-saddle}} \end{Bmatrix} \tag{2.7}$$

for $a_{110} > 0$ and $a_{220} > 0$;

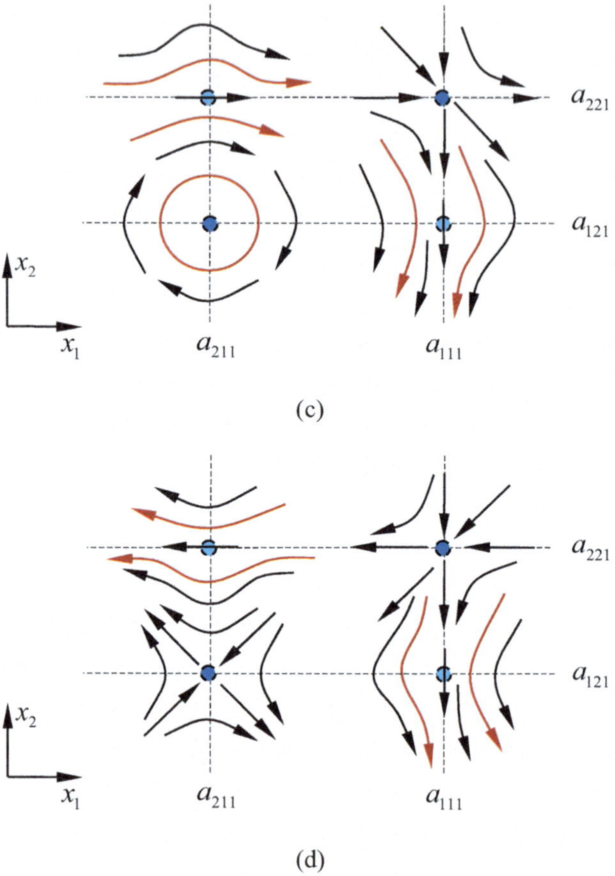

Fig. 2.1 (continued)

$$\begin{Bmatrix} (a_{211}, a_{121}) & (a_{111}, a_{121}) \\ (a_{211}, a_{221}) & (a_{111}, a_{221}) \end{Bmatrix} = \left\{ \begin{array}{cc} \underbrace{(\text{DP}_+,\text{DP}_-)}_{\text{CCW centere}} & \underbrace{(\text{DP:DP,pF})}_{\text{HB to HS flow }(+)} \\ \underbrace{(\text{pF,UP:UP})}_{\text{HB to HS flow }(+)} & \underbrace{(\text{US,US})}_{\text{double-saddle}} \end{array} \right\} \tag{2.8}$$

for $a_{110} < 0$ and $a_{220} > 0$;

$$\begin{Bmatrix} (a_{211}, a_{121}) & (a_{111}, a_{121}) \\ (a_{211}, a_{221}) & (a_{111}, a_{221}) \end{Bmatrix} = \left\{ \begin{array}{cc} \underbrace{(\text{DP}_-,\text{DP}_+)}_{\text{CW centere}} & \underbrace{(\text{DP:DP,nF})}_{\text{HB to HS flow }(-)} \\ \underbrace{(\text{nF,UP:UP})}_{\text{HB to HS flow }(-)} & \underbrace{(\text{LS,LS})}_{\text{double-saddle}} \end{array} \right\} \tag{2.9}$$

for $a_{110} > 0$ and $a_{220} < 0$;

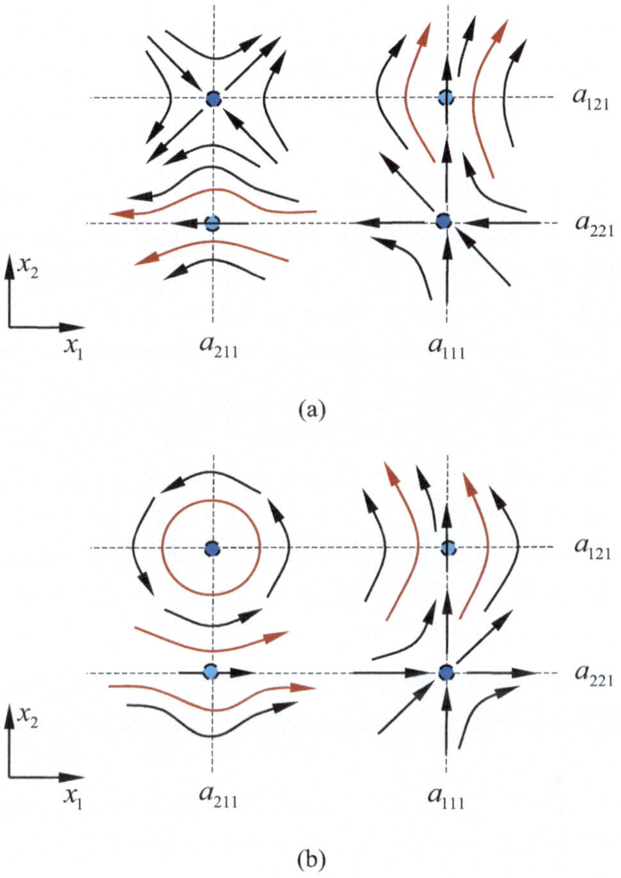

Fig. 2.2 Phase portraits ($a_{221} < a_{121}$ and $a_{211} < a_{111}$) for two-dimensional systems on the x_1-direction with $x_1^* = a_{111}, a_{211}$ and on the x_2-direction with $x_2^* = a_{121}, a_{221}$. (**a**) ($a_{110} > 0$, $a_{220} > 0$), (**b**) ($a_{110} < 0$, $a_{220} > 0$), (**c**) ($a_{110} > 0$, $a_{220} < 0$), and (**d**) ($a_{110} < 0$, $a_{220} < 0$)

$$\begin{Bmatrix} (a_{211}, a_{121}) & (a_{111}, a_{121}) \\ (a_{211}, a_{221}) & (a_{111}, a_{221}) \end{Bmatrix} = \left\{ \begin{matrix} \underbrace{(\text{UP}_-,\text{UP}_-)}_{\text{negative saddle}} & \underbrace{(\text{UP:UP,nF})}_{\text{HS to HB flow }(-)} \\ \underbrace{(\text{pF,DP:DP})}_{\text{HS to HB flow }(+)} & \underbrace{(\text{US,LS})}_{\text{double-saddle}} \end{matrix} \right\} \tag{2.10}$$

for $a_{110} < 0$ and $a_{220} < 0$.

The double-saddle bifurcations of the upper-saddle and lower-saddle are presented in Fig. 2.2 for saddle-sink and saddle-source and for the appearing bifurcation of saddles, sink, and source. The hyperbolic-secant to hyperbolic flow bifurcations are

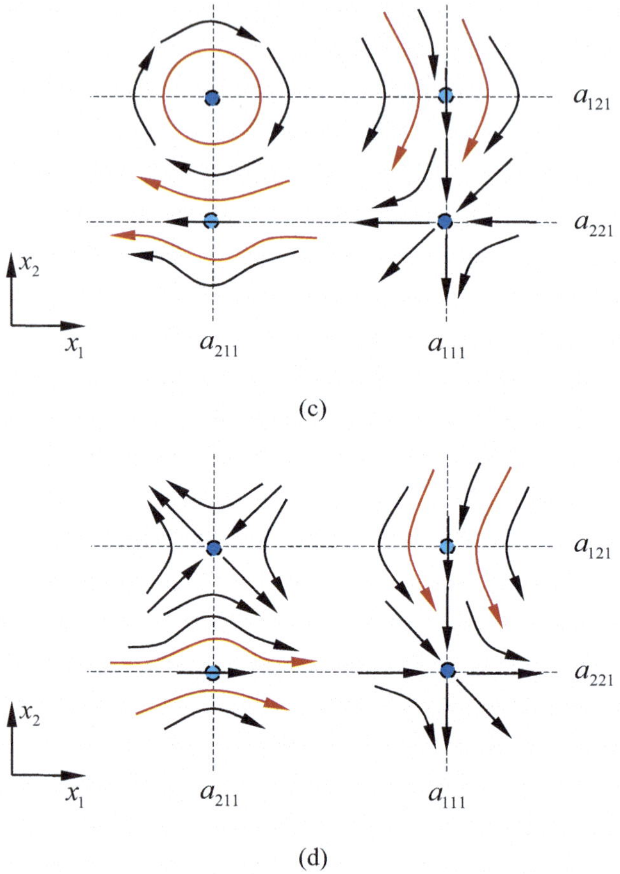

Fig. 2.2 (continued)

also presented for the appearing bifurcations of the hyperbolic and hyperbolic-secant flows.

The phase portraits of such a dynamical singular system with $a_{211} > a_{111}$ can be similarly completed as an excise by readers.

2.2 Switching Bifurcations

Consider three switching bifurcations with two second-order singularities: (i) the infinite-equilibriums $x_2^* = a_{121} = a_{221}$ and $a_{111} \neq a_{211}$, (ii) the infinite-equilibriums $x_1^* = a_{111} = a_{211}$ and $a_{121} \neq a_{221}$, (iii) the double-switching infinite-equilibriums of $x_1^* = a_{111} = a_{211}$ and $x_2^* = a_{121} = a_{221}$.

2.2.1 *Horizontal Up-Down Saddle Infinite Equilibriums*

Consider a dynamical system for $a_{121} = a_{221}$ and $a_{111} \neq a_{211}$ with up-down and down-up upper-saddle and lower-saddle infinite-equilibriums as

$$\begin{aligned} \dot{x}_1 &= a_{110}(x_1 - a_{111})^2(x_2 - a_{121}), \\ \dot{x}_2 &= a_{220}(x_1 - a_{211})(x_2 - a_{221})^2, \end{aligned} \tag{2.11}$$

and the corresponding first integral manifold ($a_{121} = a_{221}$ and $a_{111} \neq a_{211}$) is

$$\begin{aligned} &\ln \frac{|x_1 - a_{211}|}{|x_{10} - a_{211}|} - (a_{111} - a_{211})\left(\frac{1}{x_1 - a_{111}} - \frac{1}{x_{10} - a_{111}}\right) \\ &= \frac{a_{110}}{a_{220}} \ln \frac{|x_2 - a_{221}|}{|x_{20} - a_{221}|}. \end{aligned} \tag{2.12}$$

(i) The equilibriums of $x_1^* = a_{111}, a_{211}$ and $x_2^* = a_{121}, a_{221}$ for $a_{221} = a_{121}$ and $a_{211} < a_{111}$ are

$$\begin{Bmatrix} (a_{211}, a_{221}) \\ (a_{111}, a_{221}) \end{Bmatrix} = \begin{Bmatrix} \underbrace{({}_{\text{UD}}\text{LS}:{}_{\text{DU}}\text{US}, \text{DP}_-:\text{UP}_+)}_{\text{hyperbolic lower-to-upper saddle}} \\ \underbrace{(\text{DU}:\text{DU},\text{US})}_{\text{down-up upper-saddle}} \end{Bmatrix} \text{for } a_{110} > 0 \text{ and } a_{220} > 0; \tag{2.13}$$

$$\begin{Bmatrix} (a_{211}, a_{221}) \\ (a_{111}, a_{221}) \end{Bmatrix} = \begin{Bmatrix} \underbrace{({}_{\text{DU}}\text{LS}:{}_{\text{UD}}\text{US}, \text{UP}_+:\text{PP}_-)}_{\text{hyperbolic-secant lower-to-upper saddle}} \\ \underbrace{(\text{UD}:\text{UD},\text{US})}_{\text{up-down upper-saddle}} \end{Bmatrix} \text{for } a_{110} < 0 \text{ and } a_{220} > 0; \tag{2.14}$$

$$\begin{Bmatrix} (a_{211}, a_{221}) \\ (a_{111}, a_{221}) \end{Bmatrix} = \begin{Bmatrix} \underbrace{({}_{\text{DU}}\text{US}:{}_{\text{UD}}\text{LS}, \text{UP}_-:\text{PP}_+,)}_{\text{hyperbolic-secant upper-to-lower saddle}} \\ \underbrace{(\text{UD}:\text{UD},\text{LS})}_{\text{up-down lower-saddle}} \end{Bmatrix} \text{for } a_{110} > 0 \text{ and } a_{220} < 0; \tag{2.15}$$

$$\begin{Bmatrix} (a_{211}, a_{221}) \\ (a_{111}, a_{221}) \end{Bmatrix} = \begin{Bmatrix} \underbrace{({}_{\text{UD}}\text{US}:{}_{\text{DU}}\text{LS}, \text{DP}_+:\text{UP}_-)}_{\text{hyperbolic upper-to-lower saddle}} \\ \underbrace{(\text{DU}:\text{DU},\text{LS})}_{\text{down-up lower-saddle}} \end{Bmatrix} \text{for } a_{110} < 0 \text{ and } a_{220} < 0; \tag{2.16}$$

and the up-down and down-up upper saddle and lower saddle infinite equilibriums of $x_2^* = a_{121} = a_{221}$ are

$$\begin{aligned}
&(\bar{x}_1, a_{221}) = \underbrace{(\text{UD,LS})}_{\text{up-down lower-saddle}} && \text{if } \bar{x}_1 \in (-\infty, a_{211}),\\
&(\bar{x}_1, a_{221}) = \underbrace{(\text{DU,US})}_{\text{down-up upper-saddle}} && \text{if } \bar{x}_1 \in (a_{211}, a_{111}),\\
&(\bar{x}_1, a_{221}) = \underbrace{(\text{DU,US})}_{\text{down-up upper-saddle}} && \text{if } \bar{x}_1 \in (a_{111}, \infty),\\
&\text{for } a_{110} > 0 \text{ and } a_{220} > 0;
\end{aligned} \tag{2.17}$$

$$\begin{aligned}
&(\bar{x}_1, a_{221}) = \underbrace{(\text{DU,LS})}_{\text{down-up upper-saddle}} && \text{if } \bar{x}_1 \in (-\infty, a_{211}),\\
&(\bar{x}_1, a_{221}) = \underbrace{(\text{UD,US})}_{\text{up-down upper-saddle}} && \text{if } \bar{x}_1 \in (a_{211}, a_{111}),\\
&(\bar{x}_1, a_{221}) = \underbrace{(\text{UD,US})}_{\text{up-down upper-saddle}} && \text{if } \bar{x}_1 \in (a_{111}, \infty),\\
&\text{for } a_{110} < 0 \text{ and } a_{220} > 0;
\end{aligned} \tag{2.18}$$

$$\begin{aligned}
&(\bar{x}_1, a_{221}) = \underbrace{(\text{DU,US})}_{\text{down-up upper-saddle}} && \text{if } \bar{x}_1 \in (-\infty, a_{211}),\\
&(\bar{x}_1, a_{221}) = \underbrace{(\text{UD,LS})}_{\text{up-down lower-saddle}} && \text{if } \bar{x}_1 \in (a_{211}, a_{111}),\\
&(\bar{x}_1, a_{221}) = \underbrace{(\text{UD,LS})}_{\text{up-down lower-saddle}} && \text{if } \bar{x}_1 \in (a_{111}, \infty),\\
&\text{for } a_{110} < 0 \text{ and } a_{220} > 0;
\end{aligned} \tag{2.19}$$

$$\begin{aligned}
&(\bar{x}_1, a_{221}) = \underbrace{(\text{UD,US})}_{\text{up-down upper-saddle}} && \text{if } \bar{x}_1 \in (-\infty, a_{211})\\
&(\bar{x}_1, a_{221}) = \underbrace{(\text{DU,LS})}_{\text{down-up lower-saddle}} && \text{if } \bar{x}_1 \in (a_{211}, a_{111}),\\
&(\bar{x}_1, a_{221}) = \underbrace{(\text{DU,LS})}_{\text{down-up lower-saddle}} && \text{if } \bar{x}_1 \in (a_{111}, \infty),\\
&\text{for } a_{110} < 0 \text{ and } a_{220} < 0;
\end{aligned} \tag{2.20}$$

as presented in Fig. 2.3a–d for the switching bifurcations of the double saddles and hyperbolic-to-hyperbolic secant flows.

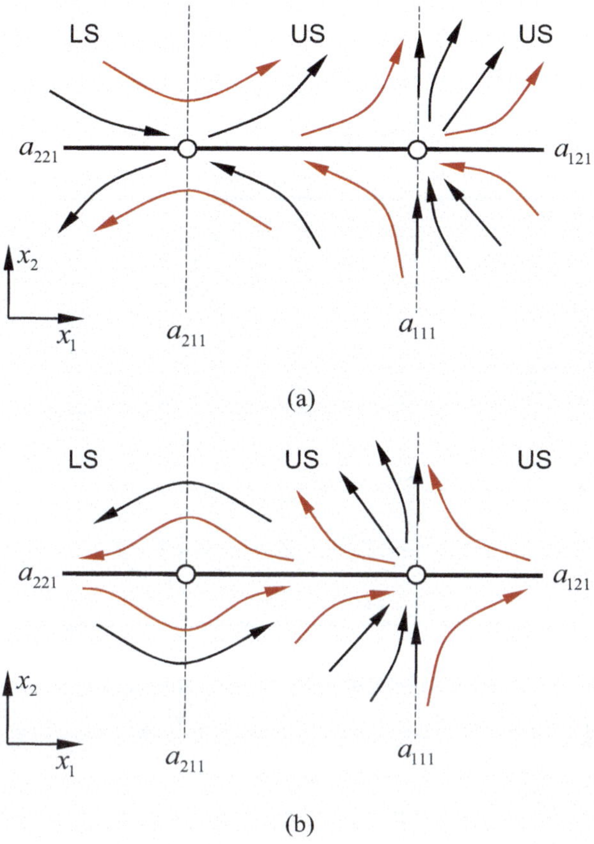

Fig. 2.3 Phase portraits of infinite equilibrium ($a_{221} = a_{121}$ and $a_{211} < a_{111}$) for two-dimensional systems on the x_1-direction with $x_1^* = a_{111}$ and on the x_2-direction with $x_2^* = a_{121} = a_{221}$. The infinite equilibriums: (**a**) ($a_{110} > 0$, $a_{220} > 0$), (**b**) ($a_{110} < 0$, $a_{220} > 0$), (**c**) ($a_{110} > 0$, $a_{220} < 0$), and (**d**) ($a_{110} < 0$, $a_{220} < 0$)

(ii) The equilibriums of $x_1^* = a_{111}, a_{211}$ and $x_2^* = a_{121}, a_{221}$ for $a_{221} = a_{121}$ and $a_{111} < a_{211}$ are

$$\begin{Bmatrix} (a_{111}, a_{221}) \\ (a_{211}, a_{221}) \end{Bmatrix} = \left\{ \begin{array}{c} \underbrace{(\text{UD:UD,LS})}_{\text{up-down lower-saddle}} \\ \underbrace{({}_{\text{UD}}\text{LS:}{}_{\text{DU}}\text{US, DP}_{-}\text{:UP}_{+})}_{\text{hyperbolic lower-to-upper saddle}} \end{array} \right\} \text{for } a_{110} > 0 \text{ and } a_{220} > 0; \quad (2.21)$$

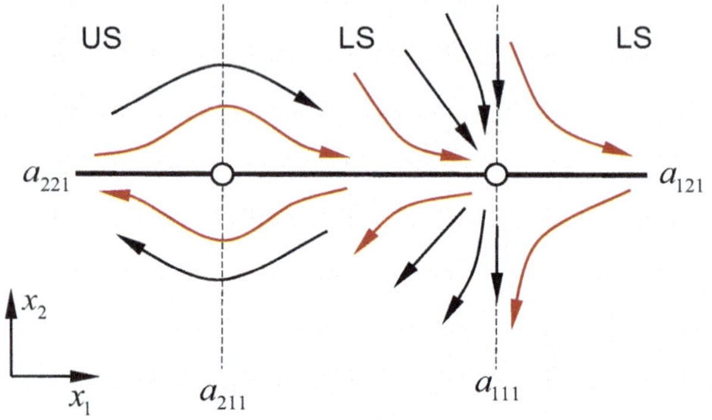

(c)

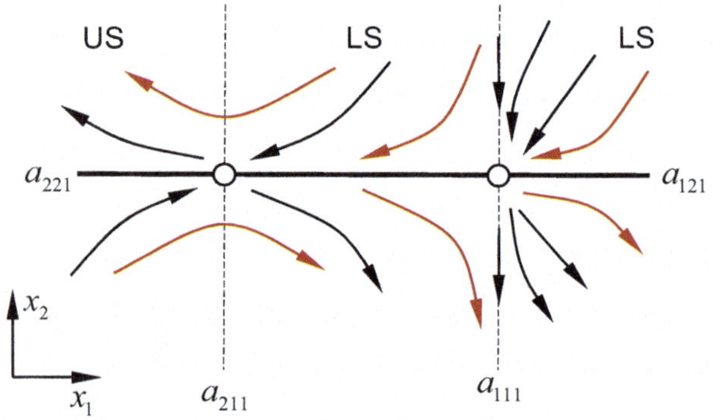

(d)

Fig. 2.3 (continued)

$$\left\{\begin{matrix}(a_{111}, a_{221})\\(a_{211}, a_{221})\end{matrix}\right\} = \left\{\begin{matrix}\underbrace{(\text{DU:DU,LS})}_{\text{down-up lower-saddle}}\\\underbrace{({}_{\text{DU}}\text{LS:}{}_{\text{UD}}\text{US,UP}_{+}\text{:DP}_{-})}_{\text{hyperbolic-secant lower-to-upper saddle}}\end{matrix}\right\} \text{for } a_{110} < 0 \text{ and } a_{220} > 0; \tag{2.22}$$

$$\left\{\begin{matrix}(a_{111}, a_{221})\\(a_{211}, a_{221})\end{matrix}\right\} = \left\{\begin{matrix}\underbrace{(\text{DU:DU,US})}_{\text{down-up upper-saddle}}\\\underbrace{({}_{\text{DU}}\text{US:}{}_{\text{UD}}\text{LS, UP}_{-}\text{:DP}_{+})}_{\text{hyperbolic-secant upper-to-lower saddle}}\end{matrix}\right\} \text{for } a_{110} > 0 \text{ and } a_{220} < 0; \tag{2.23}$$

$$\begin{Bmatrix}(a_{111},a_{221})\\(a_{211},a_{221})\end{Bmatrix}=\left\{\begin{matrix}\underbrace{(\text{UD:UD,US})}_{\text{up-down upper-saddle}}\\\underbrace{({}_{\text{UD}}\text{US:}{}_{\text{DU}}\text{LS,DP}_{+}\text{:UP}_{-})}_{\text{hyperbolic upper-to-lower saddle}}\end{matrix}\right\}\text{for } a_{110}<0 \text{ and } a_{220}<0; \tag{2.24}$$

and the up-down and down-up upper saddle and lower saddle infinite equilibriums of $x_2^*=a_{121}=a_{221}$ are

$$\begin{aligned}&(\bar{x}_1,a_{221})=\underbrace{(\text{UD,LS})}_{\text{up-down lower-saddle}}\quad\text{if } \bar{x}_1\in(-\infty,a_{111}),\\&(\bar{x}_1,a_{221})=\underbrace{(\text{UD,LS})}_{\text{up-down lower-saddle}}\quad\text{if } \bar{x}_1\in(a_{111},a_{211}),\\&(\bar{x}_1,a_{221})=\underbrace{(\text{DU,US})}_{\text{down-up upper-saddle}}\quad\text{if } \bar{x}_1\in(a_{211},\infty),\\&\text{for } a_{110}>0 \text{ and } a_{220}>0;\end{aligned} \tag{2.25}$$

$$\begin{aligned}&(\bar{x}_1,a_{221})=\underbrace{(\text{DU,LS})}_{\text{down-up lower-saddle}}\quad\text{if } \bar{x}_1\in(-\infty,a_{111}),\\&(\bar{x}_1,a_{221})=\underbrace{(\text{DU,LS})}_{\text{down-up upper-saddle}}\quad\text{if } \bar{x}_1\in(a_{111},a_{211}),\\&(\bar{x}_1,a_{221})=\underbrace{(\text{UD,US})}_{\text{up-down upper-saddle}}\quad\text{if } \bar{x}_1\in(a_{211},\infty),\\&\text{for } a_{110}<0 \text{ and } a_{220}>0;\end{aligned} \tag{2.26}$$

$$\begin{aligned}&(\bar{x}_1,a_{221})=\underbrace{(\text{DU,US})}_{\text{down-up upper-saddle}}\quad\text{if } \bar{x}_1\in(-\infty,a_{111}),\\&(\bar{x}_1,a_{221})=\underbrace{(\text{DU,US})}_{\text{down-up upper-saddle}}\quad\text{if } \bar{x}_1\in(a_{111},a_{211}),\\&(\bar{x}_1,a_{221})=\underbrace{(\text{UD,LS})}_{\text{up-down lower-saddle}}\quad\text{if } \bar{x}_1\in(a_{211},\infty),\\&\text{for } a_{110}<0 \text{ and } a_{220}>0;\end{aligned} \tag{2.27}$$

$$\begin{aligned}&(\bar{x}_1,a_{221})=\underbrace{(\text{UD,US})}_{\text{up-down upper-saddle}}\quad\text{if } \bar{x}_1\in(-\infty,a_{111})\\&(\bar{x}_1,a_{221})=\underbrace{(\text{UD,US})}_{\text{up-down upper-saddle}}\quad\text{if } \bar{x}_1\in(a_{111},a_{211}),\\&(\bar{x}_1,a_{221})=\underbrace{(\text{DU,LS})}_{\text{down-up lower-saddle}}\quad\text{if } \bar{x}_1\in(a_{211},\infty),\\&\text{for } a_{110}<0 \text{ and } a_{220}<0;\end{aligned} \tag{2.28}$$

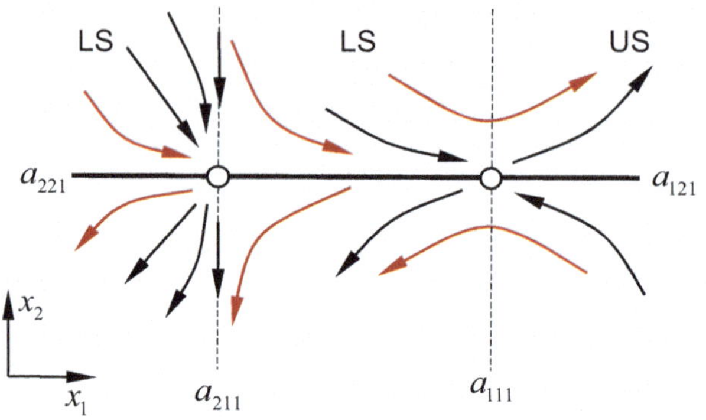

(a)

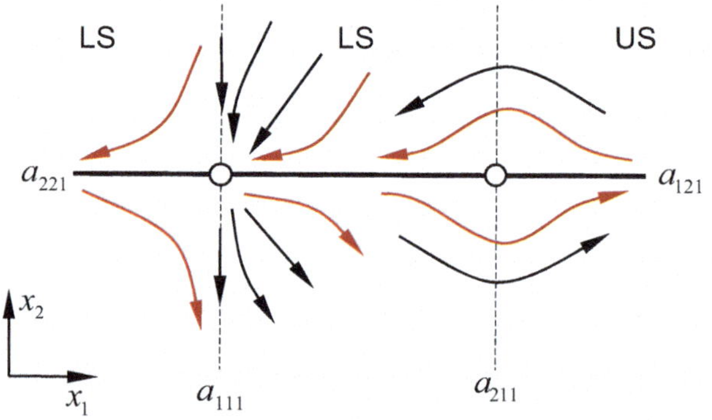

(b)

Fig. 2.4 Phase portraits of infinite equilibrium ($a_{221} = a_{121}$ and $a_{211} < a_{111}$) for two-dimensional systems on the x_1-direction with $x_1^* = a_{111}$ and on the x_2-direction with $x_2^* = a_{121} = a_{221}$. The infinite equilibriums: (**a**) ($a_{110} > 0$, $a_{220} > 0$), (**b**) ($a_{110} < 0$, $a_{220} > 0$), (**c**) ($a_{110} > 0$, $a_{220} < 0$), and (**d**) ($a_{110} < 0$, $a_{220} < 0$)

as presented in Fig. 2.4a–d for the switching bifurcations of the double saddles and hyperbolic-to-hyperbolic secant flows.

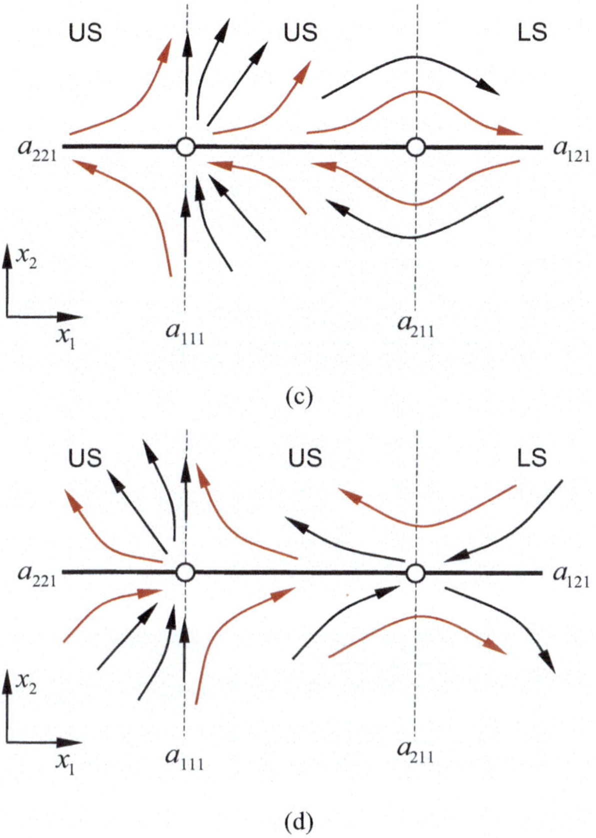

Fig. 2.4 (continued)

2.2.2 *Vertical Up-Down Saddle Infinite Equilibriums*

Consider a dynamical system as

$$\begin{aligned}\dot{x}_1 &= a_{110}(x_1 - a_{111})^2(x_2 - a_{121}),\\ \dot{x}_2 &= a_{220}(x_1 - a_{211})(x_2 - a_{221})^2,\end{aligned} \tag{2.29}$$

and the corresponding first integral manifold for $a_{111} = a_{211}$, $a_{121} \neq a_{221}$ is

$$\begin{aligned}\ln \frac{|\, x_1 - a_{111} \,|}{|\, x_{10} - a_{111} \,|} = \frac{a_{110}}{a_{220}} \Big[&\ln \frac{|\, x_2 - a_{221} \,|}{|\, x_{20} - a_{221} \,|} \\ &- (a_{221} - a_{121}) (\frac{1}{x_2 - a_{221}} - \frac{1}{x_{20} - a_{221}}) \Big].\end{aligned} \tag{2.30}$$

(i) The equilibriums of $x_1^* = a_{111}, a_{211}$ and $x_2^* = a_{121}, a_{221}$ for $a_{111} = a_{211}$ and $a_{221} < a_{121}$ are

$$\begin{Bmatrix} (a_{111}, a_{221}) \\ (a_{111}, a_{121}) \end{Bmatrix} = \begin{Bmatrix} \underbrace{(\text{US,DU:DU})}_{\text{down-up lower-saddle}} \\ \underbrace{(\text{DP}_-\text{:UP}_+, {}_{\text{UD}}\text{LS:}_{\text{DU}}\text{US})}_{\text{hyperbolic lower-to-upper saddle}} \end{Bmatrix} \text{ for } a_{110} > 0 \text{ and } a_{220} > 0; \quad (2.31)$$

$$\begin{Bmatrix} (a_{111}, a_{221}) \\ (a_{111}, a_{121}) \end{Bmatrix} = \begin{Bmatrix} \underbrace{(\text{LS,UD:UD})}_{\text{up-down lower-saddle}} \\ \underbrace{(\text{UP}_-\text{:DP}_+, {}_{\text{DU}}\text{US:}_{\text{UD}}\text{LS})}_{\text{hyperbolic-secant upper-to-lower saddle}} \end{Bmatrix} \text{ for } a_{110} < 0 \text{ and } a_{220} > 0; \quad (2.32)$$

$$\begin{Bmatrix} (a_{111}, a_{221}) \\ (a_{111}, a_{121}) \end{Bmatrix} = \begin{Bmatrix} \underbrace{(\text{US,UD:UD})}_{\text{up-down upper-saddle}} \\ \underbrace{(\text{UP}_+\text{:DP}_-, {}_{\text{DU}}\text{LS:}_{\text{UD}}\text{US})}_{\text{hyperbolic-secant lower-to-upper saddle}} \end{Bmatrix} \text{ for } a_{110} > 0 \text{ and } a_{220} < 0; \quad (2.33)$$

$$\begin{Bmatrix} (a_{111}, a_{121}) \\ (a_{111}, a_{221}) \end{Bmatrix} = \begin{Bmatrix} \underbrace{(\text{LS,DU:DU})}_{\text{down-up lower-saddle}} \\ \underbrace{(\text{DP}_+\text{:UP}_-, {}_{\text{UD}}\text{US:}_{\text{DU}}\text{LS})}_{\text{hyperbolic upper-to-lower saddle}} \end{Bmatrix} \text{ for } a_{110} < 0 \text{ and } a_{220} < 0; \quad (2.34)$$

and the infinite equilibriums of $x_1^* = a_{111} = a_{211}$ are

$$\begin{aligned} (a_{111}, \bar{x}_2) &= \underbrace{(\text{LS,UD})}_{\text{up-down lower-saddle}} \quad \text{if } \bar{x}_2 \in (-\infty, a_{121}), \\ (a_{111}, \bar{x}_2) &= \underbrace{(\text{US,DU})}_{\text{down-up lower-saddle}} \quad \text{if } \bar{x}_2 \in (a_{121}, a_{221}), \\ (a_{111}, \bar{x}_2) &= \underbrace{(\text{US,DU})}_{\text{down-up upper-saddle}} \quad \text{if } \bar{x}_2 \in (a_{221}, \infty), \\ &\text{for } a_{110} > 0 \text{ and } a_{220} > 0; \end{aligned} \quad (2.35)$$

$$\left.\begin{aligned}
(a_{111},\bar{x}_2) &= \underbrace{(\text{US,DU})}_{\text{down-up upper-saddle}} && \text{if } \bar{x}_2 \in (-\infty, a_{121}),\\
(a_{111},\bar{x}_2) &= \underbrace{(\text{LS,UD})}_{\text{up-down lower-saddle}} && \text{if } \bar{x}_2 \in (a_{121}, a_{221}),\\
(a_{111},\bar{x}_2) &= \underbrace{(\text{LS,UD})}_{\text{up-down lower-saddle}} && \text{if } \bar{x}_2 \in (a_{221}, \infty),
\end{aligned}\right. \tag{2.36}$$

for $a_{110} < 0$ and $a_{220} > 0$;

$$\left.\begin{aligned}
(a_{111},\bar{x}_2) &= \underbrace{(\text{LS,DU})}_{\text{down-up lower-saddle}} && \text{if } \bar{x}_2 \in (-\infty, a_{121}),\\
(a_{111},\bar{x}_2) &= \underbrace{(\text{US,UD})}_{\text{up-down upper-saddle}} && \text{if } \bar{x}_2 \in (a_{121}, a_{221}),\\
(a_{111},\bar{x}_2) &= \underbrace{(\text{US,UD})}_{\text{up-down upper-saddle}} && \text{if } \bar{x}_2 \in (a_{221}, \infty),
\end{aligned}\right. \tag{2.37}$$

for $a_{110} > 0$ and $a_{220} < 0$;

$$\left.\begin{aligned}
(a_{111},\bar{x}_2) &= \underbrace{(\text{US,UD})}_{\text{up-down upper-saddle}} && \text{if } \bar{x}_2 \in (-\infty, a_{121}),\\
(a_{111},\bar{x}_2) &= \underbrace{(\text{LS,DU})}_{\text{down-up lower-saddle}} && \text{if } \bar{x}_2 \in (a_{121}, a_{221}),\\
(a_{111},\bar{x}_2) &= \underbrace{(\text{LS,DU})}_{\text{down-up lower-saddle}} && \text{if } \bar{x}_2 \in (a_{221}, \infty),
\end{aligned}\right. \tag{2.38}$$

for $a_{110} < 0$ and $a_{220} < 0$.

The infinite-equilibrium of $x_1^* = a_{111} = a_{211}$ is up-down and down-up, upper saddles and lower saddles, as shown in Fig. 2.5. The hyperbolic and hyperbolic secant, lower-to-upper saddle, and upper-to-lower saddle are illustrated.

(ii) The equilibriums of $x_1^* = a_{111}, a_{211}$ and $x_2^* = a_{121}, a_{221}$ for $a_{111} = a_{211}$ and $a_{121} < a_{221}$ are

$$\left\{\begin{matrix}(a_{111}, a_{121})\\(a_{111}, a_{221})\end{matrix}\right\} = \left\{\begin{matrix}\underbrace{(\text{DP}_-\text{:UP}_+, {}_{\text{UD}}\text{LS:}_{\text{DU}}\text{US})}_{\text{hyperbolic lower-to-upper saddle}}\\ \underbrace{(\text{LS,UD:UD})}_{\text{up-down lower-saddle}}\end{matrix}\right\} \text{ for } a_{110} > 0 \text{ and } a_{220} > 0; \tag{2.39}$$

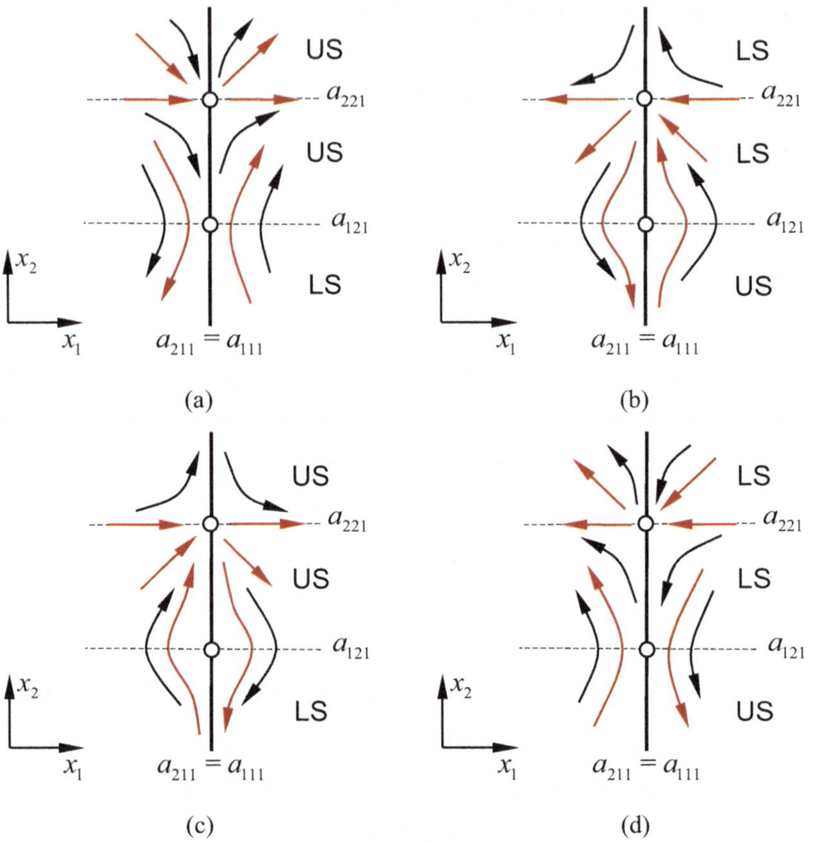

Fig. 2.5 Phase portraits ($a_{111} = a_{211}$ and $a_{121} < a_{221}$) for two-dimensional systems on the x_1-direction with $x_1^* = a_{111}$ and on the x_2-direction with $x_2^* = a_{121}, a_{221}$. (**a**) ($a_{110} > 0$, $a_{220} > 0$), (**b**) ($a_{110} < 0$, $a_{220} > 0$), (**c**) ($a_{110} > 0$, $a_{220} < 0$), and (**d**) ($a_{110} < 0$, $a_{220} < 0$)

$$\begin{Bmatrix}(a_{111}, a_{121})\\(a_{111}, a_{221})\end{Bmatrix} = \left\{\begin{array}{c}\underbrace{(\mathrm{UP}_{-}{:}\mathrm{DP}_{+},{}_{\mathrm{DU}}\mathrm{US}{:}_{\mathrm{UD}}\mathrm{LS})}_{\text{hyperbolic-secant upper-to-lower saddle}}\\ \underbrace{(\mathrm{US,DU{:}DU})}_{\text{down-up upper-saddle}}\end{array}\right\} \text{ for } a_{110} < 0 \text{ and } a_{220} > 0; \tag{2.40}$$

$$\begin{Bmatrix}(a_{111}, a_{121})\\(a_{111}, a_{221})\end{Bmatrix} = \left\{\begin{array}{c}\underbrace{(\mathrm{UP}_{+}{:}\mathrm{DP}_{-},{}_{\mathrm{DU}}\mathrm{LS}{:}_{\mathrm{UD}}\mathrm{US})}_{\text{hyperbolic-secant lower-to-upper saddle}}\\ \underbrace{(\mathrm{LS,DU{:}DU})}_{\text{down-up lower-saddle}}\end{array}\right\} \text{ for } a_{110} > 0 \text{ and } a_{220} < 0; \tag{2.41}$$

$$\begin{Bmatrix} (a_{111}, a_{121}) \\ (a_{111}, a_{221}) \end{Bmatrix} = \left\{ \begin{array}{c} \underbrace{(\mathrm{DP}_{+}{:}\mathrm{UP}_{-}, {}_{\mathrm{UD}}\mathrm{US}{:}_{\mathrm{DU}}\mathrm{LS})}_{\text{hyperbolic upper-to-lower saddle}} \\ \underbrace{(\mathrm{US},\mathrm{UD}{:}\mathrm{UD})}_{\text{up-down upper-saddle}} \end{array} \right\} \text{ for } a_{110} < 0 \text{ and } a_{220} < 0; \quad (2.42)$$

and the infinite equilibriums of $x_1^* = a_{111} = a_{211}$ are

$$\begin{aligned} (a_{111}, \bar{x}_2) &= \underbrace{(\mathrm{LS},\mathrm{UD})}_{\text{up-down lower-saddle}} \quad \text{if } \bar{x}_2 \in (-\infty, a_{221}), \\ (a_{111}, \bar{x}_2) &= \underbrace{(\mathrm{LS},\mathrm{UD})}_{\text{up-down lower-saddle}} \quad \text{if } \bar{x}_2 \in (a_{221}, a_{121}), \\ (a_{111}, \bar{x}_2) &= \underbrace{(\mathrm{US},\mathrm{DU})}_{\text{down-up upper-saddle}} \quad \text{if } \bar{x}_2 \in (a_{121}, \infty), \\ &\text{for } a_{110} > 0 \text{ and } a_{220} > 0; \end{aligned} \quad (2.43)$$

$$\begin{aligned} (a_{111}, \bar{x}_2) &= \underbrace{(\mathrm{US},\mathrm{DU})}_{\text{down-up upper-saddle}} \quad \text{if } \bar{x}_2 \in (-\infty, a_{221}), \\ (a_{111}, \bar{x}_2) &= \underbrace{(\mathrm{US},\mathrm{DU})}_{\text{down-up upper-saddle}} \quad \text{if } \bar{x}_2 \in (a_{221}, a_{121}), \\ (a_{111}, \bar{x}_2) &= \underbrace{(\mathrm{LS},\mathrm{UD})}_{\text{up-down lower-saddle}} \quad \text{if } \bar{x}_2 \in (a_{121}, \infty), \\ &\text{for } a_{110} < 0 \text{ and } a_{220} > 0; \end{aligned} \quad (2.44)$$

$$\begin{aligned} (a_{111}, \bar{x}_2) &= \underbrace{(\mathrm{LS},\mathrm{DU})}_{\text{down-up lower-saddle}} \quad \text{if } \bar{x}_2 \in (-\infty, a_{221}), \\ (a_{111}, \bar{x}_2) &= \underbrace{(\mathrm{LS},\mathrm{DU})}_{\text{down-up lower-saddle}} \quad \text{if } \bar{x}_2 \in (a_{221}, a_{121}), \\ (a_{111}, \bar{x}_2) &= \underbrace{(\mathrm{US},\mathrm{UD})}_{\text{up-down upper-saddle}} \quad \text{if } \bar{x}_2 \in (a_{121}, \infty), \\ &\text{for } a_{110} > 0 \text{ and } a_{220} < 0; \end{aligned} \quad (2.45)$$

$$\begin{aligned} (a_{111}, \bar{x}_2) &= \underbrace{(\mathrm{US},\mathrm{UD})}_{\text{up-down upper-saddle}} \quad \text{if } \bar{x}_2 \in (-\infty, a_{221}), \\ (a_{111}, \bar{x}_2) &= \underbrace{(\mathrm{US},\mathrm{UD})}_{\text{up-down upper-saddle}} \quad \text{if } \bar{x}_2 \in (a_{221}, a_{121}), \\ (a_{111}, \bar{x}_2) &= \underbrace{(\mathrm{LS},\mathrm{DU})}_{\text{down-up lower-saddle}} \quad \text{if } \bar{x}_2 \in (a_{121}, \infty), \\ &\text{for } a_{110} < 0 \text{ and } a_{220} < 0. \end{aligned} \quad (2.46)$$

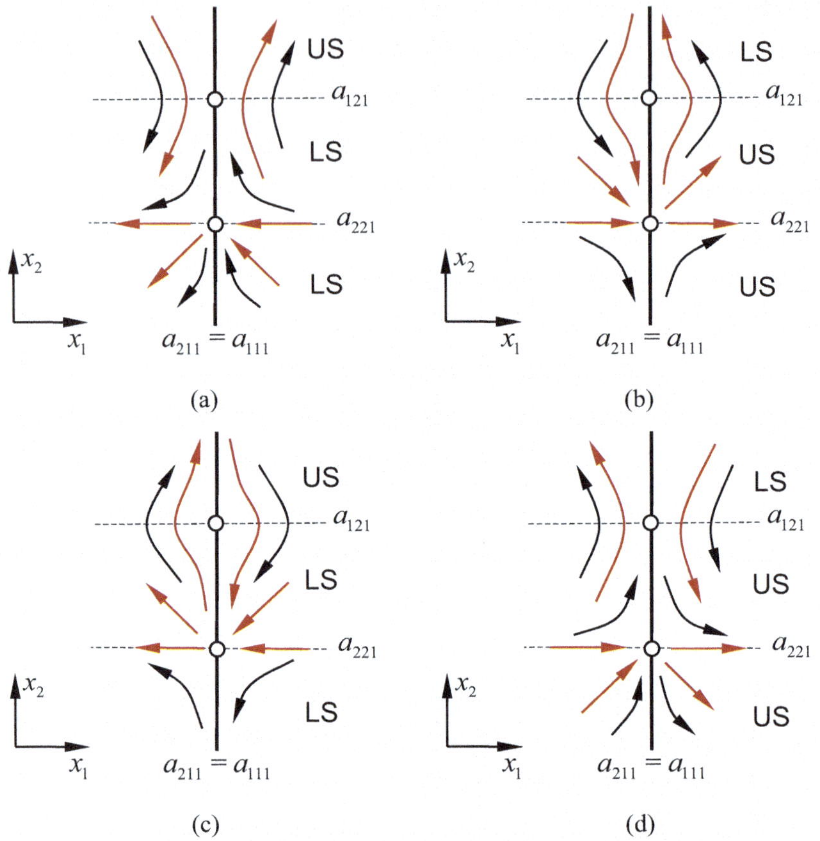

Fig. 2.6 Phase portraits ($a_{111} = a_{211}$ and $a_{221} < a_{121}$) for two-dimensional systems on the x_1-direction with $x_1^* = a_{111}$ and on the x_2-direction with $x_2^* = a_{121}, a_{221}$. (**a**) ($a_{110} > 0$, $a_{220} > 0$), (**b**) ($a_{110} < 0$, $a_{220} > 0$), (**c**) ($a_{110} > 0$, $a_{220} < 0$), and (**d**) ($a_{110} < 0$, $a_{220} < 0$)

The infinite equilibrium of $x_1^* = a_{111} = a_{211}$ is up-down and down-up, upper saddles and lower saddles, as shown in Fig. 2.6. The hyperbolic and hyperbolic secant, lower-to-upper saddle, and upper-to-lower saddle are also illustrated.

2.2.3 Double-Switching Bifurcations and Infinite Equilibriums

Consider a dynamical system as

$$\begin{aligned}\dot{x}_1 &= a_{110}(x_1 - a_{111})^2(x_2 - a_{121}),\\ \dot{x}_2 &= a_{220}(x_1 - a_{211})\,(x_2 - a_{221})^2,\end{aligned} \tag{2.47}$$

and the corresponding first integral manifold for $a_{111} = a_{211}$and $a_{121} = a_{221}$ is

$$\ln \frac{|x_1 - a_{111}|}{|x_{10} - a_{111}|} = \frac{a_{110}}{a_{220}} \ln \frac{|x_2 - a_{221}|}{|x_{20} - a_{221}|}. \tag{2.48}$$

The equilibriums of $x_1^* = a_{111} = a_{211}$ and $x_2^* = a_{121} = a_{221}$ at the intersection of two infinite-infinite equilibriums are

$$\left.\begin{aligned}
(a_{121}, a_{211}) &= \underbrace{(\mathrm{DU}_+, \mathrm{DU}_+)}_{\text{positive diagonal saddle}} \text{ for } a_{110} > 0 \text{ and } a_{220} > 0;\\
(a_{121}, a_{211}) &= \underbrace{(\mathrm{UD}_+, \mathrm{UD}_-)}_{\text{CCW saddle}} \text{ for } a_{110} < 0 \text{ and } a_{220} > 0;\\
(a_{121}, a_{211}) &= \underbrace{(\mathrm{UD}_-, \mathrm{UD}_+)}_{\text{CW saddle}} \text{ for } a_{110} > 0 \text{ and } a_{220} < 0;\\
(a_{121}, a_{211}) &= \underbrace{(\mathrm{DU}_-, \mathrm{DU}_-)}_{\text{negative diagonal saddle}} \text{ for } a_{110} < 0 \text{ and } a_{220} < 0;
\end{aligned}\right\} \tag{2.49}$$

and the corresponding infinite equilibrium of $x_1^* = a_{121} = a_{221}$ is given by

$$\begin{aligned}
(\bar{x}_1, a_{121}) &= \underbrace{(\mathrm{UD,LS})}_{\text{up-down lower-saddle}} \quad \text{if } \bar{x}_1 \in (-\infty, a_{111}),\\
(\bar{x}_1, a_{121}) &= \underbrace{(\mathrm{DU,US})}_{\text{down-up upper-saddle}} \quad \text{if } \bar{x}_1 \in (a_{111}, \infty)\\
&\text{for } a_{110} > 0 \text{ and } a_{220} > 0;
\end{aligned} \tag{2.50}$$

$$\begin{aligned}
(\bar{x}_1, a_{121}) &= \underbrace{(\mathrm{DU,LS})}_{\text{down-up lower-saddle}} \quad \text{if } \bar{x}_1 \in (-\infty, a_{111}),\\
(\bar{x}_1, a_{121}) &= \underbrace{(\mathrm{UD,US})}_{\text{up-down upper-saddle}} \quad \text{if } \bar{x}_1 \in (a_{111}, \infty)\\
&\text{for } a_{110} < 0 \text{ and } a_{220} > 0;
\end{aligned} \tag{2.51}$$

$$\begin{aligned}
(\bar{x}_1, a_{121}) &= \underbrace{(\mathrm{DU,US})}_{\text{down-up upper-saddle}} \quad \text{if } \bar{x}_1 \in (-\infty, a_{111}),\\
(\bar{x}_1, a_{121}) &= \underbrace{(\mathrm{UD,LS})}_{\text{up-down lower-saddle}} \quad \text{if } \bar{x}_1 \in (a_{111}, \infty)\\
&\text{for } a_{110} > 0 \text{ and } a_{220} < 0;
\end{aligned} \tag{2.52}$$

$$
\begin{aligned}
&(\bar{x}_1, a_{121}) = \underbrace{\text{(UD,US)}}_{\text{up-down upper-saddle}} && \text{if } \bar{x}_1 \in (-\infty, a_{111}),\\
&(\bar{x}_1, a_{121}) = \underbrace{\text{(DU,LS)}}_{\text{down-up lower-saddle}} && \text{if } \bar{x}_1 \in (a_{111}, \infty);\\
&\text{for } a_{110} < 0 \text{ and } a_{220} < 0,
\end{aligned}
\tag{2.53}
$$

and the infinite equilibrium of $x_1^* = a_{111} = a_{211}$ is given as

$$
\begin{aligned}
&(a_{111}, \bar{x}_2) = \underbrace{\text{(UD,LS)}}_{\text{up-down lower-saddle}} && \text{if } \bar{x}_2 \in (-\infty, a_{121}),\\
&(a_{111}, \bar{x}_1) = \underbrace{\text{(DU,US)}}_{\text{down-up upper-saddle}} && \text{if } \bar{x}_2 \in (a_{121}, \infty)\\
&\text{for } a_{110} > 0 \text{ and } a_{220} > 0;
\end{aligned}
\tag{2.54}
$$

$$
\begin{aligned}
&(a_{111}, \bar{x}_1) = \underbrace{\text{(DU,US)}}_{\text{down-up upper-saddle}} && \text{if } \bar{x}_2 \in (-\infty, a_{121}),\\
&(a_{111}, \bar{x}_2) = \underbrace{\text{(UD,LS)}}_{\text{up-down lower-saddle}} && \text{if } \bar{x}_2 \in (a_{121}, \infty)\\
&\text{for } a_{110} < 0 \text{ and } a_{220} > 0;
\end{aligned}
\tag{2.55}
$$

$$
\begin{aligned}
&(a_{111}, \bar{x}_1) = \underbrace{\text{(DU,LS)}}_{\text{down-up lower-saddle}} && \text{if } \bar{x}_2 \in (-\infty, a_{121}),\\
&(a_{111}, \bar{x}_2) = \underbrace{\text{(UD,US)}}_{\text{up-down upper-saddle}} && \text{if } \bar{x}_2 \in (a_{121}, \infty)\\
&\text{for } a_{110} > 0 \text{ and } a_{220} < 0;
\end{aligned}
\tag{2.56}
$$

$$
\begin{aligned}
&(a_{111}, \bar{x}_2) = \underbrace{\text{(UD,US)}}_{\text{up-down upper-saddle}} && \text{if } \bar{x}_2 \in (-\infty, a_{121}),\\
&(a_{111}, \bar{x}_1) = \underbrace{\text{(DU,LS)}}_{\text{down-up lower-saddle}} && \text{if } \bar{x}_2 \in (a_{121}, \infty)\\
&\text{for } a_{110} < 0 \text{ and } a_{220} < 0.
\end{aligned}
\tag{2.57}
$$

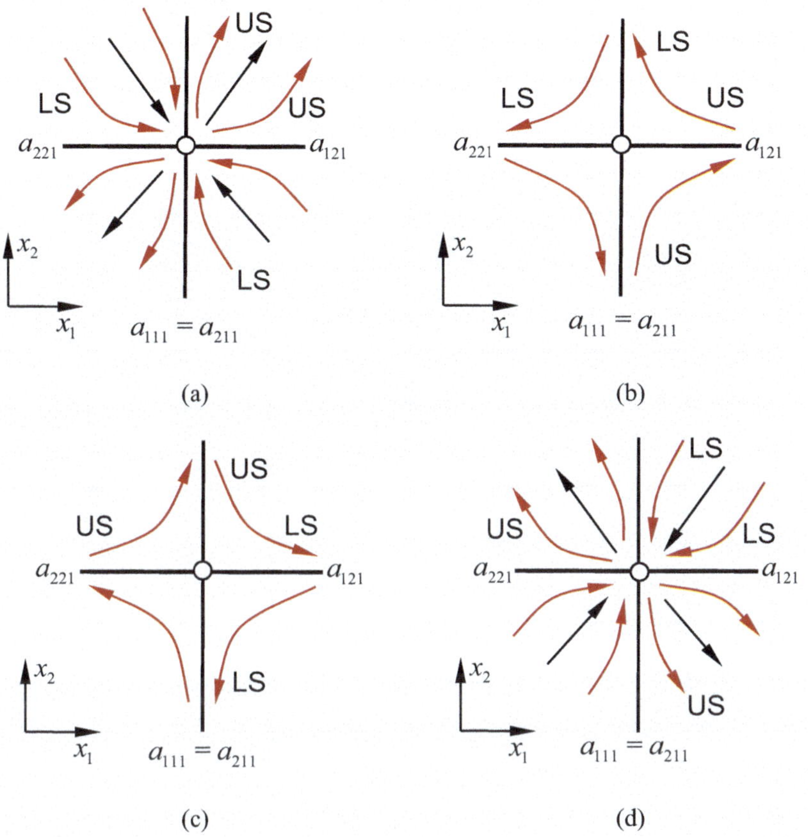

Fig. 2.7 Phase portraits of infinite equilibrium ($a_{221} = a_{121}$ and $a_{111} = a_{211}$) for two-dimensional systems on the x_1-direction with $x_1^* = a_{111} = a_{211}$ and on the x_2-direction with $x_2^* = a_{121} = a_{221}$. The infinite equilibriums: (**a**) ($a_{110} > 0$, $a_{220} > 0$), (**b**) ($a_{110} < 0$, $a_{220} > 0$), (**c**) ($a_{110} > 0$, $a_{220} < 0$), and (**d**) ($a_{110} < 0$, $a_{220} < 0$)

Such two infinite equilibriums and the corresponding intersected points are presented in phase space, as shown in Fig. 2.7. The positive diagonal saddle, counterclockwise saddle, clockwise saddle, and negative diagonal saddle are presented, which are for double-switching bifurcations.

Chapter 3
Vertically Paralleled Saddle-Source and Saddle-Sink

In this chapter, the vertically paralleled saddle-source and saddle-sink with hyperbolic flow singularity exist in equilibrium networks, and the center or saddle with separated hyperbolic and hyperbolic-secant flows are paralleled. The hyperbolic flow singularity is for the appearing bifurcations of the paralleled hyperbolic and hyperbolic-secant flows. The switching bifurcations for center (saddle) with paralleled hyperbolic (hyperbolic secant) flow and saddle source (sink) with connected hyperbolic singular flows are discussed. In addition, the switching bifurcations of the saddle source (sink) with connected saddle source (sink) and the saddle and center with paralleled hyperbolic singular flows are discussed through infinite equilibriums.

3.1 (3 × 2) Equilibrium Matrices with Singularity

Consider a dynamical system as

$$\begin{aligned}\dot{x}_1 &= a_{110}(x_1 - a_{111})^2(x_2 - a_{121}),\\ \dot{x}_2 &= a_{220}(x_1 - a_{211})(x_2 - a_{221})(x_2 - a_{222}),\end{aligned} \tag{3.1}$$

and the corresponding first integral manifold is

$$\begin{aligned}&\ln\frac{|x_1 - a_{111}|}{|x_{10} - a_{111}|} - (a_{111} - a_{211})\left(\frac{1}{x_1 - a_{111}} - \frac{1}{x_{10} - a_{111}}\right)\\ &= \frac{a_{110}}{a_{220}}\left(\frac{a_{221} - a_{121}}{a_{221} - a_{222}}\ln\frac{|x_2 - a_{221}|}{|x_{20} - a_{221}|} + \frac{a_{222} - a_{121}}{a_{222} - a_{221}}\ln\frac{|x_2 - a_{222}|}{|x_{20} - a_{222}|}\right).\end{aligned} \tag{3.2}$$

A. C. J. Luo, *Two-Dimensional Two-Product Cubic Systems, Vol. II*,
https://doi.org/10.1007/978-3-031-48491-9_3

(i) The equilibriums of $x_1^* = a_{111}, a_{211}$ and $x_2^* = a_{121}, a_{221}, a_{222}$ for $a_{211} < a_{111}$ and $a_{121} < a_{221}$ are

$$\left\{\begin{matrix} (a_{211}, a_{222}) & (a_{111}, a_{222}) \\ (a_{211}, a_{221}) & (a_{111}, a_{221}) \\ (a_{211}, a_{121}) & (a_{111}, a_{121}) \end{matrix}\right\} = \left\{\begin{matrix} \underbrace{(\text{pF,DP:UP})}_{\text{hyperbolic flow }(+)} & \underbrace{(\text{US,SO})}_{\text{upper-saddle source}} \\ \underbrace{(\text{pF,UP:DP})}_{\text{hyperbolic-secant flow }(+)} & \underbrace{(\text{US,SI})}_{\text{upper-saddle sink}} \\ \underbrace{(\text{UP}_+,\text{UP}_+)}_{\text{positive saddle}} & \underbrace{(\text{UP:UP,pF})}_{\text{HS to HB flow }(+)} \end{matrix}\right\} \tag{3.3}$$

for $a_{110} > 0$ and $a_{220} > 0$;

$$\left\{\begin{matrix} (a_{211}, a_{222}) & (a_{111}, a_{222}) \\ (a_{211}, a_{221}) & (a_{111}, a_{221}) \\ (a_{211}, a_{121}) & (a_{111}, a_{121}) \end{matrix}\right\} = \left\{\begin{matrix} \underbrace{(\text{nF,UP:DP})}_{\text{hyperbolic-secant flow }(-)} & \underbrace{(\text{LS,SO})}_{\text{lower-saddle source}} \\ \underbrace{(\text{nF,DP:UP})}_{\text{hyperbolic flow }(-)} & \underbrace{(\text{LS,SI})}_{\text{lower-saddle sink}} \\ \underbrace{(\text{DP}_+,\text{DP}_-)}_{\text{CCW center}} & \underbrace{(\text{DP:DP,pF})}_{\text{HB to HS flow }(+)} \end{matrix}\right\} \tag{3.4}$$

for $a_{110} < 0$ and $a_{220} > 0$;

$$\left\{\begin{matrix} (a_{211}, a_{222}) & (a_{111}, a_{222}) \\ (a_{211}, a_{221}) & (a_{111}, a_{221}) \\ (a_{211}, a_{121}) & (a_{111}, a_{121}) \end{matrix}\right\} = \left\{\begin{matrix} \underbrace{(\text{pF,UP:DP})}_{\text{hyperbolic-secant flow }(+)} & \underbrace{(\text{US,SI})}_{\text{upper-saddle sink}} \\ \underbrace{(\text{pF,DP:UP})}_{\text{hyperbolic flow }(+)} & \underbrace{(\text{US,SO})}_{\text{upper-saddle source}} \\ \underbrace{(\text{DP}_-,\text{DP}_+)}_{\text{CW center}} & \underbrace{(\text{DP:DP,nF})}_{\text{HB to HS flow }(-)} \end{matrix}\right\} \tag{3.5}$$

for $a_{110} > 0$ and $a_{220} < 0$;

$$\left\{\begin{matrix} (a_{211}, a_{222}) & (a_{111}, a_{222}) \\ (a_{211}, a_{221}) & (a_{111}, a_{221}) \\ (a_{211}, a_{121}) & (a_{111}, a_{121}) \end{matrix}\right\} = \left\{\begin{matrix} \underbrace{(\text{nF,DP:UP})}_{\text{hyperbolic flow }(-)} & \underbrace{(\text{LS,SI})}_{\text{lower-saddle sink}} \\ \underbrace{(\text{nF,UP:DP})}_{\text{hyperbolic-secant flow }(-)} & \underbrace{(\text{LS,SO})}_{\text{lower-saddle source}} \\ \underbrace{(\text{UP}_-,\text{UP}_-)}_{\text{negative saddle}} & \underbrace{(\text{UP:UP,nF})}_{\text{HS to HB flow }(-)} \end{matrix}\right\} \tag{3.6}$$

for $a_{110} < 0$ and $a_{220} < 0$;

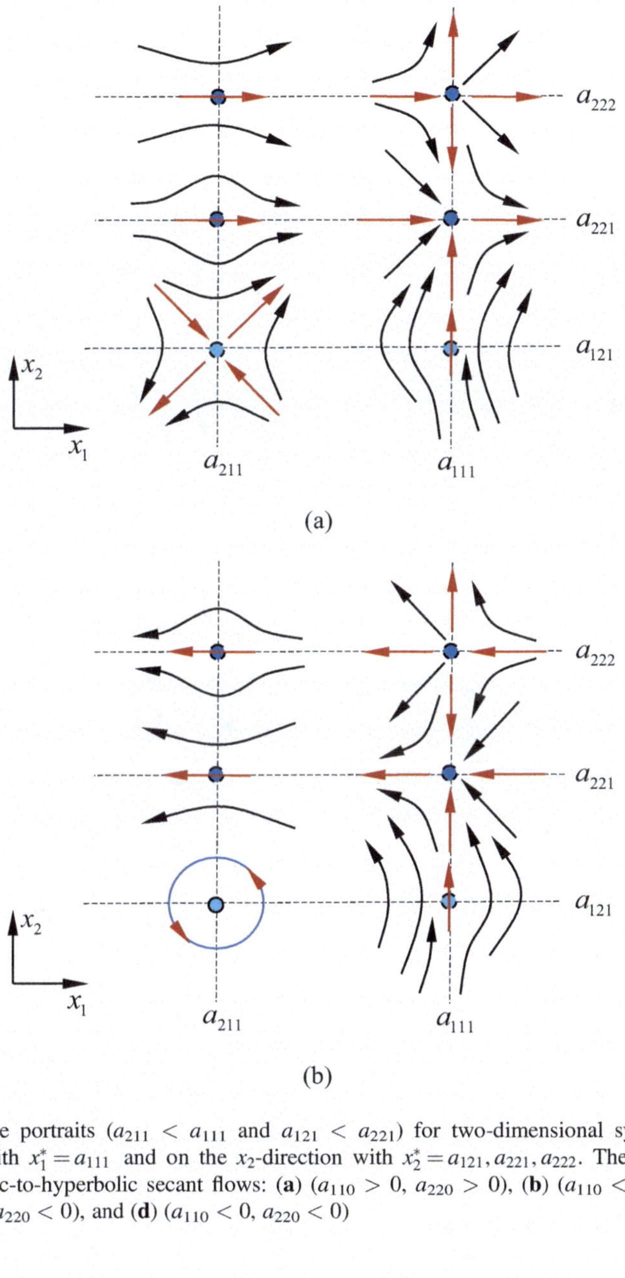

Fig. 3.1 Phase portraits ($a_{211} < a_{111}$ and $a_{121} < a_{221}$) for two-dimensional systems on the x_1-direction with $x_1^* = a_{111}$ and on the x_2-direction with $x_2^* = a_{121}, a_{221}, a_{222}$. The saddle nodes with hyperbolic-to-hyperbolic secant flows: (**a**) ($a_{110} > 0$, $a_{220} > 0$), (**b**) ($a_{110} < 0$, $a_{220} > 0$), (**c**) ($a_{110} > 0$, $a_{220} < 0$), and (**d**) ($a_{110} < 0$, $a_{220} < 0$)

as shown in Fig. 3.1. The saddle sink and saddle source are presented for the appearance of saddle and sink and for saddle and source. The hyperbolic secant-to-hyperbolic flows are for appearances of hyperbolic and hyperbolic secant

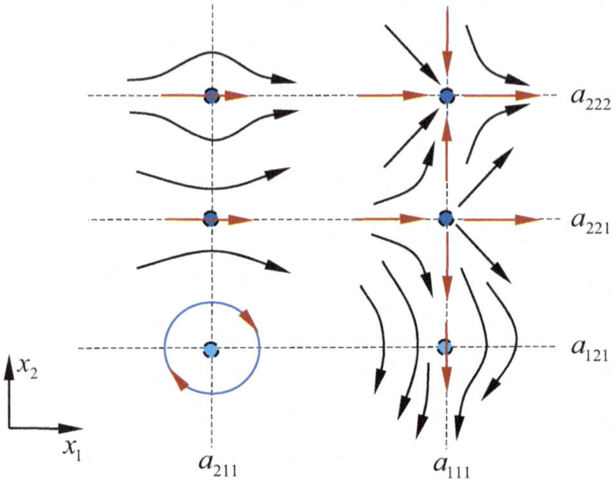

(c)

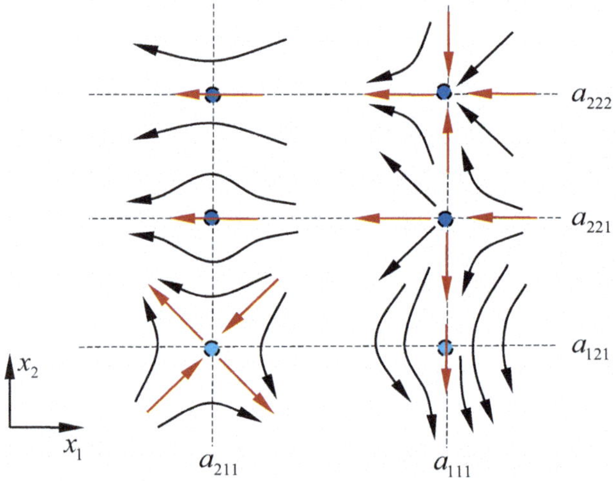

(d)

Fig. 3.1 (continued)

flows. In addition, the hyperbolic and hyperbolic secant flows are presented with positive and negative saddles or clockwise and counterclockwise centers, respectively.

(ii) The equilibriums of $x_1^* = a_{111}, a_{211}$ and $x_2^* = a_{121}, a_{221}, a_{222}$ for $a_{121} \in (a_{221}, a_{222})$ and $a_{211} < a_{111}$ are

$$\begin{Bmatrix} (a_{211}, a_{222}) & (a_{111}, a_{222}) \\ (a_{211}, a_{121}) & (a_{111}, a_{121}) \\ (a_{211}, a_{221}) & (a_{111}, a_{221}) \end{Bmatrix} = \begin{Bmatrix} \underbrace{(\text{pF,DP:UP})}_{\text{hyperbolic flow }(+)} & \underbrace{(\text{US,SO})}_{\text{upper-saddle source}} \\ \underbrace{(\text{DP}_{-},\text{DP}_{+})}_{\text{CW center}} & \underbrace{(\text{DP:DP,nF})}_{\text{HB to HS flow }(-)} \\ \underbrace{(\text{nF,DP:UP})}_{\text{hyperbolic flow }(-)} & \underbrace{(\text{LS,SI})}_{\text{lower-saddle sink}} \end{Bmatrix} \tag{3.7}$$

for $a_{110} > 0$ and $a_{220} > 0$;

$$\begin{Bmatrix} (a_{211}, a_{222}) & (a_{111}, a_{222}) \\ (a_{211}, a_{121}) & (a_{111}, a_{121}) \\ (a_{211}, a_{221}) & (a_{111}, a_{221}) \end{Bmatrix} = \begin{Bmatrix} \underbrace{(\text{nF,UP:DP})}_{\text{hyperbolic-secant flow }(-)} & \underbrace{(\text{LS,SO})}_{\text{lower-saddle source}} \\ \underbrace{(\text{UP}_{-},\text{UP}_{-})}_{\text{negative saddle}} & \underbrace{(\text{UP:UP,nF})}_{\text{HS to HB flow }(-)} \\ \underbrace{(\text{pF,UP:DP})}_{\text{hyperbolic-secant flow }(+)} & \underbrace{(\text{US,SI})}_{\text{upper-saddle sink}} \end{Bmatrix} \tag{3.8}$$

for $a_{110} < 0$ and $a_{220} > 0$;

$$\begin{Bmatrix} (a_{211}, a_{222}) & (a_{111}, a_{222}) \\ (a_{211}, a_{121}) & (a_{111}, a_{121}) \\ (a_{211}, a_{221}) & (a_{111}, a_{221}) \end{Bmatrix} = \begin{Bmatrix} \underbrace{(\text{pF,UP:DP})}_{\text{hyperbolic-secant flow }(+)} & \underbrace{(\text{US,SI})}_{\text{upper-saddle sink}} \\ \underbrace{(\text{UP}_{+},\text{UP}_{+})}_{\text{positive saddle}} & \underbrace{(\text{UP:UP,pF})}_{\text{HS to HB flow }(+)} \\ \underbrace{(\text{nF,UP:DP})}_{\text{hyperbolic-secant flow }(-)} & \underbrace{(\text{LS,SO})}_{\text{lower-saddle source}} \end{Bmatrix} \tag{3.9}$$

for $a_{110} > 0$ and $a_{220} < 0$;

$$\begin{Bmatrix} (a_{211}, a_{222}) & (a_{111}, a_{222}) \\ (a_{211}, a_{121}) & (a_{111}, a_{121}) \\ (a_{211}, a_{221}) & (a_{111}, a_{221}) \end{Bmatrix} = \begin{Bmatrix} \underbrace{(\text{nF,DP:UP})}_{\text{hyperbolic flow }(-)} & \underbrace{(\text{LS,SI})}_{\text{lower-saddle sink}} \\ \underbrace{(\text{DP}_{+},\text{DP}_{-})}_{\text{CCW center}} & \underbrace{(\text{DP:DP,pF})}_{\text{HB to HS flow }(+)} \\ \underbrace{(\text{pF,DP:UP})}_{\text{hyperbolic flow }(+)} & \underbrace{(\text{US,SO})}_{\text{upper-saddle source}} \end{Bmatrix} \tag{3.10}$$

for $a_{110} < 0$ and $a_{220} < 0$;

as shown in Fig. 3.2. In the equilibrium matrices, the saddle sink and saddle source are presented for the appearance of saddle and sink and for saddle and source. The

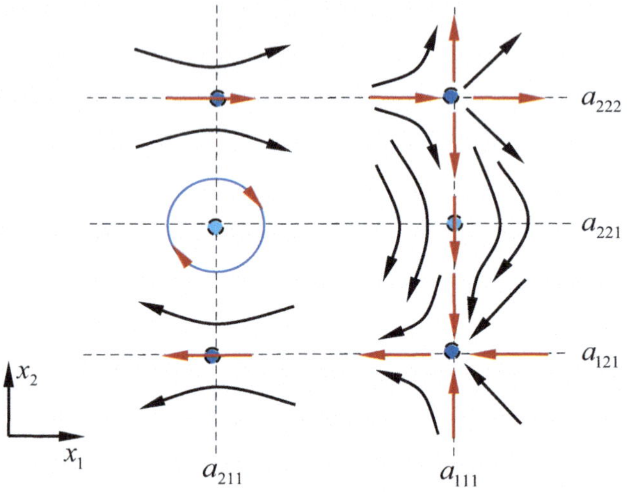

(a)

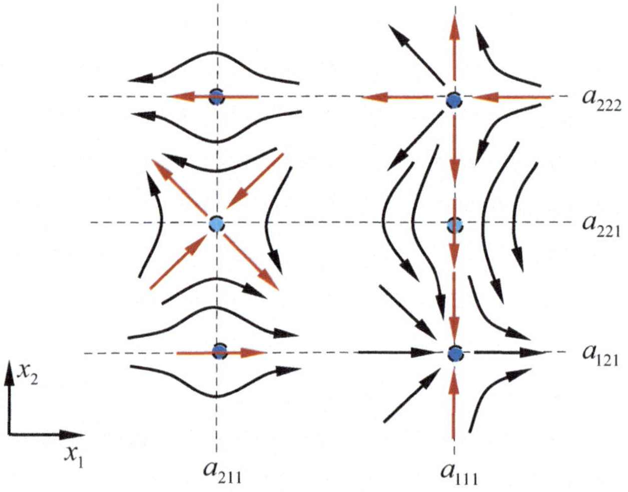

(b)

Fig. 3.2 Phase portraits ($a_{121} \in (a_{221}, a_{222})$ and $a_{211} < a_{111}$) for two-dimensional systems on the x_1-direction with $x_1^* = a_{111}$ and on the x_2-direction with $x_2^* = a_{121}, a_{221}, a_{222}$. The saddle nodes with hyperbolic-to-hyperbolic secant flows: (**a**) ($a_{110} > 0$, $a_{220} > 0$), (**b**) ($a_{110} < 0$, $a_{220} > 0$), (**c**) ($a_{110} > 0$, $a_{220} < 0$), and (**d**) ($a_{110} < 0$, $a_{220} < 0$)

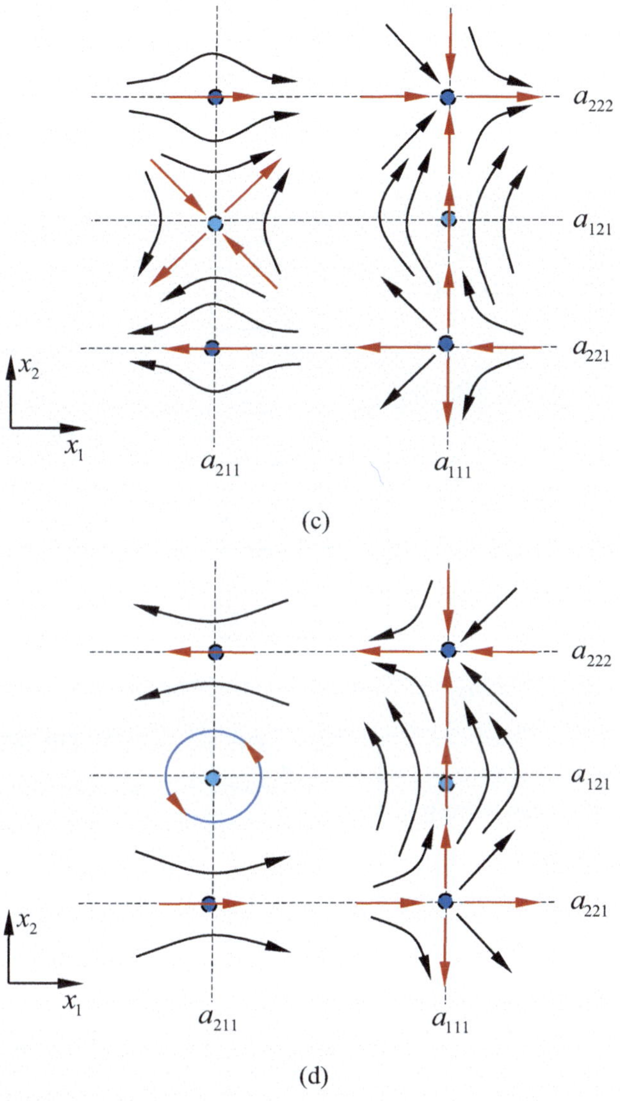

Fig. 3.2 (continued)

hyperbolic secant-to-hyperbolic flows are for appearances of hyperbolic and hyperbolic secant flows. In addition, the hyperbolic flows exist with clockwise and counterclockwise centers. The hyperbolic secant flows exist with positive and negative saddles.

(iii) The equilibriums of $x_1^* = a_{111}$ and $x_2^* = a_{121}, a_{221}, a_{222}$ for $a_{222} < a_{121}$ and $a_{211} < a_{111}$ are

$$\begin{Bmatrix} (a_{211}, a_{121}) & (a_{111}, a_{121}) \\ (a_{211}, a_{222}) & (a_{111}, a_{222}) \\ (a_{211}, a_{221}) & (a_{111}, a_{221}) \end{Bmatrix} = \begin{Bmatrix} \underbrace{(\mathrm{UP}_+,\mathrm{UP}_+)}_{\text{positive saddle}} & \underbrace{(\mathrm{UP}:\mathrm{UP},\mathrm{pF})}_{\text{HS to HB flow }(+)} \\ \underbrace{(\mathrm{nF},\mathrm{UP}:\mathrm{DP})}_{\text{hyperbolic-secant flow }(-)} & \underbrace{(\mathrm{LS},\mathrm{SO})}_{\text{lower-saddle source}} \\ \underbrace{(\mathrm{nF},\mathrm{DP}:\mathrm{UP})}_{\text{hyperbolic flow }(-)} & \underbrace{(\mathrm{LS},\mathrm{SI})}_{\text{lower-saddle sink}} \end{Bmatrix} \tag{3.11}$$

for $a_{110} > 0$ and $a_{220} > 0$;

$$\begin{Bmatrix} (a_{211}, a_{121}) & (a_{111}, a_{121}) \\ (a_{211}, a_{222}) & (a_{111}, a_{222}) \\ (a_{211}, a_{221}) & (a_{111}, a_{221}) \end{Bmatrix} = \begin{Bmatrix} \underbrace{(\mathrm{DP}_+,\mathrm{DP}_-)}_{\text{CCW center}} & \underbrace{(\mathrm{DP}:\mathrm{DP},\mathrm{pF})}_{\text{HB to HS flow }(+)} \\ \underbrace{(\mathrm{pF},\mathrm{DP}:\mathrm{UP})}_{\text{hyperbolic flow }(+)} & \underbrace{(\mathrm{US},\mathrm{SO})}_{\text{upper-saddle source}} \\ \underbrace{(\mathrm{pF},\mathrm{UP}:\mathrm{DP})}_{\text{hyperbolic-secant flow }(+)} & \underbrace{(\mathrm{US},\mathrm{SI})}_{\text{upper-saddle sink}} \end{Bmatrix} \tag{3.12}$$

for $a_{110} < 0$ and $a_{220} > 0$;

$$\begin{Bmatrix} (a_{211}, a_{121}) & (a_{111}, a_{121}) \\ (a_{211}, a_{222}) & (a_{111}, a_{222}) \\ (a_{211}, a_{221}) & (a_{111}, a_{221}) \end{Bmatrix} = \begin{Bmatrix} \underbrace{(\mathrm{DP}_-,\mathrm{DP}_+)}_{\text{CW center}} & \underbrace{(\mathrm{DP}:\mathrm{DP},\mathrm{nF})}_{\text{HB to HS flow }(-)} \\ \underbrace{(\mathrm{nF},\mathrm{DP}:\mathrm{UP})}_{\text{hyperbolic flow }(-)} & \underbrace{(\mathrm{LS},\mathrm{SI})}_{\text{lower-saddle sink}} \\ \underbrace{(\mathrm{nF},\mathrm{UP}:\mathrm{DP})}_{\text{hyperbolic-secant flow }(-)} & \underbrace{(\mathrm{LS},\mathrm{SI})}_{\text{lower-saddle source}} \end{Bmatrix} \tag{3.13}$$

for $a_{110} > 0$ and $a_{220} < 0$;

$$\begin{Bmatrix} (a_{211}, a_{121}) & (a_{111}, a_{121}) \\ (a_{211}, a_{222}) & (a_{111}, a_{222}) \\ (a_{211}, a_{221}) & (a_{111}, a_{221}) \end{Bmatrix} = \begin{Bmatrix} \underbrace{(\mathrm{UP}_-,\mathrm{UP}_-)}_{\text{negative saddle}} & \underbrace{(\mathrm{UP}:\mathrm{UP},\mathrm{nF})}_{\text{HS to HB flow }(-)} \\ \underbrace{(\mathrm{pF},\mathrm{UP}:\mathrm{DP})}_{\text{hyperbolic-secant flow }(+)} & \underbrace{(\mathrm{US},\mathrm{SI})}_{\text{upper-saddle sink}} \\ \underbrace{(\mathrm{pF},\mathrm{DP}:\mathrm{UP})}_{\text{hyperbolic flow }(+)} & \underbrace{(\mathrm{US},\mathrm{SO})}_{\text{upper-saddle source}} \end{Bmatrix} \tag{3.14}$$

for $a_{110} < 0$ and $a_{220} < 0$;

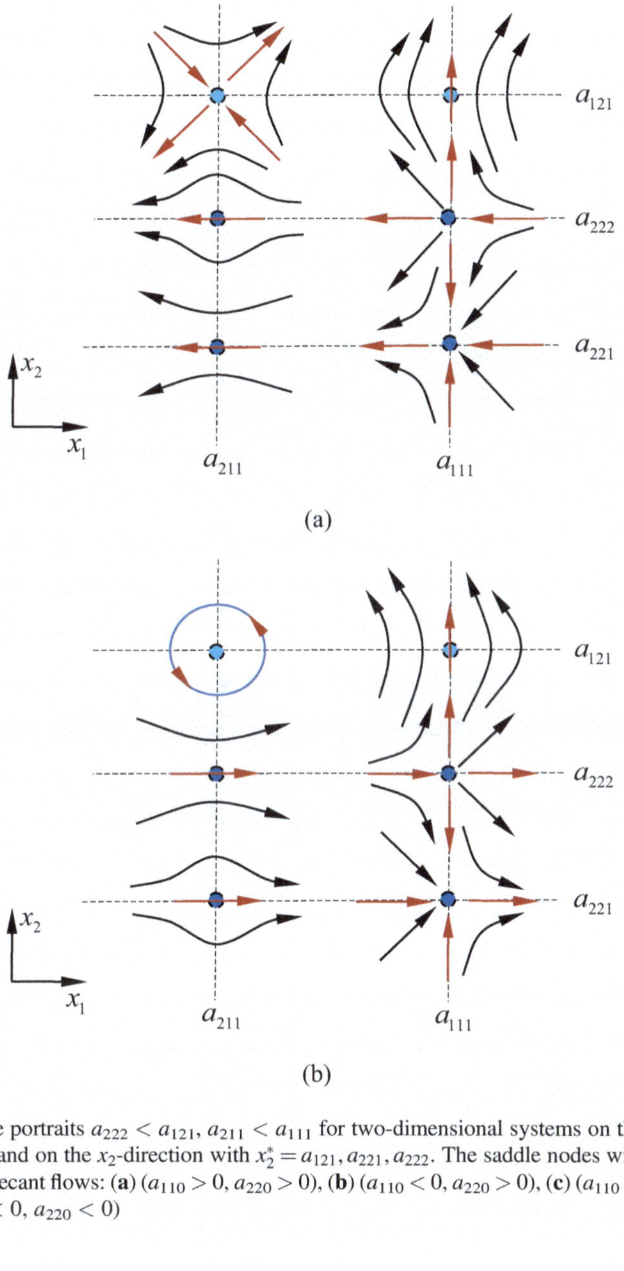

Fig. 3.3 Phase portraits $a_{222} < a_{121}$, $a_{211} < a_{111}$ for two-dimensional systems on the x_1-direction with $x_1^* = a_{111}$ and on the x_2-direction with $x_2^* = a_{121}, a_{221}, a_{222}$. The saddle nodes with hyperbolic-to-hyperbolic secant flows: (**a**) ($a_{110} > 0$, $a_{220} > 0$), (**b**) ($a_{110} < 0$, $a_{220} > 0$), (**c**) ($a_{110} > 0$, $a_{220} < 0$), and (**d**) ($a_{110} < 0$, $a_{220} < 0$)

as shown in Fig. 3.3. Similar to Fig. 3.1, the saddle sink and saddle source are presented for the appearance of saddle and sink and for saddle and source. The hyperbolic secant and hyperbolic flows are for appearances of hyperbolic and hyperbolic secant flows. In addition, the hyperbolic and hyperbolic secant flows are presented with positive and negative saddles or clockwise and counterclockwise centers, respectively.

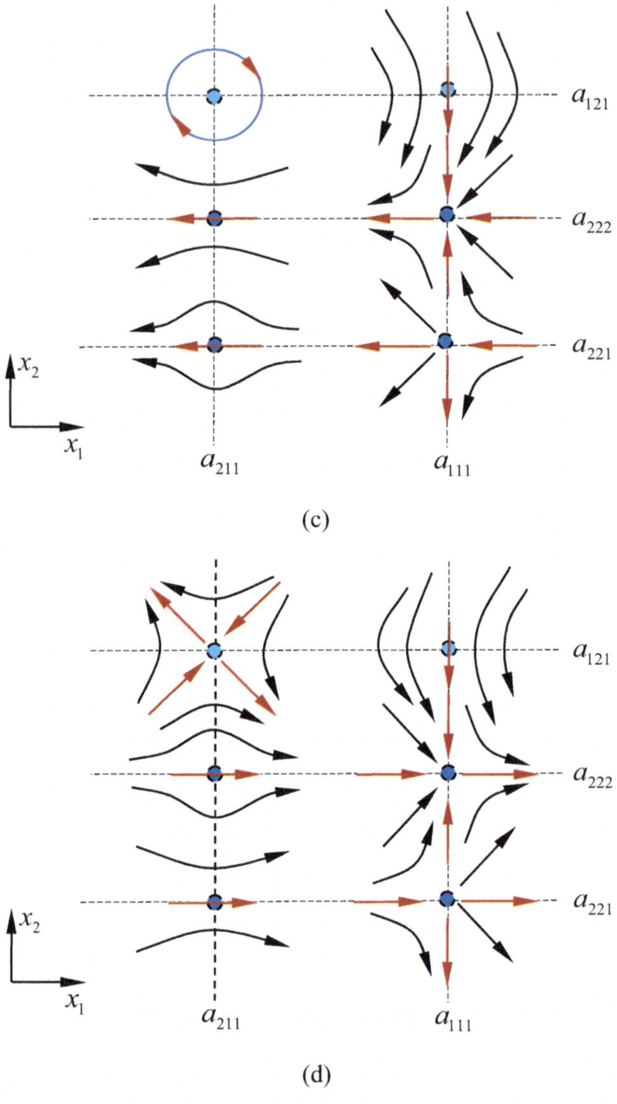

Fig. 3.3 (continued)

3.2 Infinite-Equilibriums and Switching Dynamics

Except for the appearing bifurcation, the corresponding switching bifurcation occurs through infinite equilibriums. The switching bifurcations with single and double infinite-equilibriums are presented.

3.2.1 Inflection Source and Sink Infinite Equilibriums

For the above two cases in Sect. 3.1, the corresponding switching bifurcations are discussed through the inflection source and sink infinite-equilibriums.

(i) Consider a dynamical system for $a_{221} = a_{121}$ as

$$\begin{aligned} \dot{x}_1 &= a_{110}(x_1 - a_{111})^2(x_2 - a_{121}), \\ \dot{x}_2 &= a_{220}(x_1 - a_{211})(x_2 - a_{221})(x_2 - a_{222}), \end{aligned} \tag{3.15}$$

and the corresponding first integral manifold for $a_{221} = a_{121}$ is

$$\ln \frac{|x_1 - a_{111}|}{|x_{10} - a_{111}|} - (a_{111} - a_{211})(\frac{1}{x_1 - a_{111}} - \frac{1}{x_{10} - a_{111}}) = \frac{a_{110}}{a_{220}} \ln \frac{|x_2 - a_{222}|}{|x_{20} - a_{222}|}. \tag{3.16}$$

The infinite equilibriums of $x_2^* = a_{221} = a_{121}$ are

$$\begin{Bmatrix} (a_{211}, a_{222}) & (a_{111}, a_{222}) \\ (a_{121}, a_{211}) & (a_{111}, a_{221}) \end{Bmatrix} = \begin{Bmatrix} \underbrace{(\text{pF,DP:UP})}_{\text{hyperbolic flow }(-)} & \underbrace{(\text{US,SO})}_{\text{upper-saddle source}} \\ \underbrace{(\text{LS,DP})}_{\text{down-parabola lower-saddle}} & \underbrace{(\text{DI:DI,SI})}_{\text{decreasing-inflection sink}} \end{Bmatrix} \tag{3.17}$$

for $a_{110} > 0, a_{220} > 0$,

$$\begin{Bmatrix} (a_{211}, a_{222}) & (a_{111}, a_{222}) \\ (a_{121}, a_{211}) & (a_{111}, a_{221}) \end{Bmatrix} = \begin{Bmatrix} \underbrace{(\text{nF,UP:DP})}_{\text{hyperbolic-secant flow }(-)} & \underbrace{(\text{LS,SO})}_{\text{lower-saddle source}} \\ \underbrace{(\text{LS,UP})}_{\text{up-parabola lower-saddle}} & \underbrace{(\text{II:II,SI})}_{\text{incerasing-inflection sink}} \end{Bmatrix} \tag{3.18}$$

for $a_{110} < 0, a_{220} > 0$,

$$\begin{Bmatrix} (a_{211}, a_{222}) & (a_{111}, a_{222}) \\ (a_{121}, a_{211}) & (a_{111}, a_{221}) \end{Bmatrix} = \begin{Bmatrix} \underbrace{(\text{pF,UP:DP})}_{\text{hyperbolic-secant flow }(+)} & \underbrace{(\text{US,SI})}_{\text{upper-saddle sink}} \\ \underbrace{(\text{US,UP})}_{\text{up-parabola upper-saddle}} & \underbrace{(\text{II:II,SO})}_{\text{incerasing-inflection source}} \end{Bmatrix}$$

for $a_{110} > 0, a_{220} < 0$,

(3.19)

$$\begin{Bmatrix} (a_{211}, a_{222}) & (a_{111}, a_{222}) \\ (a_{121}, a_{211}) & (a_{111}, a_{221}) \end{Bmatrix} = \begin{Bmatrix} \underbrace{(\text{nF,DP:UP})}_{\text{hyperbolic flow }(-)} & \underbrace{(\text{LS,SI})}_{\text{lower-saddle sink}} \\ \underbrace{(\text{US,DP})}_{\text{down-parabola upper-saddle}} & \underbrace{(\text{DI:DI,SO})}_{\text{decerasing-inflection source}} \end{Bmatrix}$$
$$\text{for } a_{110} < 0, a_{220} < 0; \tag{3.20}$$

and the infinite equilibrium of $x_2^* = a_{121} = a_{221}$ is summarized as follows:

$$\begin{aligned} (\bar{x}_1, a_{221}) &= \underbrace{(\text{II,SO})}_{\text{increasing-inflection source}} \quad \text{if } \bar{x}_1 \in (-\infty, a_{211}) \\ (\bar{x}_1, a_{221}) &= \underbrace{(\text{DI,SI})}_{\text{decreasing-inflection sink}} \quad \text{if } \bar{x}_1 \in (a_{211}, a_{111}), \\ (\bar{x}_1, a_{221}) &= \underbrace{(\text{DI,SI})}_{\text{decreasing-inflection sink}} \quad \text{if } \bar{x}_1 \in (a_{111}, \infty), \\ &\text{for } a_{110} > 0, a_{220} > 0; \end{aligned} \tag{3.21}$$

$$\begin{aligned} (\bar{x}_1, a_{221}) &= \underbrace{(\text{DI,SO})}_{\text{decreasing-inflection source}} \quad \text{if } \bar{x}_1 \in (-\infty, a_{211}) \\ (\bar{x}_1, a_{221}) &= \underbrace{(\text{II,SI})}_{\text{decreasing-inflection sink}} \quad \text{if } \bar{x}_1 \in (a_{211}, a_{111}), \\ (\bar{x}_1, a_{221}) &= \underbrace{(\text{II,SI})}_{\text{decreasing-inflection sink}} \quad \text{if } \bar{x}_1 \in (a_{111}, \infty), \\ &\text{for } a_{110} < 0, a_{220} > 0; \end{aligned} \tag{3.22}$$

$$\begin{aligned} (\bar{x}_1, a_{221}) &= \underbrace{(\text{DI,SI})}_{\text{decreasing-inflection sink}} \quad \text{if } \bar{x}_1 \in (-\infty, a_{211}) \\ (\bar{x}_1, a_{221}) &= \underbrace{(\text{II,SO})}_{\text{increasing-inflection source}} \quad \text{if } \bar{x}_1 \in (a_{211}, a_{111}), \\ (\bar{x}_1, a_{221}) &= \underbrace{(\text{II,SO})}_{\text{increasing-inflection source}} \quad \text{if } \bar{x}_1 \in (a_{111}, \infty), \\ &\text{for } a_{110} > 0, a_{220} < 0; \end{aligned} \tag{3.23}$$

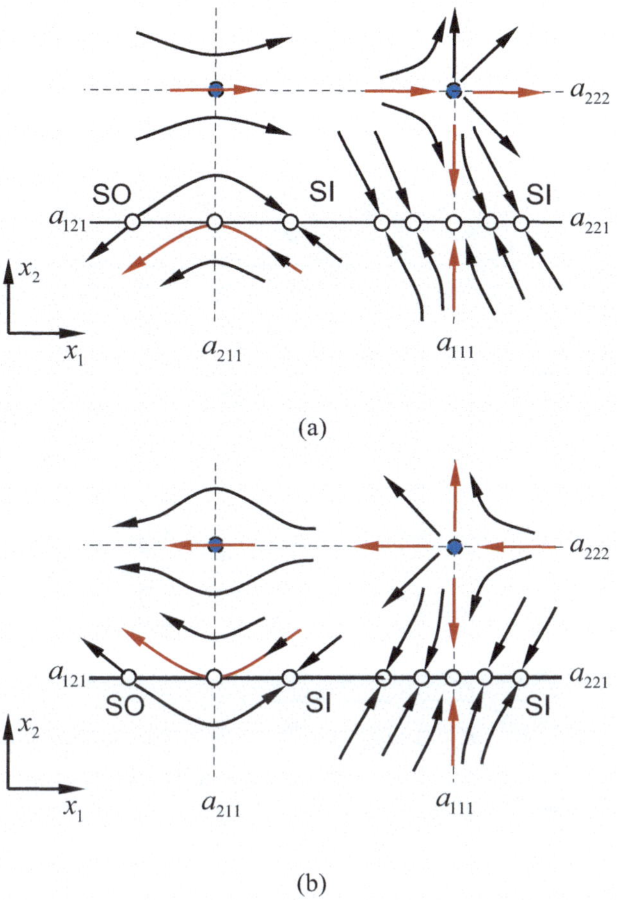

Fig. 3.4 Phase portraits ($a_{121} = a_{221}$ and $a_{211} < a_{111}$) for two-dimensional systems on the x_1-direction with $x_1^* = a_{111}$ and on the x_2-direction with $x_2^* = a_{121}, a_{221}, a_{222}$. The inflection sink and source infinite-equilibriums with saddle-source and saddle-sink: **(a)** ($a_{110} > 0$, $a_{220} > 0$), **(b)** ($a_{110} < 0$, $a_{220} > 0$), **(c)** ($a_{110} > 0$, $a_{220} < 0$), and **(d)** ($a_{110} < 0$, $a_{220} < 0$)

$$
\begin{aligned}
&(\bar{x}_1, a_{221}) = \underbrace{(\text{II,SI})}_{\text{increasing-inflection sink}} \quad \text{if } \bar{x}_1 \in (-\infty, a_{211})\\
&(\bar{x}_1, a_{221}) = \underbrace{(\text{DI,SO})}_{\text{decreasing-inflection source}} \quad \text{if } \bar{x}_1 \in (a_{211}, a_{111}),\\
&(\bar{x}_1, a_{221}) = \underbrace{(\text{DI,SO})}_{\text{decreasing-inflection source}} \quad \text{if } \bar{x}_1 \in (a_{111}, \infty),\\
&\text{for } a_{110} < 0, a_{220} < 0.
\end{aligned}
\tag{3.24}
$$

The inflection sink and source infinite equilibriums of $x_2^* = a_{121} = a_{221}$ are the switching bifurcations of saddle-sink and saddle-source with hyperbolic-to-hyperbolic-secant flows for $a_{121} < a_{221}$ and $a_{221} < a_{121} < a_{222}$, as shown in Fig. 3.4. The inflection infinite-equilibrium is below the saddle-sink or saddle-source.

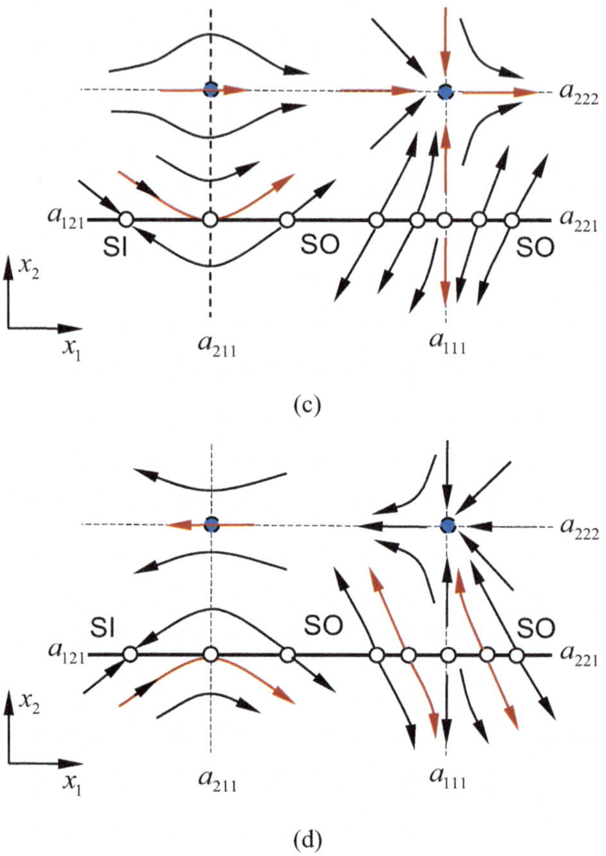

Fig. 3.4 (continued)

(ii) Consider a dynamical system for $a_{222} = a_{121}$ as

$$\begin{aligned}\dot{x}_1 &= a_{110}(x_1 - a_{111})^2(x_2 - a_{121}),\\ \dot{x}_2 &= a_{220}(x_1 - a_{211})(x_2 - a_{221})(x_2 - a_{222}),\end{aligned} \tag{3.25}$$

and the corresponding first integral manifold for $a_{221} = a_{121}$ is

$$\ln \frac{|x_1 - a_{111}|}{|x_{10} - a_{111}|} - (a_{111} - a_{211})\left(\frac{1}{x_1 - a_{111}} - \frac{1}{x_{10} - a_{111}}\right) = \frac{a_{110}}{a_{220}} \ln \frac{|x_2 - a_{221}|}{|x_{20} - a_{221}|}. \tag{3.26}$$

The infinite equilibriums of $x_2^* = a_{222} = a_{121}$ are

$$\left\{\begin{matrix}(a_{211},a_{211}) & (a_{111},a_{222})\\ (a_{211},a_{221}) & (a_{111},a_{221})\end{matrix}\right\} = \left\{\begin{matrix}\underbrace{(\text{US,UP})}_{\text{up-parabola upper-saddle}} & \underbrace{(\text{II:II,SO})}_{\text{incerasing-inflection source}}\\ \underbrace{(\text{nF,DP:UP})}_{\text{hyperbolic flow }(-)} & \underbrace{(\text{LS,SI})}_{\text{lower-saddle sink}}\end{matrix}\right\} \tag{3.27}$$

for $a_{110} > 0, a_{220} > 0$,

$$\left\{\begin{matrix}(a_{211},a_{211}) & (a_{111},a_{222})\\ (a_{211},a_{221}) & (a_{111},a_{221})\end{matrix}\right\} = \left\{\begin{matrix}\underbrace{(\text{US,DP})}_{\text{down-parabola upper-saddle}} & \underbrace{(\text{DI:DI,SO})}_{\text{decerasing-inflection source}}\\ \underbrace{(\text{pF,UP:DP})}_{\text{hyperbolic-secant flow }(+)} & \underbrace{(\text{US,SI})}_{\text{upper-saddle sink}}\end{matrix}\right\}$$

for $a_{110} < 0, a_{220} > 0$,

(3.28)

$$\left\{\begin{matrix}(a_{211},a_{211}) & (a_{111},a_{222})\\ (a_{211},a_{221}) & (a_{111},a_{221})\end{matrix}\right\} = \left\{\begin{matrix}\underbrace{(\text{LS,DP})}_{\text{down-parabola lower-saddle}} & \underbrace{(\text{DI:DI,SI})}_{\text{decerasing-inflection sink}}\\ \underbrace{(\text{nF,UP:DP})}_{\text{hyperbolic-secant flow }(-)} & \underbrace{(\text{LS,SO})}_{\text{upper-saddle source}}\end{matrix}\right\} \tag{3.29}$$

for $a_{110} > 0, a_{220} < 0$,

$$\left\{\begin{matrix}(a_{211},a_{211}) & (a_{111},a_{222})\\ (a_{211},a_{221}) & (a_{111},a_{221})\end{matrix}\right\} = \left\{\begin{matrix}\underbrace{(\text{LS,UP})}_{\text{up-parabola lower-saddle}} & \underbrace{(\text{II:II,SI})}_{\text{incerasing-inflection sink}}\\ \underbrace{(\text{pF,DP:UP})}_{\text{hyperbolic flow }(+)} & \underbrace{(\text{US,SO})}_{\text{upper-saddle source}}\end{matrix}\right\} \tag{3.30}$$

for $a_{110} < 0, a_{220} < 0$,

and the infinite equilibrium of $x_2^* = a_{121} = a_{222}$ is summarized as follows:

$$\begin{aligned}(\bar{x}_1, a_{222}) &= \underbrace{(\text{DI,SI})}_{\text{decreasing-inflection sink}} && \text{if } \bar{x}_1 \in (-\infty, a_{211}),\\ (\bar{x}_1, a_{222}) &= \underbrace{(\text{II,SO})}_{\text{increasing-inflection source}} && \text{if } \bar{x}_1 \in (a_{211}, a_{111}),\\ (\bar{x}_1, a_{222}) &= \underbrace{(\text{II,SO})}_{\text{increasing-inflection source}} && \text{if } \bar{x}_1 \in (a_{111}, \infty),\end{aligned} \tag{3.31}$$

for $a_{110} > 0, a_{220} > 0$;

$$
\begin{aligned}
&(\bar{x}_1, a_{222}) = \underbrace{(\text{II,SI})}_{\text{decreasing-inflection sink}} \quad \text{if } \bar{x}_1 \in (-\infty, a_{211}),\\
&(\bar{x}_1, a_{222}) = \underbrace{(\text{DI,SO})}_{\text{decreasing-inflection source}} \quad \text{if } \bar{x}_1 \in (a_{211}, a_{111}),\\
&(\bar{x}_1, a_{222}) = \underbrace{(\text{DI,SO})}_{\text{decreasing-inflection source}} \quad \text{if } \bar{x}_1 \in (a_{111}, \infty),\\
&\text{for } a_{110} < 0, a_{220} > 0;
\end{aligned}
\tag{3.32}
$$

$$
\begin{aligned}
&(\bar{x}_1, a_{222}) = \underbrace{(\text{II,SO})}_{\text{decreasing-inflection source}} \quad \text{if } \bar{x}_1 \in (-\infty, a_{211}),\\
&(\bar{x}_1, a_{222}) = \underbrace{(\text{DI,SI})}_{\text{decreasing-inflection sink}} \quad \text{if } \bar{x}_1 \in (a_{211}, a_{111}),\\
&(\bar{x}_1, a_{222}) = \underbrace{(\text{DI,SI})}_{\text{decreasing-inflection sink}} \quad \text{if } \bar{x}_1 \in (a_{111}, \infty),\\
&\text{for } a_{110} > 0, a_{220} < 0;
\end{aligned}
\tag{3.33}
$$

$$
\begin{aligned}
&(\bar{x}_1, a_{222}) = \underbrace{(\text{DI,SO})}_{\text{decreasing-inflection source}} \quad \text{if } \bar{x}_1 \in (-\infty, a_{211}),\\
&(\bar{x}_1, a_{222}) = \underbrace{(\text{II,SI})}_{\text{increasing-inflection sink}} \quad \text{if } \bar{x}_1 \in (a_{211}, a_{111}),\\
&(\bar{x}_1, a_{222}) = \underbrace{(\text{II,SI})}_{\text{increasing-inflection sink}} \quad \text{if } \bar{x}_1 \in (a_{111}, \infty),\\
&\text{for } a_{110} < 0, a_{220} < 0.
\end{aligned}
\tag{3.34}
$$

The inflection source and sink infinite equilibriums of $x_2^* = a_{121} = a_{222}$ are for the switching bifurcations of saddle sink and saddle source with hyperbolic-to-hyperbolic secant flows for $a_{121} < a_{222}$ and $a_{121} > a_{222}$, as presented in Fig. 3.5. The inflection infinite equilibrium is above the saddle sink or saddle source in the series of equilibriums.

3.2.2 *Up-Down and Down-Up Saddle Infinite Equilibriums*

Consider a dynamical system as

$$
\begin{aligned}
\dot{x}_1 &= a_{110}(x_1 - a_{111})^2 (x_2 - a_{121}),\\
\dot{x}_2 &= a_{220}(x_1 - a_{211})(x_2 - a_{221})(x_2 - a_{222}),
\end{aligned}
\tag{3.35}
$$

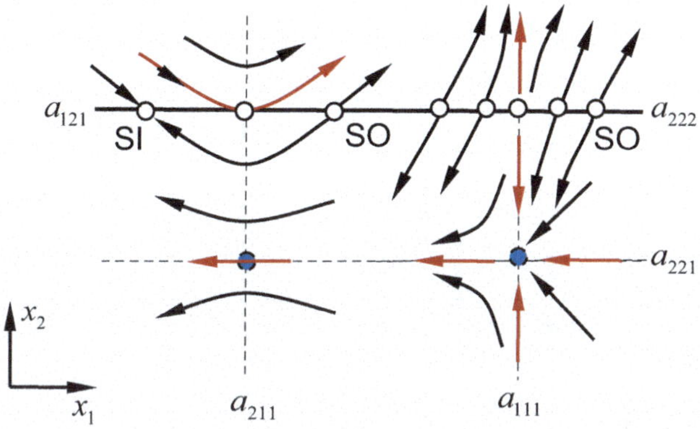

(a)

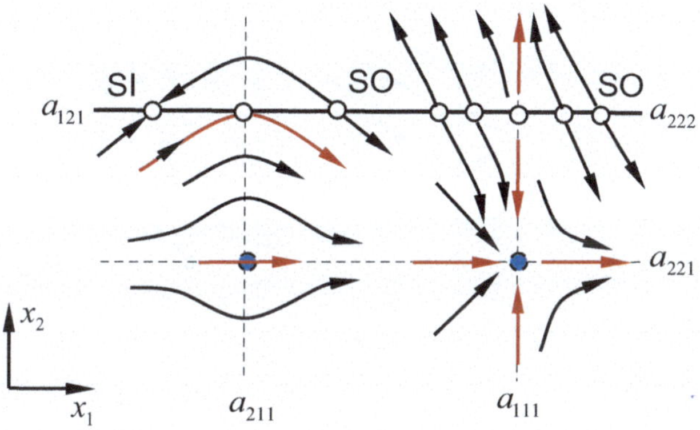

(b)

Fig. 3.5 Phase portraits ($a_{121} = a_{222}$ and $a_{211} < a_{111}$) for two-dimensional systems on the x_1-direction with $x_1^* = a_{111}$ and on the x_2-direction with $x_2^* = a_{121}, a_{221}, a_{222}$. The inflection sink and source infinite equilibriums with saddle source and saddle sink: (**a**) ($a_{110} > 0$, $a_{220} > 0$), (**b**) ($a_{110} < 0$, $a_{220} > 0$), (**c**) ($a_{110} > 0$, $a_{220} < 0$), and (**d**) ($a_{110} < 0$, $a_{220} < 0$)

and the corresponding first integral manifold for $a_{111} = a_{211}$ and $a_{121} \neq a_{221}$, a_{222} is

$$\ln \frac{|x_1 - a_{111}|}{|x_{10} - a_{111}|} = \frac{a_{110}}{a_{220}} \left(\frac{a_{221} - a_{121}}{a_{221} - a_{222}} \ln \frac{|x_2 - a_{221}|}{|x_{20} - a_{221}|} + \frac{a_{222} - a_{121}}{a_{222} - a_{221}} \ln \frac{|x_2 - a_{222}|}{|x_{20} - a_{222}|} \right). \tag{3.36}$$

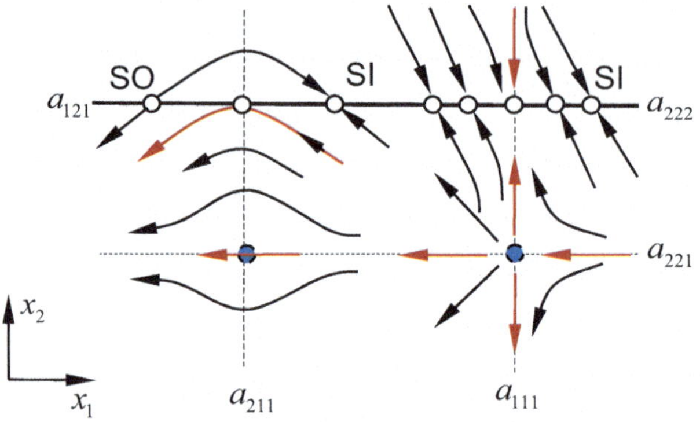

(c)

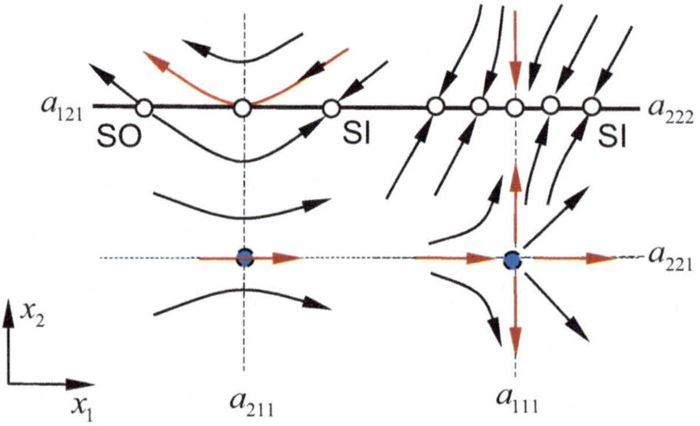

(d)

Fig. 3.5 (continued)

(i) The equilibriums of $x_1^* = a_{111} = a_{211}$ and $x_2^* = a_{121}, a_{221}, a_{222}$ for $a_{121} < a_{221}$ are

$$\begin{Bmatrix} (a_{111}, a_{222}) \\ (a_{111}, a_{221}) \\ (a_{111}, a_{121}) \end{Bmatrix} = \begin{Bmatrix} \underbrace{(\text{US,UD:DU})}_{\text{up-down-to-down-up upper-saddle}} \\ \underbrace{(\text{US,DU:UD})}_{\text{down-up-to-up-down upper-saddle}} \\ \underbrace{(\text{DP}_{-}:\text{UP}_{+}, {}_{\text{UD}}\text{LS:}_{\text{DU}}\text{US})}_{\text{hyperbolic lower-to-upper saddle}} \end{Bmatrix} \text{ for } a_{110} > 0 \text{ and } a_{220} > 0; \quad (3.37)$$

$$\begin{Bmatrix}(a_{111},a_{222})\\(a_{111},a_{221})\\(a_{111},a_{121})\end{Bmatrix}=\begin{Bmatrix}\underbrace{(\text{LS,DU:UD})}_{\text{down-up-to-up-down lower-saddle}}\\\underbrace{(\text{LS,UD:DU})}_{\text{up-down-to-down-up lower-saddle}}\\\underbrace{(\text{UP}_{-}\text{:DP}_{+},{}_{\text{DU}}\text{US:}_{\text{UD}}\text{LS})}_{\text{hyperbolic-secant upper-to-lower saddle}}\end{Bmatrix}\text{ for }a_{110}<0\text{ and }a_{220}>0;\tag{3.38}$$

$$\begin{Bmatrix}(a_{111},a_{222})\\(a_{111},a_{221})\\(a_{111},a_{121})\end{Bmatrix}=\begin{Bmatrix}\underbrace{(\text{US,DU:UD})}_{\text{down-up-to-up-down upper-saddle}}\\\underbrace{(\text{US,UD:DU})}_{\text{up-down-to-down-up upper-saddle}}\\\underbrace{(\text{UP}_{+}\text{:DP}_{-},{}_{\text{DU}}\text{LS:}_{\text{UD}}\text{US})}_{\text{hyperbolic-secant lower-to-upper saddle}}\end{Bmatrix}\text{ for }a_{110}>0\text{ and }a_{220}<0;\tag{3.39}$$

$$\begin{Bmatrix}(a_{111},a_{222})\\(a_{111},a_{221})\\(a_{111},a_{121})\end{Bmatrix}=\begin{Bmatrix}\underbrace{(\text{LS,UD:DU})}_{\text{up-down-to-down-up lower-saddle}}\\\underbrace{(\text{LS,DU:UD})}_{\text{down-up-to-up-down lower-saddle}}\\\underbrace{(\text{DP}_{+}\text{:UP}_{-},{}_{\text{UD}}\text{US:}_{\text{DU}}\text{LS})}_{\text{hyperbolic upper-to-lower saddle}}\end{Bmatrix}\text{ for }a_{110}<0\text{ and }a_{220}<0;\tag{3.40}$$

as shown in Fig. 3.6. The up-down and down-up upper-saddle and lower-saddle infinite-equilibriums are presented. The corresponding infinite-equilibrium of $x_2^* = a_{121} = a_{222}$ is summarized in Table 3.1.

(ii) The equilibriums of $x_1^* = a_{111} = a_{211}$ and $x_2^* = a_{121}, a_{221}, a_{222}$ for $a_{121} \in (a_{221}, a_{222})$ are

$$\begin{Bmatrix}(a_{111},a_{222})\\(a_{111},a_{121})\\(a_{111},a_{221})\end{Bmatrix}=\begin{Bmatrix}\underbrace{(\text{US,UD:DU})}_{\text{up-down-to-down-up upper-saddle}}\\\underbrace{(\text{UP}_{+}\text{:DP}_{-},{}_{\text{DU}}\text{LS:}_{\text{UD}}\text{US})}_{\text{hyperbolic-secant lower-to-upper saddle}}\\\underbrace{(\text{LS,UD:DU})}_{\text{up-down-to-down-up lower-saddle}}\end{Bmatrix}\text{ for }a_{110}>0\text{ and }a_{220}>0;\tag{3.41}$$

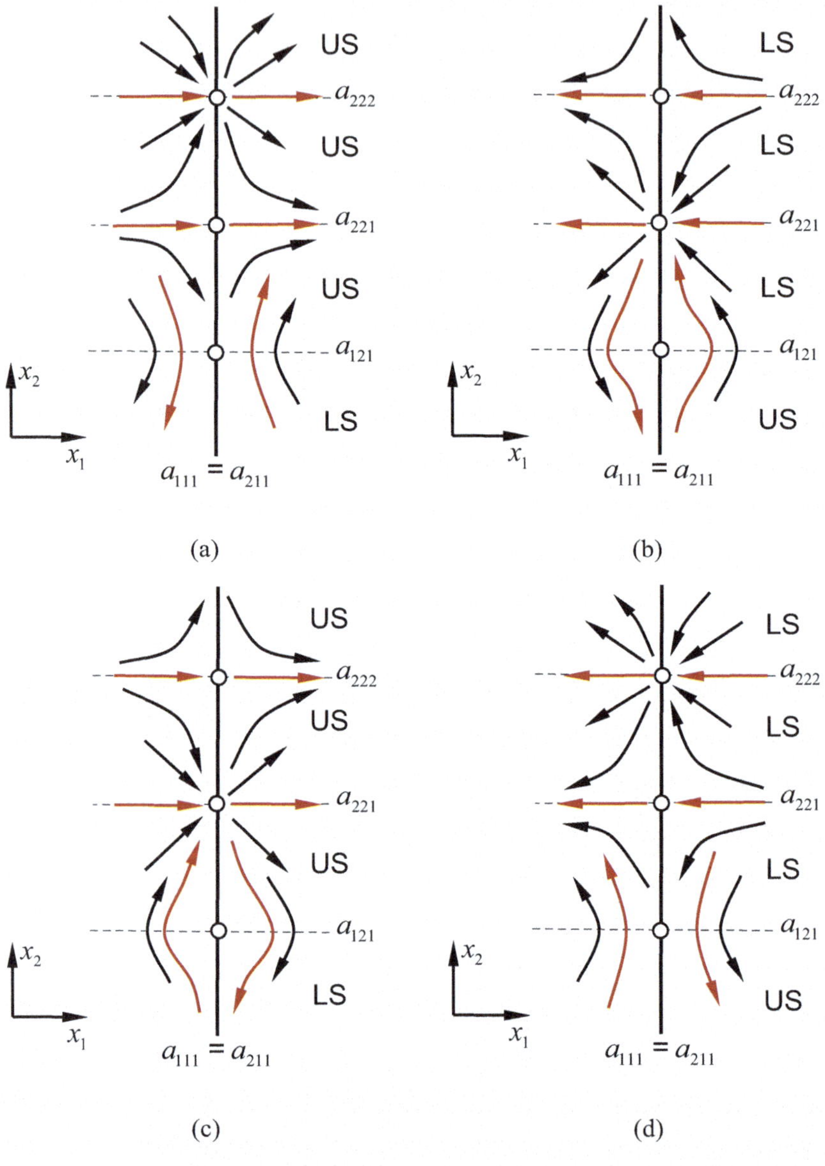

Fig. 3.6 Phase portraits ($a_{111} = a_{211}$ and $a_{121} < a_{221}$) for two-dimensional systems on the x_1-direction with $x_1^* = a_{111}, a_{211}$ and on the x_2-direction with $x_2^* = a_{121}, a_{221}, a_{222}$. Down-up and up-down infinite equilibriums: (**a**) ($a_{110} > 0$, $a_{220} > 0$), (**b**) ($a_{110} < 0$, $a_{220} > 0$), (**c**) ($a_{110} > 0$, $a_{220} < 0$), and (**d**) ($a_{110} < 0$, $a_{220} < 0$)

Table 3.1 Down-up and up-down infinite-equilibrium of $x_1^* = a_{111} = a_{211}$ ($a_{111} = a_{211}$ and $a_{121} < a_{221}$)

$(a_{111}, \bar{x}_2)$	$\bar{x}_2$			
(a_{110}, a_{220})	$(-\infty, a_{121})$	(a_{121}, a_{221})	(a_{221}, a_{222})	(a_{222}, ∞)
(+, +)	(LS,UD)	(US,DU)	(US,UD)	(US,DU)
(−, +)	(US,DU)	(LS,UD)	(LS,DU)	(LS,UD)
(+, −)	(LS,DU)	(US,UD)	(US,DU)	(US,UD)
(−, −)	(US,UD)	(LS,DU)	(LS,UD)	(LS,DU)

$$\begin{Bmatrix} (a_{111}, a_{222}) \\ (a_{111}, a_{121}) \\ (a_{111}, a_{221}) \end{Bmatrix} = \begin{Bmatrix} \underbrace{(\mathrm{LS,DU:UD})}_{\text{down-up-to-up-down lower-saddle}} \\ \underbrace{(\mathrm{DP}_+ : \mathrm{UP}_-, {}_{\mathrm{UD}}\mathrm{US}{:}_{\mathrm{DU}}\mathrm{LS})}_{\text{hyperbolic upper-to-lower saddle}} \\ \underbrace{(\mathrm{US,DU:UD})}_{\text{down-up-to-up-down upper-saddle}} \end{Bmatrix} \text{ for } a_{110} < 0 \text{ and } a_{220} > 0; \tag{3.42}$$

$$\begin{Bmatrix} (a_{111}, a_{222}) \\ (a_{111}, a_{121}) \\ (a_{111}, a_{221}) \end{Bmatrix} = \begin{Bmatrix} \underbrace{(\mathrm{US,DU:UD})}_{\text{down-up-to-up-down upper-saddle}} \\ \underbrace{(\mathrm{DP}_- : \mathrm{UP}_+, {}_{\mathrm{UD}}\mathrm{LS}{:}_{\mathrm{DU}}\mathrm{US})}_{\text{hyperbolic lower-to-upper saddle}} \\ \underbrace{(\mathrm{LS,DU:UD})}_{\text{down-up-to-up-down lower-saddle}} \end{Bmatrix} \text{ for } a_{110} > 0 \text{ and } a_{220} < 0; \tag{3.43}$$

$$\begin{Bmatrix} (a_{111}, a_{222}) \\ (a_{111}, a_{121}) \\ (a_{111}, a_{221}) \end{Bmatrix} = \begin{Bmatrix} \underbrace{(\mathrm{LS,UD:DU})}_{\text{up-down-to-down-up lower-saddle}} \\ \underbrace{(\mathrm{UP}_- : \mathrm{DP}_+, {}_{\mathrm{DU}}\mathrm{US}{:}_{\mathrm{UD}}\mathrm{LS})}_{\text{hyperbolic-secant upper-to-lower saddle}} \\ \underbrace{(\mathrm{US,UD:DU})}_{\text{up-down-to-down-up upper-saddle}} \end{Bmatrix} \text{ for } a_{110} < 0 \text{ and } a_{220} < 0; \tag{3.44}$$

as shown in Fig. 3.7. The up-down and down-up upper-saddle and lower-saddle infinite-equilibriums are presented. The corresponding infinite-equilibrium of $x_2^* = a_{121} = a_{222}$ is summarized in Table 3.2.

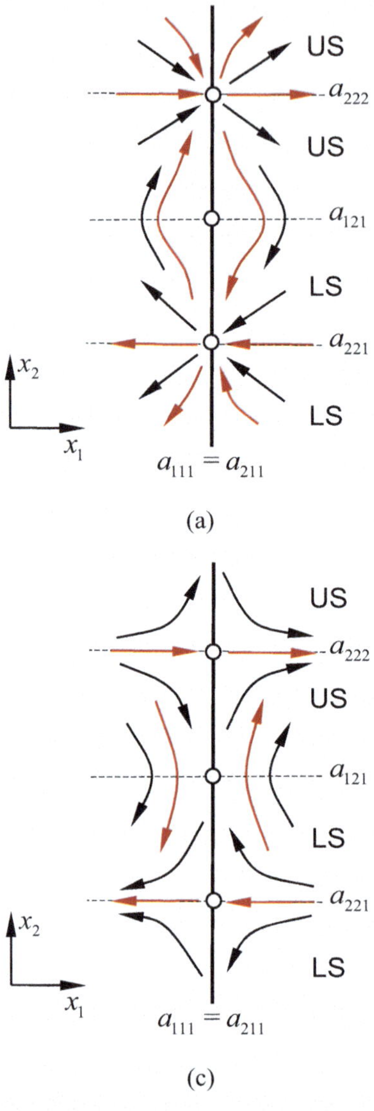

Fig. 3.7 Phase portraits ($a_{111} = a_{211}$ and $a_{121} \in (a_{221}, a_{222})$ for two-dimensional systems on the x_1-direction with $x_1^* = a_{111} = a_{211}$ and on the x_2-direction with $x_2^* = a_{121}, a_{221}, a_{222}$. Infinite equilibriums: **(a)** ($a_{110} > 0$, $a_{220} > 0$), **(b)** ($a_{110} < 0$, $a_{220} > 0$), **(c)** ($a_{110} > 0$, $a_{220} < 0$), and **(d)** ($a_{110} < 0$, $a_{220} < 0$)

Table 3.2 Down-up and up-down infinite-equilibriums of $x_1^* = a_{111} = a_{211}$ ($a_{111} = a_{211}$ and $a_{121} \in (a_{221}, a_{222})$)

$(a_{111}, \bar{x}_2)$	$\bar{x}_2$			
(a_{110}, a_{220})	$(-\infty, a_{221})$	(a_{221}, a_{121})	(a_{121}, a_{222})	(a_{222}, ∞)
(+, +)	(LS,UD)	(LS,DU)	(US,UD)	(US,DU)
$(-, +)$	(US,DU)	(US,UD)	(LS,DU)	(LS,UD)
$(+, -)$	(LS,DU)	(LS,UD)	(US,DU)	(US,UD)
$(-, -)$	(US,UD)	(US,DU)	(LS,UD)	(LS,DU)

(iii) The equilibriums of $x_1^* = a_{111} = a_{211}$ and $x_2^* = a_{121}, a_{221}, a_{222}$ for $a_{222} < a_{121}$ are

$$\begin{Bmatrix} (a_{111}, a_{121}) \\ (a_{111}, a_{222}) \\ (a_{111}, a_{221}) \end{Bmatrix} = \begin{Bmatrix} \underbrace{(\mathrm{DP}_{-}:\mathrm{UP}_{+},\ {}_{\mathrm{UD}}\mathrm{LS}:{}_{\mathrm{DU}}\mathrm{US})}_{\text{hyperbolic lower-to-upper saddle}} \\ \underbrace{(\mathrm{LS},\mathrm{DU}:\mathrm{UD})}_{\text{down-up-to-up-down lower-saddle}} \\ \underbrace{(\mathrm{LS},\mathrm{UD}:\mathrm{DU})}_{\text{up-down-to-down-up lower-saddle}} \end{Bmatrix} \text{ for } a_{110} > 0 \text{ and } a_{220} > 0; \tag{3.45}$$

$$\begin{Bmatrix} (a_{111}, a_{121}) \\ (a_{111}, a_{222}) \\ (a_{111}, a_{221}) \end{Bmatrix} = \begin{Bmatrix} \underbrace{(\mathrm{UP}_{-}:\mathrm{DP}_{+},\ {}_{\mathrm{DU}}\mathrm{US}:{}_{\mathrm{UD}}\mathrm{LS})}_{\text{hyperbolic-secant upper-to-lower saddle}} \\ \underbrace{(\mathrm{US},\mathrm{UD}:\mathrm{DU})}_{\text{up-down-to-down-up upper-saddle}} \\ \underbrace{(\mathrm{US},\mathrm{DU}:\mathrm{UD})}_{\text{down-up-to-up-down upper-saddle}} \end{Bmatrix} \text{ for } a_{110} < 0 \text{ and } a_{220} > 0; \tag{3.46}$$

$$\begin{Bmatrix} (a_{111}, a_{121}) \\ (a_{111}, a_{222}) \\ (a_{111}, a_{221}) \end{Bmatrix} = \begin{Bmatrix} \underbrace{(\mathrm{UP}_{+}:\mathrm{DP}_{-},\ {}_{\mathrm{DU}}\mathrm{LS}:{}_{\mathrm{UD}}\mathrm{US})}_{\text{hyperbolic-secant lower-to-upper saddle}} \\ \underbrace{(\mathrm{LS},\mathrm{UD}:\mathrm{DU})}_{\text{up-down-to-down-up lower-saddle}} \\ \underbrace{(\mathrm{LS},\mathrm{DU}:\mathrm{UD})}_{\text{down-up-to-up-down lower-saddle}} \end{Bmatrix} \text{ for } a_{110} > 0 \text{ and } a_{220} < 0; \tag{3.47}$$

$$\left\{\begin{array}{l}(a_{111},a_{121})\\(a_{111},a_{222})\\(a_{111},a_{221})\end{array}\right\}=\left\{\begin{array}{c}\underbrace{(\mathrm{DP}_{+}:\mathrm{UP}_{-},{}_{\mathrm{UD}}\mathrm{US}:{}_{\mathrm{DU}}\mathrm{LS})}_{\text{hyperbolic upper-to-lower saddle}}\\\underbrace{(\mathrm{US,DU:UD})}_{\text{down-up-to-up-down upper-saddle}}\\\underbrace{(\mathrm{US,UD:DU})}_{\text{up-down-to-down-up upper-saddle}}\end{array}\right\}\text{ for } a_{110}<0 \text{ and } a_{220}<0; \tag{3.48}$$

as shown in Fig. 3.8. The up-down and down-up upper-saddle and lower-saddle infinite-equilibriums are presented. The corresponding infinite equilibriums of $x_2^*=a_{121}=a_{222}$ are summarized in Table 3.3.

3.2.3 Double-Switching Bifurcations

Consider a dynamical system as

$$\begin{aligned}\dot{x}_1&=a_{110}(x_1-a_{111})^2(x_2-a_{121}),\\\dot{x}_2&=a_{220}(x_1-a_{211})(x_2-a_{221})(x_2-a_{222}),\end{aligned} \tag{3.49}$$

and the corresponding first integral manifold for $a_{111}=a_{211}$ and $a_{121}=a_{22s_1}$ (s_1, $s_2\in\{1,2\}$, $s_1\neq s_2$) is

$$\ln\frac{|x_1-a_{111}|}{|x_{10}-a_{111}|}=\frac{a_{110}}{a_{220}}\ln\frac{|x_2-a_{22s_2}|}{|x_{20}-a_{22s_2}|}. \tag{3.50}$$

(i) The equilibriums of $x_1^*=a_{111}=a_{221}$ and $x_2^*=a_{221}=a_{121}$ are

$$\left\{\begin{array}{l}(a_{111},a_{222})\\(a_{121},a_{211})\end{array}\right\}=\left\{\begin{array}{c}\underbrace{(\mathrm{US,UD:DU})}_{\text{up-down-to-down-up upper-saddle}}\\\underbrace{({}_{\mathrm{LS:US}}\mathrm{UD},{}_{\mathrm{II}}\mathrm{SO}:{}_{\mathrm{DI}}\mathrm{SI})}_{\text{up-down inflection source-to-sink}}\end{array}\right\}\text{ for } a_{110}>0, a_{220}>0, \tag{3.51}$$

$$\left\{\begin{array}{l}(a_{111},a_{222})\\(a_{121},a_{211})\end{array}\right\}=\left\{\begin{array}{c}\underbrace{(\mathrm{LS,DU:UD})}_{\text{down-up-to-up-down lower-saddle}}\\\underbrace{({}_{\mathrm{US:LS}}\mathrm{DU},{}_{\mathrm{DI}}\mathrm{SO}:{}_{\mathrm{II}}\mathrm{SI})}_{\text{up-down inflection source-to-sink}}\end{array}\right\}\text{ for } a_{110}<0, a_{220}>0, \tag{3.52}$$

$$\left\{\begin{array}{l}(a_{111},a_{222})\\(a_{121},a_{211})\end{array}\right\}=\left\{\begin{array}{c}\underbrace{(\mathrm{US,DU:UD})}_{\text{down-up-to-up-down upper-saddle}}\\\underbrace{({}_{\mathrm{LS:US}}\mathrm{DU},{}_{\mathrm{DI}}\mathrm{SI}:{}_{\mathrm{II}}\mathrm{SO})}_{\text{up-down inflection sink-to-source}}\end{array}\right\}\text{ for } a_{110}>0, a_{220}<0, \tag{3.53}$$

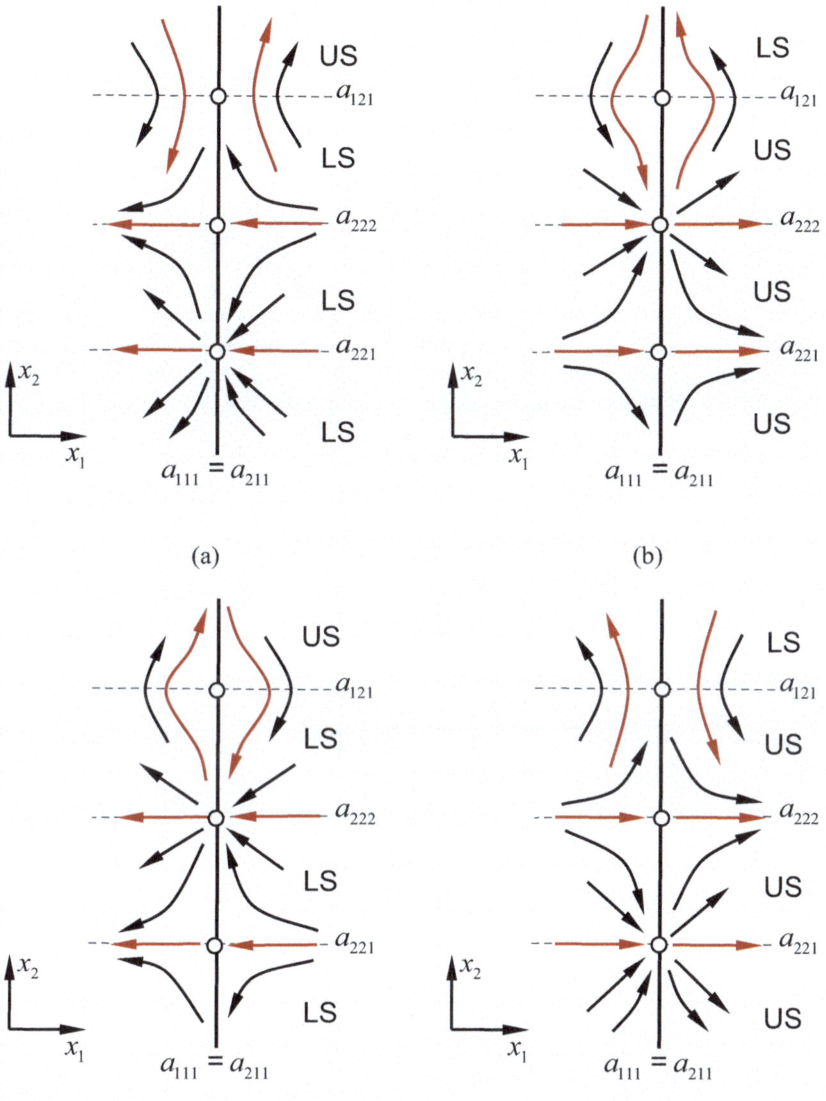

Fig. 3.8 Phase portraits ($a_{111} = a_{211}$ and $a_{222} < a_{121}$) for two-dimensional systems on the x_1-direction with $x_1^* = a_{111} = a_{211}$ and on the x_2-direction with $x_2^* = a_{121}, a_{221}, a_{222}$. Infinite equilibriums: (**a**) ($a_{110} > 0$, $a_{220} > 0$), (**b**) ($a_{110} < 0$, $a_{220} > 0$), (**c**) ($a_{110} > 0$, $a_{220} < 0$), and (**d**) ($a_{110} < 0$, $a_{220} < 0$)

$$\begin{Bmatrix} (a_{111}, a_{222}) \\ (a_{111}, a_{221}) \end{Bmatrix} = \left\{ \begin{matrix} \underbrace{(\text{LS,UD:DU})}_{\text{up-down-to-down-up lower-saddle}} \\ \underbrace{({}_{\text{US:LS}}\text{UD}, {}_{\text{II}}\text{SI}:{}_{\text{DI}}\text{SO})}_{\text{up-down inflection sink-to-source}} \end{matrix} \right\} \text{ for } a_{110} < 0, a_{220} < 0. \tag{3.54}$$

Table 3.3 Down-up and up-down infinite-equilibrium of $x_1^* = a_{111} = a_{211}$ ($a_{111} = a_{211}$ and $a_{222} < a_{121}$)

$(a_{111}, \bar{x}_2)$	$\bar{x}_2$			
(a_{110}, a_{220})	$(-\infty, a_{221})$	(a_{221}, a_{222})	(a_{222}, a_{121})	(a_{121}, ∞)
$(+,+)$	(LS,UD)	(LS,DU)	(LS,UD)	(US,DU)
$(-,+)$	(US,DU)	(US,UD)	(US,DU)	(LS,UD)
$(+,-)$	(LS,DU)	(LS,UD)	(LS,DU)	(US,UD)
$(-,-)$	(US,UD)	(US,DU)	(US,UD)	(LS,DU)

In Fig. 3.9, the up-down and down-up upper saddle and lower saddle infinite equilibriums of $x_1^* = a_{111} = a_{211}$ are presented in the horizontal direction, and the inflection sink and source infinite equilibriums of $x_2^* = a_{121} = a_{221}$ are presented. The two infinite equilibriums are summarized in Table 3.4.

(ii) The equilibriums of $x_1^* = a_{111} = a_{221}$ and $x_2^* = a_{222} = a_{121}$ are

$$\begin{Bmatrix} (a_{121}, a_{211}) \\ (a_{111}, a_{221}) \end{Bmatrix} = \begin{Bmatrix} \underbrace{({}_{\text{LS:US}}\text{DU}, {}_{\text{DI}}\text{SI}:{}_{\text{II}}\text{SO})}_{\text{up-down inflection sink-to-source}} \\ \underbrace{(\text{LS,UD:DU})}_{\text{up-down-to-down-up lower-saddle}} \end{Bmatrix} \text{ for } a_{110} > 0, a_{220} > 0, \tag{3.55}$$

$$\begin{Bmatrix} (a_{121}, a_{211}) \\ (a_{111}, a_{221}) \end{Bmatrix} = \begin{Bmatrix} \underbrace{({}_{\text{US:LS}}\text{UD}, {}_{\text{II}}\text{SI}:{}_{\text{DI}}\text{SI})}_{\text{up-down inflection sink-to-source}} \\ \underbrace{(\text{US,DU:UD})}_{\text{down-up-to-up-down upper-saddle}} \end{Bmatrix} \text{ for } a_{110} < 0, a_{220} > 0, \tag{3.56}$$

$$\begin{Bmatrix} (a_{121}, a_{211}) \\ (a_{111}, a_{221}) \end{Bmatrix} = \begin{Bmatrix} \underbrace{({}_{\text{LS:US}}\text{UD}, {}_{\text{II}}\text{SO}:{}_{\text{DI}}\text{SI})}_{\text{up-down inflection source-to-sink}} \\ \underbrace{(\text{LS,DU:UD})}_{\text{down-up-to-up-down lower-saddle}} \end{Bmatrix} \text{ for } a_{110} > 0, a_{220} < 0, \tag{3.57}$$

$$\begin{Bmatrix} (a_{111}, a_{222}) \\ (a_{111}, a_{221}) \end{Bmatrix} = \begin{Bmatrix} \underbrace{({}_{\text{US:LS}}\text{DU}, {}_{\text{DI}}\text{SO}:{}_{\text{II}}\text{SI})}_{\text{down-up inflection source-to-sink}} \\ \underbrace{(\text{US,UD:DU})}_{\text{up-down-to-down-up upper-saddle}} \end{Bmatrix} \text{ for } a_{110} < 0, a_{220} < 0. \tag{3.58}$$

In Fig. 3.10, the up-down and down-up upper saddle and lower saddle infinite equilibriums of $x_1^* = a_{111} = a_{211}$ are presented in the horizontal direction, and the inflection sink and source infinite equilibriums of $x_2^* = a_{121} = a_{221}$ are presented. The two infinite equilibriums are summarized in Table 3.5.

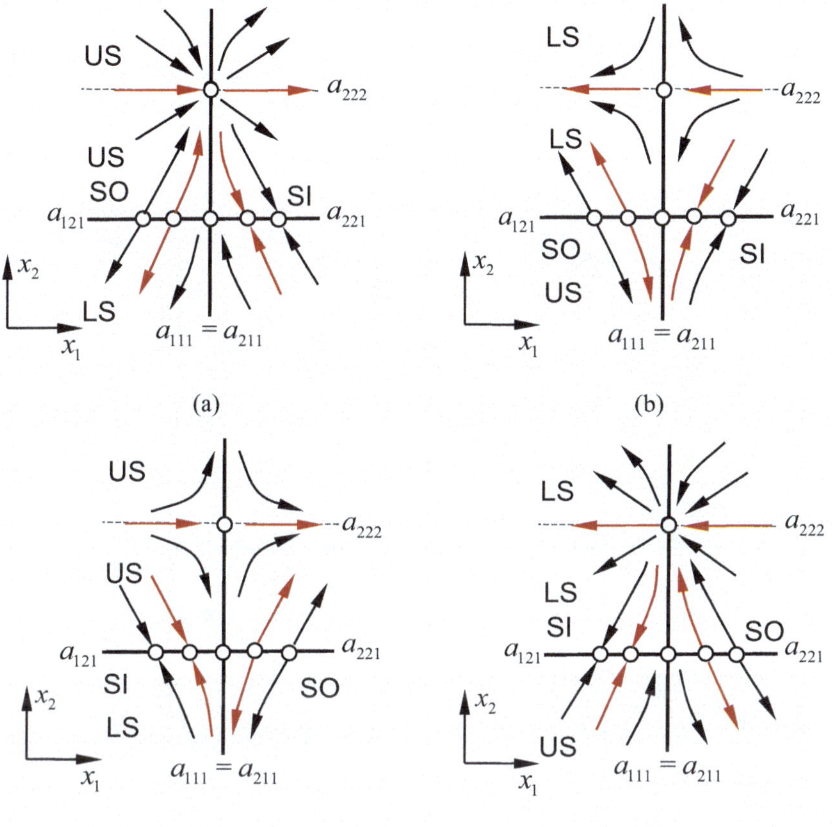

Fig. 3.9 Phase portraits ($a_{111} = a_{211}$ and $a_{121} = a_{221}$) for two-dimensional systems on the x_1-direction with $x_1^* = a_{111}, a_{211}$ and on the x_2-direction with $x_2^* = a_{121}, a_{221}, a_{222}$. Two infinite equilibriums: **(a)** ($a_{110} > 0, a_{220} > 0$), **(b)** ($a_{110} < 0, a_{220} > 0$), **(c)** ($a_{110} > 0, a_{220} < 0$), and **(d)** ($a_{110} < 0$, $a_{220} < 0$)

Table 3.4 Inflection and down-up and up-down infinite-equilibriums ($a_{111} = a_{211}$ and $a_{221} = a_{121}$)

	$\bar{x}_2$ for $(a_{111}, \bar{x}_2)$			$\bar{x}_1$ for $(\bar{x}_2, a_{221})$	
(a_{110}, a_{220})	$(-\infty, a_{221})$	(a_{221}, a_{222})	(a_{222}, ∞)	$(-\infty, a_{111})$	(a_{111}, ∞)
(+, +)	(LS,UD)	(US,UD)	(US,DU)	(II,SO)	(DI,SI)
(−, +)	(US,DU)	(LS,DU)	(LS,UD)	(DI,SO)	(II,SI)
(+, −)	(LS,DU)	(US,DU)	(US,UD)	(DI,SI)	(II,SO)
(−, −)	(US,UD)	(LS,UD)	(LS,DU)	(II,SI)	(DI,SO)

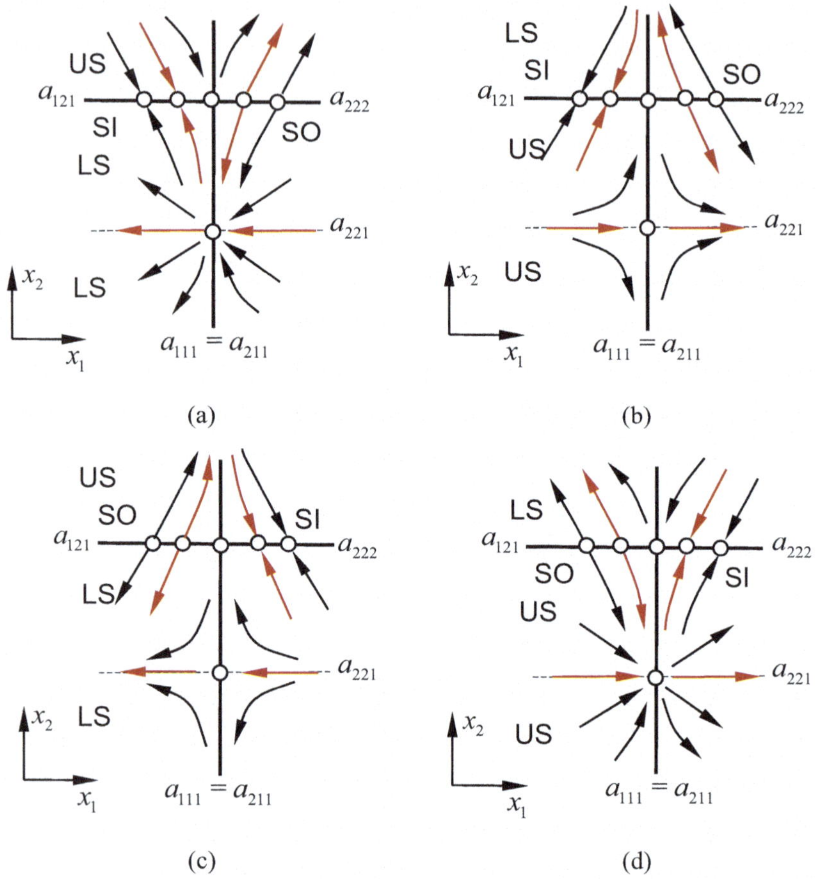

Fig. 3.10 Phase portraits ($a_{111} = a_{211}$ and $a_{121} = a_{221}$) for two-dimensional systems on the x_1-direction with $x_1^* = a_{111}, a_{211}$ and on the x_2-direction with $x_2^* = a_{121}, a_{221}, a_{222}$. Two infinite equilibriums: (**a**) ($a_{110} > 0, a_{220} > 0$), (**b**) ($a_{110} < 0, a_{220} > 0$), (**c**) ($a_{110} > 0, a_{220} < 0$), and (**d**) ($a_{110} < 0$, $a_{220} < 0$)

Table 3.5 Inflection and down-up and up-down infinite-equilibriums ($a_{111} = a_{211}$ and $a_{222} = a_{121}$)

	$\bar{x}_2$ for $(a_{111}, \bar{x}_2)$			$\bar{x}_1$ for $(\bar{x}_2, a_{221})$	
(a_{110}, a_{220})	$(-\infty, a_{221})$	(a_{221}, a_{222})	(a_{222}, ∞)	$(-\infty, a_{111})$	(a_{111}, ∞)
(+, +)	(LS,UD)	(LS,DU)	(US,DU)	(DI,SI)	(II,SO)
(−, +)	(US,DU)	(US,UD)	(LS,UD)	(II,SI)	(DI,SO)
(+, −)	(LS,DU)	(LS,UD)	(US,UD)	(II,SO)	(DI,SI)
(−, −)	(US,UD)	(US,DU)	(LS,DU)	(DI,SO)	(II,SI)

Chapter 4
Horizontally Paralleled Saddle-Source and Saddle-Sink

In this chapter, the horizontally paralleled saddle-source and saddle-sink with hyperbolic flow singularity exist in equilibrium networks, and the center or saddle with separated hyperbolic and hyperbolic-secant flows are paralleled. The hyperbolic flow singularity is for the appearing bifurcation of the paralleled hyperbolic and hyperbolic secant flows. The switching bifurcations for center (saddle) with paralleled hyperbolic (hyperbolic-secant) flow and saddle source (sink) with connected hyperbolic singular flows are discussed. In addition, the switching bifurcations of the saddle source (sink) with connected saddle source (sink) and the saddle and center with paralleled hyperbolic singular flows are discussed through infinite-equilibriums.

4.1 (2 × 3)-Equilibrium Matrices with Singularity

Consider a dynamical system as

$$\begin{aligned} \dot{x}_1 &= a_{110}(x_1 - a_{111})(x_1 - a_{112})(x_2 - a_{121}), \\ \dot{x}_2 &= a_{220}(x_1 - a_{211})(x_2 - a_{221})^2, \end{aligned} \tag{4.1}$$

and the corresponding first integral manifold is

$$\begin{aligned} &\frac{a_{112} - a_{211}}{a_{112} - a_{111}} \ln \frac{|x_1 - a_{112}|}{|x_{10} - a_{112}|} + \frac{a_{111} - a_{211}}{a_{111} - a_{211}} \ln \frac{|x_1 - a_{111}|}{|x_{10} - a_{111}|} \\ &= \frac{a_{110}}{a_{220}} \left[\ln \frac{|x_2 - a_{221}|}{|x_{20} - a_{221}|} - (a_{221} - a_{121}) \left(\frac{1}{x_2 - a_{221}} - \frac{1}{x_{20} - a_{221}} \right) \right]. \end{aligned} \tag{4.2}$$

(i) For the case of $a_{211} < a_{111}$ and $a_{121} < a_{221}$, the saddle-nodes and hyperbolic flows of $x_1^* = a_{111}, a_{112}, a_{211}$ with $x_2^* = a_{121}, a_{221}$ have the following properties:

A. C. J. Luo, *Two-Dimensional Two-Product Cubic Systems, Vol. II*,
https://doi.org/10.1007/978-3-031-48491-9_4

$$\left\{\begin{array}{cc} (a_{112},a_{121}) & (a_{112},a_{221}) \\ (a_{111},a_{121}) & (a_{111},a_{221}) \\ (a_{121},a_{211}) & (a_{211},a_{221}) \end{array}\right\} = \left\{\begin{array}{cc} \underbrace{(\text{DP:UP,pF})}_{\text{hyperbolic flow }(+)} & \underbrace{(\text{SO,US})}_{\text{upper-saddle source}} \\ \underbrace{(\text{UP:DP,pF})}_{\text{hyperbolic-secant flow }(+)} & \underbrace{(\text{SI,US})}_{\text{upper-saddle sink}} \\ \underbrace{(\text{UP}_{+},\text{UP}_{+})}_{\text{positive saddle}} & \underbrace{(\text{pF,UP:UP})}_{\text{HS-to-HB flow }(+)} \end{array}\right\} \quad (4.3)$$

for $a_{110} > 0$ and $a_{220} > 0$;

$$\left\{\begin{array}{cc} (a_{112},a_{121}) & (a_{112},a_{221}) \\ (a_{111},a_{121}) & (a_{111},a_{221}) \\ (a_{121},a_{211}) & (a_{211},a_{221}) \end{array}\right\} = \left\{\begin{array}{cc} \underbrace{(\text{UP:DP,pF})}_{\text{hyperbolic-secant flow }(+)} & \underbrace{(\text{SI,US})}_{\text{upper-saddle sink}} \\ \underbrace{(\text{DP:UP,pF})}_{\text{hyperbolic flow }(+)} & \underbrace{(\text{SO,US})}_{\text{upper-saddle source}} \\ \underbrace{(\text{DP}_{+},\text{DP}_{-})}_{\text{CCW center}} & \underbrace{(\text{nF,DP:DP, })}_{\text{HB-to-HS flow }(-)} \end{array}\right\} \quad (4.4)$$

for $a_{110} < 0$ and $a_{220} > 0$;

$$\left\{\begin{array}{cc} (a_{112},a_{121}) & (a_{112},a_{221}) \\ (a_{111},a_{121}) & (a_{111},a_{221}) \\ (a_{121},a_{211}) & (a_{211},a_{221}) \end{array}\right\} = \left\{\begin{array}{cc} \underbrace{(\text{UP:DP,nF})}_{\text{hyperbolic-secant flow }(-)} & \underbrace{(\text{SO,LS})}_{\text{lower-saddle source}} \\ \underbrace{(\text{DP:UP,nF})}_{\text{hyperbolic flow }(-)} & \underbrace{(\text{SI,LS})}_{\text{lower-saddle sink}} \\ \underbrace{(\text{DP}_{-},\text{DP}_{+})}_{\text{CW center}} & \underbrace{(\text{pF,DP:DP})}_{\text{HB-to-HS flow }(+)} \end{array}\right\} \quad (4.5)$$

for $a_{110} > 0$ and $a_{220} < 0$;

$$\left\{\begin{array}{cc} (a_{112},a_{121}) & (a_{112},a_{221}) \\ (a_{111},a_{121}) & (a_{111},a_{221}) \\ (a_{121},a_{211}) & (a_{211},a_{221}) \end{array}\right\} = \left\{\begin{array}{cc} \underbrace{(\text{DP:UP,nF})}_{\text{hyperbolic flow }(-)} & \underbrace{(\text{SI,LS})}_{\text{lower-saddle sink}} \\ \underbrace{(\text{UP:DP,nF})}_{\text{hyperbolic-secant flow }(-)} & \underbrace{(\text{SO,LS})}_{\text{lower-saddle source}} \\ \underbrace{(\text{UP}_{-},\text{UP}_{-})}_{\text{negative saddle}} & \underbrace{(\text{nF,UP:UP})}_{\text{HS-to-HB flow }(-)} \end{array}\right\} \quad (4.6)$$

for $a_{110} < 0$ and $a_{220} < 0$.

The saddle sink and saddle source with hyperbolic and hyperbolic secant flows are presented in Fig. 4.1. The saddle sink and saddle source are the appearing bifurcation of saddle and sink and the appearing bifurcation of saddle and source, respectively. The saddle sink and saddle source are above the hyperbolic and hyperbolic secant flows.

(ii) For $a_{211} < a_{111}$ and $a_{221} < a_{121}$, the equilibriums of $x_1^* = a_{111}, a_{112}, a_{211}$ with $x_2^* = a_{121}, a_{221}$ have the following properties:

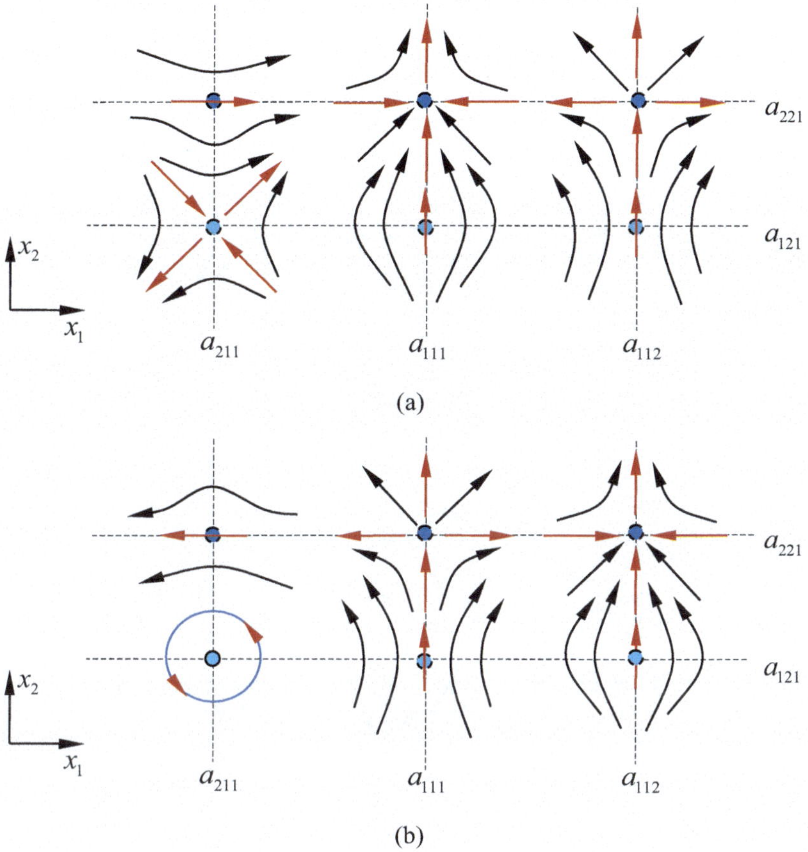

Fig. 4.1 Phase portraits ($a_{121} < a_{221}, a_{211} < a_{111}$) for two-dimensional systems on the x_1-direction with $x_1^* = a_{111}, a_{112}, a_{211}$ and on the x_2-direction with $x_2^* = a_{121}, a_{221}$. (2 × 3)-equilibrium matrices: (**a**) ($a_{110} > 0$, $a_{220} > 0$), (**b**) ($a_{110} < 0$, $a_{220} > 0$), (**c**) ($a_{110} > 0$, $a_{220} < 0$), and (**d**) ($a_{110} < 0$, $a_{220} < 0$)

$$
\begin{Bmatrix}
(a_{112}, a_{221}) & (a_{112}, a_{121}) \\
(a_{111}, a_{221}) & (a_{111}, a_{121}) \\
(a_{211}, a_{221}) & (a_{121}, a_{211})
\end{Bmatrix}
=
\begin{Bmatrix}
\underbrace{(\text{SI,US})}_{\text{upper-saddle sink}} & \underbrace{(\text{DP:UP,pF})}_{\text{hyperbolic flow } (+)} \\
\underbrace{(\text{SO,US})}_{\text{upper-saddle source}} & \underbrace{(\text{UP:DP,pF})}_{\text{hyperbolic-secant flow } (+)} \\
\underbrace{(\text{nF,DP:DP})}_{\text{HB-to-HS flow } (-)} & \underbrace{(\text{UP}_+,\text{UP}_+)}_{\text{positive saddle}}
\end{Bmatrix}
\tag{4.7}
$$

for $a_{110} > 0$ and $a_{220} > 0$;

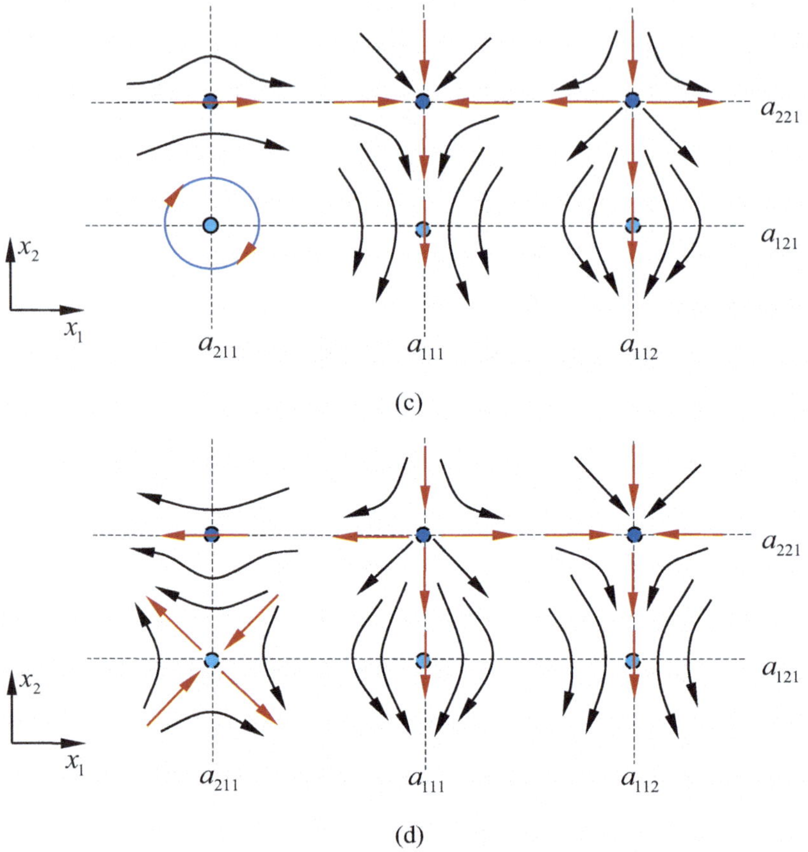

Fig. 4.1 (continued)

$$\begin{Bmatrix} (a_{112}, a_{221}) & (a_{112}, a_{121}) \\ (a_{111}, a_{221}) & (a_{111}, a_{121}) \\ (a_{211}, a_{221}) & (a_{121}, a_{211}) \end{Bmatrix} = \left\{ \begin{array}{cc} \underbrace{(\text{SO,US})}_{\text{upper-saddle source}} & \underbrace{(\text{UP:DP,pF})}_{\text{hyperbolic-secant flow }(+)} \\ \underbrace{(\text{SI,US})}_{\text{upper-saddle sink}} & \underbrace{(\text{DP:UP,pF})}_{\text{hyperbolic flow }(+)} \\ \underbrace{(\text{pF,UP:UP})}_{\text{HS-to-HB flow }(+)} & \underbrace{(\text{DP}_+,\text{DP}_-)}_{\text{CCW center}} \end{array} \right\} \tag{4.8}$$

for $a_{110} < 0$ and $a_{220} > 0$;

$$\begin{Bmatrix} (a_{112},a_{221}) & (a_{112},a_{121}) \\ (a_{111},a_{221}) & (a_{111},a_{121}) \\ (a_{211},a_{221}) & (a_{121},a_{211}) \end{Bmatrix} = \begin{Bmatrix} \underbrace{(\text{SI,LS})}_{\text{lower-saddle sink}} & \underbrace{(\text{UP:DP,nF})}_{\text{hyperbolic-secant flow }(-)} \\ \underbrace{(\text{SO,LS})}_{\text{lower-saddle source}} & \underbrace{(\text{DP:UP,nF})}_{\text{hyperbolic flow }(-)} \\ \underbrace{(\text{nF,UP:UP})}_{\text{HS-to-HB flow }(-)} & \underbrace{(\text{DP}_{-},\text{DP}_{+})}_{\text{CW center}} \end{Bmatrix} \tag{4.9}$$

for $a_{110} > 0$ and $a_{220} < 0$;

$$\begin{Bmatrix} (a_{112},a_{221}) & (a_{112},a_{121}) \\ (a_{111},a_{221}) & (a_{111},a_{121}) \\ (a_{211},a_{221}) & (a_{121},a_{211}) \end{Bmatrix} = \begin{Bmatrix} \underbrace{(\text{SO,LS})}_{\text{lower-saddle source}} & \underbrace{(\text{DP:UP,nF})}_{\text{hyperbolic flow }(-)} \\ \underbrace{(\text{SI,LS})}_{\text{lower-saddle sink}} & \underbrace{(\text{UP:DP,nF})}_{\text{hyperbolic-secant flow }(-)} \\ \underbrace{(\text{pF,DP:DP})}_{\text{HB-to-HS flow }(+)} & \underbrace{(\text{UP}_{-},\text{UP}_{-})}_{\text{negative saddle}} \end{Bmatrix} \tag{4.10}$$

for $a_{110} < 0$ and $a_{220} < 0$.

The saddle sink and saddle source with hyperbolic and hyperbolic secant flows are presented in Fig. 4.2. The saddle sink and saddle source are the appearing bifurcation of saddle and sink and the appearing bifurcation of saddle and source, respectively. The saddle sink and saddle source are below the hyperbolic and hyperbolic secant flows. The hyperbolic singular flows are the appearing bifurcation flows for hyperbolic and hyperbolic secant flows.

(iii) For the case of $a_{211} \in (a_{111}, a_{112})$ and $a_{121} < a_{221}$, the saddle nodes and hyperbolic flows of $x_1^* = a_{111}, a_{112}, a_{211}$ with $x_2^* = a_{121}, a_{221}$ have the following properties:

$$\begin{Bmatrix} (a_{112},a_{121}) & (a_{112},a_{221}) \\ (a_{121},a_{211}) & (a_{211},a_{221}) \\ (a_{111},a_{121}) & (a_{111},a_{221}) \end{Bmatrix} = \begin{Bmatrix} \underbrace{(\text{DP:UP,pF})}_{\text{hyperbolic flow }(+)} & \underbrace{(\text{SO,US})}_{\text{upper-saddle source}} \\ \underbrace{(\text{DP}_{+},\text{DP}_{-})}_{\text{CCW center}} & \underbrace{(\text{nF,DP:DP})}_{\text{HB-to-HS flow }(-)} \\ \underbrace{(\text{DP:UP,nF})}_{\text{hyperbolic flow }(-)} & \underbrace{(\text{SI,LS})}_{\text{lower-saddle sink}} \end{Bmatrix} \tag{4.11}$$

for $a_{110} > 0$ and $a_{220} > 0$;

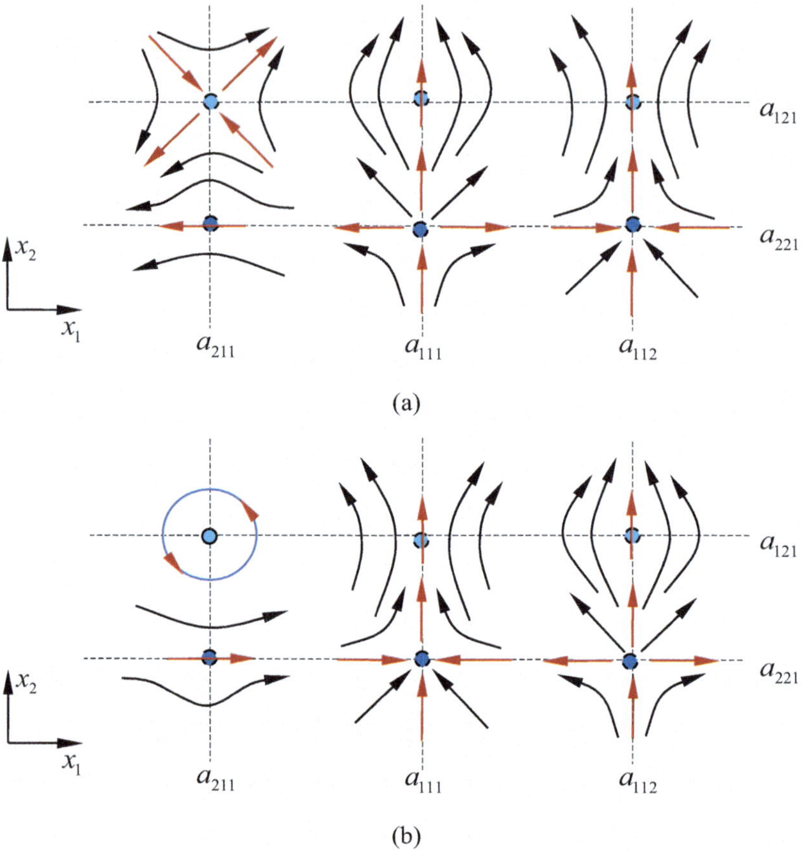

Fig. 4.2 Phase portraits ($a_{221} < a_{121}$, $a_{211} < a_{111}$) for two-dimensional systems on the x_1-direction with $x_1^* = a_{111}, a_{112}, a_{211}$ and on the x_2-direction with $x_2^* = a_{121}, a_{221}$. (2 × 3)-equilibrium matrices: (**a**) ($a_{110} > 0$, $a_{220} > 0$), (**b**) ($a_{110} < 0$, $a_{220} > 0$), (**c**) ($a_{110} > 0$, $a_{220} < 0$), and (**d**) ($a_{110} < 0$, $a_{220} < 0$)

$$\begin{Bmatrix} (a_{112}, a_{121}) & (a_{112}, a_{221}) \\ (a_{121}, a_{211}) & (a_{211}, a_{221}) \\ (a_{111}, a_{121}) & (a_{111}, a_{221}) \end{Bmatrix} = \begin{Bmatrix} \underbrace{(\text{UP:DP,pF})}_{\text{hyperbolic-secant flow }(+)} & \underbrace{(\text{SI,US})}_{\text{upper-saddle sink}} \\ \underbrace{(\text{UP}_+,\text{UP}_+)}_{\text{positive saddle}} & \underbrace{(\text{pF,UP:UP})}_{\text{HS-to-HB flow }(+)} \\ \underbrace{(\text{UP:DP,nF})}_{\text{hyperbolic-secant flow }(-)} & \underbrace{(\text{SO,LS})}_{\text{lower-saddle source}} \end{Bmatrix} \tag{4.12}$$

for $a_{110} < 0$ and $a_{220} > 0$;

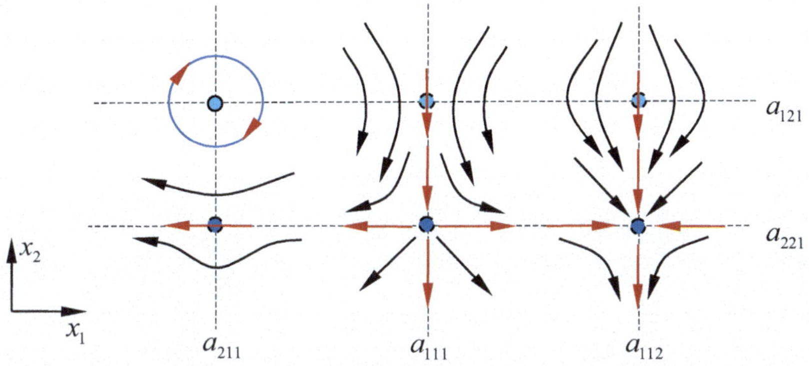

(c)

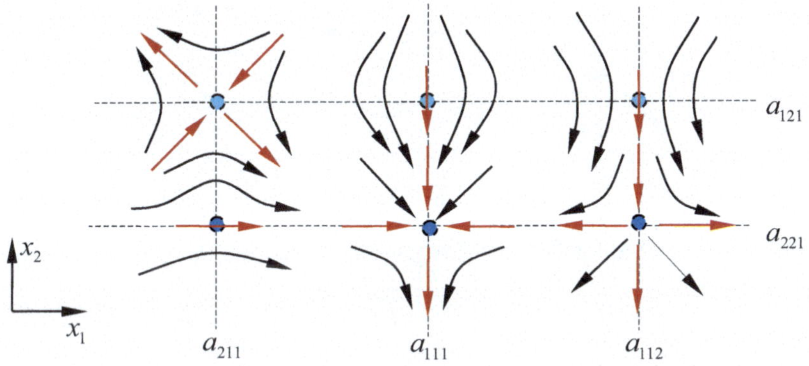

(d)

Fig. 4.2 (continued)

$$\left\{\begin{matrix}(a_{112},a_{121}) & (a_{112},a_{221})\\(a_{121},a_{211}) & (a_{211},a_{221})\\(a_{111},a_{121}) & (a_{111},a_{221})\end{matrix}\right\}=\left\{\begin{matrix}\underbrace{(\text{UP:DP,nF})}_{\text{hyperbolic-secant flow }(-)} & \underbrace{(\text{SO,LS})}_{\text{lower-saddle source}}\\\underbrace{(\text{UP}_{-},\text{UP}_{-})}_{\text{negative saddle}} & \underbrace{(\text{nF,UP:UP})}_{\text{HS-to-HB flow }(-)}\\\underbrace{(\text{UP:DP,pF})}_{\text{hyperbolic-secant flow }(+)} & \underbrace{(\text{SI,US})}_{\text{upper-saddle sink}}\end{matrix}\right\}\tag{4.13}$$

for $a_{110} > 0$ and $a_{220} < 0$;

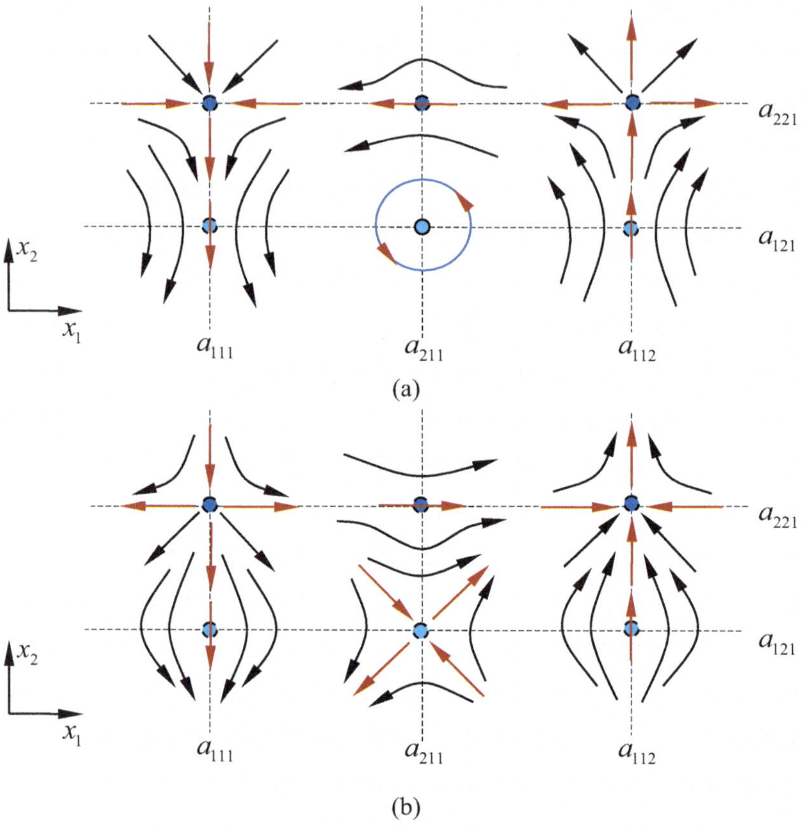

Fig. 4.3 Phase portraits ($a_{121} < a_{221}$, $a_{211} \in (a_{111}, a_{112})$) for two-dimensional systems on the x_1-direction with $x_1^* = a_{111}, a_{112}, a_{211}$ and on the x_2-direction with $x_2^* = a_{121}, a_{221}$. (2×3)-equilibrium matrices: (**a**) ($a_{110} > 0, a_{220} > 0$), (**b**) ($a_{110} < 0, a_{220} > 0$), (**c**) ($a_{110} > 0, a_{220} < 0$), and (**d**) ($a_{110} < 0$, $a_{220} < 0$)

$$\begin{Bmatrix} (a_{112}, a_{121}) & (a_{112}, a_{221}) \\ (a_{121}, a_{211}) & (a_{211}, a_{221}) \\ (a_{111}, a_{121}) & (a_{111}, a_{221}) \end{Bmatrix} = \begin{Bmatrix} \underbrace{(\text{DP:UP,nF})}_{\text{hyperbolic flow } (-)} & \underbrace{(\text{SI,LS})}_{\text{lower-saddle sink}} \\ \underbrace{(\text{DP}_-,\text{DP}_+)}_{\text{CW center}} & \underbrace{(\text{pF,DP:DP})}_{\text{HB-to-HS flow } (+)} \\ \underbrace{(\text{DP:UP,pF})}_{\text{hyperbolic flow } (+)} & \underbrace{(\text{SO,US})}_{\text{upper-saddle source}} \end{Bmatrix} \tag{4.14}$$

for $a_{110} < 0$ and $a_{220} < 0$.

The saddle sink and saddle source with hyperbolic and hyperbolic secant flows are presented in Fig. 4.3. The saddle sink and saddle source are the appearing bifurcation of saddle and sink and the appearing bifurcation of saddle and source, respectively.

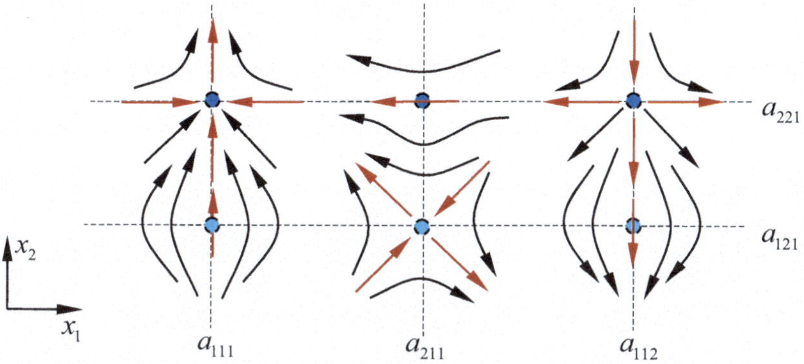

(c)

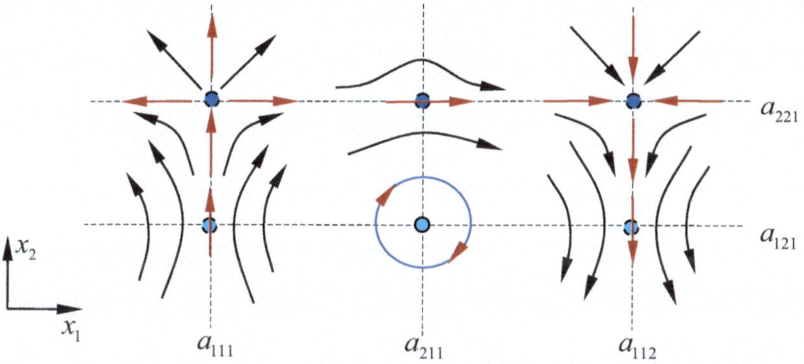

(d)

Fig. 4.3 (continued)

The saddle sink and saddle source are above the hyperbolic and hyperbolic secant flows.

(iv) For $a_{221} < a_{121}$ and $a_{211} \in (a_{111}, a_{112})$, the equilibriums of $x_1^* = a_{111}, a_{112}, a_{211}$ with $x_2^* = a_{121}, a_{221}$ have the following properties:

$$\begin{Bmatrix} (a_{112}, a_{221}) & (a_{112}, a_{121}) \\ (a_{211}, a_{221}) & (a_{121}, a_{211}) \\ (a_{111}, a_{221}) & (a_{111}, a_{121}) \end{Bmatrix} = \begin{Bmatrix} \underbrace{(\text{SI,US})}_{\text{upper-saddle sink}} & \underbrace{(\text{DP:UP,pF})}_{\text{hyperbolic flow }(+)} \\ \underbrace{(\text{pF,UP:UP})}_{\text{HS-to-HB flow }(+)} & \underbrace{(\text{DP}_+,\text{DP}_-)}_{\text{CCW center}} \\ \underbrace{(\text{SO,LS})}_{\text{lower-saddle source}} & \underbrace{(\text{DP:UP,nF})}_{\text{hyperbolic flow }(-)} \end{Bmatrix} \tag{4.15}$$

for $a_{110} > 0$ and $a_{220} > 0$;

$$\left\{\begin{matrix} (a_{112}, a_{221}) & (a_{112}, a_{121}) \\ (a_{211}, a_{221}) & (a_{121}, a_{211}) \\ (a_{111}, a_{221}) & (a_{111}, a_{121}) \end{matrix}\right\} = \left\{\begin{matrix} \underbrace{\text{(SO,US)}}_{\text{upper-saddle source}} & \underbrace{\text{(UP:DP,pF)}}_{\text{hyperbolic-secant flow }(+)} \\ \underbrace{\text{(nF,DP:DP)}}_{\text{HB-to-HS flow }(-)} & \underbrace{(\text{UP}_+,\text{UP}_+)}_{\text{positive saddle}} \\ \underbrace{\text{(SI,LS)}}_{\text{lower-saddle sink}} & \underbrace{\text{(UP:DP,nF)}}_{\text{hyperbolic-secant flow }(-)} \end{matrix}\right\} \tag{4.16}$$

for $a_{110} < 0$ and $a_{220} > 0$;

$$\left\{\begin{matrix} (a_{112}, a_{221}) & (a_{112}, a_{121}) \\ (a_{211}, a_{221}) & (a_{121}, a_{211}) \\ (a_{111}, a_{221}) & (a_{111}, a_{121}) \end{matrix}\right\} = \left\{\begin{matrix} \underbrace{\text{(SI,LS)}}_{\text{lower-saddle sink}} & \underbrace{\text{(UP:DP,nF)}}_{\text{hyperbolic-secant flow }(-)} \\ \underbrace{\text{(pF,DP:DP)}}_{\text{HB-to-HS flow }(+)} & \underbrace{(\text{UP}_-,\text{ UP}_-)}_{\text{neagative saddle}} \\ \underbrace{\text{(SO,US)}}_{\text{upper-saddle source}} & \underbrace{\text{(UP:DP,pF)}}_{\text{hyperbolic-secant flow }(+)} \end{matrix}\right\} \tag{4.17}$$

for $a_{110} > 0$ and $a_{220} < 0$;

$$\left\{\begin{matrix} (a_{112}, a_{221}) & (a_{112}, a_{121}) \\ (a_{211}, a_{221}) & (a_{121}, a_{211}) \\ (a_{111}, a_{221}) & (a_{111}, a_{121}) \end{matrix}\right\} = \left\{\begin{matrix} \underbrace{\text{(SO,LS)}}_{\text{lower-saddle source}} & \underbrace{\text{(DP:UP,nF)}}_{\text{hyperbolic flow }(-)} \\ \underbrace{\text{(nF,UP:UP)}}_{\text{HS-to-HB flow }(-)} & \underbrace{(\text{DP}_-,\text{DP}_+)}_{\text{CW center}} \\ \underbrace{\text{(SI,US)}}_{\text{upper-saddle sink}} & \underbrace{\text{(DP:UP,pF)}}_{\text{hyperbolic flow }(+)} \end{matrix}\right\} \tag{4.18}$$

for $a_{110} < 0$ and $a_{220} < 0$.

The saddle-sink and saddle-source with hyperbolic and hyperbolic-secant flows are presented in Fig. 4.4. The saddle-sink and saddle-source are the appearing bifurcation of saddle and sink and the appearing bifurcation of saddle and source, respectively. The saddle-sink and saddle-source are below the hyperbolic and hyperbolic-secant flows.

(v) For the case of $a_{112} < a_{211}$ and $a_{121} < a_{221}$, the saddle nodes and hyperbolic flows of $x_1^* = a_{111}, a_{112}, a_{211}$ with $x_2^* = a_{121}, a_{221}$ have the following properties:

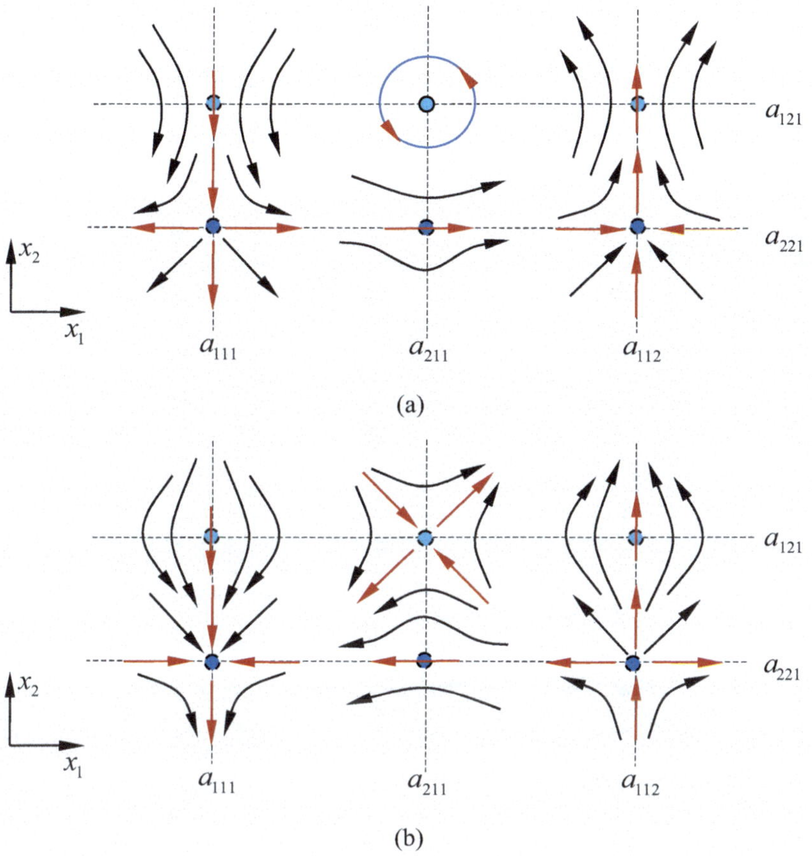

Fig. 4.4 Phase portraits ($a_{221} < a_{121}$, $a_{211} \in (a_{111}, a_{112})$) for two-dimensional systems on the x_1-direction with $x_1^* = a_{111}, a_{112}, a_{211}$ and on the x_2-direction with $x_2^* = a_{121}, a_{221}$. (2 × 3)-equilibrium matrices: **(a)** ($a_{110} > 0, a_{220} > 0$), **(b)** ($a_{110} < 0, a_{220} > 0$), **(c)** ($a_{110} > 0, a_{220} < 0$), and **(d)** ($a_{110} < 0$, $a_{220} < 0$)

$$\begin{Bmatrix} (a_{121}, a_{211}) & (a_{211}, a_{221}) \\ (a_{112}, a_{121}) & (a_{112}, a_{221}) \\ (a_{111}, a_{121}) & (a_{111}, a_{221}) \end{Bmatrix} = \begin{Bmatrix} \underbrace{(\text{DP}_-,\text{DP}_+)}_{\text{CW center}} & \underbrace{(\text{pF},\text{DP:DP})}_{\text{HB-to-HS flow }(+)} \\ \underbrace{(\text{DP:UP},\text{pF})}_{\text{hyperbolic flow }(+)} & \underbrace{(\text{SO},\text{US})}_{\text{upper-saddle source}} \\ \underbrace{(\text{UP:DP},\text{pF})}_{\text{hyperbolic-secant flow }(+)} & \underbrace{(\text{SI},\text{US})}_{\text{upper-saddle sink}} \end{Bmatrix} \tag{4.19}$$

for $a_{110} > 0$ and $a_{220} > 0$;

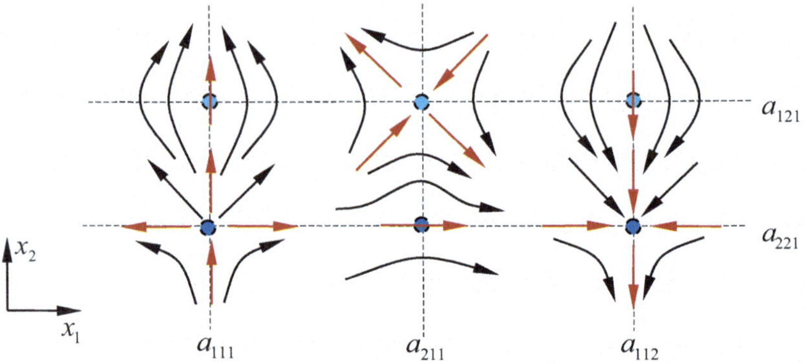

(c)

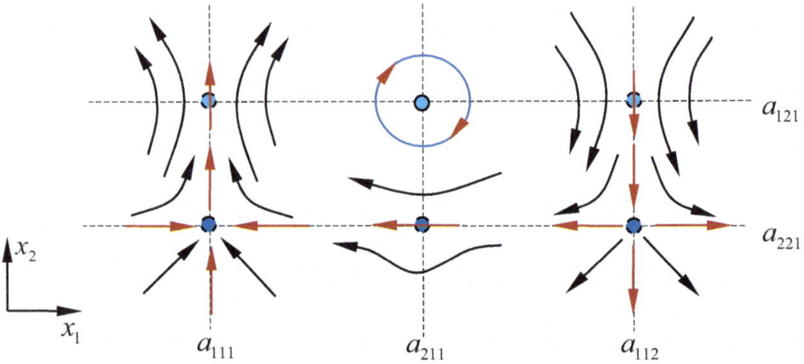

(d)

Fig. 4.4 (continued)

$$\begin{Bmatrix} (a_{121}, a_{211}) & (a_{211}, a_{221}) \\ (a_{112}, a_{121}) & (a_{112}, a_{221}) \\ (a_{111}, a_{121}) & (a_{111}, a_{221}) \end{Bmatrix} = \begin{Bmatrix} \underbrace{(\text{UP}_-, \text{UP}_-)}_{\text{negative saddle}} & \underbrace{(\text{nF,UP:UP})}_{\text{HS-to-HB flow }(-)} \\ \underbrace{(\text{UP:DP,pF})}_{\text{hyperbolic-secant flow }(+)} & \underbrace{(\text{SI,US})}_{\text{upper-saddle sink}} \\ \underbrace{(\text{DP:UP,pF})}_{\text{hyperbolic flow }(+)} & \underbrace{(\text{SO,US})}_{\text{upper-saddle source}} \end{Bmatrix} \tag{4.20}$$

for $a_{110} < 0$ and $a_{220} > 0$;

$$\begin{Bmatrix} (a_{121}, a_{211}) & (a_{211}, a_{221}) \\ (a_{112}, a_{121}) & (a_{112}, a_{221}) \\ (a_{111}, a_{121}) & (a_{111}, a_{221}) \end{Bmatrix} = \begin{Bmatrix} \underbrace{(\text{UP}_+, \text{UP}_+)}_{\text{positive saddle}} & \underbrace{(\text{pF,UP:UP})}_{\text{HS-to-HB flow }(+)} \\ \underbrace{(\text{UP:DP,nF})}_{\text{hyperbolic-secant flow }(-)} & \underbrace{(\text{SO,LS})}_{\text{lower-saddle source}} \\ \underbrace{(\text{DP:UP,nF})}_{\text{hyperbolic flow }(-)} & \underbrace{(\text{SI,LS})}_{\text{lower-saddle sink}} \end{Bmatrix} \tag{4.21}$$

for $a_{110} > 0$ and $a_{220} < 0$;

$$\begin{Bmatrix} (a_{121}, a_{211}) & (a_{211}, a_{221}) \\ (a_{112}, a_{121}) & (a_{112}, a_{221}) \\ (a_{111}, a_{121}) & (a_{111}, a_{221}) \end{Bmatrix} = \begin{Bmatrix} \underbrace{(\mathrm{DP}_{+},\mathrm{DP}_{-})}_{\text{CCW center}} & \underbrace{(\mathrm{nF},\mathrm{DP}:\mathrm{DP})}_{\text{HB-to-HS flow }(-)} \\ \underbrace{(\mathrm{DP}:\mathrm{UP},\mathrm{nF})}_{\text{hyperbolic flow }(-)} & \underbrace{(\mathrm{SI},\mathrm{LS})}_{\text{lower-saddle sink}} \\ \underbrace{(\mathrm{UP}:\mathrm{DP},\mathrm{nF})}_{\text{hyperbolic-secant flow }(-)} & \underbrace{(\mathrm{SO},\mathrm{LS})}_{\text{lower-saddle source}} \end{Bmatrix} \tag{4.22}$$

for $a_{110} < 0$ and $a_{220} < 0$.

The horizontally paralleled saddle-sink and saddle-source with connected hyperbolic singular flows plus two separated hyperbolic and hyperbolic-secant flows and saddle and center in equilibrium matrices are presented in Fig. 4.5. The saddle-sink and saddle-source are the appearing bifurcation of saddle and sink and the appearing bifurcation of saddle and source, respectively. The saddle-sink and saddle-source are above the hyperbolic and hyperbolic-secant flows.

(vi) For $a_{221} < a_{121}$ and $a_{112} < a_{211}$, the equilibriums of $x_1^* = a_{111}, a_{112}, a_{211}$ with $x_2^* = a_{121}, a_{221}$ have the following properties:

$$\begin{Bmatrix} (a_{211}, a_{221}) & (a_{121}, a_{211}) \\ (a_{112}, a_{221}) & (a_{112}, a_{121}) \\ (a_{111}, a_{221}) & (a_{111}, a_{121}) \end{Bmatrix} = \begin{Bmatrix} \underbrace{(\mathrm{nF},\mathrm{UP}:\mathrm{UP})}_{\text{HS-to-HB flow }(-)} & \underbrace{(\mathrm{DP}_{-},\mathrm{DP}_{+})}_{\text{CW center}} \\ \underbrace{(\mathrm{SI},\mathrm{US})}_{\text{upper-saddle sink}} & \underbrace{(\mathrm{DP}:\mathrm{UP},\mathrm{pF})}_{\text{hyperbolic flow }(+)} \\ \underbrace{(\mathrm{SO},\mathrm{US})}_{\text{upper-saddle source}} & \underbrace{(\mathrm{UP}:\mathrm{DP},\mathrm{pF})}_{\text{hyperbolic-secant flow }(+)} \end{Bmatrix} \tag{4.23}$$

for $a_{110} > 0$ and $a_{220} > 0$;

$$\begin{Bmatrix} (a_{211}, a_{221}) & (a_{121}, a_{211}) \\ (a_{112}, a_{221}) & (a_{112}, a_{121}) \\ (a_{111}, a_{221}) & (a_{111}, a_{121}) \end{Bmatrix} = \begin{Bmatrix} \underbrace{(\mathrm{pF},\mathrm{DP}:\mathrm{DP})}_{\text{HB-to-HS flow }(+)} & \underbrace{(\mathrm{UP}_{-},\mathrm{UP}_{-})}_{\text{negative saddle}} \\ \underbrace{(\mathrm{SO},\mathrm{US})}_{\text{upper-saddle source}} & \underbrace{(\mathrm{UP}:\mathrm{DP},\mathrm{pF})}_{\text{hyperbolic-secant flow }(+)} \\ \underbrace{(\mathrm{SI},\mathrm{US})}_{\text{upper-saddle sink}} & \underbrace{(\mathrm{DP}:\mathrm{UP},\mathrm{pF})}_{\text{hyperbolic flow }(+)} \end{Bmatrix} \tag{4.24}$$

for $a_{110} < 0$ and $a_{220} > 0$;

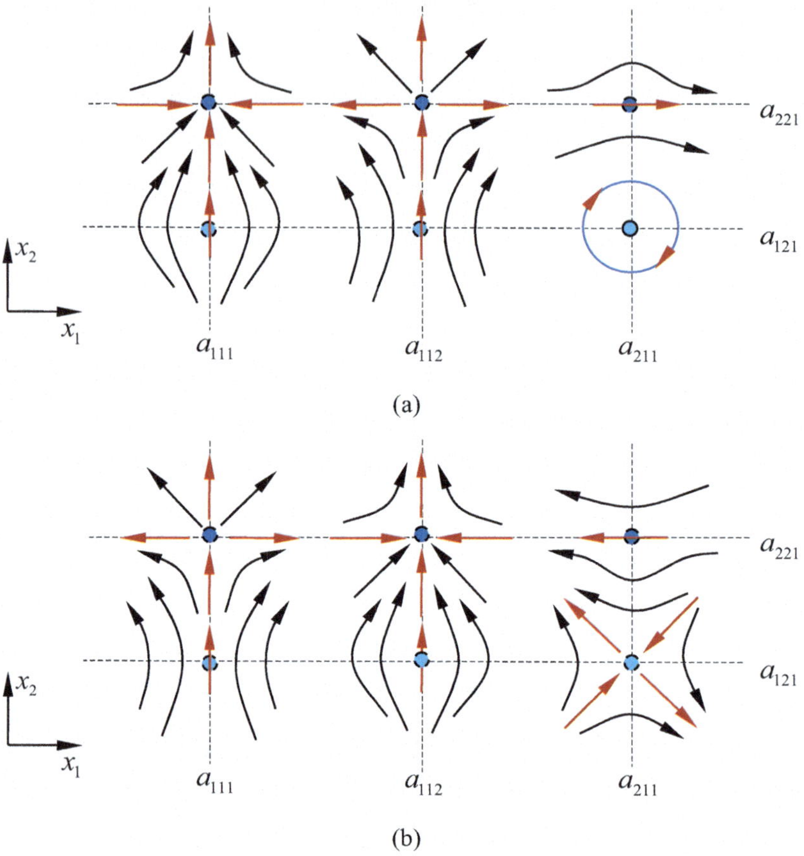

Fig. 4.5 Phase portraits ($a_{121} < a_{221}$, $a_{112} < a_{211}$) for two-dimensional systems on the x_1-direction with $x_1^* = a_{111}, a_{112}, a_{211}$ and on the x_2-direction with $x_2^* = a_{121}, a_{221}$. (2 × 3)-equilibrium matrices: (**a**) ($a_{110} > 0$, $a_{220} > 0$), (**b**) ($a_{110} < 0$, $a_{220} > 0$), (**c**) ($a_{110} > 0$, $a_{220} < 0$), and (**d**) ($a_{110} < 0$, $a_{220} < 0$)

$$\begin{Bmatrix} (a_{211}, a_{221}) & (a_{121}, a_{211}) \\ (a_{112}, a_{221}) & (a_{112}, a_{121}) \\ (a_{111}, a_{221}) & (a_{111}, a_{121}) \end{Bmatrix} = \begin{Bmatrix} \underbrace{(\text{nF,DP:DP})}_{\text{HB-to-HS flow }(-)} & \underbrace{(\text{UP}_+,\text{UP}_+)}_{\text{positive saddle}} \\ \underbrace{(\text{SI,LS})}_{\text{lower-saddle sink}} & \underbrace{(\text{UP:DP,nF})}_{\text{hyperbolic-secant flow }(-)} \\ \underbrace{(\text{SO,LS})}_{\text{lower-saddle source}} & \underbrace{(\text{DP:UP,nF})}_{\text{hyperbolic flow }(-)} \end{Bmatrix} \tag{4.25}$$

for $a_{110} > 0$ and $a_{220} < 0$;

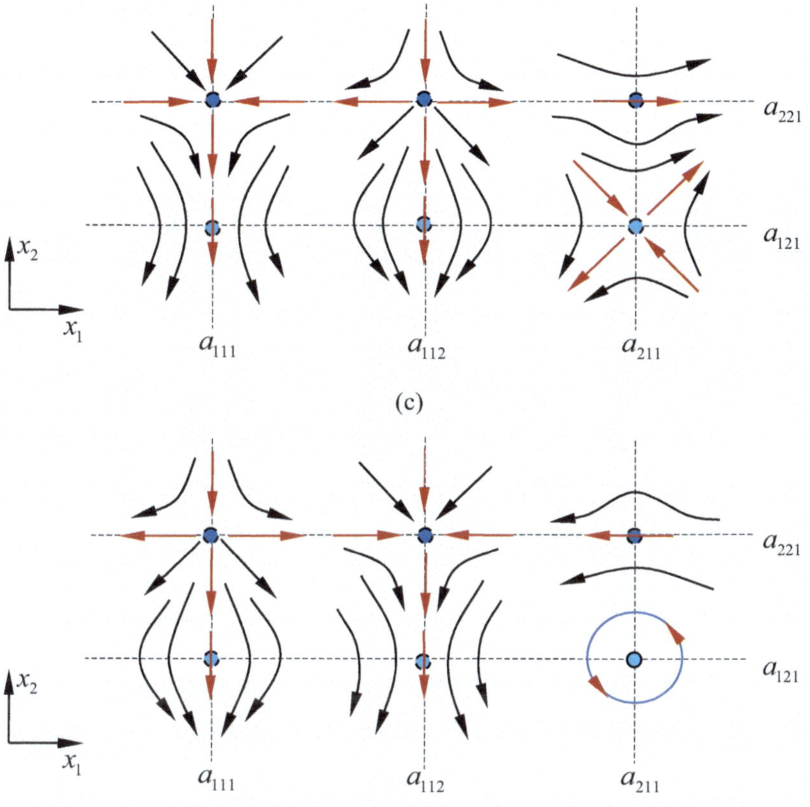

(d)

Fig. 4.5 (continued)

$$\begin{Bmatrix} (a_{211}, a_{221}) & (a_{121}, a_{211}) \\ (a_{112}, a_{221}) & (a_{112}, a_{121}) \\ (a_{111}, a_{221}) & (a_{111}, a_{121}) \end{Bmatrix} = \begin{Bmatrix} \underbrace{(\text{pF,UP:UP})}_{\text{HS-to-HB flow }(+)} & \underbrace{(\text{DP}_{+},\text{DP}_{-})}_{\text{CCW center}} \\ \underbrace{(\text{SO,LS})}_{\text{lower-saddle source}} & \underbrace{(\text{DP:UP,nF})}_{\text{hyperbolic flow }(-)} \\ \underbrace{(\text{SI,LS})}_{\text{lower-saddle sink}} & \underbrace{(\text{UP:DP,nF})}_{\text{hyperbolic-secant flow }(-)} \end{Bmatrix} \tag{4.26}$$

for $a_{110} < 0$ and $a_{220} < 0$.

The saddle-sink and saddle-source with hyperbolic and hyperbolic-secant flows are presented in Fig. 4.6. The saddle-sink and saddle-source are the appearing bifurcation of saddle and sink and the appearing bifurcation of saddle and source, respectively. The saddle-sink and saddle-source are below the hyperbolic and hyperbolic-secant flows.

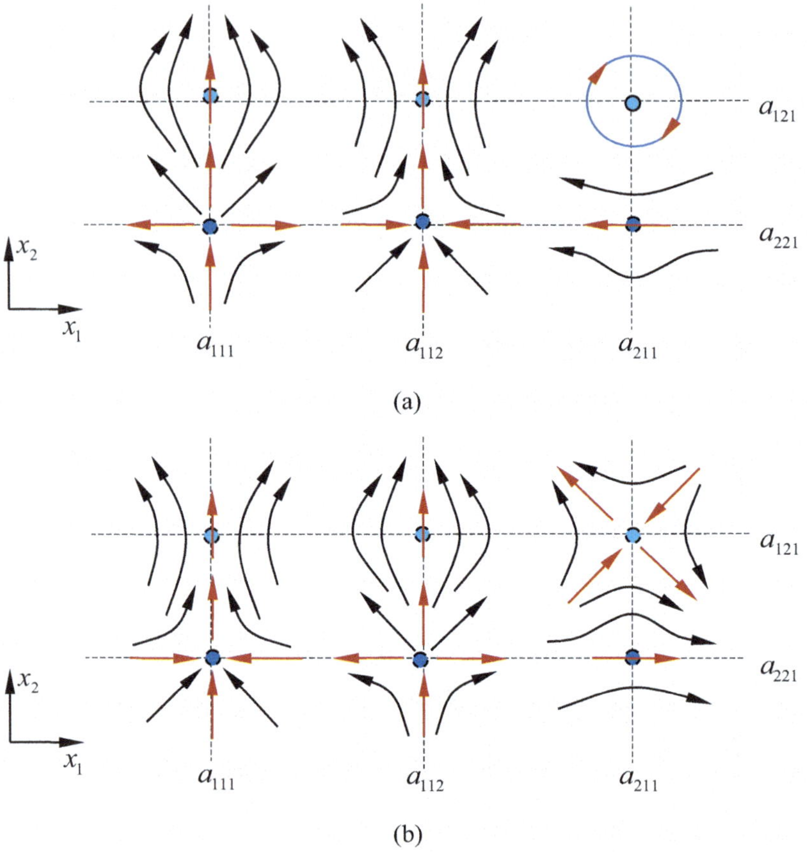

Fig. 4.6 Phase portraits ($a_{221} < a_{121}, a_{112} < a_{211}$) for two-dimensional systems on the x_1-direction with $x_1^* = a_{111}, a_{112}, a_{211}$ and on the x_2-direction with $x_2^* = a_{121}, a_{221}$. (2×3)-equilibrium matrices: **(a)** ($a_{110} > 0$, $a_{220} > 0$), **(b)** ($a_{110} < 0$, $a_{220} > 0$), **(c)** ($a_{110} > 0$, $a_{220} < 0$), and **(d)** ($a_{110} < 0$, $a_{220} < 0$)

4.2 Infinite-Equilibriums and Switching Bifurcation Dynamics

In this section, the up-down and down-up upper-saddle and lower-saddle infinite-equilibriums in the horizontal direction are discussed first for the switching bifurcations. The up-down and down-up saddle infinite-equilibriums are for the switching of a hyperbolic flow with connected saddle-nodes and the saddle and center with a paralleled hyperbolic singular flow. The inflection-source (sink) infinite-equilibriums in the vertical direction are discussed, which are for the switching of the saddle and center with paralleled hyperbolic flows and the saddle-nodes with paralleled hyperbolic singular flow. The double-switching bifurcations in two directions are also presented.

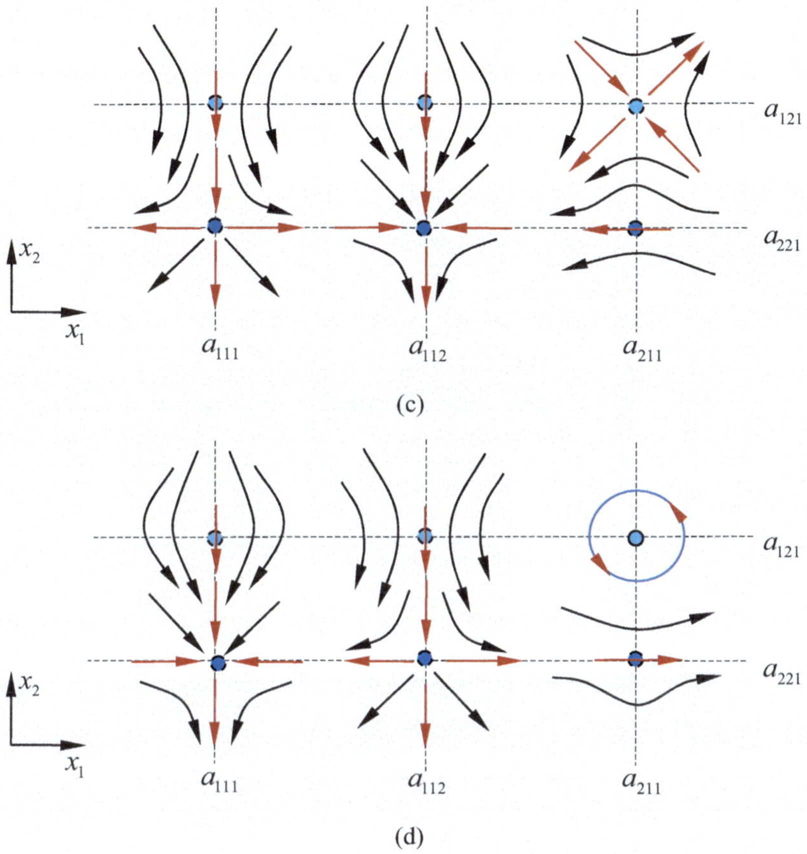

Fig. 4.6 (continued)

4.2.1 Up-Down and Down-Up Saddle Infinite Equilibriums

Consider a dynamical system with up-down and down-up upper saddle and lower saddle infinite equilibrium as

$$\begin{aligned}\dot{x}_1 &= a_{110}(x_1 - a_{111})(x_1 - a_{112})(x_2 - a_{121}),\\ \dot{x}_2 &= a_{220}(x_1 - a_{211})(x_2 - a_{221})^2.\end{aligned} \tag{4.27}$$

For $a_{121} = a_{221}$, the corresponding first integral manifold is

$$\frac{a_{112} - a_{211}}{a_{112} - a_{111}} \ln \frac{|x_1 - a_{112}|}{|x_{10} - a_{112}|} + \frac{a_{111} - a_{211}}{a_{111} - a_{112}} \ln \frac{|x_1 - a_{111}|}{|x_{10} - a_{111}|} = \frac{a_{110}}{a_{220}} \ln \frac{|x_2 - a_{221}|}{|x_{20} - a_{221}|}. \tag{4.28}$$

(i) The infinite-equilibriums of $x_2^* = a_{121} = a_{221}$ with $x_1^* = a_{111}, a_{112}, a_{211}$ $(a_{211} < a_{111})$ have the following properties:

$$\left.\begin{array}{l}(a_{112}, a_{221})\\(a_{111}, a_{221})\\(a_{211}, a_{221})\end{array}\right\} = \left\{\begin{array}{c}\underbrace{(\text{UD:DU,US})}_{\text{up-down-to-down-up upper-saddle}}\\\underbrace{(\text{DU:UD,US})}_{\text{down-up-to-up-down upper-saddle}}\\\underbrace{({}_{\text{UD}}\text{LS:}{}_{\text{DU}}\text{US, DP}_{-}\text{:UP}_{+})}_{\text{hyperbolic lower-to-upper saddle}}\end{array}\right\} \text{ for } a_{110} > 0, a_{220} > 0, \tag{4.29}$$

$$\left.\begin{array}{l}(a_{112}, a_{221})\\(a_{111}, a_{221})\\(a_{211}, a_{221})\end{array}\right\} = \left\{\begin{array}{c}\underbrace{(\text{DU:UD,US})}_{\text{down-up-to-up-down upper-saddle}}\\\underbrace{(\text{UD:DU,US})}_{\text{up-down-to-down-up upper-saddle}}\\\underbrace{({}_{\text{DU}}\text{LS:}{}_{\text{UD}}\text{US, UP}_{-}\text{: DP}_{+})}_{\text{hyperbolic-secant lower-to-upper saddle}}\end{array}\right\} \text{ for } a_{110} < 0, a_{220} > 0, \tag{4.30}$$

$$\left.\begin{array}{l}(a_{112}, a_{221})\\(a_{111}, a_{221})\\(a_{211}, a_{221})\end{array}\right\} = \left\{\begin{array}{c}\underbrace{(\text{DU:UD,LS})}_{\text{down-up-to-up-down lower-saddle}}\\\underbrace{(\text{UD:DU,LS})}_{\text{up-down-to-down-up lower-saddle}}\\\underbrace{({}_{\text{DU}}\text{US:}{}_{\text{UD}}\text{LS, UP}_{+}\text{: DP}_{-})}_{\text{hyperbolic-secant upper-to-lower saddle}}\end{array}\right\} \text{ for } a_{110} > 0, a_{220} < 0, \tag{4.31}$$

$$\left.\begin{array}{l}(a_{112}, a_{221})\\(a_{111}, a_{221})\\(a_{211}, a_{221})\end{array}\right\} = \left\{\begin{array}{c}\underbrace{(\text{UD:DU,LS})}_{\text{up-down-to-down-up lower-saddle}}\\\underbrace{(\text{DU:UD,LS})}_{\text{down-up-to-up-down lower-saddle}}\\\underbrace{({}_{\text{UD}}\text{US:}{}_{\text{DU}}\text{LS, DP}_{+}\text{: UP}_{-})}_{\text{hyperbolic upper-to-lower saddle}}\end{array}\right\} \text{ for } a_{110} < 0, a_{220} < 0. \tag{4.32}$$

The down-up and up-down upper-saddle and lower-saddle infinite-equilibriums of $x_2^* = a_{121} = a_{221}$ are presented in Fig. 4.7. The hyperbolic and hyperbolic-secant, lower-to-upper saddle, and upper-to-lower saddle are illustrated. The infinite-equilibrium is for the switching bifurcation for $a_{121} < a_{221}$ and $a_{121} > a_{221}$. The infinite-equilibrium of $x_2^* = a_{121} = a_{222}$ is summarized in Table 4.1.

(ii) The infinite-equilibriums of $x_2^* = a_{121} = a_{221}$ with $x_1^* = a_{111}, a_{112}, a_{211}$ $(a_{211} \in (a_{111}, a_{112}))$ have the following properties:

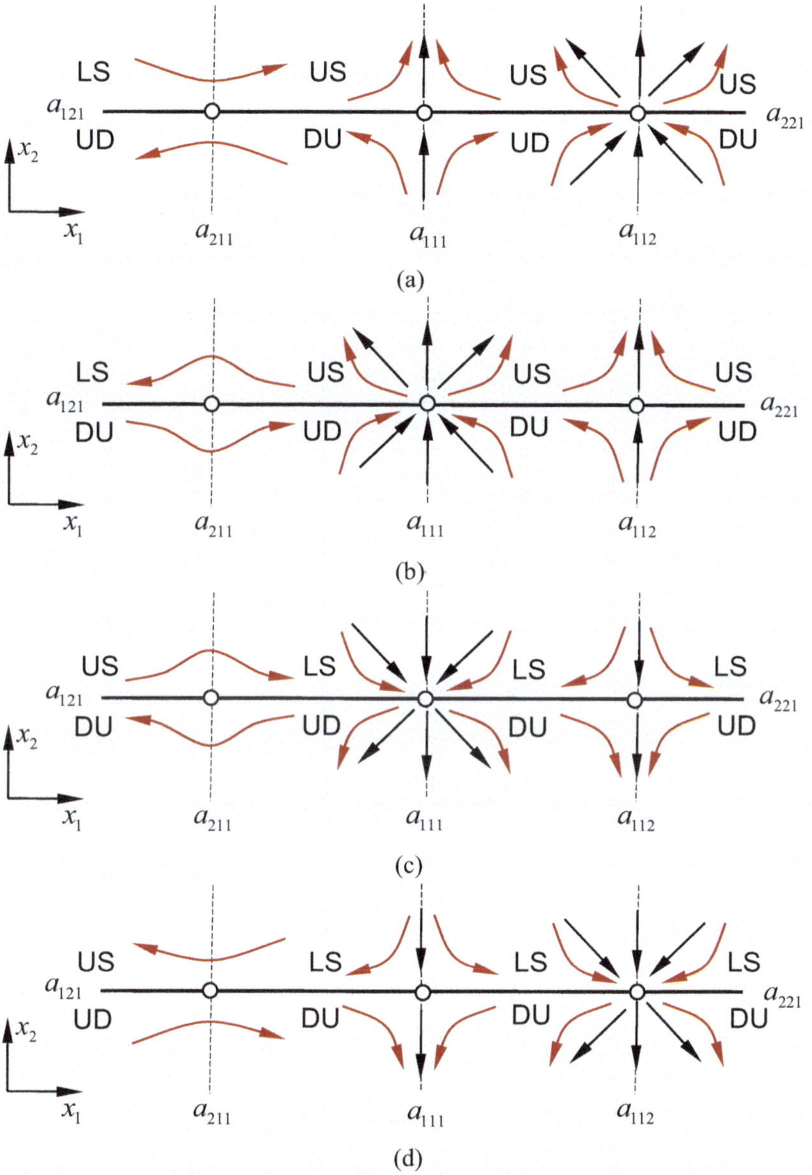

Fig. 4.7 Phase portraits ($a_{211} < a_{111}$ $a_{121} = a_{221}$) for two-dimensional systems on the x_1-direction with $x_1^* = a_{111}, a_{112}, a_{211}$ and on the x_2-direction with $x_2^* = a_{121}, a_{221}$. The up-down and down-up, upper-saddle, and lower-saddle infinite-equilibriums: (**a**) ($a_{110} > 0$, $a_{220} > 0$), (**b**) ($a_{110} < 0$, $a_{220} > 0$), (**c**) ($a_{110} > 0$, $a_{220} < 0$), and (**d**) ($a_{110} < 0$, $a_{220} < 0$)

Table 4.1 Down-up and up-down infinite equilibriums of $x_1^* = a_{221} = a_{121}$ $(a_{211} = a_{121}$ and $a_{211} < a_{111})$

$(\bar{x}_1, a_{221})$	$\bar{x}_1$			
(a_{110}, a_{220})	$(-\infty, a_{211})$	(a_{211}, a_{111})	(a_{111}, a_{112})	(a_{112}, ∞)
$(+,+)$	(UD,LS)	(DU,US)	(UD,US)	(DU,US)
$(-,+)$	(DU,LS)	(UD,US)	(DU,US)	(UD, US)
$(+,-)$	(DU,US)	(UD,LS)	(DU,LS)	(UD,LS)
$(-,-)$	(UD,US)	(DU,LS)	(UD,LS)	(DU,LS)

$$\left\{\begin{matrix}(a_{112}, a_{221})\\(a_{211}, a_{221})\\(a_{111}, a_{221})\end{matrix}\right\} = \left\{\begin{matrix}\underbrace{(\text{UD:DU,US})}_{\text{up-down-to-down-up upper-saddle}}\\\underbrace{({}_{\text{DU}}\text{US:}{}_{\text{UD}}\text{LS, UP}_+\text{: DP}_-)}_{\text{hyperbolic lower-to-upper saddle}}\\\underbrace{(\text{UD:DU,LS})}_{\text{up-down-to-down-up lower-saddle}}\end{matrix}\right\} \text{ for } a_{110} > 0, a_{220} > 0, \tag{4.33}$$

$$\left\{\begin{matrix}(a_{112}, a_{221})\\(a_{211}, a_{221})\\(a_{111}, a_{221})\end{matrix}\right\} = \left\{\begin{matrix}\underbrace{(\text{DU:UD,US})}_{\text{down-up-to-up-down upper-saddle}}\\\underbrace{({}_{\text{UD}}\text{LS:}{}_{\text{DU}}\text{US, DP}_-\text{ :UP}_+)}_{\text{hyperbolic lower-to-upper saddle}}\\\underbrace{(\text{DU:UD,LS})}_{\text{down-up-to-up-down lower-saddle}}\end{matrix}\right\} \text{ for } a_{110} < 0, a_{220} > 0, \tag{4.34}$$

$$\left\{\begin{matrix}(a_{112}, a_{221})\\(a_{211}, a_{221})\\(a_{111}, a_{221})\end{matrix}\right\} = \left\{\begin{matrix}\underbrace{(\text{DU:UD,LS})}_{\text{down-up-to-up-down lower-saddle}}\\\underbrace{({}_{\text{UD}}\text{US:}{}_{\text{DU}}\text{LS, DP}_+\text{: UP}_-)}_{\text{hyperbolic upper-to-lower saddle}}\\\underbrace{(\text{DU:UD,US})}_{\text{down-up-to-up-down upper-saddle}}\end{matrix}\right\} \text{ for } a_{110} > 0, a_{220} < 0, \tag{4.35}$$

$$\left\{\begin{matrix}(a_{112}, a_{221})\\(a_{211}, a_{221})\\(a_{111}, a_{221})\end{matrix}\right\} = \left\{\begin{matrix}\underbrace{(\text{UD:DU,LS})}_{\text{up-down-to-down-up lower-saddle}}\\\underbrace{({}_{\text{DU}}\text{LS:}{}_{\text{UD}}\text{US, UP}_-\text{ : DP}_+)}_{\text{hyperbolic upper-to-lower saddle}}\\\underbrace{(\text{UD:DU,US})}_{\text{up-down-to-down-up upper-saddle}}\end{matrix}\right\} \text{ for } a_{110} < 0, a_{220} < 0, \tag{4.36}$$

The down-up and up-down upper-saddle and lower-saddle infinite-equilibriums of $x_2^* = a_{121} = a_{221}$ $(a_{211} \in (a_{111}, a_{112}))$ are presented in Fig. 4.8. The hyperbolic and hyperbolic-secant, lower-to-upper saddle, and upper-to-lower saddle are illustrated. The infinite-equilibrium is for the switching bifurcation of equilibriums for

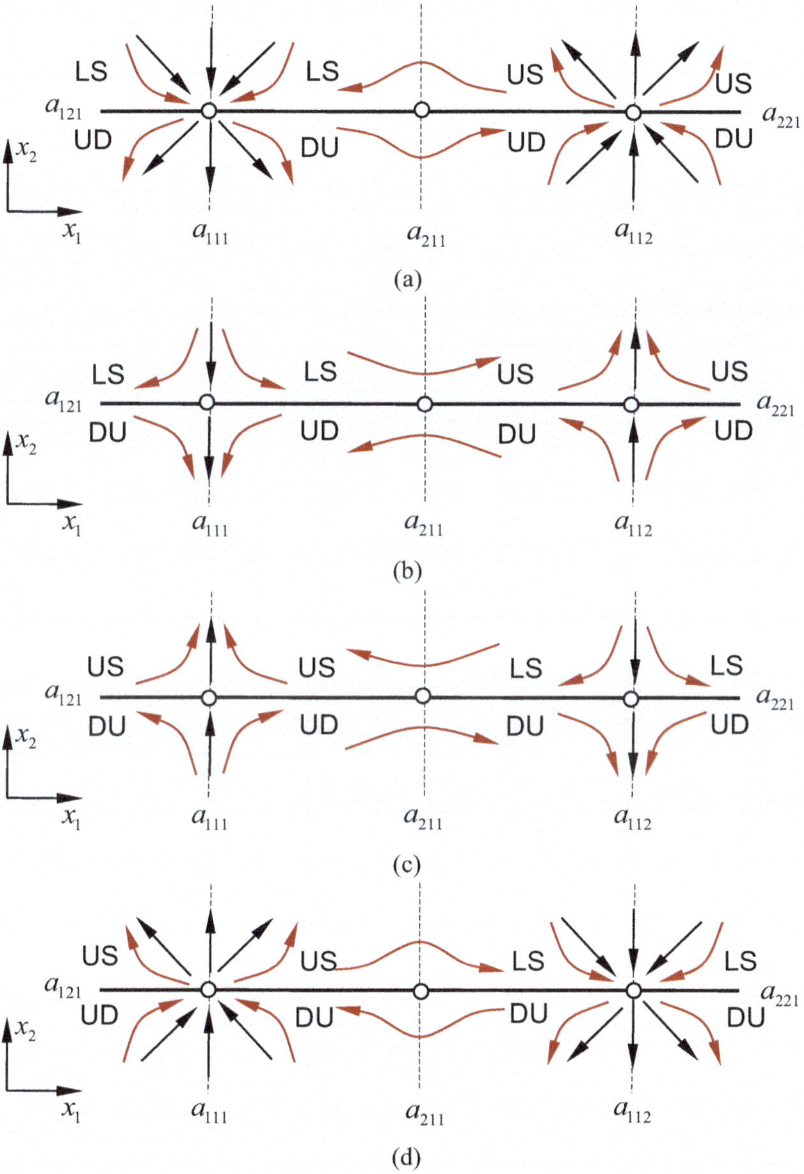

Fig. 4.8 Phase portraits ($a_{211} \in (a_{111}, a_{112})$ and $a_{121} = a_{221}$) for two-dimensional systems on the x_1-direction with $x_1^* = a_{111}, a_{112}$ and on the x_2-direction with $x_2^* = a_{121}, a_{221}$. The up-down and down-up, upper-saddle, and lower-saddle infinite-equilibriums: **(a)** ($a_{110} > 0$, $a_{220} > 0$), **(b)** ($a_{110} < 0$, $a_{220} > 0$), **(c)** ($a_{110} > 0$, $a_{220} < 0$), and **(d)** ($a_{110} < 0$, $a_{220} < 0$)

$a_{121} < a_{221}$ and $a_{121} > a_{221}$. The infinite-equilibrium of $x_2^* = a_{121} = a_{222}$ for $a_{211} \in (a_{111}, a_{112})$ is summarized in Table 4.2.

Table 4.2 Down-up and up-down infinite equilibriums of $x_1^* = a_{221} = a_{121}$ $(a_{211} = a_{121}$ and $a_{211} \in (a_{111}, a_{112}))$

$(\overline{x}_1, a_{221})$	$\overline{x}_1$			
(a_{110}, a_{220})	$(-\infty, a_{211})$	(a_{211}, a_{111})	(a_{111}, a_{112})	(a_{112}, ∞)
$(+,+)$	(UD,LS)	(DU,LS)	(UD,US)	(DU,US)
$(-,+)$	(DU,LS)	(UD,LS)	(DU,US)	(UD, US)
$(+,-)$	(DU,US)	(UD,US)	(DU,LS)	(UD,LS)
$(-,-)$	(UD,US)	(DU,US)	(UD,LS)	(DU,LS)

(iii) The infinite equilibriums of $x_2^* = a_{121} = a_{221}$ with $x_1^* = a_{111}, a_{112}, a_{211}$ $(a_{112} < a_{211})$ have the following properties:

$$\begin{Bmatrix} (a_{211}, a_{221}) \\ (a_{112}, a_{221}) \\ (a_{111}, a_{221}) \end{Bmatrix} = \begin{Bmatrix} \underbrace{({}_{\text{UD}}\text{LS:}{}_{\text{DU}}\text{US}, \text{DP}_-: \text{UP}_+)}_{\text{hyperbolic lower-to-upper saddle}} \\ \underbrace{(\text{DU:UD,LS})}_{\text{down-up-to-up-down lower-saddle}} \\ \underbrace{(\text{UD:DU,LS})}_{\text{up-down-to-down-up lower-saddle}} \end{Bmatrix} \text{ for } a_{110} > 0, a_{220} > 0, \tag{4.37}$$

$$\begin{Bmatrix} (a_{211}, a_{221}) \\ (a_{112}, a_{221}) \\ (a_{111}, a_{221}) \end{Bmatrix} = \begin{Bmatrix} \underbrace{({}_{\text{DU}}\text{LS:}{}_{\text{UD}}\text{US}, \text{DP}_+: \text{UP}_-)}_{\text{hyperbolic-secant lower-to-upper saddle}} \\ \underbrace{(\text{UD:DU,LS})}_{\text{up-down-to-down-up lower-saddle}} \\ \underbrace{(\text{DU:UD,LS})}_{\text{down-up-to-up-down lower-saddle}} \end{Bmatrix} \text{ for } a_{110} < 0, a_{220} > 0, \tag{4.38}$$

$$\begin{Bmatrix} (a_{211}, a_{221}) \\ (a_{112}, a_{221}) \\ (a_{111}, a_{221}) \end{Bmatrix} = \begin{Bmatrix} \underbrace{({}_{\text{DU}}\text{US:}{}_{\text{UD}}\text{LS}, \text{DP}_-: \text{UP}_+)}_{\text{hyperbolic-secant upper-to-lower saddle}} \\ \underbrace{(\text{UD:DU,US})}_{\text{up-down-to-down-up upper-saddle}} \\ \underbrace{(\text{DU:UD,US})}_{\text{down-up-to-up-down upper-saddle}} \end{Bmatrix} \text{ for } a_{110} > 0, a_{220} < 0, \tag{4.39}$$

$$\begin{Bmatrix} (a_{211}, a_{221}) \\ (a_{112}, a_{221}) \\ (a_{111}, a_{221}) \end{Bmatrix} = \begin{Bmatrix} \underbrace{({}_{\text{UD}}\text{US:}{}_{\text{DU}}\text{LS}, \text{DP}_+: \text{UP}_-)}_{\text{hyperbolic upper-to-lower saddle}} \\ \underbrace{(\text{DU:UD,US})}_{\text{down-up-to-up-down upper-saddle}} \\ \underbrace{(\text{UD:DU,US})}_{\text{up-down-to-down-up upper-saddle}} \end{Bmatrix} \text{ for } a_{110} < 0, a_{220} < 0. \tag{4.40}$$

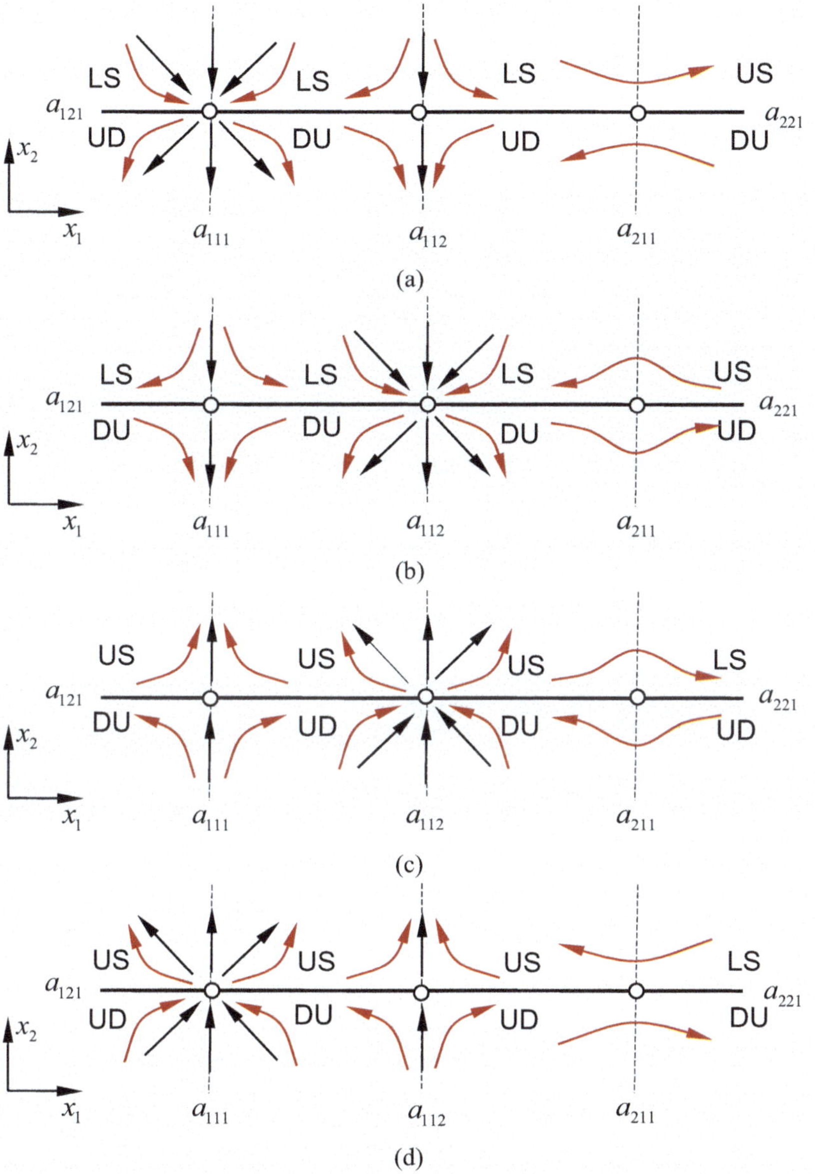

Fig. 4.9 Phase portraits ($a_{112} < a_{211}$ $a_{121} = a_{221}$) for two-dimensional systems on the x_1-direction with $x_1^* = a_{111}, a_{112}$ and on the x_2-direction with $x_2^* = a_{121}, a_{221}$. The up-down and down-up, upper-saddle, and lower-saddle infinite-equilibriums: (**a**) ($a_{110} > 0, a_{220} > 0$), (**b**) ($a_{110} < 0, a_{220} > 0$), (**c**) ($a_{110} > 0, a_{220} < 0$), and (**d**) ($a_{110} < 0, a_{220} < 0$)

The down-up and up-down infinite-equilibriums of $x_2^* = a_{121} = a_{221}$ are up-down and down-up upper-saddle and lower-saddle, as presented in Fig. 4.9. The hyperbolic and hyperbolic-secant, lower-to-upper saddle, and upper-to-lower saddle are illustrated. The infinite-equilibrium is for the switching bifurcation of equilibriums

Table 4.3 Down-up and up-down infinite equilibriums of $x_1^* = a_{221} = a_{121}$ ($a_{211} = a_{121}$ and $a_{112} < a_{211}$)

$(\bar{x}_1, a_{221})$	$\bar{x}_1$			
(a_{110}, a_{220})	$(-\infty, a_{211})$	(a_{211}, a_{111})	(a_{111}, a_{112})	(a_{112}, ∞)
(+, +)	(UD,LS)	(DU,LS)	(UD,LS)	(DU,US)
(−, +)	(DU,LS)	(UD,LS)	(DU,LS)	(UD, US)
(+, −)	(DU,US)	(UD,US)	(DU,US)	(UD,LS)
(−, −)	(UD,US)	(DU,US)	(UD,US)	(DU,LS)

for $a_{121} < a_{221}$ and $a_{121} > a_{221}$. The infinite-equilibrium of $x_2^* = a_{121} = a_{222}$ is summarized in Table 4.3.

4.2.2 *Inflection-Sink and Source Infinite-Equilibriums*

Consider a dynamical system for $a_{211} = a_{11s_1}$ ($s_1, s_2 \in \{1, 2\}, s_1 \neq s_2$) as

$$\begin{aligned}\dot{x}_1 &= a_{110}(x_1 - a_{111})(x_1 - a_{112})(x_2 - a_{121}),\\ \dot{x}_2 &= a_{220}(x_1 - a_{211})(x_2 - a_{221})^2,\end{aligned} \tag{4.41}$$

and the corresponding first integral manifold is

$$\ln\frac{|x_1 - a_{11s_2}|}{|x_{10} - a_{11s_2}|} = \frac{a_{110}}{a_{220}}\left[\ln\frac{|x_2 - a_{221}|}{|x_{20} - a_{221}|} - (a_{221} - a_{121})(\frac{1}{x_2 - a_{221}} - \frac{1}{x_{20} - a_{221}})\right]. \tag{4.42}$$

(i) For the case of $a_{211} = a_{111}$ and $a_{121} < a_{221}$, the saddle-nodes and hyperbolic flows of $x_1^* = a_{111}, a_{112}, a_{211}$ with $x_2^* = a_{121}, a_{221}$ have the following properties:

$$\begin{Bmatrix}(a_{111}, a_{221}) & (a_{112}, a_{221})\\ (a_{121}, a_{211}) & (a_{112}, a_{121})\end{Bmatrix} = \begin{Bmatrix}\underbrace{(\text{DI:DI,SI})}_{\text{decreasing-inflection sink}} & \underbrace{(\text{SO,US})}_{\text{upper-saddle source}}\\ \underbrace{({}_{\text{II:DI}}\text{DP}, {}_{\text{SO:SI}}\text{LS})}_{\text{down-parabola lower-saddle}} & \underbrace{(\text{DP:UP,pF})}_{\text{hyperbolic flow }(+)}\end{Bmatrix} \tag{4.43}$$

for $a_{110} > 0$ and $a_{220} > 0$;

$$\begin{Bmatrix}(a_{111}, a_{221}) & (a_{112}, a_{221})\\ (a_{121}, a_{211}) & (a_{112}, a_{121})\end{Bmatrix} = \begin{Bmatrix}\underbrace{(\text{II:II,SO})}_{\text{increasing-inflection source}} & \underbrace{(\text{SI,US})}_{\text{upper-saddle sink}}\\ \underbrace{({}_{\text{DI:II}}\text{UP}, {}_{\text{SI:SO}}\text{US})}_{\text{up-parabola upper-saddle}} & \underbrace{(\text{UP:DP,pF})}_{\text{hyperbolic-secant flow }(+)}\end{Bmatrix}$$

for $a_{110} < 0$ and $a_{220} > 0$;

(4.44)

$$\left\{\begin{matrix}(a_{111},a_{221}) & (a_{112},a_{221})\\ (a_{121},a_{211}) & (a_{112},a_{121})\end{matrix}\right\} = \left\{\begin{matrix}\underbrace{\text{(II:II,SI)}}_{\text{increasing-inflection sink}} & \underbrace{\text{(SO,LS)}}_{\text{lower-saddle source}}\\ \underbrace{({}_{\text{DI:II}}\text{UP},{}_{\text{SO:SI}}\text{LS})}_{\text{up-parabola lower-saddle}} & \underbrace{\text{(UP:DP,nF)}}_{\text{hyperbolic-secant flow }(-)}\end{matrix}\right\} \tag{4.45}$$

for $a_{110} > 0$ and $a_{220} < 0$;

$$\left\{\begin{matrix}(a_{111},a_{221}) & (a_{112},a_{221})\\ (a_{121},a_{211}) & (a_{112},a_{121})\end{matrix}\right\} = \left\{\begin{matrix}\underbrace{\text{(DI:DI,SO)}}_{\text{decreasing-inflection source}} & \underbrace{\text{(SI,LS)}}_{\text{lower-saddle sink}}\\ \underbrace{({}_{\text{II:DI}}\text{DP},{}_{\text{SI:SO}}\text{US})}_{\text{down-parabola upper-saddle}} & \underbrace{\text{(DP:UP,nF)}}_{\text{hyperbolic flow }(-)}\end{matrix}\right\} \tag{4.46}$$

for $a_{110} < 0$ and $a_{220} < 0$.

The inflection-sink and source infinite-equilibriums with the saddle-sink and saddle-source with hyperbolic and hyperbolic-secant flows are presented in Fig. 4.10 for $a_{111} = a_{211}$ and $a_{121} < a_{221}$. The increasing and decreasing inflection-source and sink infinite-equilibriums of $x_1^* = a_{111} = a_{211}$ are the switching bifurcations, as listed in Table 4.4.

(ii) For the case of $a_{211} = a_{112}$ and $a_{121} < a_{221}$, the saddle-nodes and hyperbolic flows of $x_1^* = a_{111}, a_{112}, a_{211}$ with $x_2^* = a_{121}, a_{221}$ have the following properties:

$$\left\{\begin{matrix}(a_{111},a_{221}) & (a_{112},a_{221})\\ (a_{111},a_{121}) & (a_{121},a_{211})\end{matrix}\right\} = \left\{\begin{matrix}\underbrace{\text{(SI,LS)}}_{\text{lower-saddle sink}} & \underbrace{\text{(II:II,SO)}}_{\text{increasing-inflection source}}\\ \underbrace{\text{(DP:UP,nF)}}_{\text{hyperbolic flow }(-)} & \underbrace{({}_{\text{DI:II}}\text{UP},{}_{\text{SO:SI}}\text{US})}_{\text{up-parabola upper-saddle}}\end{matrix}\right\} \tag{4.47}$$

for $a_{110} > 0$ and $a_{220} > 0$;

$$\left\{\begin{matrix}(a_{111},a_{221}) & (a_{112},a_{221})\\ (a_{111},a_{121}) & (a_{121},a_{211})\end{matrix}\right\} = \left\{\begin{matrix}\underbrace{\text{(SO,LS)}}_{\text{lower-saddle source}} & \underbrace{\text{(DI:DI,SI)}}_{\text{decreasing-inflection sink}}\\ \underbrace{\text{(UP:DP,nF)}}_{\text{hyperbolic-secant flow }(-)} & \underbrace{({}_{\text{II:DI}}\text{DP},{}_{\text{SO:SI}}\text{LS})}_{\text{inflection down-parabola lower-saddle}}\end{matrix}\right\}$$

for $a_{110} < 0$ and $a_{220} > 0$;

(4.48)

$$\left\{\begin{matrix}(a_{111},a_{221}) & (a_{112},a_{221})\\ (a_{111},a_{121}) & (a_{121},a_{211})\end{matrix}\right\} = \left\{\begin{matrix}\underbrace{\text{(SI,US)}}_{\text{upper-saddle sink}} & \underbrace{\text{(DI:DI,SO)}}_{\text{decreasing-inflection source}}\\ \underbrace{\text{(UP:DP,pF)}}_{\text{hyperbolic-secant flow }(+)} & \underbrace{({}_{\text{II:DI}}\text{DP},{}_{\text{SO:SI}}\text{US})}_{\text{inflection down-parabola upper-saddle}}\end{matrix}\right\} \tag{4.49}$$

for $a_{110} > 0$ and $a_{220} < 0$;

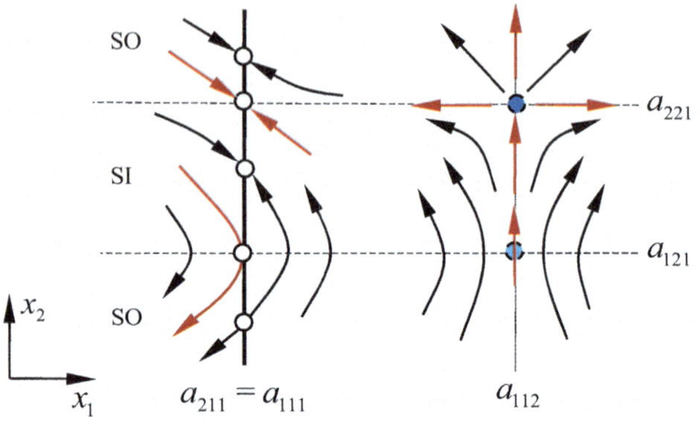

(a)

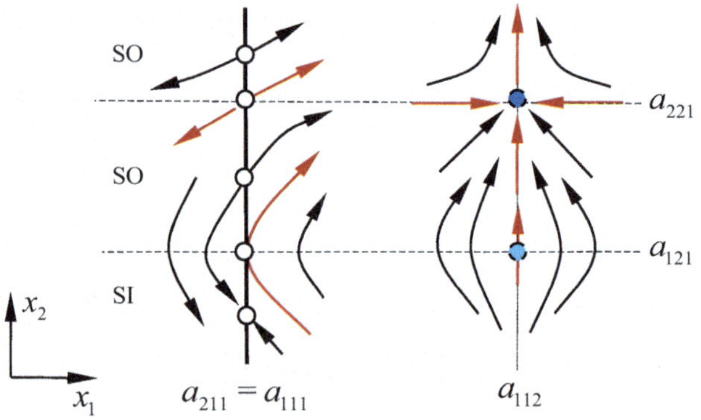

(b)

Fig. 4.10 Phase portraits ($a_{111} = a_{211}$ and $a_{121} < a_{221}$) for two-dimensional systems on the x_1-direction with $x_1^* = a_{112}$ and on the x_2-direction with $x_2^* = a_{121}, a_{221}, a_{222}$. The inflection-sink and source infinite-equilibriums with saddle source and saddle sink: (**a**) ($a_{110} > 0$, $a_{220} > 0$), (**b**) ($a_{110} < 0$, $a_{220} > 0$), (**c**) ($a_{110} > 0$, $a_{220} < 0$), and (**d**) ($a_{110} < 0$, $a_{220} < 0$)

$$\begin{Bmatrix} (a_{111}, a_{221}) & (a_{112}, a_{221}) \\ (a_{111}, a_{121}) & (a_{121}, a_{211}) \end{Bmatrix} = \begin{Bmatrix} \underbrace{(\text{SO,US})}_{\text{upper-saddle source}} & \underbrace{(\text{II:II,SI})}_{\text{increasing-inflection sink}} \\ \underbrace{({}_{\text{DP:UP}}\text{HB, pF})}_{\text{hyperbolic flow }(+)} & \underbrace{({}_{\text{DI:II}}\text{UP, }{}_{\text{SI:SO}}\text{LS})}_{\text{up-parabola lower-saddle}} \end{Bmatrix} \tag{4.50}$$

for $a_{110} < 0$ and $a_{220} < 0$.

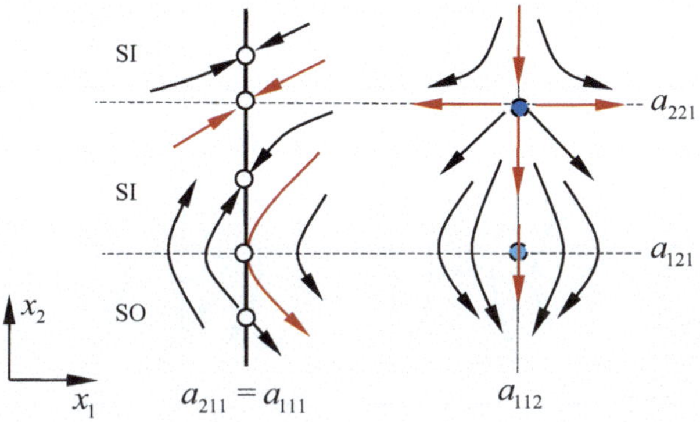

(c)

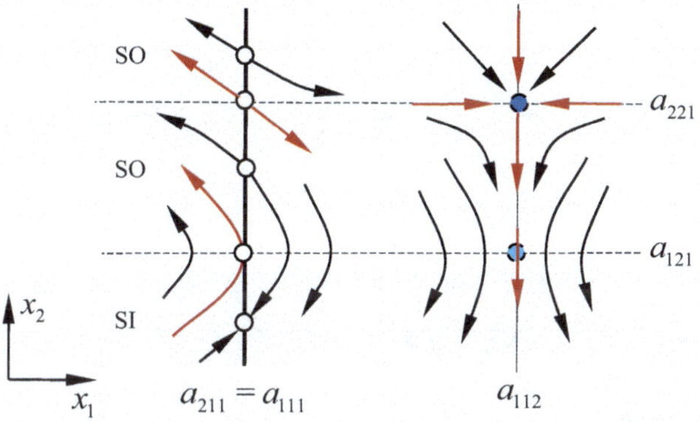

(d)

Fig. 4.10 (continued)

Table 4.4 Inflection-sink and source infinite-equilibriums of $x_1^* = a_{111} = a_{211}$ ($a_{111} = a_{211}$ and $a_{121} < a_{221}$)

$(a_{111}, \bar{x}_2)$	$\bar{x}_2$		
(a_{110}, a_{220})	$(-\infty, a_{121})$	(a_{121}, a_{221})	(a_{221}, ∞)
$(+,+)$	(SO,II)	(SI,DI)	(SI,DI)
$(-,+)$	(SI,DI)	(SO,II)	(SO,II)
$(+,-)$	(SO,DI)	(SI,II)	(SI,II)
$(-,-)$	(SI,II)	(SO,DI)	(SO,DI)

The inflection-sink and source infinite-equilibriums with the saddle-sink and saddle-source with hyperbolic and hyperbolic-secant flows are presented in Fig. 4.11 for $a_{112} = a_{211}$ and $a_{121} < a_{221}$. The increasing and decreasing inflection-source and sink infinite-equilibriums of $x_1^* = a_{112} = a_{211}$ are the switching bifurcations, as listed in Table 4.5.

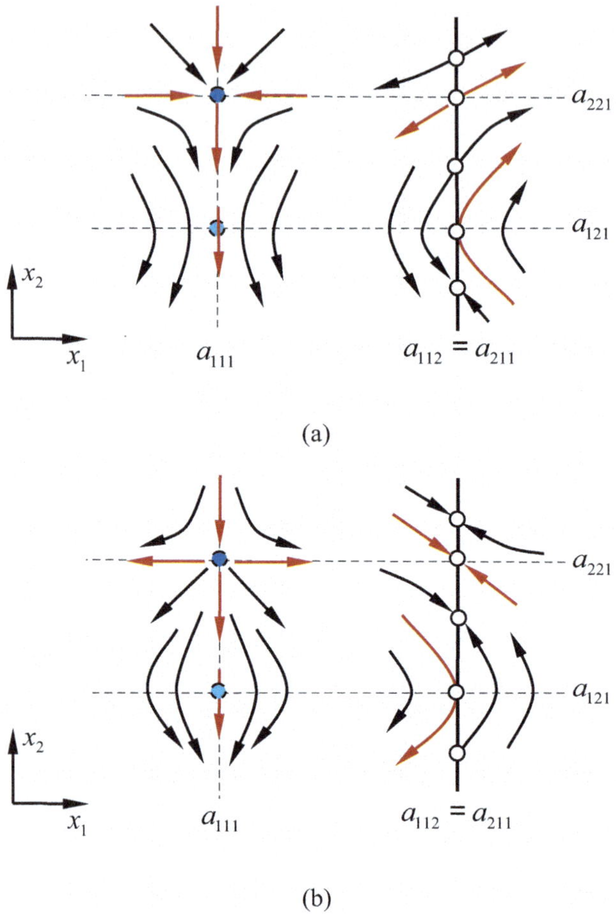

Fig. 4.11 Phase portraits ($a_{112} = a_{211}$ and $a_{121} < a_{221}$) for two-dimensional systems on the x_1-direction with $x_1^* = a_{111}$ and on the x_2-direction with $x_2^* = a_{121}, a_{221}, a_{222}$. The inflection-sink and source infinite-equilibriums with saddle-source and saddle-sink: (**a**) ($a_{110} > 0$, $a_{220} > 0$), (**b**) ($a_{110} < 0$, $a_{220} > 0$), (**c**) ($a_{110} > 0$, $a_{220} < 0$), and (**d**) ($a_{110} < 0$, $a_{220} < 0$)

4.2.3 *Double-Switching Bifurcations*

Consider a dynamical system for $a_{221} = a_{121}$ and $a_{211} = a_{11s_1}$ ($s_1, s_2 \in \{1, 2\}$, $s_1 \neq s_2$) as

$$\begin{aligned} \dot{x}_1 &= a_{110}(x_1 - a_{111})(x_1 - a_{112})(x_2 - a_{121}), \\ \dot{x}_2 &= a_{220}(x_1 - a_{211})(x_2 - a_{221})^2, \end{aligned} \tag{4.51}$$

and the corresponding first integral manifold is

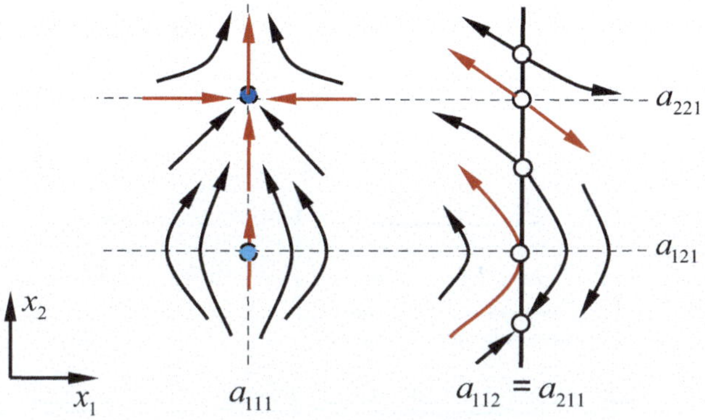

(c)

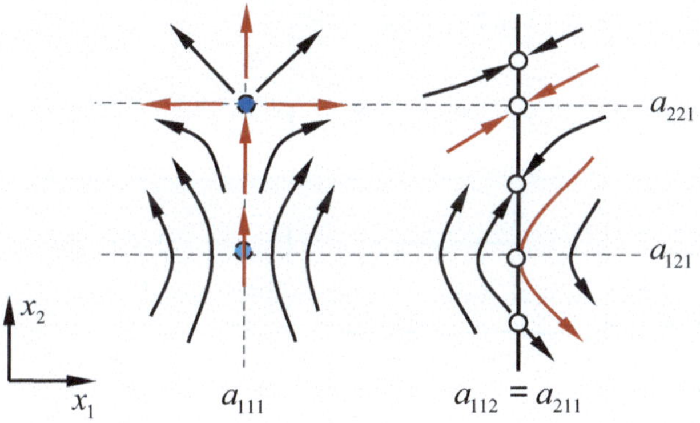

(d)

Fig. 4.11 (continued)

Table 4.5 Inflection-sink and source infinite-equilibriums of $x_1^* = a_{112} = a_{211}$ ($a_{112} = a_{211}$ and $a_{121} < a_{221}$)

$(a_{112}, \bar{x}_2)$	$\bar{x}_2$		
(a_{110}, a_{220})	$(-\infty, a_{121})$	(a_{121}, a_{221})	(a_{221}, ∞)
(+, +)	(SI,DI)	(SO,II)	(SO,II)
(−, +)	(SO,II)	(SI,DI)	(SI,DI)
(+, −)	(SI,II)	(SO,DI)	(SO,DI)
(−, −)	(SO,DI)	(SI,II)	(SI,II)

$$\ln \frac{|x_1 - a_{11s_2}|}{|x_{10} - a_{11s_2}|} = \frac{a_{110}}{a_{220}} \ln \frac{|x_2 - a_{221}|}{|x_{20} - a_{221}|}. \tag{4.52}$$

(i) For the case of $a_{211} = a_{111}$ and $a_{121} = a_{221}$, the saddle nodes and hyperbolic flows of $x_1^* = a_{111}, a_{112}, a_{211}$ with $x_2^* = a_{121}, a_{221}$ have the following properties:

$$\begin{Bmatrix}(a_{112}, a_{121})\\(a_{111}, a_{121})\end{Bmatrix} = \begin{Bmatrix}\underbrace{(\text{UD:DU,US})}_{\text{up-down-to-down-up upper-saddle}}\\ \underbrace{({}_{\text{II}}\text{SO:}{}_{\text{DI}}\text{SI}, {}_{\text{LS:US}}\text{UD})}_{\text{up-down inflection source-to-sink}}\end{Bmatrix} \text{ for } a_{110} > 0 \text{ and } a_{220} > 0; \tag{4.53}$$

$$\begin{Bmatrix}(a_{112}, a_{121})\\(a_{111}, a_{121})\end{Bmatrix} = \begin{Bmatrix}\underbrace{(\text{DU:UD,US})}_{\text{down-up-to-up-down upper-saddle}}\\ \underbrace{({}_{\text{DI}}\text{SI:}{}_{\text{II}}\text{SO}, {}_{\text{LS:US}}\text{DU})}_{\text{down-up inflection sink-to-source}}\end{Bmatrix} \text{ for } a_{110} < 0 \text{ and } a_{220} > 0; \tag{4.54}$$

$$\begin{Bmatrix}(a_{112}, a_{121})\\(a_{111}, a_{121})\end{Bmatrix} = \begin{Bmatrix}\underbrace{(\text{DU:UD,LS})}_{\text{down-up-to-up-down lower-saddle}}\\ \underbrace{({}_{\text{DI}}\text{SO:}{}_{\text{II}}\text{SI}, {}_{\text{US:LS}}\text{DU})}_{\text{down-up inflection source-to-sink}}\end{Bmatrix} \text{ for } a_{110} > 0 \text{ and } a_{220} < 0; \tag{4.55}$$

$$\begin{Bmatrix}(a_{112}, a_{121})\\(a_{111}, a_{121})\end{Bmatrix} = \begin{Bmatrix}\underbrace{(\text{UD:DU,LS})}_{\text{up-down-to-down-up lower-saddle}}\\ \underbrace{({}_{\text{II}}\text{SI:}{}_{\text{DI}}\text{SO}, {}_{\text{US:LS}}\text{UD})}_{\text{up-down inflection sink-to-source}}\end{Bmatrix} \text{ for } a_{110} < 0 \text{ and } a_{220} < 0. \tag{4.56}$$

The inflection sink and source infinite equilibriums with the saddle sink and saddle source with hyperbolic and hyperbolic secant flows are presented in Fig. 4.12 for $a_{111} = a_{211}$ and $a_{121} = a_{221}$. The increasing and decreasing inflection-source and sink infinite-equilibriums of $x_1^* = a_{111} = a_{211}$ are the switching bifurcations, and the up-down and down-up saddle infinite-equilibriums of $x_2^* = a_{121} = a_{221}$ are the switching bifurcations, as listed in Table 4.6.

(ii) For the case of $a_{211} = a_{112}$ and $a_{121} = a_{221}$, the saddle nodes and hyperbolic flows of $x_1^* = a_{111}, a_{112}, a_{211}$ with $x_2^* = a_{121}, a_{221}$ have the following properties:

$$\begin{Bmatrix}(a_{112}, a_{121})\\(a_{111}, a_{121})\end{Bmatrix} = \begin{Bmatrix}\underbrace{({}_{\text{DI}}\text{SI:}{}_{\text{II}}\text{SO}, {}_{\text{LS:US}}\text{DU})}_{\text{down-up inflection sink-to-source}}\\ \underbrace{(\text{UD:DU,LS})}_{\text{up-down-to-down-up lower-saddle}}\end{Bmatrix} \text{ for } a_{110} > 0 \text{ and } a_{220} > 0; \tag{4.57}$$

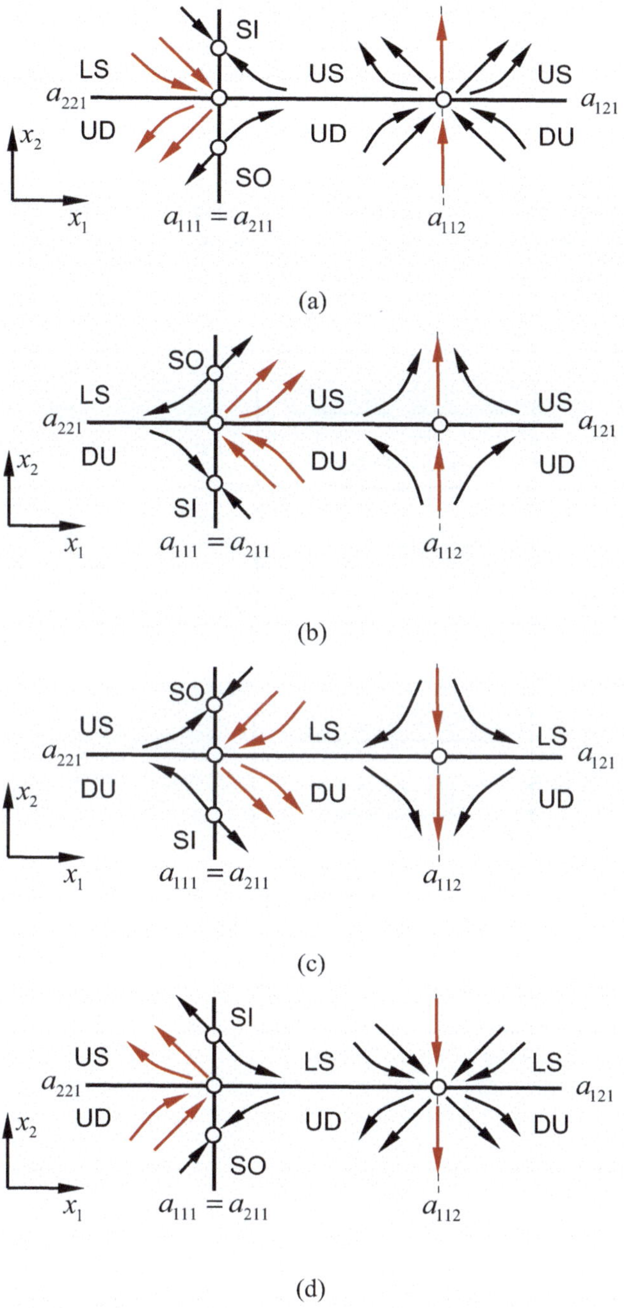

Fig. 4.12 Phase portraits ($a_{111} = a_{211}$ and $a_{121} = a_{221}$) for two-dimensional systems on the x_1-direction with $x_1^* = a_{111}, a_{112}, a_{211}$ and on the x_2-direction with $x_2^* = a_{121}, a_{221}$. The equilibrium flows: (**a**) ($a_{110} > 0$, $a_{220} > 0$), (**b**) ($a_{110} < 0$, $a_{220} > 0$), (**c**) ($a_{110} > 0$, $a_{220} < 0$), and (**d**) ($a_{110} < 0$, $a_{220} < 0$)

Table 4.6 Inflection-sink and source infinite-equilibriums of $x_1^* = a_{111} = a_{211}$ and up-down and down-up saddle infinite-equilibriums of $x_2^* = a_{121} = a_{221}$ ($a_{111} = a_{211}$ and $a_{121} = a_{221}$)

	$(\bar{x}_1, a_{221})$			$(a_{111}, \bar{x}_2)$	
	$\bar{x}_1$			$\bar{x}_2$	
(a_{110}, a_{220})	$(-\infty, a_{111})$	(a_{111}, a_{112})	(a_{112}, ∞)	$(-\infty, a_{221})$	(a_{221}, ∞)
(+, +)	(UD,LS)	(UD,US)	(DU,US)	(SO,II)	(SI,DI)
(−, +)	(DU,LS)	(DU,US)	(UD,US)	(SI,DI)	(SO,II)
(+, −)	(DU,US)	(DU,LS)	(UD,LS)	(SO,DI)	(SI,II)
(−, −)	(UD,US)	(UD,LS)	(DU,LS)	(SI,II)	(SO,DI)

$$\left\{\begin{matrix}(a_{112}, a_{121})\\(a_{111}, a_{121})\end{matrix}\right\} = \left\{\begin{matrix}\underbrace{({}_{II}SO{:}_{DI}SI,\ {}_{LS:US}UD)}_{\text{up-down inflection source-to-sink}}\\ \underbrace{(DU{:}UD,LS)}_{\text{down-up-to-up-down lower-saddle}}\end{matrix}\right\} \text{for } a_{110} < 0 \text{ and } a_{220} > 0; \tag{4.58}$$

$$\left\{\begin{matrix}(a_{112}, a_{121})\\(a_{111}, a_{121})\end{matrix}\right\} = \left\{\begin{matrix}\underbrace{({}_{II}SI{:}_{DI}SO,\ {}_{US:LS}UD)}_{\text{up-down inflection sink-to-source}}\\ \underbrace{(DU{:}UD,US)}_{\text{down-up-to-up-down upper-saddle}}\end{matrix}\right\} \text{for } a_{110} > 0 \text{ and } a_{220} < 0; \tag{4.59}$$

$$\left\{\begin{matrix}(a_{112}, a_{121})\\(a_{111}, a_{121})\end{matrix}\right\} = \left\{\begin{matrix}\underbrace{({}_{DI}SO{:}_{II}SI,\ {}_{US:LS}DU)}_{\text{down-up inflection source-to-sink}}\\ \underbrace{(UD{:}DU,US)}_{\text{up-down-to-down-up upper-saddle}}\end{matrix}\right\} \text{for } a_{110} < 0 \text{ and } a_{220} < 0. \tag{4.60}$$

The inflection sink and source infinite equilibriums with the saddle sink and saddle source with hyperbolic and hyperbolic secant flows are presented in Fig. 4.13 for $a_{112} = a_{211}$ and $a_{121} = a_{221}$. The increasing and decreasing-inflection source and sink infinite-equilibriums of $x_1^* = a_{112} = a_{211}$ are the switching bifurcations, and the up-down and down-up saddle infinite-equilibriums of $x_2^* = a_{121} = a_{221}$ are the switching bifurcations, as listed in Table 4.7.

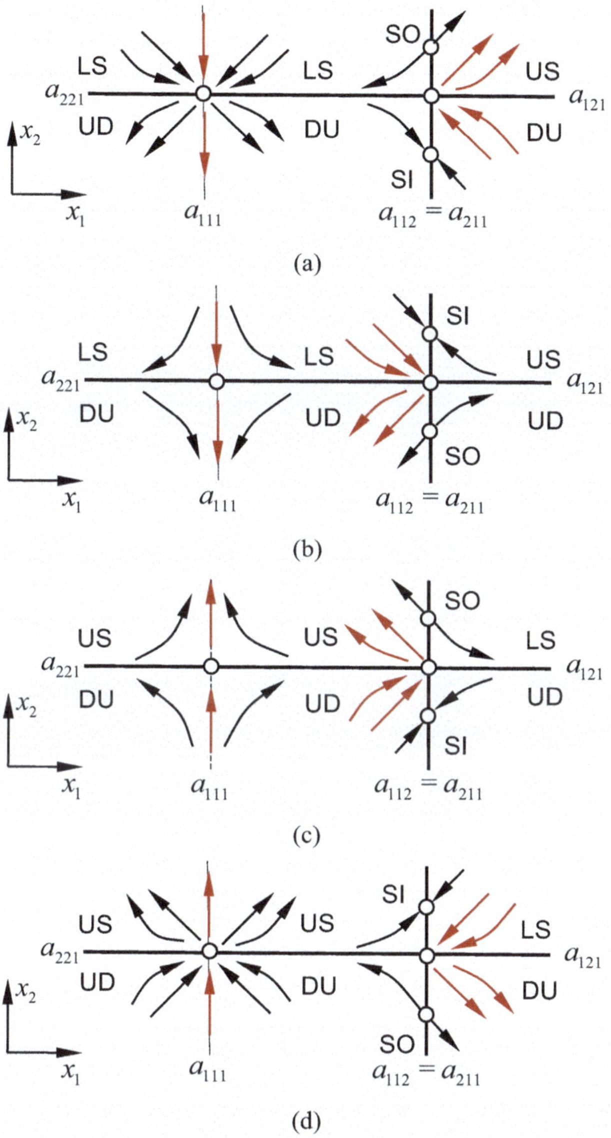

Fig. 4.13 Phase portraits ($a_{112} = a_{211}$ and $a_{121} = a_{221}$) for two-dimensional systems on the x_1-direction with $x_1^* = a_{111}$ and on the x_2-direction with $x_2^* = a_{121}, a_{221}, a_{222}$. The inflection-sink and source infinite-equilibriums with saddle-source and saddle-sink: (**a**) ($a_{110} > 0$, $a_{220} > 0$), (**b**) ($a_{110} < 0$, $a_{220} > 0$), (**c**) ($a_{110} > 0$, $a_{220} < 0$), and (**d**) ($a_{110} < 0$, $a_{220} < 0$)

Table 4.7 Inflection-source and sink infinite-equilibriums of $x_1^* = a_{112} = a_{211}$ and up-down and down-up saddle infinite-equilibriums of $x_2^* = a_{121} = a_{221}$ ($a_{112} = a_{211}$ and $a_{121} = a_{221}$)

	$(\overline{x}_1, a_{221})$			$(a_{111}, \overline{x}_2)$	
	$\overline{x}_1$			$\overline{x}_2$	
(a_{110}, a_{220})	$(-\infty, a_{111})$	(a_{111}, a_{112})	(a_{112}, ∞)	$(-\infty, a_{221},)$	(a_{221}, ∞)
(+, +)	(UD,LS)	(DU,LS)	(DU,US)	(SI,DI)	(SO,II)
(−, +)	(DU,LS)	(UD,LS)	(UD,US)	(SO,II)	(SI,DI)
(+, −)	(DU,US)	(UD,US)	(UD,LS)	(SI,II)	(SO,DI)
(−, −)	(UD,US)	(DU,US)	(DU,LS)	(SO,DI)	(SI,II)

Chapter 5
Simple Equilibrium Networks and Switching Dynamics

In this chapter, the simple equilibrium networks are presented for the cubic product system with two self-linear and crossing-quadratic product vector fields, and the corresponding inflection source and sink infinite equilibriums are for hyperbolic flows with paralleled saddle and center and for hyperbolic flows with connected saddle, source, and sink. The (3 × 3) simple equilibrium networks consist of negative and positive saddles, counterclockwise and clockwise centers, saddles, sink and source, and vertical and horizontal hyperbolic and hyperbolic-secant flows. In addition, the double-switching bifurcations are generated through the inflection-saddles formed by the two inflection-source (sink) infinite-equilibriums.

5.1 (3 × 3) Simple Equilibrium Networks with Hyperbolic Flows

Consider a dynamical system as

$$\begin{aligned}\dot{x}_1 &= a_{110}(x_1 - a_{111})(x_1 - a_{112})(x_2 - a_{121}),\\ \dot{x}_2 &= a_{220}(x_1 - a_{211})(x_2 - a_{221})(x_2 - a_{222}),\end{aligned} \tag{5.1}$$

and the corresponding first integral manifold is

$$\begin{aligned}&\frac{a_{112} - a_{211}}{a_{112} - a_{111}} \ln \frac{|\, x_1 - a_{112} \,|}{|\, x_{10} - a_{112} \,|} + \frac{a_{111} - a_{211}}{a_{111} - a_{211}} \ln \frac{|\, x_1 - a_{111} \,|}{|\, x_{10} - a_{111} \,|}\\ &\quad = \frac{a_{110}}{a_{220}} \left(\frac{a_{221} - a_{121}}{a_{221} - a_{222}} \ln \frac{|\, x_2 - a_{221} \,|}{|\, x_{20} - a_{221} \,|} + \frac{a_{222} - a_{121}}{a_{222} - a_{221}} \ln \frac{|\, x_2 - a_{222} \,|}{|\, x_{20} - a_{222} \,|} \right).\end{aligned} \tag{5.2}$$

A. C. J. Luo, *Two-Dimensional Two-Product Cubic Systems, Vol. II*,
https://doi.org/10.1007/978-3-031-48491-9_5

(i) The simple equilibrium network for $a_{211} < a_{111}$ and $a_{121} < a_{221}$ has the following properties:

$$\left\{\begin{array}{ccc} (a_{211},a_{222}) & (a_{111},a_{222}) & (a_{112},a_{222}) \\ (a_{211},a_{221}) & (a_{111},a_{221}) & (a_{112},a_{221}) \\ (a_{121},a_{211}) & (a_{111},a_{121}) & (a_{112},a_{121}) \end{array}\right\}$$
$$= \left\{\begin{array}{ccc} \underbrace{(\text{pF},\text{DP:UP})}_{\text{hyperbolic flow }(+)} & \underbrace{(\text{SI},\text{SO})}_{\text{saddle}} & \underbrace{(\text{SO},\text{SO})}_{\text{source}} \\ \underbrace{(\text{pF},\text{UP:DP})}_{\text{hyperbolic-secant flow }(+)} & \underbrace{(\text{SI},\text{SI})}_{\text{sink}} & \underbrace{(\text{SO},\text{SI})}_{\text{saddle}} \\ \underbrace{(\text{UP}_+,\text{UP}_+)}_{\text{positive saddle}} & \underbrace{(\text{UP:DP},\text{pF})}_{\text{hyperbolic-secant flow }(+)} & \underbrace{(\text{DP:UP},\text{pF})}_{\text{hyperbolic flow }(+)} \end{array}\right\}$$
for $a_{110} > 0$ and $a_{220} > 0$; (5.3)

$$\left\{\begin{array}{ccc} (a_{211},a_{222}) & (a_{111},a_{222}) & (a_{112},a_{222}) \\ (a_{211},a_{221}) & (a_{111},a_{221}) & (a_{112},a_{221}) \\ (a_{121},a_{211}) & (a_{111},a_{121}) & (a_{112},a_{121}) \end{array}\right\}$$
$$= \left\{\begin{array}{ccc} \underbrace{(\text{nF},\text{UP:DP})}_{\text{hyperbolic-secant flow }(-)} & \underbrace{(\text{SO},\text{SO})}_{\text{source}} & \underbrace{(\text{SI},\text{SO})}_{\text{saddle}} \\ \underbrace{(\text{nF},\text{DP:UP})}_{\text{hyperbolic flow }(-)} & \underbrace{(\text{SO},\text{SI})}_{\text{saddle}} & \underbrace{(\text{SI},\text{SI})}_{\text{sink}} \\ \underbrace{(\text{DP}_+,\text{DP}_-)}_{\text{CCW center}} & \underbrace{(\text{DP:UP},\text{pF})}_{\text{hyperbolic flow }(+)} & \underbrace{(\text{UP:DP},\text{pF})}_{\text{hyperbolic-secant flow }(+)} \end{array}\right\}$$
for $a_{110} < 0$ and $a_{220} > 0$; (5.4)

$$\left\{\begin{array}{ccc} (a_{211},a_{222}) & (a_{111},a_{222}) & (a_{112},a_{222}) \\ (a_{211},a_{221}) & (a_{111},a_{221}) & (a_{112},a_{221}) \\ (a_{121},a_{211}) & (a_{111},a_{121}) & (a_{112},a_{121}) \end{array}\right\}$$
$$= \left\{\begin{array}{ccc} \underbrace{(\text{pF},\text{UP:DP})}_{\text{hyperbolic-secant flow }(+)} & \underbrace{(\text{SI},\text{SI})}_{\text{sink}} & \underbrace{(\text{SO},\text{SI})}_{\text{saddle}} \\ \underbrace{(\text{pF},\text{DP:UP})}_{\text{hyperbolic flow }(+)} & \underbrace{(\text{SI},\text{SO})}_{\text{saddle}} & \underbrace{(\text{SO},\text{SO})}_{\text{source}} \\ \underbrace{(\text{DP}_-,\text{DP}_+)}_{\text{CW center}} & \underbrace{(\text{DP:UP},\text{nF})}_{\text{hyperbolic flow }(-)} & \underbrace{(\text{UP:DP},\text{nF})}_{\text{hyperbolic-secant flow }(-)} \end{array}\right\}$$
for $a_{110} > 0$ and $a_{220} < 0$; (5.5)

$$\left\{\begin{array}{ccc}(a_{211},a_{222}) & (a_{111},a_{222}) & (a_{112},a_{222})\\ (a_{211},a_{221}) & (a_{111},a_{221}) & (a_{112},a_{221})\\ (a_{121},a_{211}) & (a_{111},a_{121}) & (a_{112},a_{121})\end{array}\right\}$$

$$=\left\{\begin{array}{ccc}\underbrace{(\text{nF,DP:UP})}_{\text{hyperbolic flow }(-)} & \underbrace{(\text{SO,SI})}_{\text{saddle}} & \underbrace{(\text{SI,SI})}_{\text{sink}}\\ \underbrace{(\text{nF,UP:DP})}_{\text{hyperbolic-secant flow }(-)} & \underbrace{(\text{SO,SO})}_{\text{source}} & \underbrace{(\text{SI,SO})}_{\text{saddle}}\\ \underbrace{(\text{UP}_{-},\text{UP}_{-})}_{\text{negative saddle}} & \underbrace{(\text{UP:DP,nF})}_{\text{hyperbolic-secant flow }(-)} & \underbrace{(\text{DP:UP,nF})}_{\text{hyperbolic flow }(-)}\end{array}\right\}$$

for $a_{110} < 0$ and $a_{220} < 0$; (5.6)

as shown in Fig. 5.1. The directrix flows of $x_1 = a_{111}, a_{112}$ are positive and negative. The directrix flows of $x_1 = a_{221}, a_{222}$ are also positive and negative. The networks of saddle, sink, and source with hyperbolic and hyperbolic-secant flows are presented. The saddles, sink, and sources in the network of equilibriums are above the hyperbolic and hyperbolic-secant flows.

(ii) The simple equilibrium networks for $a_{211} < a_{111}$ and $a_{222} < a_{121}$ have the following properties:

$$\left\{\begin{array}{ccc}(a_{121},a_{211}) & (a_{111},a_{121}) & (a_{112},a_{121})\\ (a_{211},a_{222}) & (a_{111},a_{222}) & (a_{112},a_{222})\\ (a_{211},a_{221}) & (a_{111},a_{221}) & (a_{112},a_{221})\end{array}\right\}$$

$$=\left\{\begin{array}{ccc}\underbrace{(\text{UP}_{+},\text{UP}_{+})}_{\text{positive saddle}} & \underbrace{(\text{UP:PP,pF})}_{\text{hyperbolic-secant flow }(+)} & \underbrace{(\text{DP:UP,pF})}_{\text{hyperbolic flow }(+)}\\ \underbrace{(\text{nF,UP:DP})}_{\text{hyperbolic-secant flow }(-)} & \underbrace{(\text{SO,SO})}_{\text{source}} & \underbrace{(\text{SI,SO})}_{\text{saddle}}\\ \underbrace{(\text{nF,DP:UP})}_{\text{hyperbolic flow }(-)} & \underbrace{(\text{SO,SI})}_{\text{saddle}} & \underbrace{(\text{SI,SI})}_{\text{sink}}\end{array}\right\}$$

for $a_{110} > 0$ and $a_{220} > 0$; (5.7)

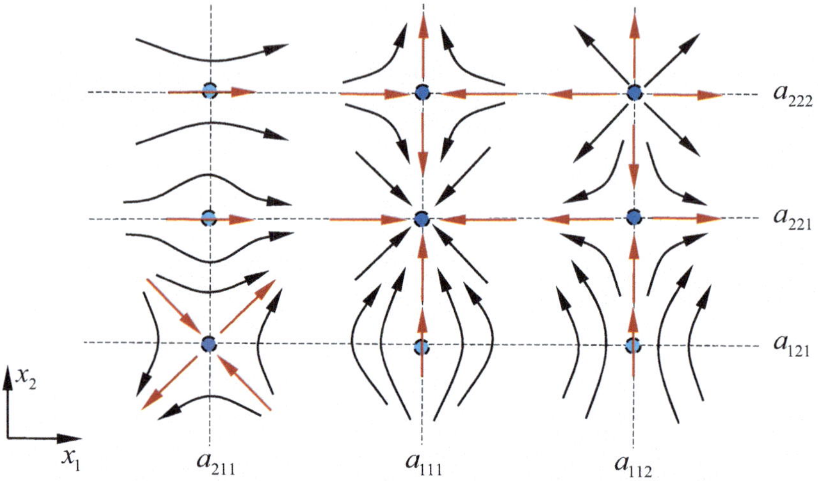

(a)

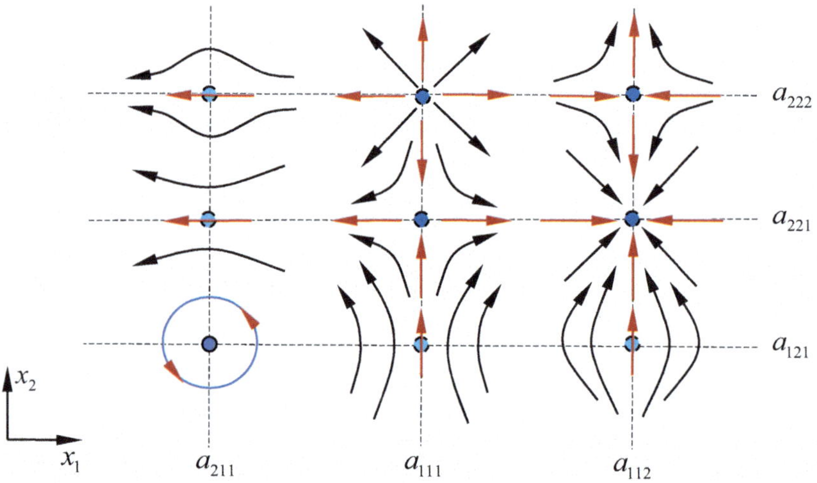

(b)

Fig. 5.1 Phase portraits ($a_{211} < a_{111}$ and $a_{121} < a_{221}$) for two-dimensional systems on the x_1-direction with $x_1^* = a_{211}, a_{111}, a_{112}$ and on the x_2-direction with $x_2^* = a_{121}, a_{221}, a_{222}$. The simple equilibriums with hyperbolic flows: (**a**) ($a_{110} > 0$, $a_{220} > 0$), (**b**) ($a_{110} < 0$, $a_{220} > 0$), (**c**) ($a_{110} > 0$, $a_{220} < 0$), and (**d**) ($a_{110} < 0$, $a_{220} < 0$)

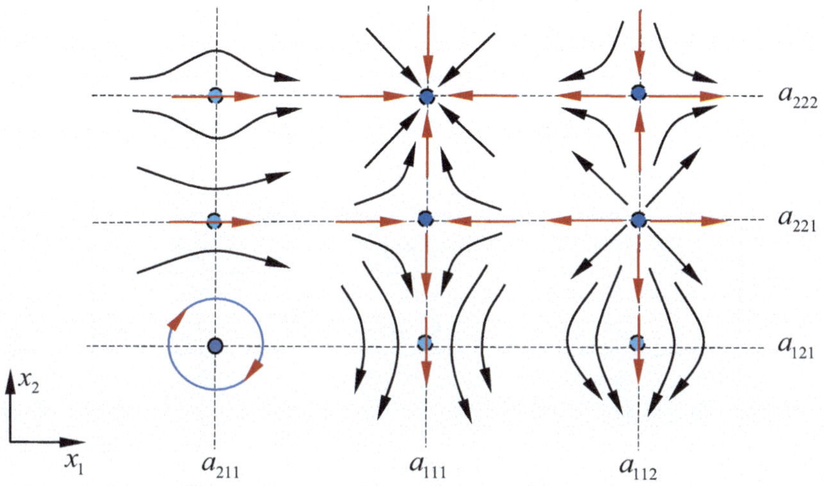

(c)

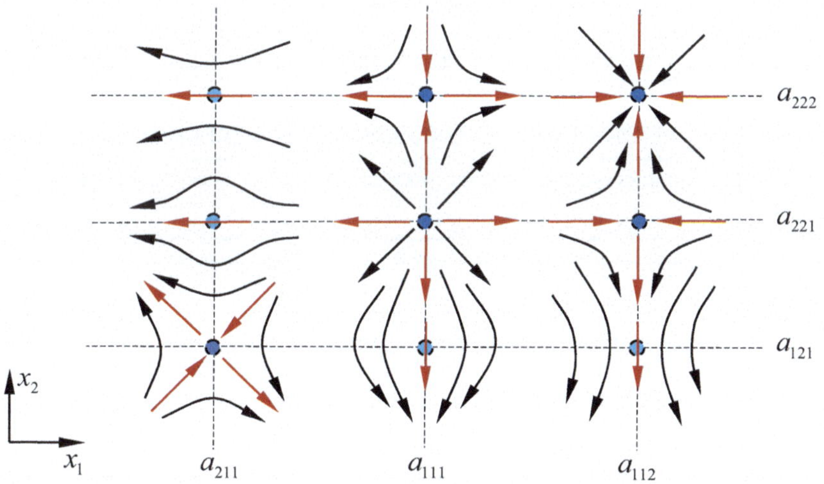

(d)

Fig. 5.1 (continued)

$$\left\{\begin{array}{ccc} (a_{121}, a_{211}) & (a_{111}, a_{121}) & (a_{112}, a_{121}) \\ (a_{211}, a_{222}) & (a_{111}, a_{222}) & (a_{112}, a_{222}) \\ (a_{211}, a_{221}) & (a_{111}, a_{221}) & (a_{112}, a_{221}) \end{array}\right\}$$

$$= \left\{\begin{array}{ccc} \underbrace{(\mathrm{DP}_{+}, \mathrm{DP}_{-})}_{\text{CCW center}} & \underbrace{(\mathrm{DP{:}UP}, \mathrm{pF})}_{\text{hyperbolic flow } (+)} & \underbrace{(\mathrm{UP{:}DP}, \mathrm{pF})}_{\text{hyperbolic-secant flow } (+)} \\ \underbrace{(\mathrm{pF}, \mathrm{DP{:}UP})}_{\text{hyperbolic flow } (+)} & \underbrace{(\mathrm{SI}, \mathrm{SO})}_{\text{saddle}} & \underbrace{(\mathrm{SO}, \mathrm{SO})}_{\text{source}} \\ \underbrace{(\mathrm{pF}, \mathrm{UP{:}DP})}_{\text{hyperbolic-secant flow } (+)} & \underbrace{(\mathrm{SI}, \mathrm{SI})}_{\text{sink}} & \underbrace{(\mathrm{SO}, \mathrm{SI})}_{\text{saddle}} \end{array}\right\}$$

for $a_{110} < 0$ and $a_{220} > 0$; (5.8)

$$\left\{\begin{array}{ccc} (a_{121}, a_{211}) & (a_{111}, a_{121}) & (a_{112}, a_{121}) \\ (a_{211}, a_{222}) & (a_{111}, a_{222}) & (a_{112}, a_{222}) \\ (a_{211}, a_{221}) & (a_{111}, a_{221}) & (a_{112}, a_{221}) \end{array}\right\}$$

$$= \left\{\begin{array}{ccc} \underbrace{(\mathrm{DP}_{-}, \mathrm{DP}_{+})}_{\text{CW center}} & \underbrace{(\mathrm{DP{:}UP}, \mathrm{nF})}_{\text{hyperbolic flow } (-)} & \underbrace{(\mathrm{UP{:}DP}, \mathrm{nF})}_{\text{hyperbolic-secant flow } (-)} \\ \underbrace{(\mathrm{nF}, \mathrm{DP{:}UP})}_{\text{hyperbolic flow } (-)} & \underbrace{(\mathrm{SO}, \mathrm{SI})}_{\text{saddle}} & \underbrace{(\mathrm{SI}, \mathrm{SI})}_{\text{sink}} \\ \underbrace{(\mathrm{nF}, \mathrm{UP{:}DP})}_{\text{hyperbolic-secant flow } (-)} & \underbrace{(\mathrm{SO}, \mathrm{SO})}_{\text{source}} & \underbrace{(\mathrm{SI}, \mathrm{SO})}_{\text{saddle}} \end{array}\right\}$$

for $a_{110} > 0$ and $a_{220} < 0$; (5.9)

$$\left\{\begin{array}{ccc} (a_{121}, a_{211}) & (a_{111}, a_{121}) & (a_{112}, a_{121}) \\ (a_{211}, a_{222}) & (a_{111}, a_{222}) & (a_{112}, a_{222}) \\ (a_{211}, a_{221}) & (a_{111}, a_{221}) & (a_{112}, a_{221}) \end{array}\right\}$$

$$= \left\{\begin{array}{ccc} \underbrace{(\mathrm{UP}_{-}, \mathrm{UP}_{-})}_{\text{negative saddle}} & \underbrace{(\mathrm{UP{:}DP}, \mathrm{nF})}_{\text{hyperbolic-secant flow } (-)} & \underbrace{(\mathrm{DP{:}UP}, \mathrm{nF})}_{\text{hyperbolic flow } (-)} \\ \underbrace{(\mathrm{pF}, \mathrm{UP{:}DP})}_{\text{hyperbolic-secant flow } (+)} & \underbrace{(\mathrm{SI}, \mathrm{SI})}_{\text{sink}} & \underbrace{(\mathrm{SO}, \mathrm{SI})}_{\text{saddle}} \\ \underbrace{(\mathrm{pF}, \mathrm{DP{:}UP})}_{\text{hyperbolic flow } (+)} & \underbrace{(\mathrm{SI}, \mathrm{SO})}_{\text{saddle}} & \underbrace{(\mathrm{SO}, \mathrm{SO})}_{\text{source}} \end{array}\right\}$$

for $a_{110} < 0$ and $a_{220} < 0$; (5.10)

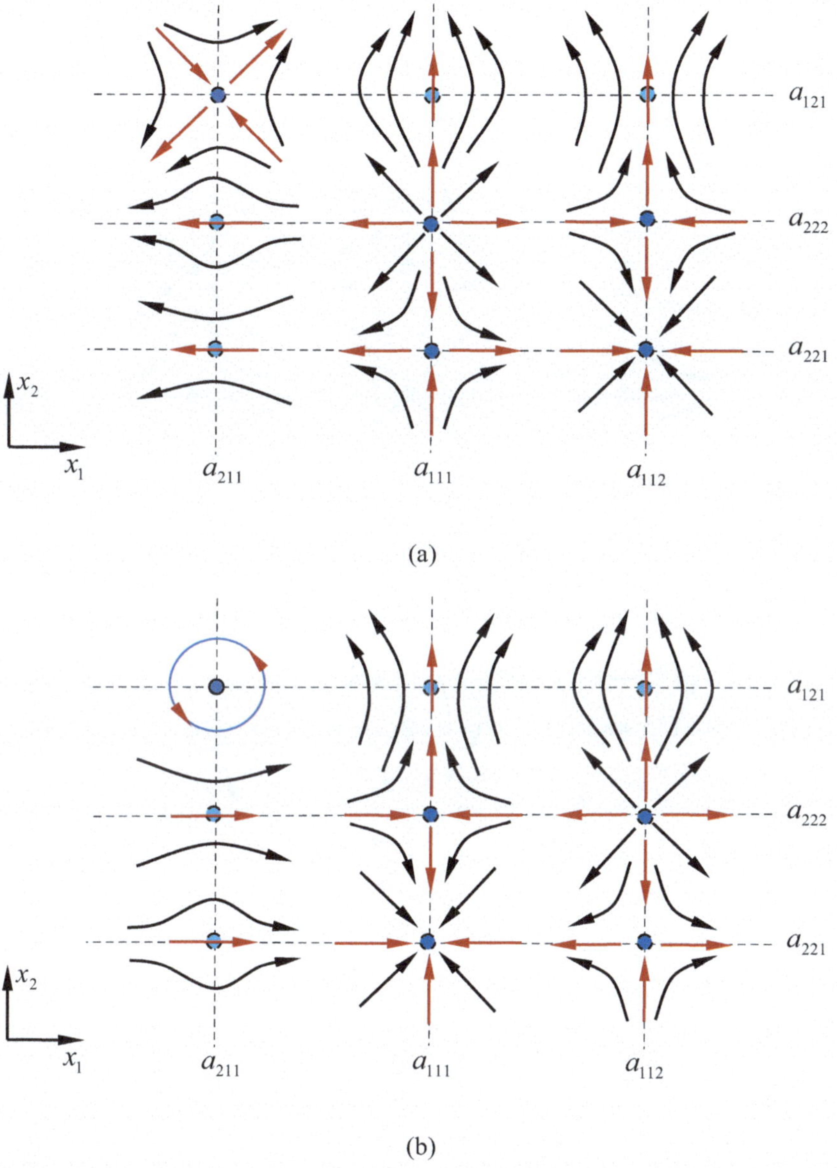

Fig. 5.2 Phase portraits ($a_{211} < a_{111}$ and $a_{222} < a_{121}$) for two-dimensional systems on the x_1-direction with $x_1^* = a_{211}, a_{111}, a_{112}$ and on the x_2-direction with $x_2^* = a_{221}, a_{222}, a_{121}$. The simple equilibriums with hyperbolic flows: **(a)** ($a_{110} > 0$, $a_{220} > 0$), **(b)** ($a_{110} < 0$, $a_{220} > 0$), **(c)** ($a_{110} > 0$, $a_{220} < 0$), and **(d)** ($a_{110} < 0$, $a_{220} < 0$)

as shown in Fig. 5.2. The directrix flows of $x_1 = a_{111}, a_{112}$ are positive and negative. The networks of saddle, sink, and source with hyperbolic and hyperbolic secant flows are presented. The saddles, sink, and sources in the network of equilibriums are *below* the hyperbolic and hyperbolic secant flows.

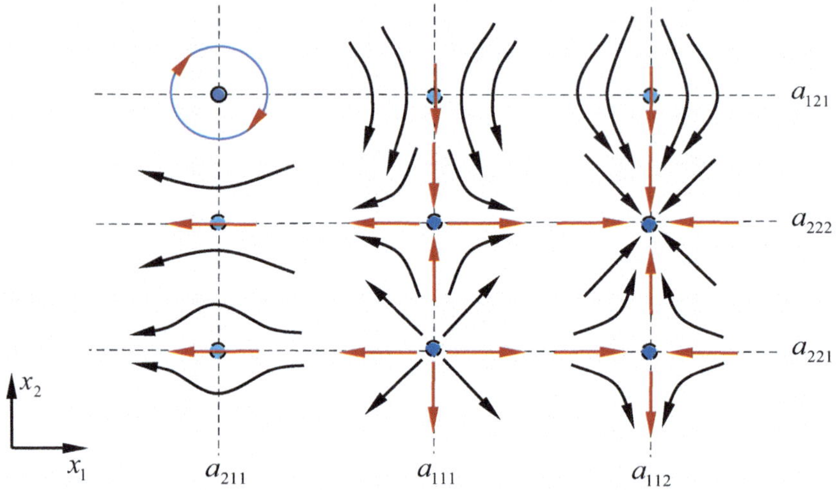

(c)

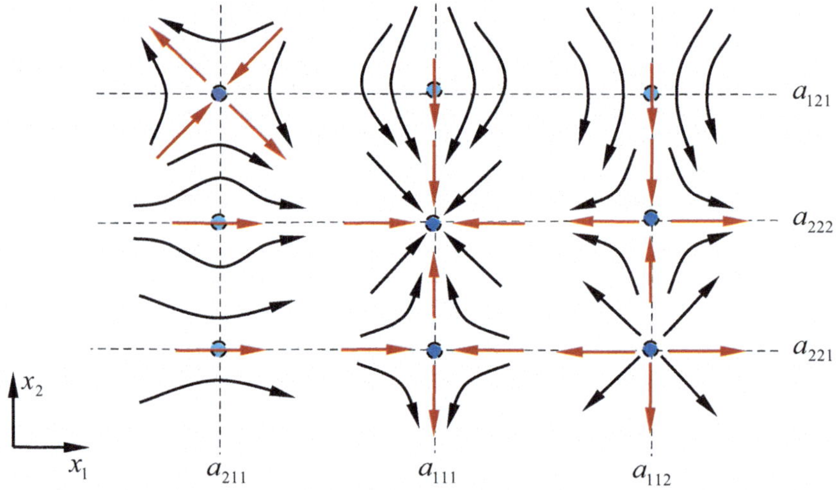

(d)

Fig. 5.2 (continued)

(iii) The simple equilibrium networks for $a_{211} < a_{111}$ and $a_{121} \in (a_{221}, a_{222})$ have the following properties:

$$\left\{\begin{array}{ccc} (a_{211}, a_{222}) & (a_{111}, a_{222}) & (a_{112}, a_{222}) \\ (a_{121}, a_{211}) & (a_{111}, a_{121}) & (a_{112}, a_{121}) \\ (a_{211}, a_{221}) & (a_{111}, a_{221}) & (a_{112}, a_{221}) \end{array}\right\}$$

$$= \left\{\begin{array}{ccc} \underbrace{(\text{pF,DP:UP})}_{\text{hyperbolic flow }(+)} & \underbrace{(\text{SI,SO})}_{\text{saddle}} & \underbrace{(\text{SO,SO})}_{\text{source}} \\ \underbrace{(\text{DP}_{-},\text{DP}_{+})}_{\text{CW center}} & \underbrace{(\text{DP:UP,nF})}_{\text{hyperbolic flow }(-)} & \underbrace{(\text{UP:DP,nF})}_{\text{hyperbolic-secant flow }(-)} \\ \underbrace{(\text{nF,DP:UP})}_{\text{hyperbolic flow }(-)} & \underbrace{(\text{SO,SI})}_{\text{saddle}} & \underbrace{(\text{SI,SI})}_{\text{sink}} \end{array}\right\}$$

for $a_{110} > 0$ and $a_{220} > 0$; (5.11)

$$\left\{\begin{array}{ccc} (a_{211}, a_{222}) & (a_{111}, a_{222}) & (a_{112}, a_{222}) \\ (a_{121}, a_{211}) & (a_{111}, a_{121}) & (a_{112}, a_{121}) \\ (a_{211}, a_{221}) & (a_{111}, a_{221}) & (a_{112}, a_{221}) \end{array}\right\}$$

$$= \left\{\begin{array}{ccc} \underbrace{(\text{nF,UP:DP})}_{\text{hyperbolic flow }(-)} & \underbrace{(\text{SO,SO})}_{\text{source}} & \underbrace{(\text{SI,SO})}_{\text{saddle}} \\ \underbrace{(\text{UP}_{-},\text{UP}_{-})}_{\text{negative saddle}} & \underbrace{(\text{UP:DP,nF})}_{\text{hyperbolic-secant flow }(-)} & \underbrace{(\text{DP:UP,nF})}_{\text{hyperbolic flow }(-)} \\ \underbrace{(\text{pF,UP:DP})}_{\text{hyperbolic-secant flow }(+)} & \underbrace{(\text{SI,SI})}_{\text{sink}} & \underbrace{(\text{SO,SI})}_{\text{saddle}} \end{array}\right\}$$

for $a_{110} < 0$ and $a_{220} > 0$; (5.12)

$$\left\{\begin{array}{ccc} (a_{211}, a_{222}) & (a_{111}, a_{222}) & (a_{112}, a_{222}) \\ (a_{121}, a_{211}) & (a_{111}, a_{121}) & (a_{112}, a_{121}) \\ (a_{211}, a_{221}) & (a_{111}, a_{221}) & (a_{112}, a_{221}) \end{array}\right\}$$

$$= \left\{\begin{array}{ccc} \underbrace{(\text{pF,UP:DP})}_{\text{hyperbolic flow }(+)} & \underbrace{(\text{SI,SI})}_{\text{sink}} & \underbrace{(\text{SO,SI})}_{\text{saddle}} \\ \underbrace{(\text{UP}_{+},\text{UP}_{+})}_{\text{positive saddle}} & \underbrace{(\text{UP:DP,pF})}_{\text{hyperbolic-secant flow }(+)} & \underbrace{(\text{DP:UP,pF})}_{\text{hyperbolic flow }(+)} \\ \underbrace{(\text{nF,UP:DP})}_{\text{hyperbolic-secant flow }(-)} & \underbrace{(\text{SO,SO})}_{\text{source}} & \underbrace{(\text{SI,SO})}_{\text{saddle}} \end{array}\right\}$$

for $a_{110} > 0$ and $a_{220} < 0$; (5.13)

$$
\begin{Bmatrix}
(a_{211}, a_{222}) & (a_{111}, a_{222}) & (a_{112}, a_{222}) \\
(a_{121}, a_{211}) & (a_{111}, a_{121}) & (a_{112}, a_{121}) \\
(a_{211}, a_{221}) & (a_{111}, a_{221}) & (a_{112}, a_{221})
\end{Bmatrix}
= \begin{Bmatrix}
\underbrace{(\text{nF,DP:UP})}_{\text{hyperbolic flow }(-)} & \underbrace{(\text{SO,SI})}_{\text{saddle}} & \underbrace{(\text{SI,SI})}_{\text{sink}} \\
\underbrace{(\text{DP}_+,\text{DP}_-)}_{\text{CCW center}} & \underbrace{(\text{DP:UP,pF})}_{\text{hyperbolic flow }(+)} & \underbrace{(\text{UP:DP,pF})}_{\text{hyperbolic-secant flow }(+)} \\
\underbrace{(\text{pF,DP:UP})}_{\text{hyperbolic flow }(+)} & \underbrace{(\text{SI,SO})}_{\text{saddle}} & \underbrace{(\text{SO,SO})}_{\text{source}}
\end{Bmatrix}
$$

for $a_{110} < 0$ and $a_{220} < 0$; (5.14)

as shown in Fig. 5.3. The directrix flows of $x_1 = a_{111}, a_{112}$ are positive and negative. The networks of saddle, sink, and source with hyperbolic and hyperbolic-secant flows are presented. The saddles, sink, and sources in the network of equilibriums are *separated by* the hyperbolic and hyperbolic-secant flows.

(iv) The simple equilibrium network for $a_{211} \in (a_{111}, a_{112})$ and $a_{121} < a_{221}$ has the following properties:

$$
\begin{Bmatrix}
(a_{111}, a_{222}) & (a_{211}, a_{222}) & (a_{112}, a_{222}) \\
(a_{111}, a_{221}) & (a_{211}, a_{221}) & (a_{112}, a_{221}) \\
(a_{111}, a_{121}) & (a_{121}, a_{211}) & (a_{112}, a_{121})
\end{Bmatrix}
= \begin{Bmatrix}
\underbrace{(\text{SI,SI})}_{\text{sink}} & \underbrace{(\text{nF,UP:DP})}_{\text{hyperbolic-secant flow }(-)} & \underbrace{(\text{SO,SO})}_{\text{source}} \\
\underbrace{(\text{SI,SO})}_{\text{saddle}} & \underbrace{(\text{nF,DP:UP})}_{\text{hyperbolic flow }(-)} & \underbrace{(\text{SO,SI})}_{\text{saddle}} \\
\underbrace{(\text{DP:UP,nF})}_{\text{hyperbolic flow }(-)} & \underbrace{(\text{DP}_+,\text{DP}_-)}_{\text{CCW center}} & \underbrace{(\text{DP:UP,pF})}_{\text{hyperbolic flow }(+)}
\end{Bmatrix}
$$

for $a_{110} > 0$ and $a_{220} > 0$; (5.15)

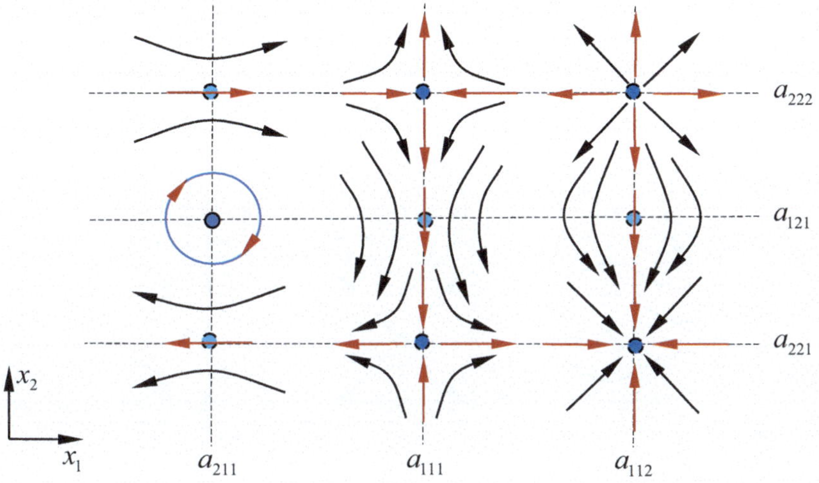

(a)

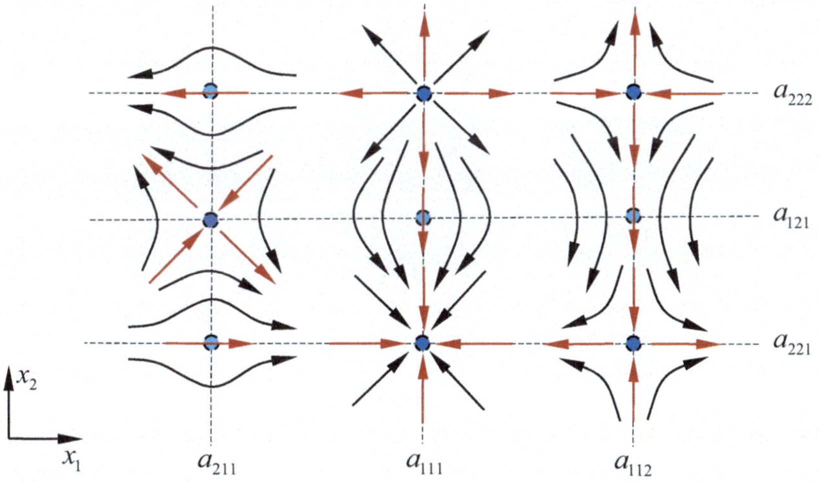

(b)

Fig. 5.3 Phase portraits ($a_{211} < a_{111}$ and $a_{121} \in (a_{221}, a_{222})$) for two-dimensional systems on the x_1-direction with $x_1^* = a_{211}, a_{111}, a_{112}$ and on the x_2-direction with $x_2^* = a_{221}, a_{121}, a_{222}$. The simple equilibriums with hyperbolic flows: **(a)** ($a_{110} > 0$, $a_{220} > 0$), **(b)** ($a_{110} < 0$, $a_{220} > 0$), **(c)** ($a_{110} > 0$, $a_{220} < 0$), and **(d)** ($a_{110} < 0$, $a_{220} < 0$)

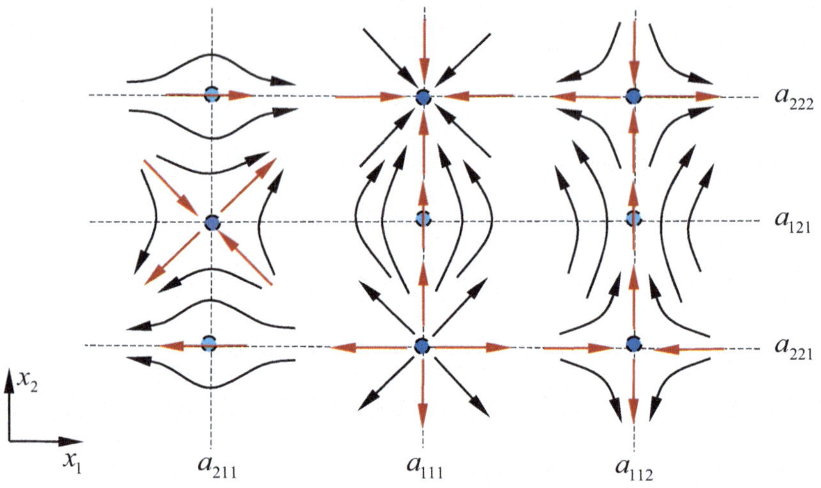

(c)

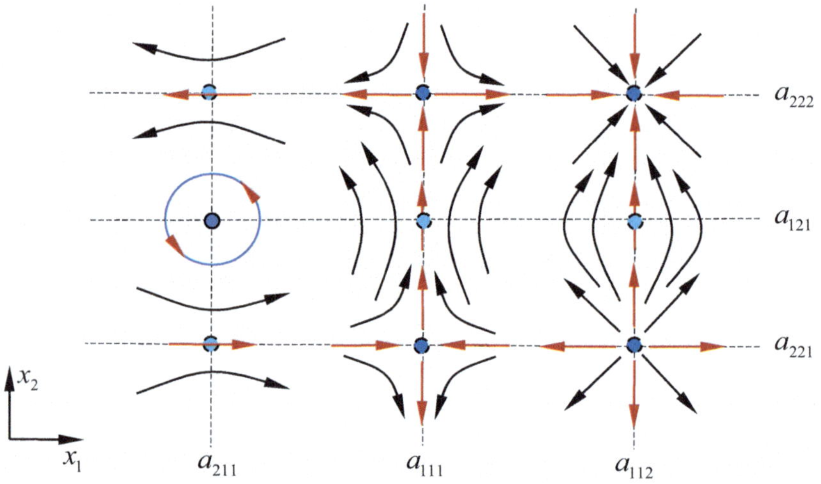

(d)

Fig. 5.3 (continued)

$$\left\{\begin{array}{ccc}(a_{111},a_{222}) & (a_{211},a_{222}) & (a_{112},a_{222})\\(a_{111},a_{221}) & (a_{211},a_{221}) & (a_{112},a_{221})\\(a_{111},a_{121}) & (a_{121},a_{211}) & (a_{112},a_{121})\end{array}\right\}$$

$$=\left\{\begin{array}{ccc}\underbrace{(\text{SO,SI})}_{\text{saddle}} & \underbrace{(\text{pF,DP:UP})}_{\text{hyperbolic flow }(+)} & \underbrace{(\text{SI,SO})}_{\text{saddle}}\\\underbrace{(\text{SO,SO})}_{\text{source}} & \underbrace{(\text{pF,UP:DP})}_{\text{hyperbolic-secant flow }(+)} & \underbrace{(\text{SI,SI})}_{\text{sink}}\\\underbrace{(\text{UP:DP,nF})}_{\text{hyperbolic-secant flow }(-)} & \underbrace{(\text{UP}_{+},\text{UP}_{+})}_{\text{positive saddle}} & \underbrace{(\text{UP:DP,pF})}_{\text{hyperbolic-secant flow }(+)}\end{array}\right\}$$

for $a_{110} < 0$ and $a_{220} > 0$; (5.16)

$$\left\{\begin{array}{ccc}(a_{111},a_{222}) & (a_{211},a_{222}) & (a_{112},a_{222})\\(a_{111},a_{221}) & (a_{211},a_{221}) & (a_{112},a_{221})\\(a_{111},a_{121}) & (a_{121},a_{211}) & (a_{112},a_{121})\end{array}\right\}$$

$$=\left\{\begin{array}{ccc}\underbrace{(\text{SI,SO})}_{\text{saddle}} & \underbrace{(\text{nF,DP:UP})}_{\text{hyperbolic flow }(-)} & \underbrace{(\text{SO,SI})}_{\text{saddle}}\\\underbrace{(\text{SI,SI})}_{\text{sink}} & \underbrace{(\text{nF,UP:DP})}_{\text{hyperbolic-secant flow }(-)} & \underbrace{(\text{SO,SO})}_{\text{source}}\\\underbrace{(\text{UP:DP,pF})}_{\text{hyperbolic-secant flow }(+)} & \underbrace{(\text{UP}_{-},\text{UP}_{-})}_{\text{negative saddle}} & \underbrace{(\text{UP:DP,nF})}_{\text{hyperbolic-secant flow }(-)}\end{array}\right\}$$

for $a_{110} > 0$ and $a_{220} < 0$; (5.17)

$$\left\{\begin{array}{ccc}(a_{111},a_{222}) & (a_{211},a_{222}) & (a_{112},a_{222})\\(a_{111},a_{221}) & (a_{211},a_{221}) & (a_{112},a_{221})\\(a_{111},a_{121}) & (a_{121},a_{211}) & (a_{112},a_{121})\end{array}\right\}$$

$$=\left\{\begin{array}{ccc}\underbrace{(\text{SO,SO})}_{\text{source}} & \underbrace{(\text{pF,UP:DP})}_{\text{hyperbolic-secant flow }(+)} & \underbrace{(\text{SI,SI})}_{\text{sink}}\\\underbrace{(\text{SO,SI})}_{\text{saddle}} & \underbrace{(\text{pF,DP:UP})}_{\text{hyperbolic flow }(+)} & \underbrace{(\text{SI,SO})}_{\text{saddle}}\\\underbrace{(\text{DP:UP,pF})}_{\text{hyperbolic flow }(+)} & \underbrace{(\text{DP}_{-},\text{DP}_{+})}_{\text{CW center}} & \underbrace{(\text{DP:UP,nF})}_{\text{hyperbolic flow }(-)}\end{array}\right\}$$

for $a_{110} < 0$ and $a_{220} < 0$; (5.18)

as shown in Fig. 5.4. The directrix flows of $x_1 = a_{111}, a_{112}$ are positive and negative. The networks of saddle, sink, and source with hyperbolic and hyperbolic-secant flows are presented. The saddles, sink, and sources in the network of equilibriums are above the hyperbolic and hyperbolic-secant flows.

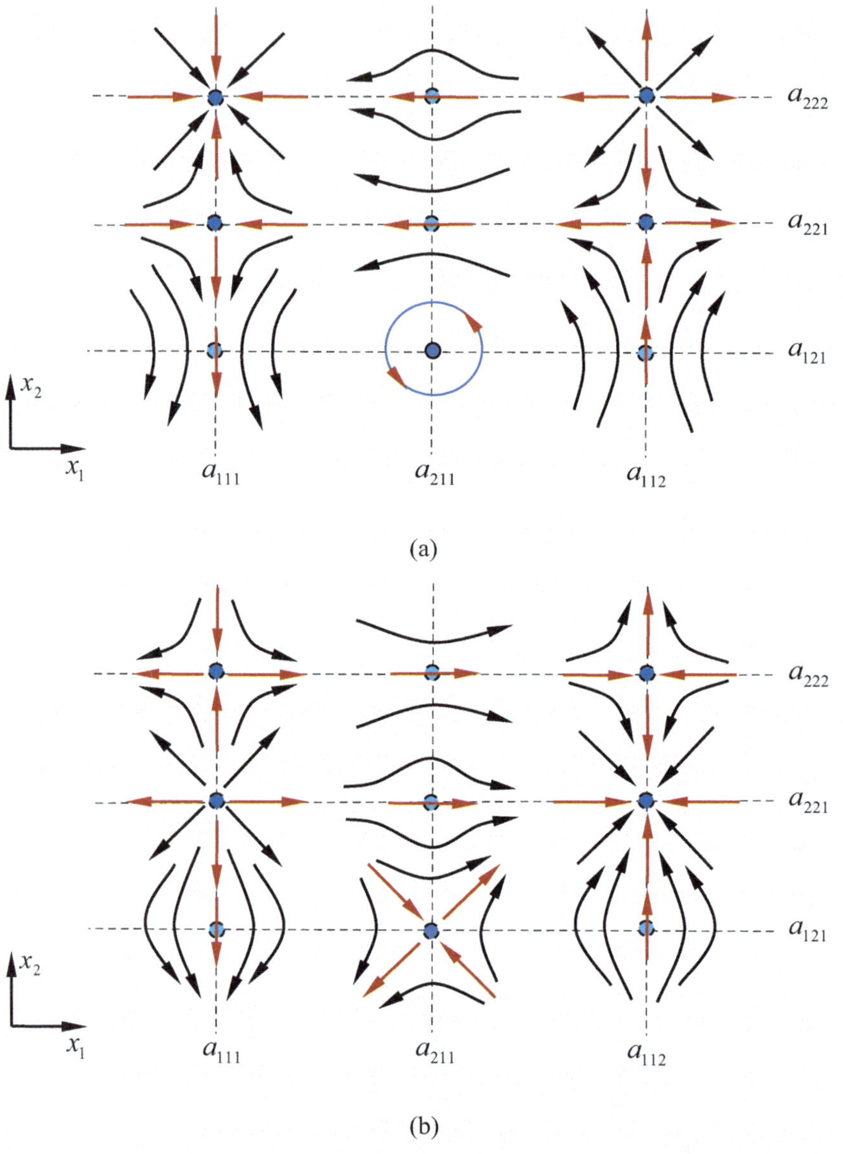

Fig. 5.4 Phase portraits ($a_{211} \in (a_{111}, a_{112})$ and $a_{121} < a_{221}$) for two-dimensional systems on the x_1-direction with $x_1^* = a_{111}, a_{211}, a_{112}$ and on the x_2-direction with $x_2^* = a_{121}, a_{221}, a_{222}$. The simple equilibriums with hyperbolic flows: **(a)** ($a_{110} > 0$, $a_{220} > 0$), **(b)** ($a_{110} < 0$, $a_{220} > 0$), **(c)** ($a_{110} > 0$, $a_{220} < 0$), and **(d)** ($a_{110} < 0$, $a_{220} < 0$)

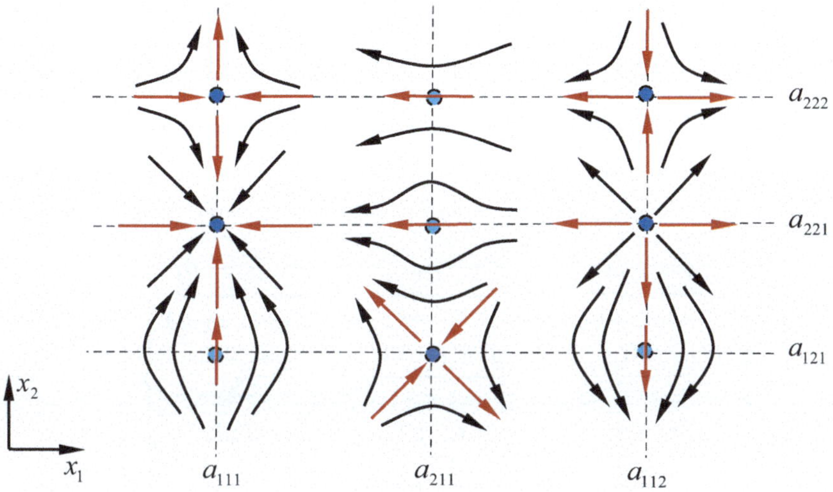

(c)

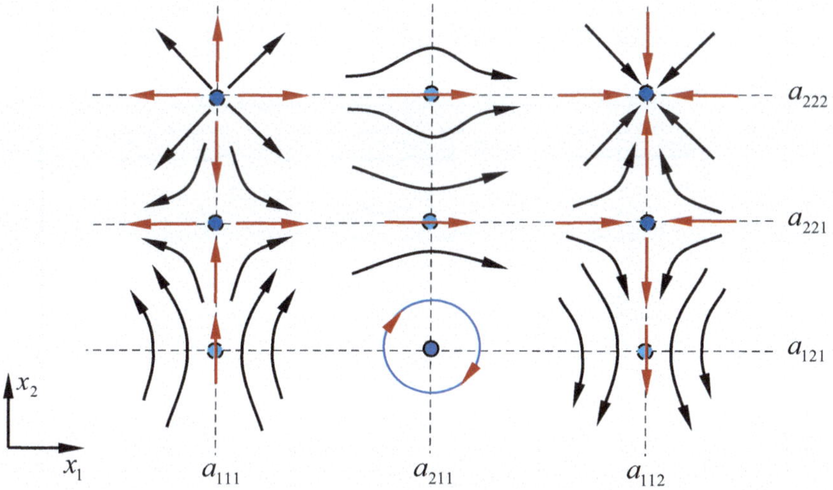

(d)

Fig. 5.4 (continued)

(v) The simple equilibrium networks for $a_{211} \in (a_{111}, a_{112})$ and $a_{222} < a_{121}$ have the following properties:

$$\begin{Bmatrix} (a_{111}, a_{121}) & (a_{121}, a_{211}) & (a_{112}, a_{121}) \\ (a_{111}, a_{222}) & (a_{211}, a_{222}) & (a_{112}, a_{222}) \\ (a_{111}, a_{221}) & (a_{211}, a_{221}) & (a_{112}, a_{221}) \end{Bmatrix}$$
$$= \begin{Bmatrix} \underbrace{(\text{DP:UP,nF})}_{\text{hyperbolic flow }(-)} & \underbrace{(\text{DP}_{+},\text{DP}_{-})}_{\text{CCW center}} & \underbrace{(\text{DP:UP,pF})}_{\text{hyperbolic flow }(+)} \\ \underbrace{(\text{SO,SI})}_{\text{saddle}} & \underbrace{(\text{pF,DP:UP})}_{\text{hyperbolic flow }(+)} & \underbrace{(\text{SI,SO})}_{\text{saddle}} \\ \underbrace{(\text{SO,SO})}_{\text{source}} & \underbrace{(\text{pF,UP:DP})}_{\text{hyperbolic-secant flow }(+)} & \underbrace{(\text{SI,SI})}_{\text{sink}} \end{Bmatrix}$$

for $a_{110} > 0$ and $a_{220} > 0$; (5.19)

$$\begin{Bmatrix} (a_{111}, a_{121}) & (a_{121}, a_{211}) & (a_{112}, a_{121}) \\ (a_{111}, a_{222}) & (a_{211}, a_{222}) & (a_{112}, a_{222}) \\ (a_{111}, a_{221}) & (a_{211}, a_{221}) & (a_{112}, a_{221}) \end{Bmatrix}$$
$$= \begin{Bmatrix} \underbrace{(\text{UP:DP,nF})}_{\text{hyperbolic-secant flow }(-)} & \underbrace{(\text{UP}_{+},\text{UP}_{+})}_{\text{positive saddle}} & \underbrace{(\text{UP:DP,pF})}_{\text{hyperbolic-secant flow }(+)} \\ \underbrace{(\text{SI,SI})}_{\text{sink}} & \underbrace{(\text{nF,UP:DP})}_{\text{hyperbolic-secant flow }(-)} & \underbrace{(\text{SO,SO})}_{\text{source}} \\ \underbrace{(\text{SI,SO})}_{\text{saddle}} & \underbrace{(\text{nF,DP:UP})}_{\text{hyperbolic flow }(-)} & \underbrace{(\text{SO,SI})}_{\text{saddle}} \end{Bmatrix}$$

for $a_{110} < 0$ and $a_{220} > 0$; (5.20)

$$\begin{Bmatrix} (a_{111}, a_{121}) & (a_{121}, a_{211}) & (a_{112}, a_{121}) \\ (a_{111}, a_{222}) & (a_{211}, a_{222}) & (a_{112}, a_{222}) \\ (a_{111}, a_{221}) & (a_{211}, a_{221}) & (a_{112}, a_{221}) \end{Bmatrix}$$
$$= \begin{Bmatrix} \underbrace{(\text{UP:DP,pF})}_{\text{hyperbolic-secant flow }(+)} & \underbrace{(\text{UP}_{-},\text{UP}_{-})}_{\text{negative saddle}} & \underbrace{(\text{UP:DP,nF})}_{\text{hyperbolic-secant flow }(-)} \\ \underbrace{(\text{SO,SO})}_{\text{source}} & \underbrace{(\text{pF,UP:DP})}_{\text{hyperbolic-secant flow }(+)} & \underbrace{(\text{SI,SI})}_{\text{sink}} \\ \underbrace{(\text{SO,SI})}_{\text{saddle}} & \underbrace{(\text{nF,DP:UP})}_{\text{hyperbolic flow }(+)} & \underbrace{(\text{SI,SO})}_{\text{saddle}} \end{Bmatrix}$$

for $a_{110} > 0$ and $a_{220} < 0$; (5.21)

$$\left\{\begin{array}{ccc}(a_{111},a_{121}) & (a_{121},a_{211}) & (a_{112},a_{121})\\ (a_{111},a_{222}) & (a_{211},a_{222}) & (a_{112},a_{222})\\ (a_{111},a_{221}) & (a_{211},a_{221}) & (a_{112},a_{221})\end{array}\right\}$$

$$=\left\{\begin{array}{ccc}\underbrace{(\mathrm{DP{:}UP,pF})}_{\text{hyperbolic flow }(+)} & \underbrace{(\mathrm{DP}_-,\mathrm{DP}_+)}_{\text{CW center}} & \underbrace{(\mathrm{DP{:}UP,nF})}_{\text{hyperbolic flow }(-)}\\ \underbrace{(\mathrm{SI,SO})}_{\text{saddle}} & \underbrace{(\mathrm{nF,DP{:}UP})}_{\text{hyperbolic flow }(-)} & \underbrace{(\mathrm{SO,SI})}_{\text{saddle}}\\ \underbrace{(\mathrm{SI,SI})}_{\text{sink}} & \underbrace{(\mathrm{nF,UP{:}DP})}_{\text{hyperbolic-secant flow }(-)} & \underbrace{(\mathrm{SO,SO})}_{\text{source}}\end{array}\right\}$$

for $a_{110} < 0$ and $a_{220} < 0$; (5.22)

as shown in Fig. 5.5. The directrix flows of $x_1 = a_{111}, a_{112}$ are positive and negative. The networks of saddle, sink, and source with hyperbolic and hyperbolic-secant flows are presented. The saddles, sink, and sources in the network of equilibriums are *below* the hyperbolic and hyperbolic secant flows.

(vi) The simple equilibrium networks for $a_{211} \in (a_{111}, a_{112})$ and $a_{121} \in (a_{221}, a_{222})$ have the following properties:

$$\left\{\begin{array}{ccc}(a_{111},a_{222}) & (a_{211},a_{222}) & (a_{112},a_{222})\\ (a_{111},a_{121}) & (a_{121},a_{211}) & (a_{112},a_{121})\\ (a_{111},a_{221}) & (a_{211},a_{221}) & (a_{112},a_{221})\end{array}\right\}$$

$$=\left\{\begin{array}{ccc}\underbrace{(\mathrm{SI,SI})}_{\text{sink}} & \underbrace{(\mathrm{nF,UP{:}DP})}_{\text{hyperbolic-secant flow }(-)} & \underbrace{(\mathrm{SO,SO})}_{\text{source}}\\ \underbrace{(\mathrm{UP{:}DP,pF})}_{\text{hyperbolic-secant flow }(+)} & \underbrace{(\mathrm{UP}_-,\mathrm{UP}_-)}_{\text{negative saddle}} & \underbrace{(\mathrm{UP{:}DP,nF})}_{\text{hyperbolic-secant flow }(-)}\\ \underbrace{(\mathrm{SO,SO})}_{\text{source}} & \underbrace{(\mathrm{pF,UP{:}DP})}_{\text{hyperbolic-secant flow }(+)} & \underbrace{(\mathrm{SI,SI})}_{\text{sink}}\end{array}\right\}$$

for $a_{110} > 0$ and $a_{220} > 0$; (5.23)

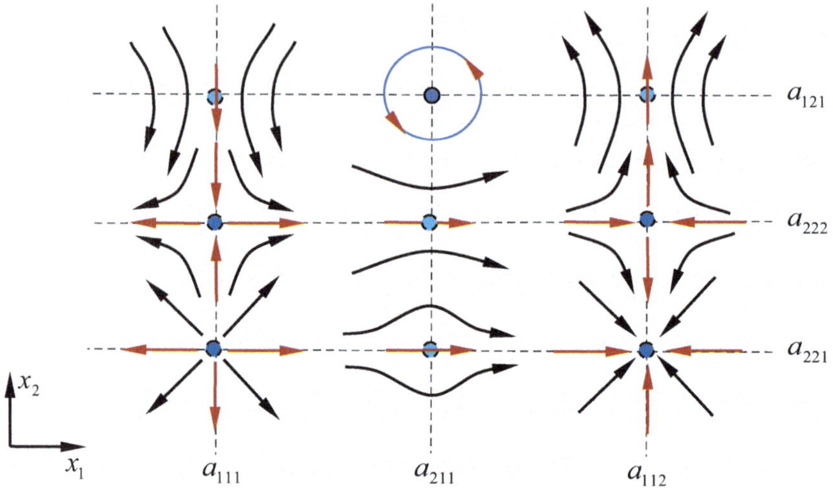

(a)

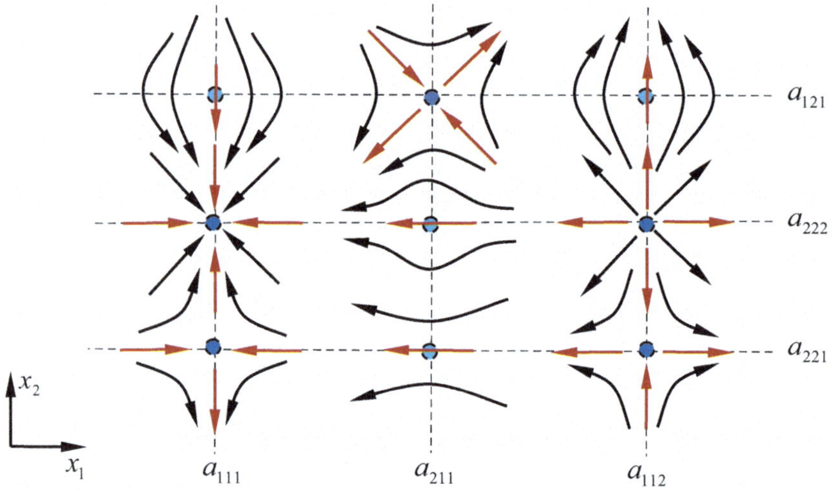

(b)

Fig. 5.5 Phase portraits ($a_{211} \in (a_{111}, a_{112})$ and $a_{222} < a_{121}$) for two-dimensional systems on the x_1-direction with $x_1^* = a_{111}, a_{211}, a_{112}$ and on the x_2-direction with $x_2^* = a_{221}, a_{222}, a_{121}$. The simple equilibriums with hyperbolic flows: (**a**) ($a_{110} > 0$, $a_{220} > 0$), (**b**) ($a_{110} < 0$, $a_{220} > 0$), (**c**) ($a_{110} > 0$, $a_{220} < 0$), and (**d**) ($a_{110} < 0$, $a_{220} < 0$)

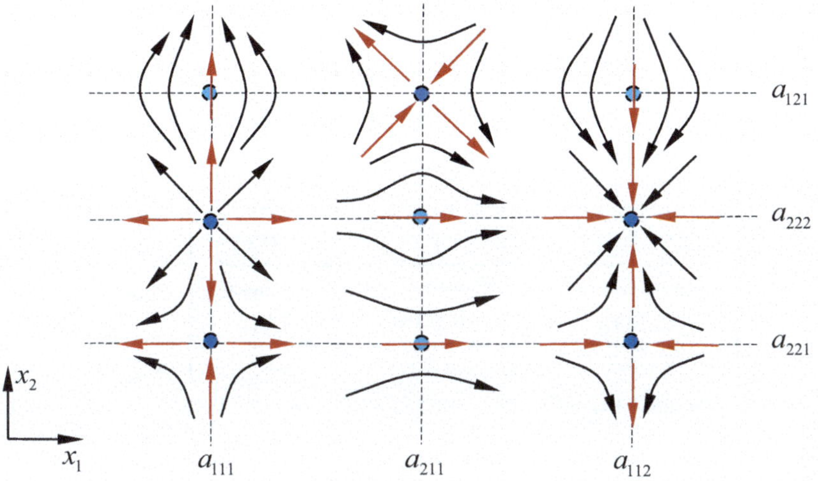

(c)

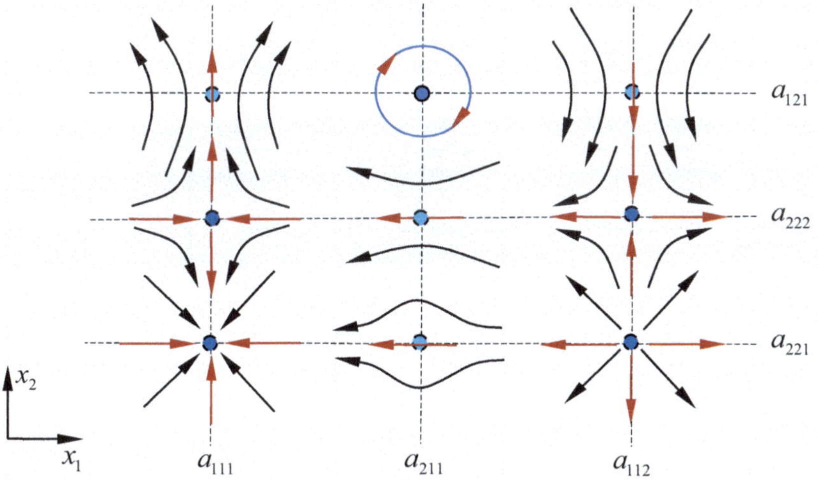

(d)

Fig. 5.5 (continued)

$$\left\{\begin{matrix} (a_{111},a_{222}) & (a_{211},a_{222}) & (a_{112},a_{222}) \\ (a_{111},a_{121}) & (a_{121},a_{211}) & (a_{112},a_{121}) \\ (a_{111},a_{221}) & (a_{211},a_{221}) & (a_{112},a_{221}) \end{matrix}\right\}$$

$$= \left\{\begin{matrix} \underbrace{(\text{SO,SI})}_{\text{saddle}} & \underbrace{(\text{pF,DP:UP})}_{\text{hyperbolic flow }(+)} & \underbrace{(\text{SI,SO})}_{\text{saddle}} \\ \underbrace{(\text{DP:UP,pF})}_{\text{hyperbolic flow }(+)} & \underbrace{(\text{DP}_{-},\text{DP}_{+})}_{\text{CW center}} & \underbrace{(\text{DP:UP,nF})}_{\text{hyperbolic flow }(-)} \\ \underbrace{(\text{SI,SO})}_{\text{saddle}} & \underbrace{(\text{nF,DP:UP})}_{\text{hyperbolic flow }(-)} & \underbrace{(\text{SO,SI})}_{\text{saddle}} \end{matrix}\right\}$$

for $a_{110} < 0$ and $a_{220} > 0$; (5.24)

$$\left\{\begin{matrix} (a_{111},a_{222}) & (a_{211},a_{222}) & (a_{112},a_{222}) \\ (a_{111},a_{121}) & (a_{121},a_{211}) & (a_{112},a_{121}) \\ (a_{111},a_{221}) & (a_{211},a_{221}) & (a_{112},a_{221}) \end{matrix}\right\}$$

$$= \left\{\begin{matrix} \underbrace{(\text{SI,SO})}_{\text{saddle}} & \underbrace{(\text{nF,DP:UP})}_{\text{hyperbolic flow }(-)} & \underbrace{(\text{SO,SI})}_{\text{saddle}} \\ \underbrace{(\text{DP:UP,nF})}_{\text{hyperbolic flow }(-)} & \underbrace{(\text{DP}_{+},\text{DP}_{-})}_{\text{CCW center}} & \underbrace{(\text{DP:UP,pF})}_{\text{hyperbolic flow }(+)} \\ \underbrace{(\text{SO,SI})}_{\text{saddle}} & \underbrace{(\text{pF,DP:UP})}_{\text{hyperbolic flow }(+)} & \underbrace{(\text{SI,SO})}_{\text{saddle}} \end{matrix}\right\}$$

for $a_{110} > 0$ and $a_{220} < 0$; (5.25)

$$\left\{\begin{matrix} (a_{111},a_{222}) & (a_{211},a_{222}) & (a_{112},a_{222}) \\ (a_{111},a_{121}) & (a_{121},a_{211}) & (a_{112},a_{121}) \\ (a_{111},a_{221}) & (a_{211},a_{221}) & (a_{112},a_{221}) \end{matrix}\right\}$$

$$= \left\{\begin{matrix} \underbrace{(\text{SO,SO})}_{\text{source}} & \underbrace{(\text{pF,UP:DP})}_{\text{hyperbolic-secant flow }(+)} & \underbrace{(\text{SI,SI})}_{\text{sink}} \\ \underbrace{(\text{UP:DP,nF})}_{\text{hyperbolic-secant flow }(-)} & \underbrace{(\text{UP}_{+},\text{UP}_{+})}_{\text{positive saddle}} & \underbrace{(\text{UP:DP,pF})}_{\text{hyperbolic-secant flow }(+)} \\ \underbrace{(\text{SI,SI})}_{\text{sink}} & \underbrace{(\text{nF,UP:DP})}_{\text{hyperbolic-secant flow }(-)} & \underbrace{(\text{SO,SO})}_{\text{source}} \end{matrix}\right\}$$

for $a_{110} < 0$ and $a_{220} < 0$; (5.26)

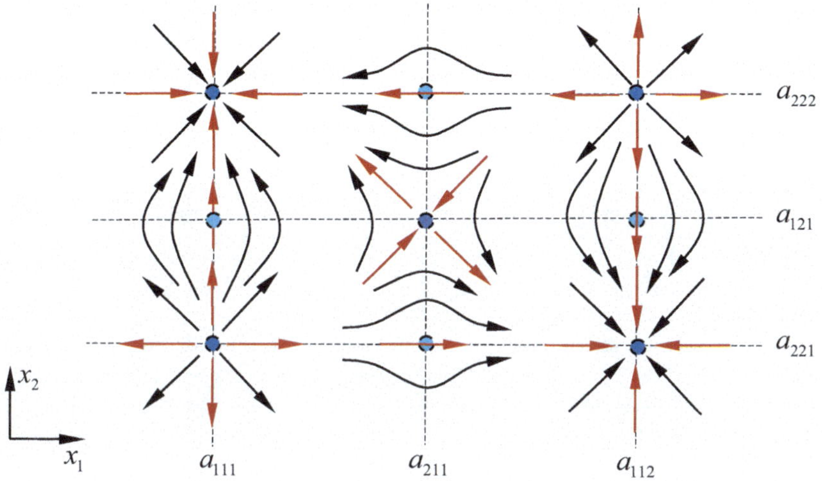

(a)

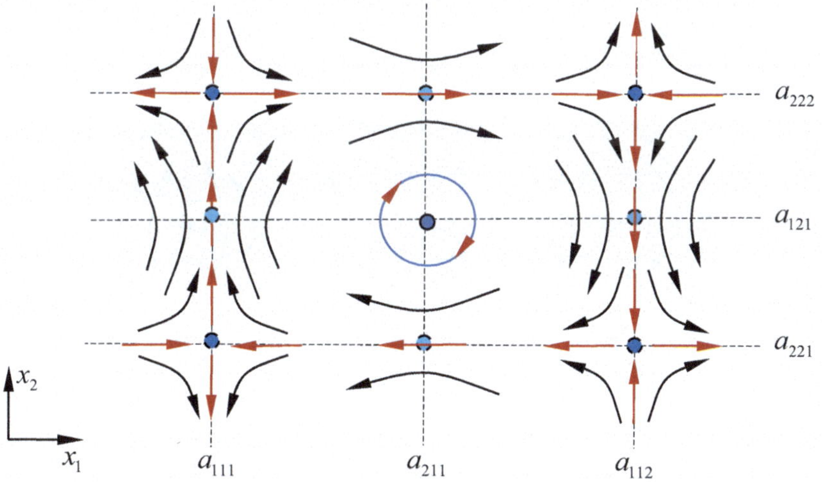

(b)

Fig. 5.6 Phase portraits ($a_{211} \in (a_{111}, a_{112})$ and $a_{121} \in (a_{221}, a_{222})$) for two-dimensional systems on the x_1-direction with $x_1^* = a_{111}, a_{211}, a_{112}$ and on the x_2-direction with $x_2^* = a_{221}, a_{121}, a_{222}$. The simple equilibriums with hyperbolic flows: (**a**) ($a_{110} > 0$, $a_{220} > 0$), (**b**) ($a_{110} < 0$, $a_{220} > 0$), (**c**) ($a_{110} > 0$, $a_{220} < 0$), and (**d**) ($a_{110} < 0$, $a_{220} < 0$)

as shown in Fig. 5.6. The directrix flows of $x_1 = a_{111}, a_{112}$ are positive and negative. The networks of saddle, sink, and source with hyperbolic and hyperbolic secant flows are presented. The saddles, sink, and sources in the network of equilibriums are *separated by* the hyperbolic and hyperbolic secant flows.

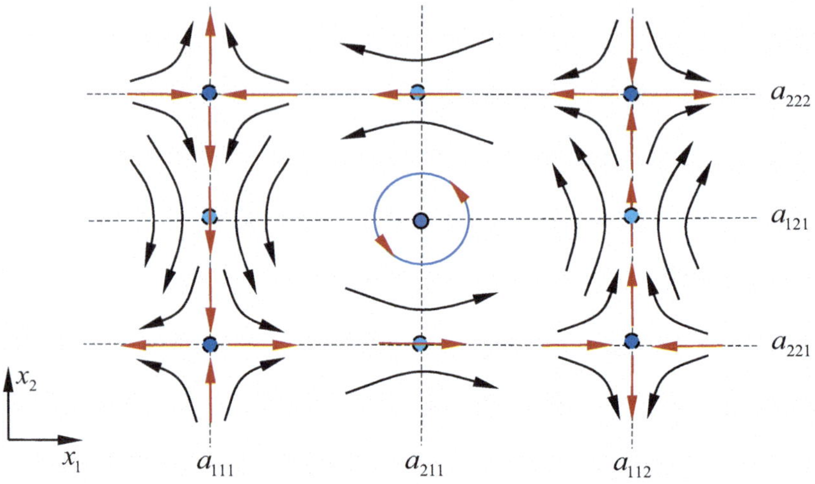

(c)

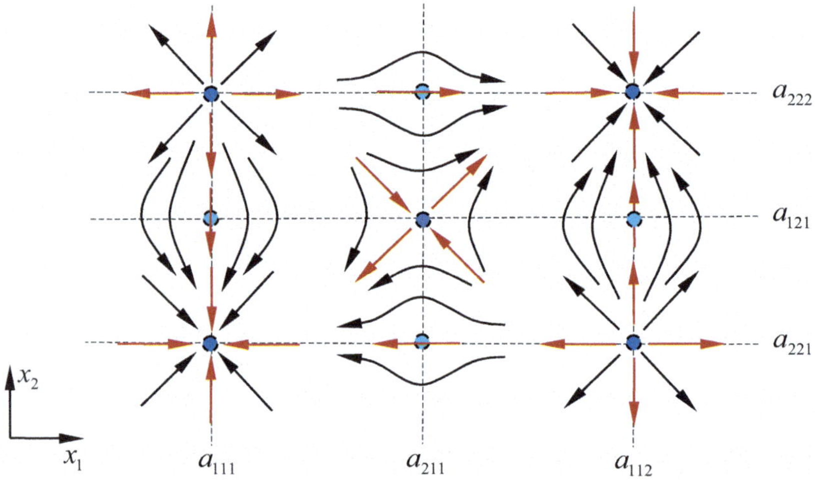

(d)

Fig. 5.6 (continued)

(vii) The simple equilibrium network for $a_{112} < a_{211}$ and $a_{121} < a_{221}$ has the following properties:

$$
\begin{gathered}
\left\{\begin{array}{ccc}
(a_{111}, a_{222}) & (a_{112}, a_{222}) & (a_{211}, a_{222}) \\
(a_{111}, a_{221}) & (a_{112}, a_{221}) & (a_{211}, a_{221}) \\
(a_{111}, a_{121}) & (a_{112}, a_{121}) & (a_{121}, a_{211})
\end{array}\right\} \\
= \left\{\begin{array}{ccc}
\underbrace{(\mathrm{SI},\mathrm{SI})}_{\text{sink}} & \underbrace{(\mathrm{SO},\mathrm{SI})}_{\text{saddle}} & \underbrace{(\mathrm{pF},\mathrm{DP}:\mathrm{UP})}_{\text{hyperbolic flow }(+)} \\
\underbrace{(\mathrm{SI},\mathrm{SO})}_{\text{saddle}} & \underbrace{(\mathrm{SO},\mathrm{SO})}_{\text{source}} & \underbrace{(\mathrm{pF},\mathrm{UP}:\mathrm{DP})}_{\text{hyperbolic-secant flow }(+)} \\
\underbrace{(\mathrm{DP}:\mathrm{UP},\mathrm{nF})}_{\text{hyperbolic flow }(-)} & \underbrace{(\mathrm{UP}:\mathrm{DP},\mathrm{nF})}_{\text{hyperbolic-secant flow }(-)} & \underbrace{(\mathrm{UP}_{+},\mathrm{UP}_{+})}_{\text{positive saddle}}
\end{array}\right\} \\
\text{for } a_{110} > 0 \text{ and } a_{220} > 0;
\end{gathered}
\tag{5.27}
$$

$$
\begin{gathered}
\left\{\begin{array}{ccc}
(a_{111}, a_{222}) & (a_{112}, a_{222}) & (a_{211}, a_{222}) \\
(a_{111}, a_{221}) & (a_{112}, a_{221}) & (a_{211}, a_{221}) \\
(a_{111}, a_{121}) & (a_{112}, a_{121}) & (a_{121}, a_{211})
\end{array}\right\} \\
= \left\{\begin{array}{ccc}
\underbrace{(\mathrm{SO},\mathrm{SI})}_{\text{saddle}} & \underbrace{(\mathrm{SI},\mathrm{SI})}_{\text{sink}} & \underbrace{(\mathrm{nF},\mathrm{UP}:\mathrm{DP})}_{\text{hyperbolic-secant flow }(-)} \\
\underbrace{(\mathrm{SO},\mathrm{SO})}_{\text{source}} & \underbrace{(\mathrm{SI},\mathrm{SO})}_{\text{saddle}} & \underbrace{(\mathrm{nF},\mathrm{DP}:\mathrm{UP})}_{\text{hyperbolic flow }(-)} \\
\underbrace{(\mathrm{UP}:\mathrm{DP},\mathrm{nF})}_{\text{hyperbolic-secant flow }(-)} & \underbrace{(\mathrm{DP}:\mathrm{UP},\mathrm{nF})}_{\text{hyperbolic flow }(-)} & \underbrace{(\mathrm{DP}_{+},\mathrm{DP}_{-})}_{\text{CCW center}}
\end{array}\right\} \\
\text{for } a_{110} < 0 \text{ and } a_{220} > 0;
\end{gathered}
\tag{5.28}
$$

$$
\begin{gathered}
\left\{\begin{array}{ccc}
(a_{111}, a_{222}) & (a_{112}, a_{222}) & (a_{211}, a_{222}) \\
(a_{111}, a_{221}) & (a_{112}, a_{221}) & (a_{211}, a_{221}) \\
(a_{111}, a_{121}) & (a_{112}, a_{121}) & (a_{121}, a_{211})
\end{array}\right\} \\
= \left\{\begin{array}{ccc}
\underbrace{(\mathrm{SI},\mathrm{SO})}_{\text{saddle}} & \underbrace{(\mathrm{SO},\mathrm{SO})}_{\text{source}} & \underbrace{(\mathrm{pF},\mathrm{UP}:\mathrm{DP})}_{\text{hyperbolic-secant flow }(+)} \\
\underbrace{(\mathrm{SI},\mathrm{SI})}_{\text{sink}} & \underbrace{(\mathrm{SO},\mathrm{SI})}_{\text{saddle}} & \underbrace{(\mathrm{pF},\mathrm{DP}:\mathrm{UP})}_{\text{hyperbolic flow }(+)} \\
\underbrace{(\mathrm{UP}:\mathrm{DP},\mathrm{pF})}_{\text{hyperbolic-secant flow }(+)} & \underbrace{(\mathrm{DP}:\mathrm{UP},\mathrm{pF})}_{\text{hyperbolic flow }(+)} & \underbrace{(\mathrm{DP}_{-},\mathrm{DP}_{+})}_{\text{CW center}}
\end{array}\right\} \\
\text{for } a_{110} > 0 \text{ and } a_{220} < 0;
\end{gathered}
\tag{5.29}
$$

$$\left\{\begin{matrix} (a_{111},a_{222}) & (a_{112},a_{222}) & (a_{211},a_{222}) \\ (a_{111},a_{221}) & (a_{112},a_{221}) & (a_{211},a_{221}) \\ (a_{111},a_{121}) & (a_{112},a_{121}) & (a_{121},a_{211}) \end{matrix}\right\}$$
$$= \left\{\begin{matrix} \underbrace{(\text{SO,SO})}_{\text{source}} & \underbrace{(\text{SI,SO})}_{\text{saddle}} & \underbrace{(\text{nF,DP:UP})}_{\text{hyperbolic flow }(-)} \\ \underbrace{(\text{SO,SI})}_{\text{saddle}} & \underbrace{(\text{SI,SI})}_{\text{sink}} & \underbrace{(\text{nF,UP:DP})}_{\text{hyperbolic-secant flow }(-)} \\ \underbrace{(\text{DP:UP,pF})}_{\text{hyperbolic flow }(+)} & \underbrace{(\text{UP:PP,pF})}_{\text{hyperbolic-secant flow }(+)} & \underbrace{(\text{UP}_-,\text{UP}_-)}_{\text{negative center}} \end{matrix}\right\}$$
$$\text{for } a_{110} < 0 \text{ and } a_{220} < 0; \tag{5.30}$$

as shown in Fig. 5.7. The directrix flows of $x_1 = a_{111}, a_{112}$ are positive and negative. The directrix flows of $x_1 = a_{221}, a_{222}$ are also positive and negative. The networks of saddle, sink, and source with hyperbolic and hyperbolic-secant flows are presented.

(viii) The simple equilibrium networks for $a_{112} < a_{211}$ and $a_{222} < a_{121}$ have the following properties:

$$\left\{\begin{matrix} (a_{111},a_{121}) & (a_{112},a_{121}) & (a_{121},a_{211}) \\ (a_{111},a_{222}) & (a_{112},a_{222}) & (a_{211},a_{222}) \\ (a_{111},a_{221}) & (a_{112},a_{221}) & (a_{211},a_{221}) \end{matrix}\right\}$$
$$= \left\{\begin{matrix} \underbrace{(\text{DP:UP,nF})}_{\text{hyperbolic flow }(-)} & \underbrace{(\text{UP:DP,nF})}_{\text{hyperbolic-secant flow }(-)} & \underbrace{(\text{UP}_+,\text{UP}_+)}_{\text{positve saddle}} \\ \underbrace{(\text{SO,SI})}_{\text{saddle}} & \underbrace{(\text{SI,SI})}_{\text{sink}} & \underbrace{(\text{nF,UP:DP})}_{\text{hyperbolic-secant flow }(-)} \\ \underbrace{(\text{SO,SO})}_{\text{source}} & \underbrace{(\text{SI,SO})}_{\text{saddle}} & \underbrace{(\text{nF,DP:UP})}_{\text{hyperbolic flow }(-)} \end{matrix}\right\}$$
$$\text{for } a_{110} > 0 \text{ and } a_{220} > 0; \tag{5.31}$$

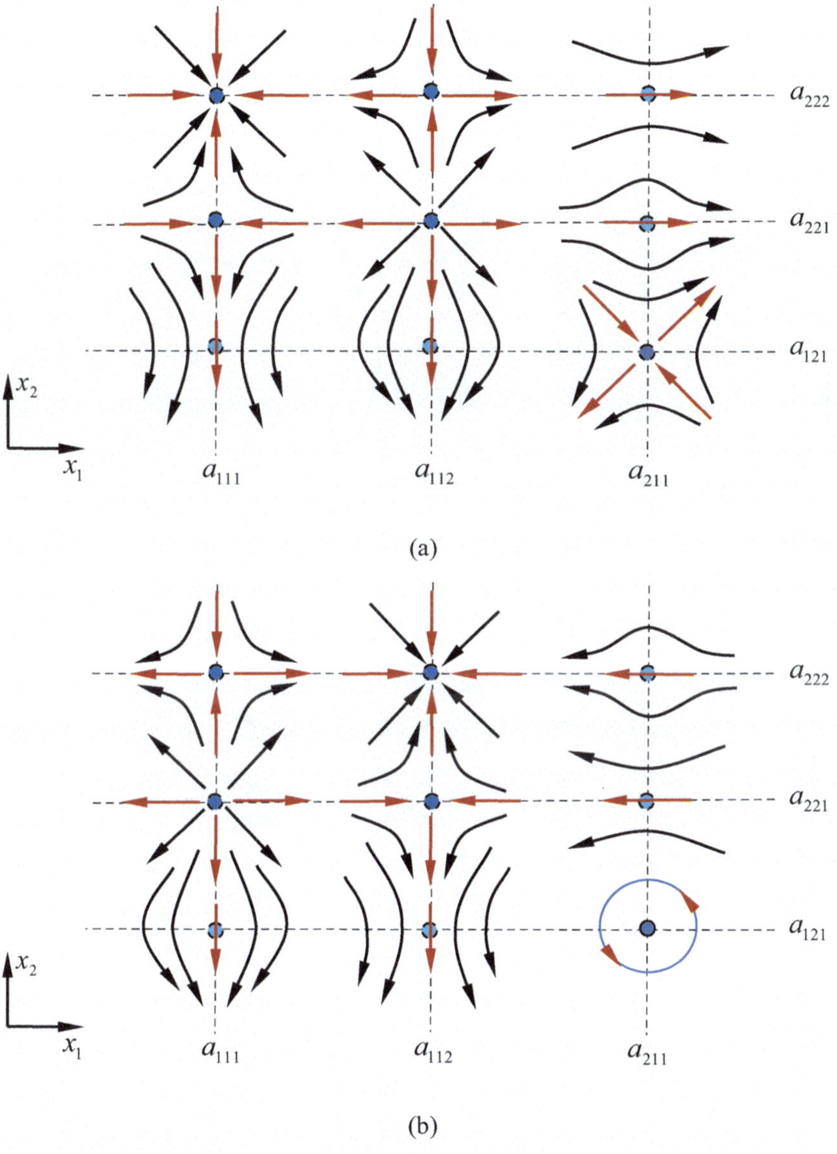

Fig. 5.7 Phase portraits ($a_{112} < a_{211}$ and $a_{121} < a_{221}$) for two-dimensional systems on the x_1-direction with $x_1^* = a_{111}, a_{112}, a_{211}$ and on the x_2-direction with $x_2^* = a_{121}, a_{221}, a_{222}$. The simple equilibriums with hyperbolic flows: (**a**) ($a_{110} > 0$, $a_{220} > 0$), (**b**) ($a_{110} < 0$, $a_{220} > 0$), (**c**) ($a_{110} > 0$, $a_{220} < 0$), and (**d**) ($a_{110} < 0$, $a_{220} < 0$)

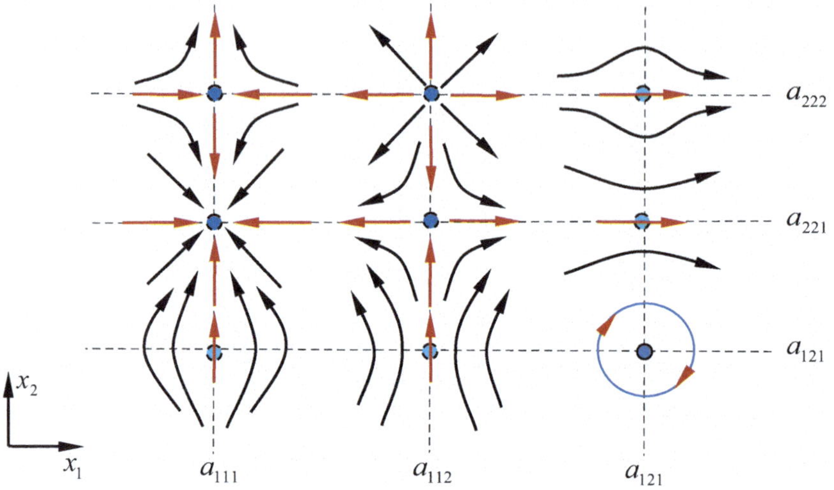

(c)

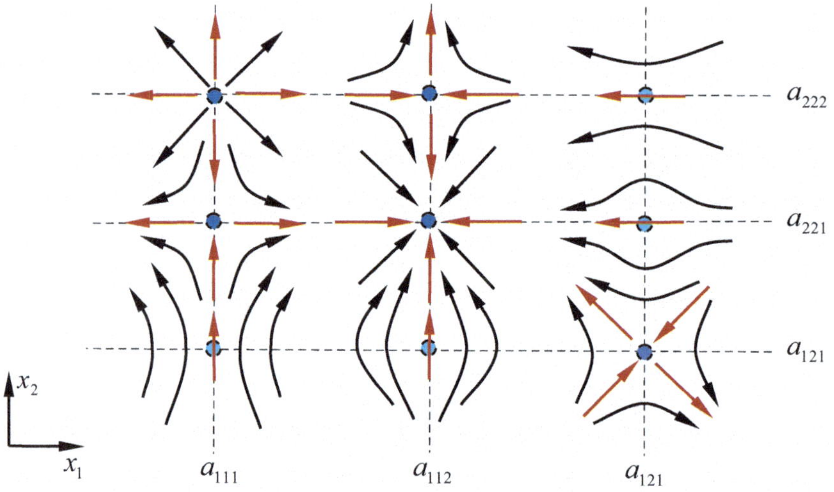

(d)

Fig. 5.7 (continued)

$$
\left\{\begin{array}{ccc}
(a_{111},a_{121}) & (a_{112},a_{121}) & (a_{121},a_{211})\\
(a_{111},a_{222}) & (a_{112},a_{222}) & (a_{211},a_{222})\\
(a_{111},a_{221}) & (a_{112},a_{221}) & (a_{211},a_{221})
\end{array}\right\}
=\left\{\begin{array}{ccc}
\underbrace{(\text{UP:DP,nF})}_{\text{hyperbolic-secant flow }(-)} & \underbrace{(\text{DP:UP,nF})}_{\text{hyperbolic flow }(-)} & \underbrace{(\text{DP}_{+},\text{DP}_{-})}_{\text{CCW center}}\\
\underbrace{(\text{SI,SI})}_{\text{sink}} & \underbrace{(\text{SO,SI})}_{\text{saddle}} & \underbrace{(\text{pF,DP:UP})}_{\text{hyperbolic flow }(+)}\\
\underbrace{(\text{SI,SO})}_{\text{saddle}} & \underbrace{(\text{SO,SO})}_{\text{source}} & \underbrace{(\text{pF,UP:DP})}_{\text{hyperbolic-secant flow }(+)}
\end{array}\right\}
$$

for $a_{110} < 0$ and $a_{220} > 0$; (5.32)

$$
\left\{\begin{array}{ccc}
(a_{111},a_{121}) & (a_{112},a_{121}) & (a_{121},a_{211})\\
(a_{111},a_{222}) & (a_{112},a_{222}) & (a_{211},a_{222})\\
(a_{111},a_{221}) & (a_{112},a_{221}) & (a_{211},a_{221})
\end{array}\right\}
=\left\{\begin{array}{ccc}
\underbrace{(\text{UP:DP,pF})}_{\text{hyperbolic-secant flow }(+)} & \underbrace{(\text{DP:UP,pF})}_{\text{hyperbolic flow }(+)} & \underbrace{(\text{DP}_{-},\text{DP}_{+})}_{\text{CW center}}\\
\underbrace{(\text{SO,SO})}_{\text{source}} & \underbrace{(\text{SI,SO})}_{\text{saddle}} & \underbrace{(\text{nF,DP:UP})}_{\text{hyperbolic flow }(-)}\\
\underbrace{(\text{SO,SI})}_{\text{saddle}} & \underbrace{(\text{SI,SI})}_{\text{sink}} & \underbrace{(\text{nF,UP:DP})}_{\text{hyperbolic-secant flow }(-)}
\end{array}\right\}
$$

for $a_{110} > 0$ and $a_{220} < 0$; (5.33)

$$
\left\{\begin{array}{ccc}
(a_{111},a_{121}) & (a_{112},a_{121}) & (a_{121},a_{211})\\
(a_{111},a_{222}) & (a_{112},a_{222}) & (a_{211},a_{222})\\
(a_{111},a_{221}) & (a_{112},a_{221}) & (a_{211},a_{221})
\end{array}\right\}
=\left\{\begin{array}{ccc}
\underbrace{(\text{DP:UP,pF})}_{\text{hyperbolic flow }(+)} & \underbrace{(\text{UP:DP,pF})}_{\text{hyperbolic-secant flow }(+)} & \underbrace{(\text{UP}_{-},\text{UP}_{-})}_{\text{negaitve saddle}}\\
\underbrace{(\text{SI,SO})}_{\text{saddle}} & \underbrace{(\text{SO,SO})}_{\text{source}} & \underbrace{(\text{pF,UP:DP})}_{\text{hyperbolic-secant flow }(+)}\\
\underbrace{(\text{SI,SI})}_{\text{sink}} & \underbrace{(\text{SO,SI})}_{\text{saddle}} & \underbrace{(\text{pF,DP:UP})}_{\text{hyperbolic flow }(+)}
\end{array}\right\}
$$

for $a_{110} < 0$ and $a_{220} < 0$; (5.34)

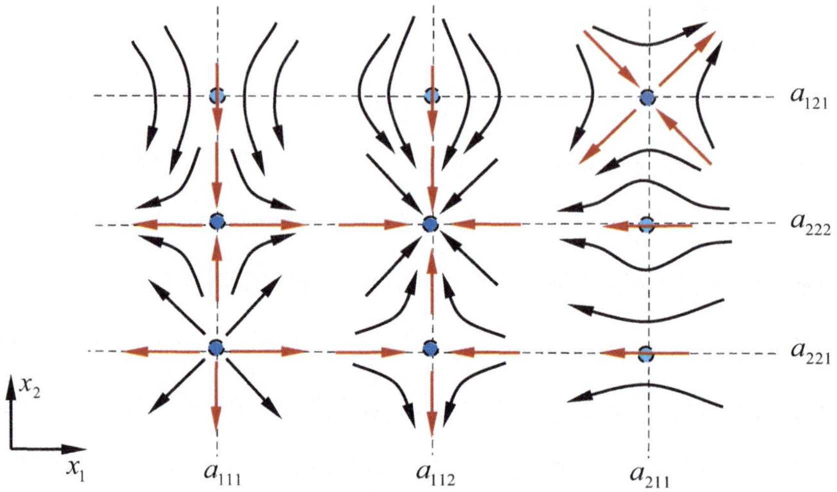

(a)

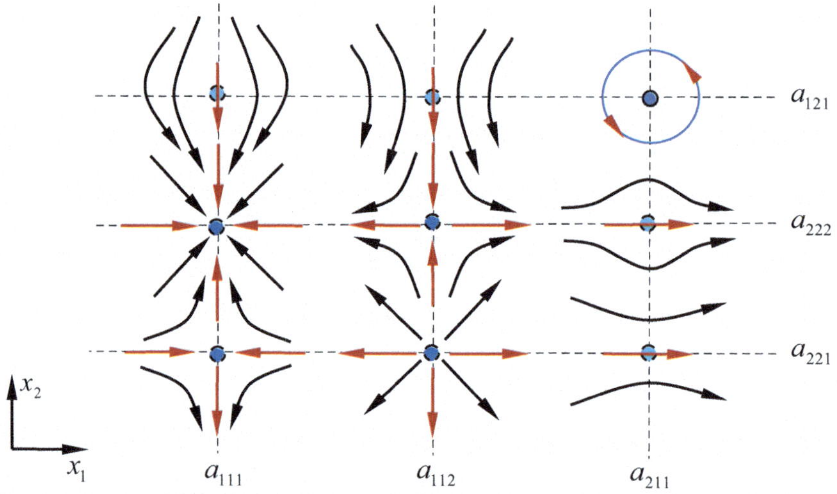

(b)

Fig. 5.8 Phase portraits ($a_{112} < a_{211}$ and $a_{222} < a_{121}$) for two-dimensional systems on the x_1-direction with $x_1^* = a_{111}, a_{112}, a_{211}$ and on the x_2-direction with $x_2^* = a_{221}, a_{222}, a_{121}$. The simple equilibriums with hyperbolic flows: (**a**) ($a_{110} > 0$, $a_{220} > 0$), (**b**) ($a_{110} < 0$, $a_{220} > 0$), (**c**) ($a_{110} > 0$, $a_{220} < 0$), and (**d**) ($a_{110} < 0$, $a_{220} < 0$)

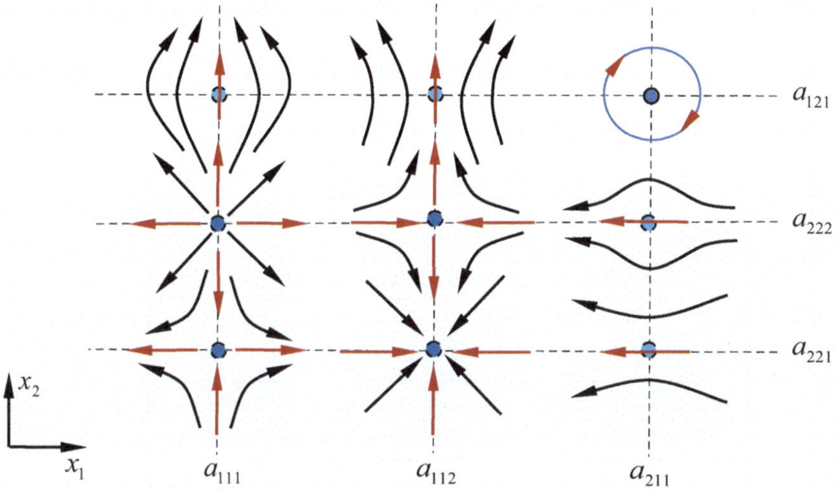

(c)

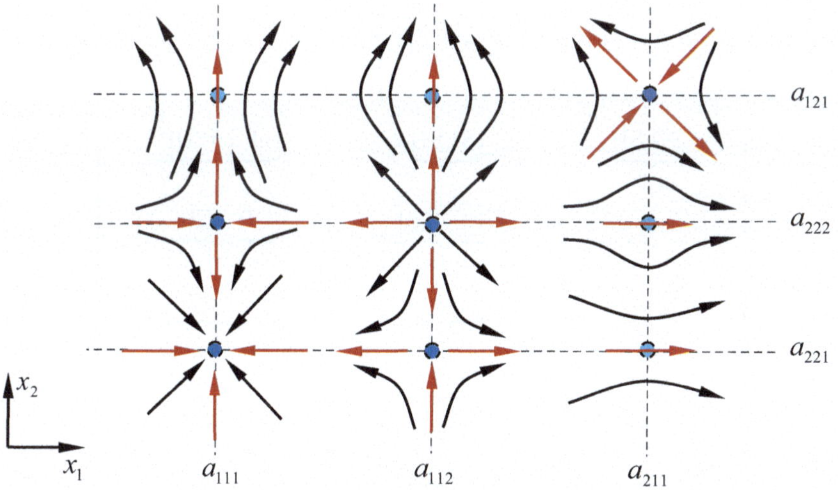

(d)

Fig. 5.8 (continued)

as shown in Fig. 5.8. The directrix flows of $x_1 = a_{111}, a_{112}$ are positive and negative. The networks of saddle, sink, and source with hyperbolic and hyperbolic secant flows are presented. The saddles, sink, and sources in the network of equilibriums are *below* the hyperbolic and hyperbolic secant flows.

(ix) The simple equilibrium networks for $a_{112} < a_{211}$ and $a_{121} \in (a_{221}, a_{222})$ have the following properties:

$$\left\{\begin{matrix} (a_{111}, a_{222}) & (a_{112}, a_{222}) & (a_{211}, a_{222}) \\ (a_{111}, a_{121}) & (a_{112}, a_{121}) & (a_{121}, a_{211}) \\ (a_{111}, a_{221}) & (a_{112}, a_{221}) & (a_{211}, a_{221}) \end{matrix}\right\}$$

$$= \left\{\begin{matrix} \underbrace{(\text{SI,SI})}_{\text{sink}} & \underbrace{(\text{SO,SI})}_{\text{saddle}} & \underbrace{(\text{pF,DP:UP})}_{\text{hyperbolic flow }(+)} \\ \underbrace{(\text{UP:DP,pF})}_{\text{hyperbolic-secant flow }(+)} & \underbrace{(\text{DP:UP,pF})}_{\text{hyperbolic flow }(+)} & \underbrace{(\text{DP}_-,\text{DP}_+)}_{\text{CW center}} \\ \underbrace{(\text{SO,SO})}_{\text{source}} & \underbrace{(\text{SI,SO})}_{\text{saddle}} & \underbrace{(\text{nF,DP:UP})}_{\text{hyperbolic flow }(-)} \end{matrix}\right\}$$

for $a_{110} > 0$ and $a_{220} > 0$; (5.35)

$$\left\{\begin{matrix} (a_{111}, a_{222}) & (a_{112}, a_{222}) & (a_{211}, a_{222}) \\ (a_{111}, a_{121}) & (a_{112}, a_{121}) & (a_{121}, a_{211}) \\ (a_{111}, a_{221}) & (a_{112}, a_{221}) & (a_{211}, a_{221}) \end{matrix}\right\}$$

$$= \left\{\begin{matrix} \underbrace{(\text{SO,SI})}_{\text{saddle}} & \underbrace{(\text{SI,SI})}_{\text{sink}} & \underbrace{(\text{nF,UP:DP})}_{\text{hyperbolic flow }(-)} \\ \underbrace{(\text{DP:UP,pF})}_{\text{hyperbolic flow }(+)} & \underbrace{(\text{UP:DP,pF})}_{\text{hyperbolic-secant flow }(+)} & \underbrace{(\text{UP}_-,\text{UP}_-)}_{\text{negative saddle}} \\ \underbrace{(\text{SI,SO})}_{\text{saddle}} & \underbrace{(\text{SO,SO})}_{\text{source}} & \underbrace{(\text{pF,UP:DP})}_{\text{hyperbolic-secant flow }(+)} \end{matrix}\right\}$$

for $a_{110} < 0$ and $a_{220} > 0$; (5.36)

$$\left\{\begin{matrix} (a_{111}, a_{222}) & (a_{112}, a_{222}) & (a_{211}, a_{222}) \\ (a_{111}, a_{121}) & (a_{112}, a_{121}) & (a_{121}, a_{211}) \\ (a_{111}, a_{221}) & (a_{112}, a_{221}) & (a_{211}, a_{221}) \end{matrix}\right\}$$

$$= \left\{\begin{matrix} \underbrace{(\text{SI,SO})}_{\text{saddle}} & \underbrace{(\text{SO,SO})}_{\text{source}} & \underbrace{(\text{pF,UP:DP})}_{\text{hyperbolic flow }(+)} \\ \underbrace{(\text{DP:UP,nF})}_{\text{hyperbolic flow }(-)} & \underbrace{(\text{UP:DP,nF})}_{\text{hyperbolic-secant flow }(-)} & \underbrace{(\text{UP}_+,\text{UP}_+)}_{\text{positive saddle}} \\ \underbrace{(\text{SO,SI})}_{\text{saddle}} & \underbrace{(\text{SI,SI})}_{\text{sink}} & \underbrace{(\text{nF,UP:DP})}_{\text{hyperbolic-secant flow }(-)} \end{matrix}\right\}$$

for $a_{110} > 0$ and $a_{220} < 0$; (5.37)

$$\begin{Bmatrix} (a_{111}, a_{222}) & (a_{112}, a_{222}) & (a_{211}, a_{222}) \\ (a_{111}, a_{121}) & (a_{112}, a_{121}) & (a_{121}, a_{211}) \\ (a_{111}, a_{221}) & (a_{112}, a_{221}) & (a_{211}, a_{221}) \end{Bmatrix}$$

$$= \begin{Bmatrix} \underbrace{(\mathrm{SO,SO})}_{\text{source}} & \underbrace{(\mathrm{SI,SO})}_{\text{saddle}} & \underbrace{(\mathrm{nF,DP:UP})}_{\text{hyperbolic flow }(-)} \\ \underbrace{(\mathrm{UP:DP,nF})}_{\text{hyperbolic-secant flow }(-)} & \underbrace{(\mathrm{DP:UP,nF})}_{\text{hyperbolic flow }(-)} & \underbrace{(\mathrm{DP_+,DP_-})}_{\text{CCW center}} \\ \underbrace{(\mathrm{SI,SI})}_{\text{sink}} & \underbrace{(\mathrm{SO,SI})}_{\text{saddle}} & \underbrace{(\mathrm{pF,DP:UP})}_{\text{hyperbolic flow }(+)} \end{Bmatrix}$$

for $a_{110} < 0$ and $a_{220} < 0$; (5.38)

as shown in Fig. 5.9. The directrix flows of $x_1 = a_{111}, a_{112}$ are positive and negative. The networks of saddle, sink, and source with hyperbolic and hyperbolic-secant flows are presented. The saddles, sink, and sources in the network of equilibriums are *separated by* the hyperbolic and hyperbolic-secant flows.

5.2 Infinite-Equilibriums and Switching Bifurcations

In this section, two types of inflection-sink and source infinite-equilibriums and one type of double inflection-sink and source infinite-equilibriums are discussed.

5.2.1 Horizontal Inflection-Sink and Source Infinite-Equilibriums

Consider a dynamical system for $a_{121} = a_{22l_1}$ $(l_1, l_2 \in \{1,2\}, l_1 \neq l_2)$ as

$$\begin{aligned} \dot{x}_1 &= a_{110}(x_1 - a_{111})(x_1 - a_{112})(x_2 - a_{121}), \\ \dot{x}_2 &= a_{220}(x_1 - a_{211})(x_2 - a_{221})(x_2 - a_{222}), \end{aligned} \tag{5.39}$$

and the corresponding first integral manifold is

$$\sum_{s_1=1, s_2 \neq s_1}^{2} \frac{a_{11s_1} - a_{211}}{a_{11s_1} - a_{11s_2}} \ln \frac{|x_1 - a_{11s_1}|}{|x_{10} - a_{11s_1}|} = \frac{a_{110}}{a_{220}} \ln \frac{|x_2 - a_{22l_2}|}{|x_{20} - a_{22l_2}|}. \tag{5.40}$$

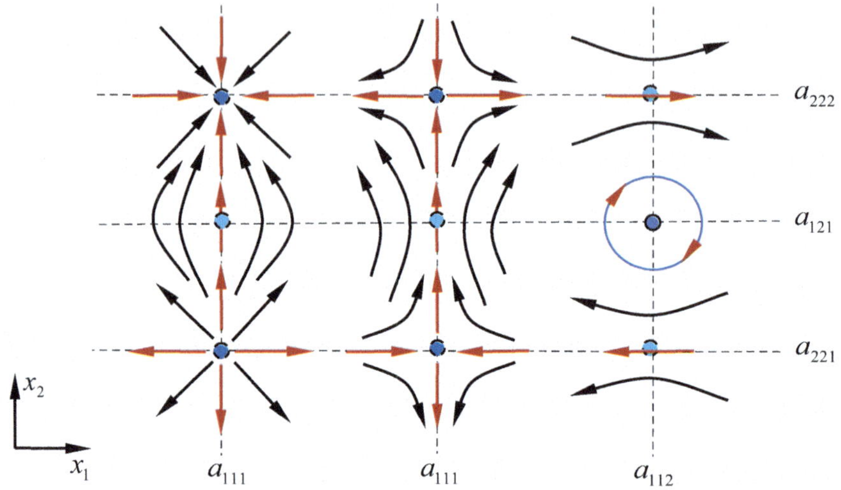

(a)

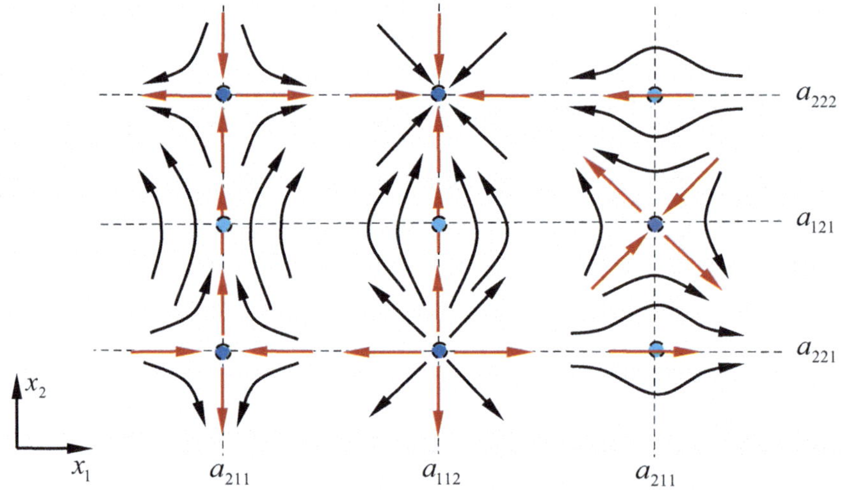

(b)

Fig. 5.9 Phase portraits ($a_{112} < a_{211}$ and $a_{121} \in (a_{221}, a_{222})$) for two-dimensional systems on the x_1-direction with $x_1^* = a_{111}, a_{112}, a_{211}$ and on the x_2-direction with $x_2^* = a_{221}, a_{121}, a_{222}$. The simple equilibriums with hyperbolic flows: (**a**) ($a_{110} > 0$, $a_{220} > 0$), (**b**) ($a_{110} < 0$, $a_{220} > 0$), (**c**) ($a_{110} > 0$, $a_{220} < 0$), and (**d**) ($a_{110} < 0$, $a_{220} < 0$)

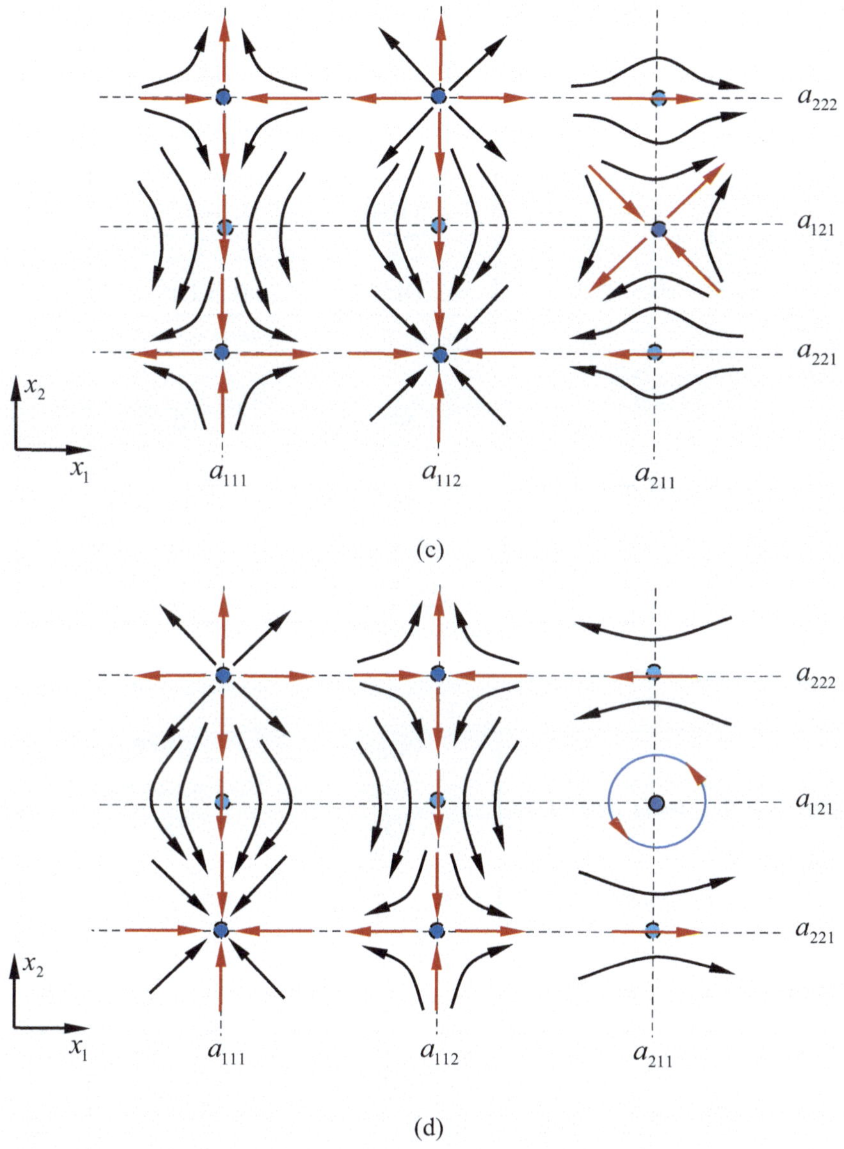

Fig. 5.9 (continued)

(i) The infinite-equilibriums of $x_2^* = a_{121} = a_{222}$ with $a_{211} < a_{111}$ have the following properties:

$$\left\{\begin{matrix} (a_{112}, a_{221}) & (a_{112}, a_{222}) \\ (a_{111}, a_{221}) & (a_{111}, a_{222}) \\ (a_{211}, a_{221}) & (a_{121}, a_{211}) \end{matrix}\right\} = \left\{\begin{matrix} \underbrace{(\text{SI,SI})}_{\text{sink}} & \underbrace{(\text{DI:II,SO})}_{\text{inflection source}} \\ \underbrace{(\text{SO,SI})}_{\text{saddle}} & \underbrace{(\text{II:DI,SO})}_{\text{inflection source}} \\ \underbrace{(\text{nF,DP:UP})}_{\text{hyperbolic flow }(-)} & \underbrace{(\text{US,UP})}_{\text{up-parabola upper-saddle}} \end{matrix}\right\} \tag{5.41}$$

for $a_{110} > 0, a_{220} > 0$,

$$\left\{\begin{matrix} (a_{112}, a_{221}) & (a_{112}, a_{222}) \\ (a_{111}, a_{221}) & (a_{111}, a_{222}) \\ (a_{211}, a_{221}) & (a_{121}, a_{211}) \end{matrix}\right\} = \left\{\begin{matrix} \underbrace{(\text{SO,SI})}_{\text{saddle}} & \underbrace{(\text{II:DI,SO})}_{\text{inflection source}} \\ \underbrace{(\text{SI,SI})}_{\text{sink}} & \underbrace{(\text{DI:II,SO})}_{\text{inflection source}} \\ \underbrace{(\text{pF,UP:DP})}_{\text{hyperbolic-secant flow }(+)} & \underbrace{(\text{US,DP})}_{\text{down-parabola upper-saddle}} \end{matrix}\right\}$$

for $a_{110} < 0, a_{220} > 0$,

(5.42)

$$\left\{\begin{matrix} (a_{112}, a_{221}) & (a_{112}, a_{222}) \\ (a_{111}, a_{221}) & (a_{111}, a_{222}) \\ (a_{211}, a_{221}) & (a_{121}, a_{211}) \end{matrix}\right\} = \left\{\begin{matrix} \underbrace{(\text{SI,SO})}_{\text{ssaddle}} & \underbrace{(\text{II:DI,SI})}_{\text{inflection sink}} \\ \underbrace{(\text{SO,SO})}_{\text{source}} & \underbrace{(\text{DI:II,SI})}_{\text{inflection sink}} \\ \underbrace{(\text{nF,UP:DP})}_{\text{hyperbolic-secant flow }(-)} & \underbrace{(\text{LS,DP})}_{\text{down-parabola lower-saddle}} \end{matrix}\right\}$$

for $a_{110} > 0, a_{220} < 0$,

(5.43)

$$\left\{\begin{matrix} (a_{112}, a_{221}) & (a_{112}, a_{222}) \\ (a_{111}, a_{221}) & (a_{111}, a_{222}) \\ (a_{211}, a_{221}) & (a_{121}, a_{211}) \end{matrix}\right\} = \left\{\begin{matrix} \underbrace{(\text{SO,SO})}_{\text{source}} & \underbrace{(\text{DI:II,SI})}_{\text{inflection sink}} \\ \underbrace{(\text{SI,SO})}_{\text{saddle}} & \underbrace{(\text{II:DI,SI})}_{\text{inflection sink}} \\ \underbrace{(\text{pF,DP:UP})}_{\text{hyperbolic flow }(+)} & \underbrace{(\text{LS,UP})}_{\text{up-parabola lower-saddle}} \end{matrix}\right\} \tag{5.44}$$

for $a_{110} < 0, a_{220} < 0$.

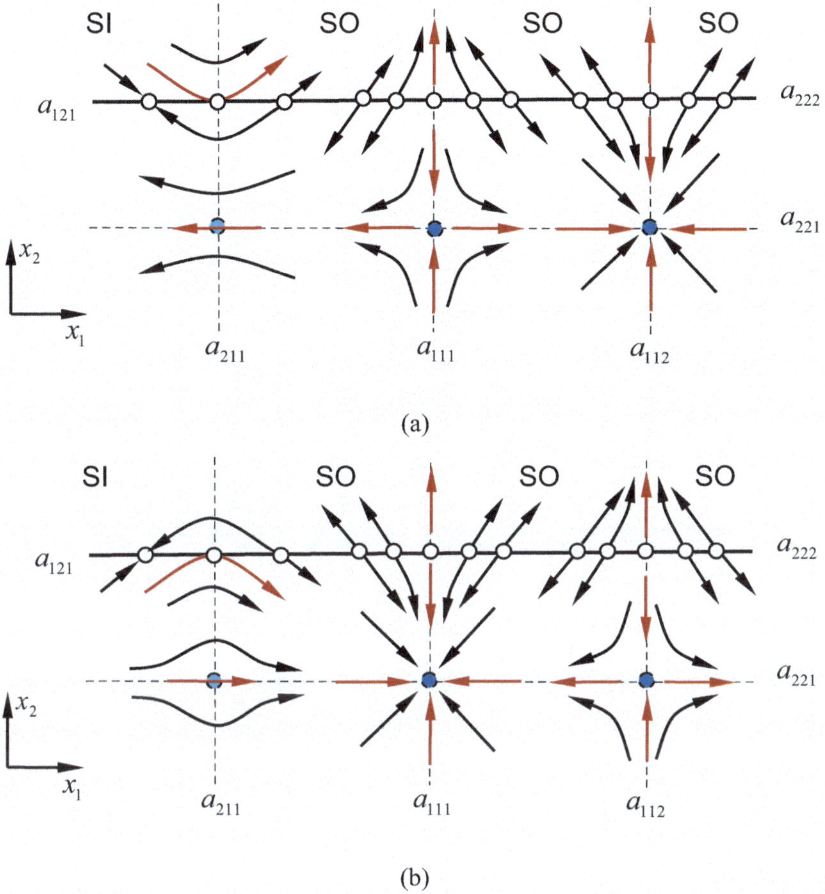

Fig. 5.10 Phase portraits ($a_{211} < a_{111}$ and $a_{121} = a_{222}$) for two-dimensional systems on the x_1-direction with $x_1^* = a_{211}, a_{111}, a_{112}$ and on the x_2-direction with $x_2^* = a_{221}, a_{121}, a_{222}$. The inflection sink and source infinite equilibriums with sink, source, and saddles plus a hyperbolic flow: (**a**) ($a_{110} > 0, a_{220} > 0$), (**b**) ($a_{110} < 0, a_{220} > 0$), (**c**) ($a_{110} > 0, a_{220} < 0$), and (**d**) ($a_{110} < 0, a_{220} < 0$)

Such infinite-equilibriums of $x_2^* = a_{121} = a_{222}$ with $a_{211} < a_{111}$ for such a dynamical system are presented in Fig. 5.10. The inflection-sink and source infinite-equilibriums are listed in Table 5.1. The infinite-equilibriums are the switching bifurcation for $a_{222} < a_{121}$ and $a_{121} \in (a_{221}, a_{222})$ with $a_{211} < a_{111}$.

(ii) The infinite-equilibriums of $x_2^* = a_{121} = a_{222}$ with $a_{211} > a_{112}$ have the following properties:

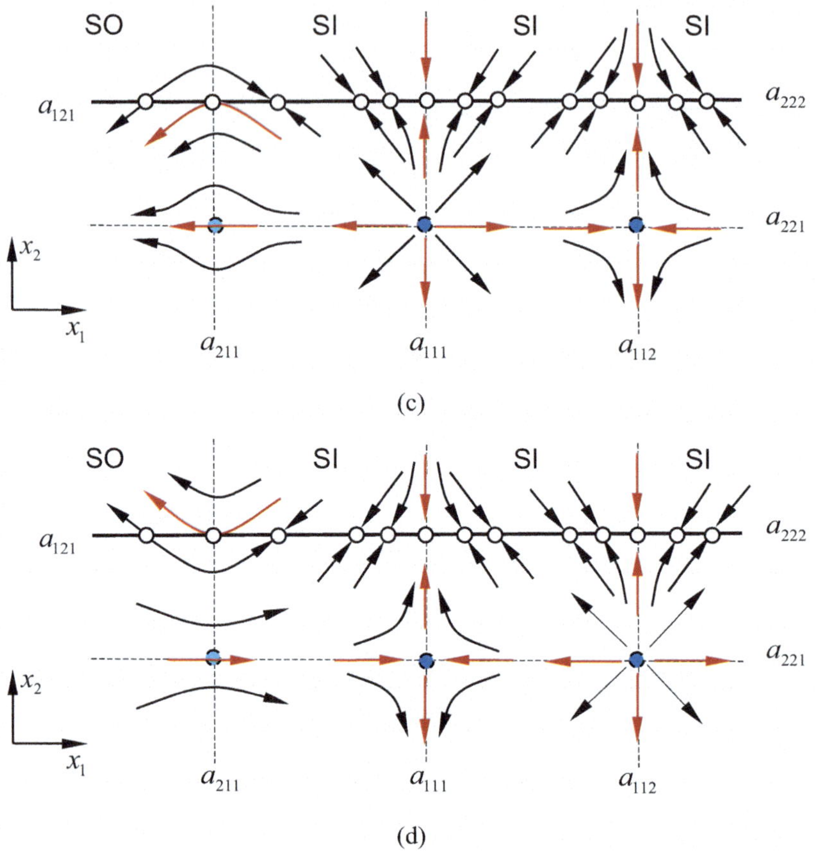

Fig. 5.10 (continued)

Table 5.1 Inflection-sink and source infinite-equilibriums of $x_2^* = a_{121} = a_{222}$ ($a_{211} < a_{111}$ and $a_{121} = a_{222}$)

	$(\bar{x}_1, a_{222}) \equiv (\bar{x}_1, a_{121})$			
	$\bar{x}_1$			
(a_{110}, a_{220})	$(-\infty, a_{211})$	(a_{211}, a_{111})	(a_{111}, a_{112})	(a_{112}, ∞)
$(+,+)$	(DI,SI)	(II,SO)	(DI,SO)	(II,SO)
$(-,+)$	(II,SI)	(DI,SO)	(II,SO)	(DI,SO)
$(+,-)$	(II,SO)	(DI,SI)	(II,SI)	(DI,SI)
$(-,-)$	(DI,SO)	(II,SI)	(DI,SI)	(II,SI)

$$\left\{\begin{matrix} (a_{211},a_{221}) & (a_{121},a_{211}) \\ (a_{112},a_{221}) & (a_{112},a_{222}) \\ (a_{111},a_{221}) & (a_{111},a_{222}) \end{matrix}\right\} = \left\{\begin{matrix} \underbrace{(\text{nF,DP:UP})}_{\text{hyperbolic flow }(-)} & \underbrace{(\text{US,UP})}_{\text{up-parabola upper-saddle}} \\ \underbrace{(\text{SI,SO})}_{\text{saddle}} & \underbrace{(\text{II:DI,SI})}_{\text{inflection sink}} \\ \underbrace{(\text{SO,SO})}_{\text{source}} & \underbrace{(\text{DI:II,SI})}_{\text{inflection sink}} \end{matrix}\right\} \tag{5.45}$$

for $a_{110} > 0, a_{220} > 0$,

$$\left\{\begin{matrix} (a_{211},a_{221}) & (a_{121},a_{211}) \\ (a_{112},a_{221}) & (a_{112},a_{222}) \\ (a_{111},a_{221}) & (a_{111},a_{222}) \end{matrix}\right\} = \left\{\begin{matrix} \underbrace{(\text{pF,UP:DP})}_{\text{hyperbolic-secant flow }(+)} & \underbrace{(\text{US,DP})}_{\text{down-parabola upper-saddle}} \\ \underbrace{(\text{SO,SO})}_{\text{source}} & \underbrace{(\text{DI:II,SI})}_{\text{inflection sink}} \\ \underbrace{(\text{SI,SO})}_{\text{saddle}} & \underbrace{(\text{II:DI,SI})}_{\text{inflection sink}} \end{matrix}\right\}$$

for $a_{110} < 0, a_{220} > 0$,

(5.46)

$$\left\{\begin{matrix} (a_{211},a_{221}) & (a_{121},a_{211}) \\ (a_{112},a_{221}) & (a_{112},a_{222}) \\ (a_{111},a_{221}) & (a_{111},a_{222}) \end{matrix}\right\} = \left\{\begin{matrix} \underbrace{(\text{nF,UP:DP})}_{\text{hyperbolic-secant flow }(-)} & \underbrace{(\text{LS,DP})}_{\text{down-parabola lower-saddle}} \\ \underbrace{(\text{SI,SI})}_{\text{sink}} & \underbrace{(\text{DI:II,SO})}_{\text{inflection source}} \\ \underbrace{(\text{SO,SI})}_{\text{saddle}} & \underbrace{(\text{II:DI,SO})}_{\text{inflection source}} \end{matrix}\right\}$$

for $a_{110} > 0, a_{220} < 0$,

(5.47)

$$\left\{\begin{matrix} (a_{211},a_{221}) & (a_{121},a_{211}) \\ (a_{112},a_{221}) & (a_{112},a_{222}) \\ (a_{111},a_{221}) & (a_{111},a_{222}) \end{matrix}\right\} = \left\{\begin{matrix} \underbrace{(\text{pF,DP:UP})}_{\text{hyperbolic flow }(+)} & \underbrace{(\text{LS,UP})}_{\text{up-parabola lower-saddle}} \\ \underbrace{(\text{SO,SI})}_{\text{saddle}} & \underbrace{(\text{II:DI,SO})}_{\text{inflection source}} \\ \underbrace{(\text{SI,SI})}_{\text{sink}} & \underbrace{(\text{DI:II,SO})}_{\text{inflection source}} \end{matrix}\right\} \tag{5.48}$$

for $a_{110} < 0, a_{220} < 0$.

Such infinite-equilibriums of $x_2^* = a_{121} = a_{222}$ with $a_{211} < a_{111}$ for such a dynamical system are presented in Fig. 5.11. The inflection-sink and source infinite-equilibriums are listed in Table 5.2. The infinite-equilibriums are the switching bifurcation for $a_{222} < a_{121}$ and $a_{121} \in (a_{221}, a_{222})$ with $a_{211} > a_{112}$.

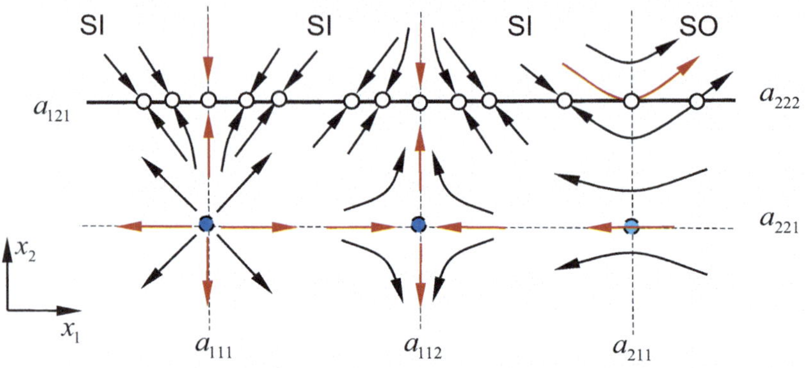

(a)

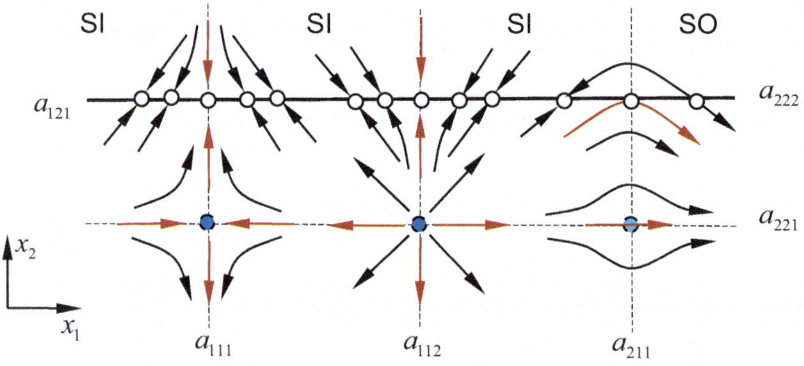

(b)

Fig. 5.11 Phase portraits ($a_{112} < a_{211}$ and $a_{121} = a_{222}$) for two-dimensional systems on the x_1-direction with $x_1^* = a_{111}, a_{112}, a_{211}$ and on the x_2-direction with $x_2^* = a_{221}, a_{121}, a_{222}$. The inflection-sink and source infinite-equilibriums with sink, source, and saddles plus a hyperbolic flow: (**a**) ($a_{110} > 0, a_{220} > 0$), (**b**) ($a_{110} < 0, a_{220} > 0$), (**c**) ($a_{110} > 0, a_{220} < 0$), and (**d**) ($a_{110} < 0, a_{220} < 0$)

(iii) The infinite-equilibriums of $x_2^* = a_{121} = a_{222}$ with $a_{211} \in (a_{111}, a_{112})$ have the following properties:

$$\left\{\begin{matrix} (a_{112}, a_{221}) & (a_{112}, a_{222}) \\ (a_{211}, a_{221}) & (a_{121}, a_{211}) \\ (a_{111}, a_{221}) & (a_{111}, a_{222}) \end{matrix}\right\} = \left\{\begin{matrix} \underbrace{(\text{SI},\text{SI})}_{\text{sink}} & \underbrace{(\text{DI:II},\text{SO})}_{\text{inflection source}} \\ \underbrace{(\text{pF},\text{UP:DP})}_{\text{hyperbolic-secant flow }(+)} & \underbrace{(\text{US},\text{DP})}_{\text{down-parabola upper-saddle}} \\ \underbrace{(\text{SO},\text{SO})}_{\text{source}} & \underbrace{(\text{DI:II},\text{SI})}_{\text{inflection sink}} \end{matrix}\right\}$$

for $a_{110} > 0, a_{220} > 0$,

(5.49)

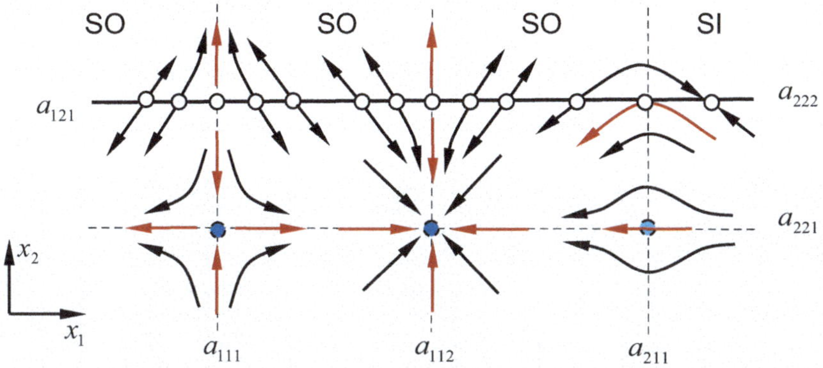

(c)

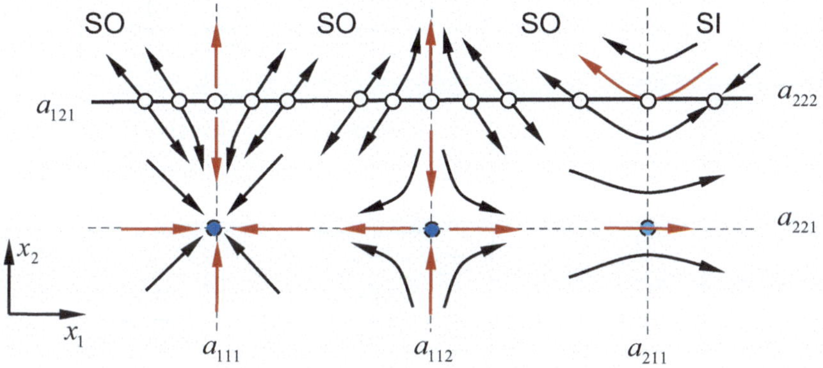

(d)

Fig. 5.11 (continued)

Table 5.2 Inflection-sink and source infinite-equilibriums of $x_2^* = a_{121} = a_{222}$ ($a_{211} > a_{112}$ and $a_{121} = a_{222}$)

	$(\bar{x}_1, a_{222}) \equiv (\bar{x}_1, a_{121})$			
	$\bar{x}_1$			
(a_{110}, a_{220})	$(-\infty, a_{211})$	(a_{211}, a_{111})	(a_{111}, a_{112})	(a_{112}, ∞)
$(+,+)$	(DI,SI)	(II,SI)	(DI,SI)	(II,SO)
$(-,+)$	(II,SI)	(DI,SI)	(II,SI)	(DI,SO)
$(+,-)$	(II,SO)	(DI,SO)	(II,SO)	(DI,SI)
$(-,-)$	(DI,SO)	(II,SO)	(DI,SO)	(II,SI)

$$\left\{\begin{matrix} (a_{112},a_{221}) & (a_{112},a_{222}) \\ (a_{211},a_{221}) & (a_{121},a_{211}) \\ (a_{111},a_{221}) & (a_{111},a_{222}) \end{matrix}\right\} = \left\{\begin{matrix} \underbrace{(\text{SO,SI})}_{\text{saddle}} & \underbrace{(\text{II:DI,SO})}_{\text{inflection source}} \\ \underbrace{(\text{nF,DP:UP})}_{\text{hyperbolic flow }(-)} & \underbrace{(\text{US,UP})}_{\text{up-parabola upper-saddle}} \\ \underbrace{(\text{SI,SO})}_{\text{saddle}} & \underbrace{(\text{II:DI,SI})}_{\text{inflection sink}} \end{matrix}\right\} \tag{5.50}$$

for $a_{110} < 0, a_{220} > 0$,

$$\left\{\begin{matrix} (a_{112},a_{221}) & (a_{112},a_{222}) \\ (a_{211},a_{221}) & (a_{121},a_{211}) \\ (a_{111},a_{221}) & (a_{111},a_{222}) \end{matrix}\right\} = \left\{\begin{matrix} \underbrace{(\text{SI,SO})}_{\text{saddle}} & \underbrace{(\text{II:DI,SI})}_{\text{inflection sink}} \\ \underbrace{(\text{pF,DP:UP})}_{\text{hyperbolic flow }(+)} & \underbrace{(\text{LS,UP})}_{\text{up-parabola lower-saddle}} \\ \underbrace{(\text{SO,SI})}_{\text{saddle}} & \underbrace{(\text{II:DI,SO})}_{\text{inflection source}} \end{matrix}\right\} \tag{5.51}$$

for $a_{110} > 0, a_{220} < 0$,

$$\left\{\begin{matrix} (a_{112},a_{221}) & (a_{112},a_{222}) \\ (a_{211},a_{221}) & (a_{121},a_{211}) \\ (a_{111},a_{221}) & (a_{111},a_{222}) \end{matrix}\right\} = \left\{\begin{matrix} \underbrace{(\text{SO,SO})}_{\text{source}} & \underbrace{(\text{DI:II,SI})}_{\text{inflection sink}} \\ \underbrace{(\text{nF,UP:DP})}_{\text{hyperbolic-secant flow }(-)} & \underbrace{(\text{LS,DP})}_{\text{down-parabola lower-saddle}} \\ \underbrace{(\text{SI,SI})}_{\text{sink}} & \underbrace{(\text{DI:II,SO})}_{\text{inflection source}} \end{matrix}\right\}$$

for $a_{110} < 0, a_{220} < 0$. (5.52)

Such infinite-equilibriums of $x_2^* = a_{121} = a_{222}$ with $a_{211} < a_{111}$ for such a dynamical system are presented in Fig. 5.12. The inflection-sink and source infinite-equilibriums are listed in Table 5.3. The infinite-equilibriums are the switching bifurcation for $a_{222} < a_{121}$ and $a_{121} \in (a_{221}, a_{222})$ with $a_{211} \in (a_{111}, a_{112})$.

(iv) The infinite-equilibriums of $x_2^* = a_{121} = a_{221}$ with $a_{211} < a_{111}$ have the following properties:

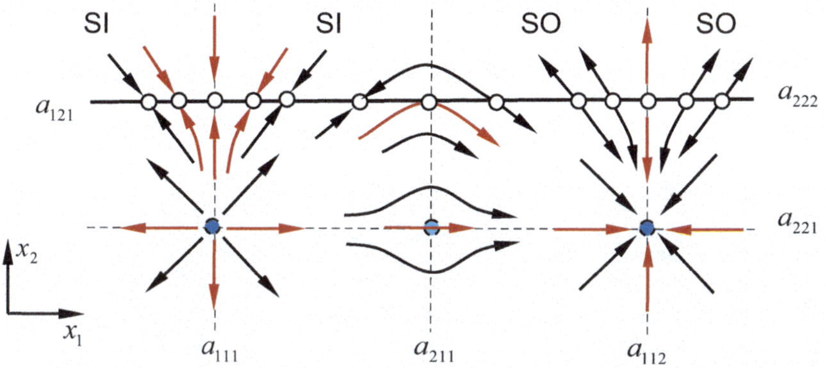

(a)

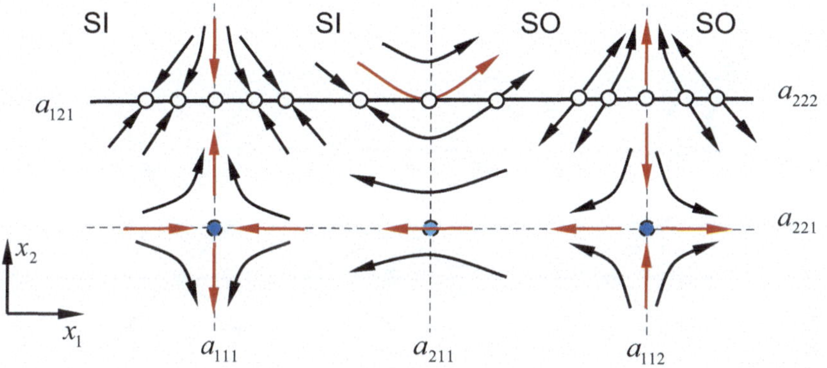

(b)

Fig. 5.12 Phase portraits ($a_{211} \in (a_{111}, a_{112})$ and $a_{121} = a_{222}$) for two-dimensional systems on the x_1-direction with $x_1^* = a_{111}, a_{211}, a_{112}$ and on the x_2-direction with $x_2^* = a_{221}, a_{121}, a_{222}$. The inflection-sink and source infinite-equilibriums with sink, source, and saddles plus a hyperbolic flow: (**a**) ($a_{110} > 0$, $a_{220} > 0$), (**b**) ($a_{110} < 0$, $a_{220} > 0$), (**c**) ($a_{110} > 0$, $a_{220} < 0$), and (**d**) ($a_{110} < 0$, $a_{220} < 0$)

$$\left\{\begin{matrix} (a_{112}, a_{221}) & (a_{112}, a_{222}) \\ (a_{111}, a_{221}) & (a_{111}, a_{222}) \\ (a_{121}, a_{211}) & (a_{211}, a_{222}) \end{matrix}\right\} = \left\{\begin{matrix} \underbrace{(\text{II:DI,SI})}_{\text{inflection sink}} & \underbrace{(\text{SO,SO})}_{\text{source}} \\ \underbrace{(\text{DI:II,SI})}_{\text{inflection sink}} & \underbrace{(\text{SI,SO})}_{\text{saddle}} \\ \underbrace{(\text{LS,DP})}_{\text{down-parabola lower-saddle}} & \underbrace{(\text{pF,DP:UP})}_{\text{hyperbolic flow }(+)} \end{matrix}\right\} \tag{5.53}$$

for $a_{110} > 0, a_{220} > 0$,

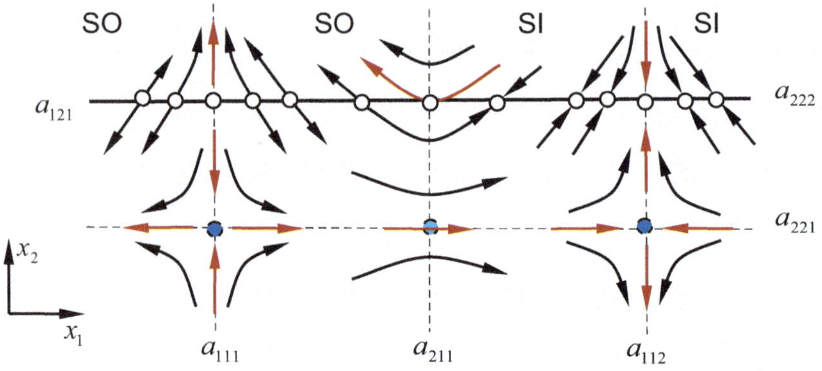

(c)

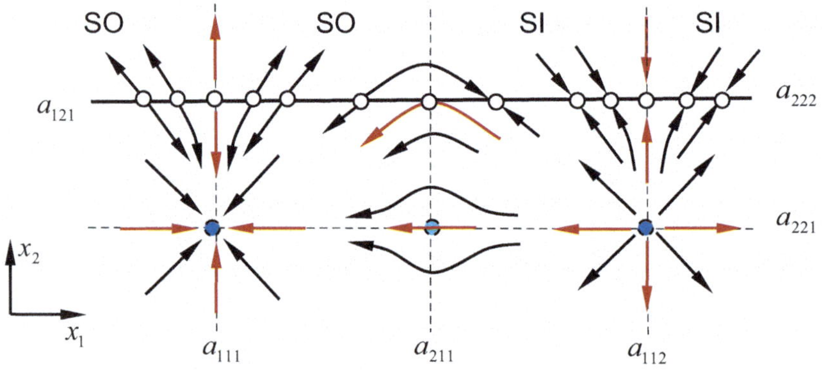

(d)

Fig. 5.12 (continued)

Table 5.3 Inflection-sink and source infinite-equilibriums of $x_2^* = a_{121} = a_{222}$ ($a_{211} \in (a_{111}, a_{112})$ and $a_{121} = a_{222}$)

	$(\bar{x}_1, a_{222}) \equiv (\bar{x}_1, a_{121})$			
	$\bar{x}_1$			
(a_{110}, a_{220})	$(-\infty, a_{211})$	(a_{211}, a_{111})	(a_{111}, a_{112})	(a_{112}, ∞)
$(+,+)$	(DI,SI)	(II,SI)	(DI,SO)	(II,SO)
$(-,+)$	(II,SI)	(DI,SI)	(II,SO)	(DI,SO)
$(+,-)$	(II,SO)	(DI,SO)	(II,SI)	(DI,SI)
$(-,-)$	(DI,SO)	(II,SO)	(DI,SI)	(II,SI)

$$\left\{\begin{array}{cc}(a_{112},a_{221}) & (a_{112},a_{222})\\ (a_{111},a_{221}) & (a_{111},a_{222})\\ (a_{121},a_{211}) & (a_{211},a_{222})\end{array}\right\}=\left\{\begin{array}{cc}\underbrace{(\text{DI:II,SI})}_{\text{inflection sink}} & \underbrace{(\text{SI,SO})}_{\text{saddle}}\\ \underbrace{(\text{II:DI,SI})}_{\text{inflection sink}} & \underbrace{(\text{SO,SO})}_{\text{source}}\\ \underbrace{(\text{LS,UP})}_{\text{up-parabola lower-saddle}} & \underbrace{(\text{nF,DP:UP})}_{\text{hyperbolic-secant flow }(-)}\end{array}\right\}$$

for $a_{110} < 0, a_{220} > 0$,

(5.54)

$$\left\{\begin{array}{cc}(a_{112},a_{221}) & (a_{112},a_{222})\\ (a_{111},a_{221}) & (a_{111},a_{222})\\ (a_{121},a_{211}) & (a_{211},a_{222})\end{array}\right\}=\left\{\begin{array}{cc}\underbrace{(\text{DI:II,SO})}_{\text{inflection source}} & \underbrace{(\text{SO,SI})}_{\text{saddle}}\\ \underbrace{(\text{II:DI,SO})}_{\text{inflection source}} & \underbrace{(\text{SI,SI})}_{\text{sink}}\\ \underbrace{(\text{US,UP})}_{\text{up-parabola upper-saddle}} & \underbrace{(\text{pF,UP:DP})}_{\text{hyperbolic-secant flow }(+)}\end{array}\right\}$$

for $a_{110} > 0, a_{220} < 0$,

(5.55)

$$\left\{\begin{array}{cc}(a_{112},a_{221}) & (a_{112},a_{222})\\ (a_{111},a_{221}) & (a_{111},a_{222})\\ (a_{121},a_{211}) & (a_{211},a_{222})\end{array}\right\}=\left\{\begin{array}{cc}\underbrace{(\text{II:DI,SO})}_{\text{inflection source}} & \underbrace{(\text{SI,SI})}_{\text{sink}}\\ \underbrace{(\text{DI:II,SO})}_{\text{inflection source}} & \underbrace{(\text{SO,SI})}_{\text{saddle}}\\ \underbrace{(\text{US,DP})}_{\text{down-parabola upper-saddle}} & \underbrace{(\text{nF,DP:UP})}_{\text{hyperbolic flow }(-)}\end{array}\right\} \tag{5.56}$$

for $a_{110} < 0, a_{220} < 0$.

Such infinite-equilibriums of $x_2^* = a_{121} = a_{221}$ with $a_{211} < a_{111}$ for such a dynamical system are presented in Fig. 5.13. The inflection-sink and source infinite-equilibriums are listed in Table 5.4. The infinite-equilibriums are the switching bifurcation for $a_{121} < a_{221}$ and $a_{121} \in (a_{221}, a_{222})$ with $a_{211} < a_{111}$.

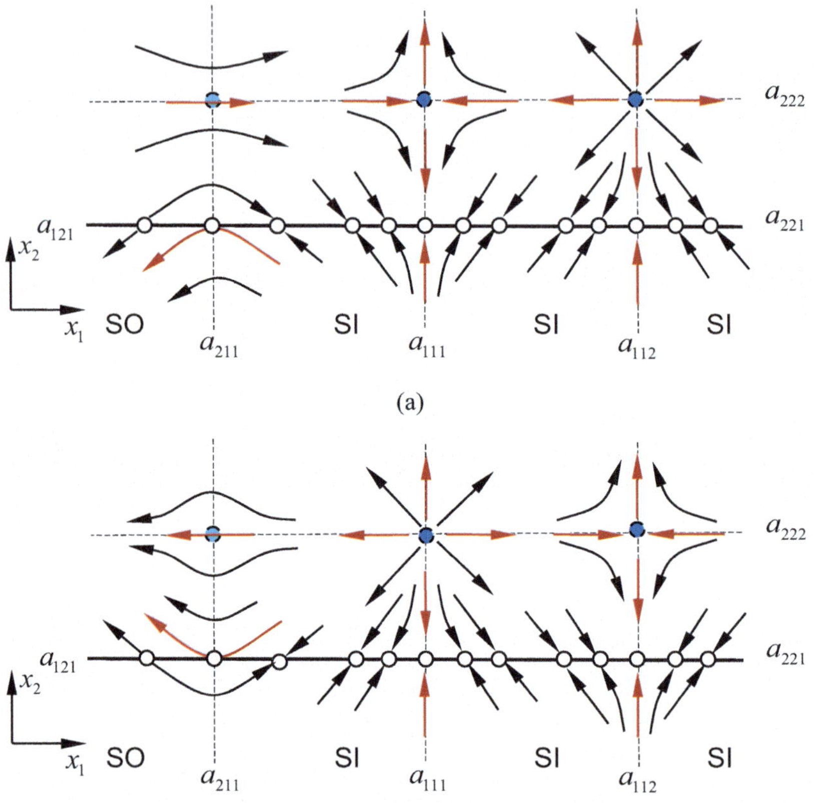

(b)

Fig. 5.13 Phase portraits ($a_{211} < a_{111}$ and $a_{121} = a_{221}$) for two-dimensional systems on the x_1-direction with $x_1^* = a_{211}, a_{111}, a_{112}$ and on the x_2-direction with $x_2^* = a_{121}, a_{221}, a_{222}$. The inflection-sink and source infinite-equilibriums with sink, source, and saddles plus a hyperbolic flow: (**a**) ($a_{110} > 0$, $a_{220} > 0$), (**b**) ($a_{110} < 0$, $a_{220} > 0$), (**c**) ($a_{110} > 0$, $a_{220} < 0$), and (**d**) ($a_{110} < 0$, $a_{220} < 0$)

(v) The infinite equilibriums of $x_2^* = a_{121} = a_{221}$ with $a_{211} > a_{112}$ have the following properties:

$$\left\{\begin{array}{cc} (a_{121}, a_{211}) & (a_{211}, a_{222}) \\ (a_{112}, a_{221}) & (a_{112}, a_{222}) \\ (a_{111}, a_{221}) & (a_{111}, a_{222}) \end{array}\right\} = \left\{\begin{array}{cc} \underbrace{(\text{LS,DP})}_{\text{down-parabola lower-saddle}} & \underbrace{(\text{pF,DP:UP})}_{\text{hyperbolic flow }(+)} \\ \underbrace{(\text{DI:II,SO})}_{\text{inflection source}} & \underbrace{(\text{SO,SI})}_{\text{saddle}} \\ \underbrace{(\text{II:DI,SO})}_{\text{inflection source}} & \underbrace{(\text{SI,SI})}_{\text{sink}} \end{array}\right\} \tag{5.57}$$

for $a_{110} > 0, a_{220} > 0$,

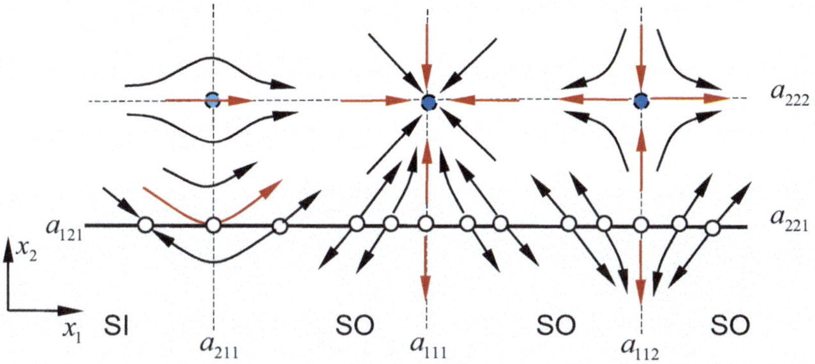

(c)

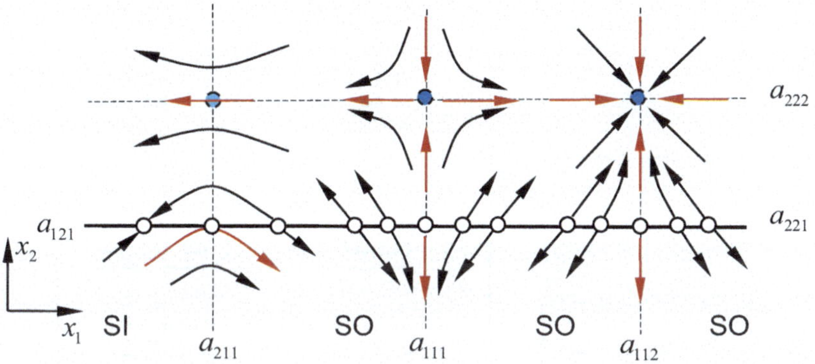

(d)

Fig. 5.13 (continued)

Table 5.4 Inflection-sink and source infinite-equilibriums of $x_2^* = a_{121} = a_{221}$ $(a_{211} < a_{111}$ and $a_{121} = a_{221})$

	$(\bar{x}_1, a_{221}) \equiv (\bar{x}_1, a_{121})$			
	$\bar{x}_1$			
(a_{110}, a_{220})	$(-\infty, a_{211})$	(a_{211}, a_{111})	(a_{111}, a_{112})	(a_{112}, ∞)
$(+,+)$	(II,SO)	(DI,SI)	(II,SI)	(DI,SI)
$(-,+)$	(DI,SO)	(II,SI)	(DI,SI)	(II,SI)
$(+,-)$	(DI,SI)	(II,SO)	(DI,SO)	(II,SO)
$(-,-)$	(II,SI)	(DI,SO)	(II,SO)	(DI,SO)

$$\left\{\begin{array}{ll}(a_{121},a_{211}) & (a_{211},a_{222})\\(a_{112},a_{221}) & (a_{112},a_{222})\\(a_{111},a_{221}) & (a_{111},a_{222})\end{array}\right\}=\left\{\begin{array}{cc}\underbrace{(\mathrm{LS,UP})}_{\text{up-parabola lower-saddle}} & \underbrace{(\mathrm{nF,UP:DP})}_{\text{hyperbolic-secant flow }(-)}\\\underbrace{(\mathrm{II:DI,SO})}_{\text{inflection source}} & \underbrace{(\mathrm{SI,SI})}_{\text{sink}}\\\underbrace{(\mathrm{DI:II,SO})}_{\text{inflection source}} & \underbrace{(\mathrm{SO,SI})}_{\text{saddle}}\end{array}\right\}$$

$$\text{for } a_{110}<0, a_{220}>0, \tag{5.58}$$

$$\left\{\begin{array}{ll}(a_{121},a_{211}) & (a_{211},a_{222})\\(a_{112},a_{221}) & (a_{112},a_{222})\\(a_{111},a_{221}) & (a_{111},a_{222})\end{array}\right\}=\left\{\begin{array}{cc}\underbrace{(\mathrm{US,UP})}_{\text{up-parabola upper-saddle}} & \underbrace{(\mathrm{pF,UP:DP})}_{\text{hyperbolic-secant flow }(+)}\\\underbrace{(\mathrm{II:DI,SI})}_{\text{inflection sink}} & \underbrace{(\mathrm{SO,SO})}_{\text{sink}}\\\underbrace{(\mathrm{DI:II,SI})}_{\text{inflection sink}} & \underbrace{(\mathrm{SI,SO})}_{\text{saddle}}\end{array}\right\}$$

$$\text{for } a_{110}>0, a_{220}<0, \tag{5.59}$$

$$\left\{\begin{array}{ll}(a_{121},a_{211}) & (a_{211},a_{222})\\(a_{112},a_{221}) & (a_{112},a_{222})\\(a_{111},a_{221}) & (a_{111},a_{222})\end{array}\right\}=\left\{\begin{array}{cc}\underbrace{(\mathrm{US,DP})}_{\text{down-parabola upper-saddle}} & \underbrace{(\mathrm{nF,DP:UP})}_{\text{hyperbolic flow }(-)}\\\underbrace{(\mathrm{DI:II,SI})}_{\text{inflection sink}} & \underbrace{(\mathrm{SI,SO})}_{\text{saddle}}\\\underbrace{(\mathrm{II:DI,SI})}_{\text{inflection sink}} & \underbrace{(\mathrm{SO,SO})}_{\text{source}}\end{array}\right\} \tag{5.60}$$

$$\text{for } a_{110}<0, a_{220}<0.$$

Such infinite-equilibriums of $x_2^*=a_{121}=a_{221}$ with $a_{211}>a_{112}$ for such a dynamical system are presented in Fig. 5.14. The inflection-sink and source infinite-equilibriums are listed in Table 5.5. The infinite-equilibriums are the switching bifurcation for $a_{121}<a_{221}$ and $a_{121}\in(a_{221},a_{222})$ with $a_{211}>a_{112}$.

(vi) The infinite-equilibriums of $x_2^*=a_{121}=a_{221}$ with $a_{211}\in(a_{111},a_{112})$ have the following properties:

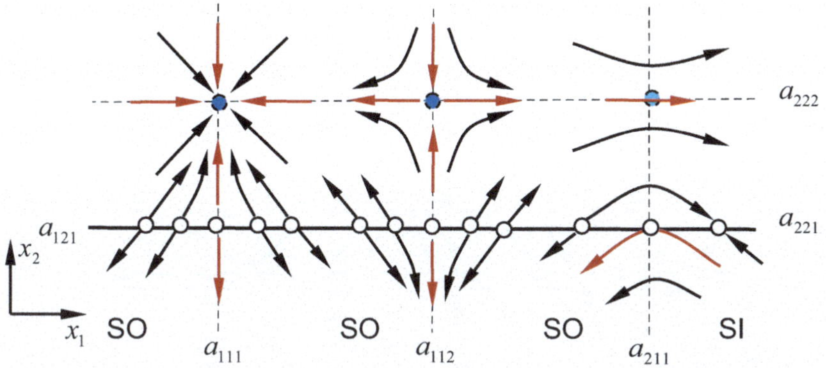

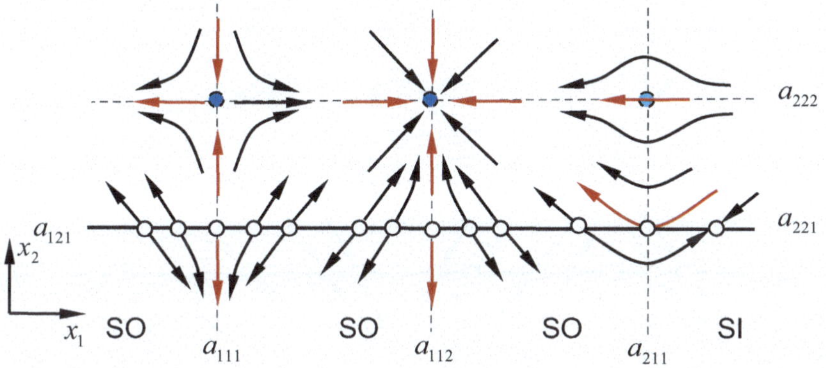

Fig. 5.14 Phase portraits ($a_{112} < a_{211}$ and $a_{121} = a_{221}$) for two-dimensional systems on the x_1-direction with $x_1^* = a_{111}, a_{112}, a_{211}$ and on the x_2-direction with $x_2^* = a_{121}, a_{221}, a_{222}$. The inflection-sink and source infinite-equilibriums with sink, source, and saddles plus a hyperbolic flow: (**a**) ($a_{110} > 0, a_{220} > 0$), (**b**) ($a_{110} < 0, a_{220} > 0$), (**c**) ($a_{110} > 0, a_{220} < 0$), and (**d**) ($a_{110} < 0, a_{220} < 0$)

$$
\left\{ \begin{matrix} (a_{112}, a_{221}) & (a_{112}, a_{222}) \\ (a_{121}, a_{211}) & (a_{211}, a_{222}) \\ (a_{111}, a_{221}) & (a_{111}, a_{222}) \end{matrix} \right\} = \left\{ \begin{matrix} \underbrace{(\text{II:DI,SI})}_{\text{inflection sink}} & \underbrace{(\text{SO,SO})}_{\text{source}} \\ \underbrace{(\text{LS,UP})}_{\text{up-parabola lower-saddle}} & \underbrace{(\text{nF,UP:DP})}_{\text{hyperbolic-secant flow }(-)} \\ \underbrace{(\text{II:DI,SO})}_{\text{inflection source}} & \underbrace{(\text{SI,SI})}_{\text{sink}} \end{matrix} \right\}
$$

for $a_{110} > 0, a_{220} > 0,$

(5.61)

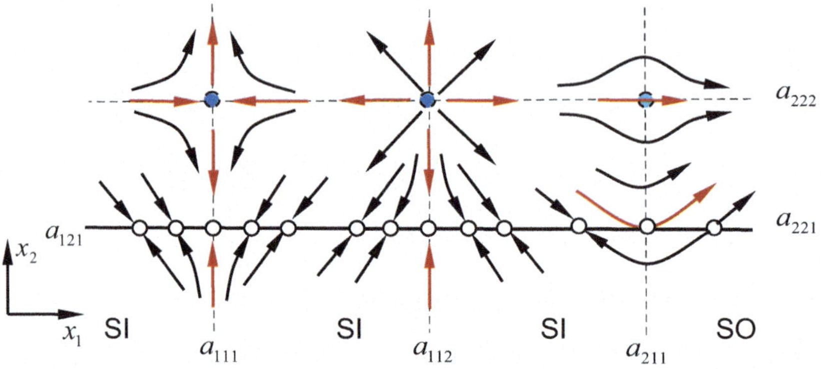

(c)

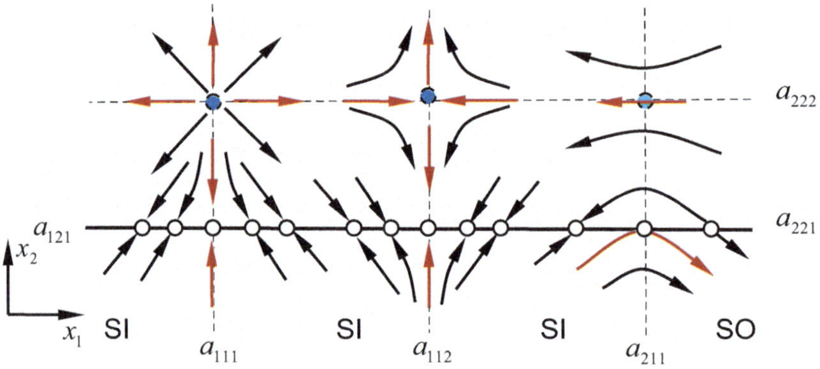

(d)

Fig. 5.14 (continued)

Table 5.5 Inflection-sink and source infinite-equilibriums of $x_2^* = a_{121} = a_{221}$ ($a_{112} < a_{211}$ and $a_{121} = a_{221}$)

	$(\bar{x}_1, a_{221}) \equiv (\bar{x}_1, a_{121})$			
	$\bar{x}_1$			
(a_{110}, a_{220})	$(-\infty, a_{211})$	(a_{211}, a_{111})	(a_{111}, a_{112})	(a_{112}, ∞)
$(+,+)$	(II,SO)	(DI,SO)	(II,SO)	(DI,SI)
$(-,+)$	(DI,SO)	(II,SO)	(DI,SO)	(II,SI)
$(+,-)$	(DI,SI)	(II,SI)	(DI,SI)	(II,SO)
$(-,-)$	(II,SI)	(DI,SI)	(II,SI)	(DI,SO)

$$\left\{\begin{array}{ll}(a_{112}, a_{221}) & (a_{112}, a_{222}) \\ (a_{121}, a_{211}) & (a_{211}, a_{222}) \\ (a_{111}, a_{221}) & (a_{111}, a_{222})\end{array}\right\} = \left\{\begin{array}{cc}\underbrace{\text{(DI:II,SI)}}_{\text{inflection sink}} & \underbrace{\text{(SI,SO)}}_{\text{saddle}} \\ \underbrace{\text{(LS,DP)}}_{\text{parabola lower-saddle}} & \underbrace{\text{(pF,DP:UP)}}_{\text{hyperbolic flow }(+)} \\ \underbrace{\text{(DI:II,SO)}}_{\text{inflection source}} & \underbrace{\text{(SO,SI)}}_{\text{saddle}}\end{array}\right\} \tag{5.62}$$

for $a_{110} < 0, a_{220} > 0$,

$$\left\{\begin{array}{ll}(a_{112}, a_{221}) & (a_{112}, a_{222}) \\ (a_{121}, a_{211}) & (a_{211}, a_{222}) \\ (a_{111}, a_{221}) & (a_{111}, a_{222})\end{array}\right\} = \left\{\begin{array}{cc}\underbrace{\text{(DI:II,SO)}}_{\text{inflection source}} & \underbrace{\text{(SO,SI)}}_{\text{saddle}} \\ \underbrace{\text{(US,DP)}}_{\text{down-parabola upper-saddle}} & \underbrace{\text{(nF,DP:UP)}}_{\text{hyperbolic flow }(-)} \\ \underbrace{\text{(DI:II,SI)}}_{\text{inflection sink}} & \underbrace{\text{(SI,SO)}}_{\text{saddle}}\end{array}\right\} \tag{5.63}$$

for $a_{110} > 0, a_{220} < 0$,

$$\left\{\begin{array}{ll}(a_{112}, a_{221}) & (a_{112}, a_{222}) \\ (a_{121}, a_{211}) & (a_{211}, a_{222}) \\ (a_{111}, a_{221}) & (a_{111}, a_{222})\end{array}\right\} = \left\{\begin{array}{cc}\underbrace{\text{(II:DI,SO)}}_{\text{inflection source}} & \underbrace{\text{(SI,SI)}}_{\text{sink}} \\ \underbrace{\text{(US,UP)}}_{\text{up-parabola upper-saddle}} & \underbrace{\text{(pF,UP:DP)}}_{\text{hyperbolic-secant flow }(+)} \\ \underbrace{\text{(II:DI,SI)}}_{\text{inflection sink}} & \underbrace{\text{(SO,SO)}}_{\text{source}}\end{array}\right\}$$

for $a_{110} < 0, a_{220} < 0$.

$$\tag{5.64}$$

Such infinite-equilibriums of $x_2^* = a_{121} = a_{221}$ with $a_{211} \in (a_{111}, a_{112})$ for such a dynamical system are presented in Fig. 5.15. The inflection-sink and source infinite-equilibriums are listed in Table 5.6. The infinite-equilibriums are the switching bifurcation for $a_{121} < a_{221}$ and $a_{121} \in (a_{221}, a_{222})$ with $a_{211} \in (a_{111}, a_{112})$.

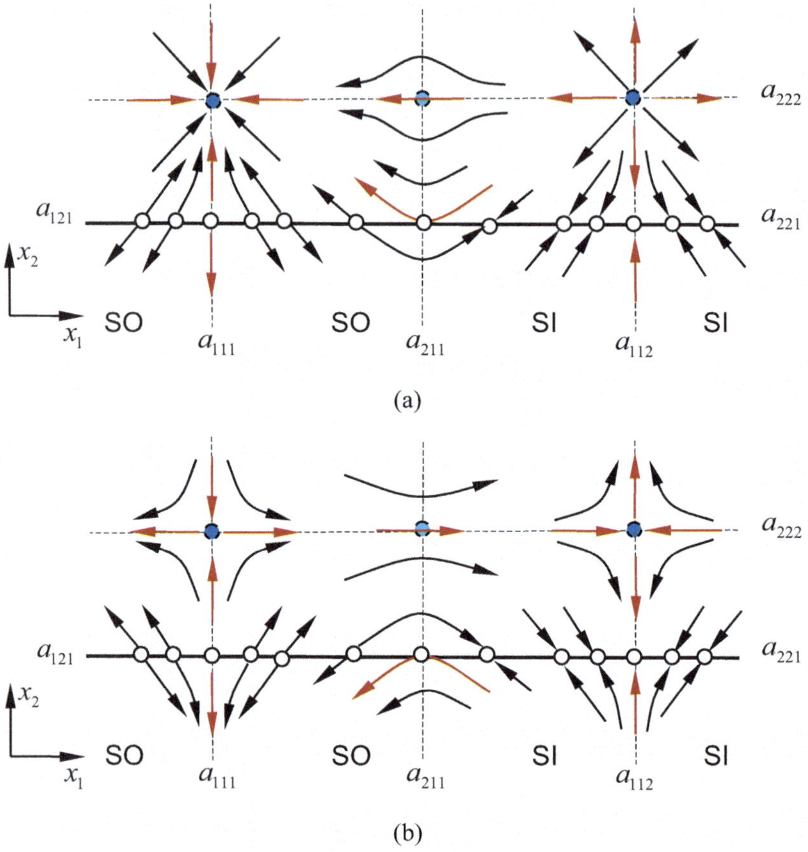

Fig. 5.15 Phase portraits ($a_{211} \in (a_{111}, a_{112})$ and $a_{121} = a_{221}$) for two-dimensional systems on the x_1-direction with $x_1^* = a_{111}, a_{211}, a_{112}$ and on the x_2-direction with $x_2^* = a_{121} = a_{221}, a_{222}$. The inflection-sink and source infinite-equilibriums with sink, source, and saddles: (**a**) ($a_{110} > 0$, $a_{220} > 0$), (**b**) ($a_{110} < 0$, $a_{220} > 0$), (**c**) ($a_{110} > 0$, $a_{220} < 0$), and (**d**) ($a_{110} < 0$, $a_{220} < 0$)

5.2.2 *Vertical Inflection-Source and Sink Infinite-Equilibriums*

Consider a dynamical system for $a_{211} = a_{11s_1}$ ($s_1, s_2 \in \{1, 2\}$, $s_1 \neq s_2$) as

$$\begin{aligned} \dot{x}_1 &= a_{110}(x_1 - a_{111})(x_1 - a_{112})(x_2 - a_{121}), \\ \dot{x}_2 &= a_{220}(x_1 - a_{211})(x_2 - a_{221})(x_2 - a_{222}), \end{aligned} \tag{5.65}$$

and the corresponding first integral manifold is

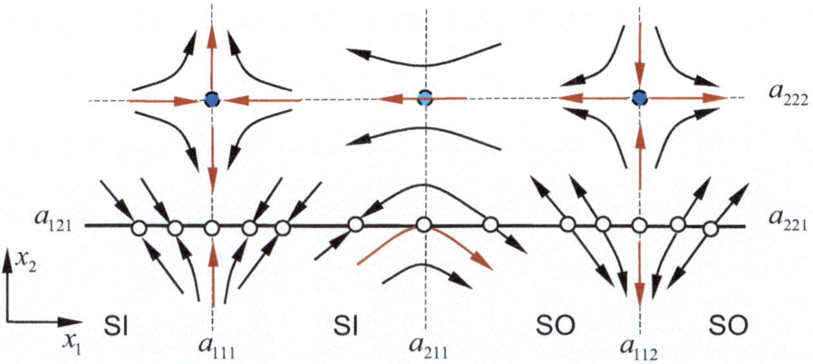

(c)

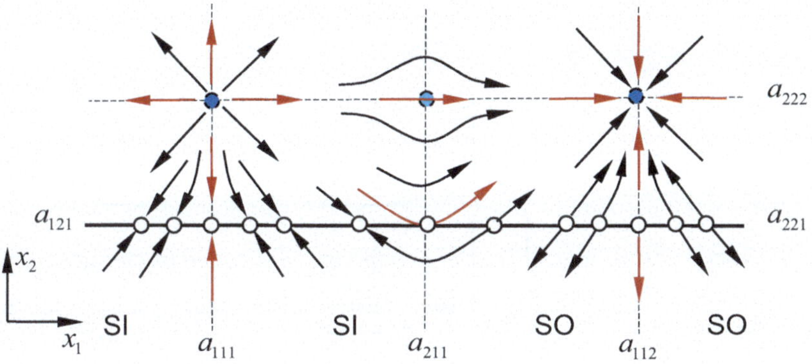

(d)

Fig. 5.15 (continued)

Table 5.6 Inflection-sink and source infinite-equilibriums of $x_2^* = a_{121} = a_{221}$ ($a_{211} \in (a_{111}, a_{112})$ and $a_{121} = a_{221}$)

	$(\overline{x}_1, a_{221}) \equiv (\overline{x}_1, a_{121})$			
	$\overline{x}_1$			
(a_{110}, a_{220})	$(-\infty, a_{211})$	(a_{211}, a_{111})	(a_{111}, a_{112})	(a_{112}, ∞)
$(+,+)$	(II,SO)	(DI,SO)	(II,SI)	(DI,SI)
$(-,+)$	(DI,SO)	(II,SO)	(DI,SI)	(II,SI)
$(+,-)$	(DI,SI)	(II,SI)	(DI,SO)	(II,SO)
$(-,-)$	(II,SI)	(DI,SI)	(II,SO)	(DI,SO)

$$\ln \frac{|x_1 - a_{11s_2}|}{|x_{10} - a_{11s_2}|} = \frac{a_{110}}{a_{220}} \sum_{l_1=1,l_2\neq l_1}^{2} \frac{a_{22l_1} - a_{121}}{a_{22l_1} - a_{22l_2}} \ln \frac{|x_2 - a_{22l_1}|}{|x_{20} - a_{22l_1}|}. \tag{5.66}$$

(i) The infinite-equilibriums of $x_1^* = a_{211} = a_{111}$ with $a_{121} < a_{221}$ have the following properties:

$$\left\{\begin{array}{cc} (a_{111}, a_{222}) & (a_{112}, a_{222}) \\ (a_{111}, a_{221}) & (a_{112}, a_{221}) \\ (a_{121}, a_{211}) & (a_{112}, a_{121}) \end{array}\right\} = \left\{\begin{array}{cc} \underbrace{\text{(SI,II:DI)}}_{\text{inflection source}} & \underbrace{\text{(SO,SO)}}_{\text{source}} \\ \underbrace{\text{(SI,DI:II)}}_{\text{inflection source}} & \underbrace{\text{(SO,SI)}}_{\text{saddle}} \\ \underbrace{\text{(DP,LS)}}_{\text{down-parabola lower-saddle}} & \underbrace{\text{(DP:UP,pF)}}_{\text{hyperbolic flow (+)}} \end{array}\right\} \tag{5.67}$$

for $a_{110} > 0, a_{220} > 0$,

$$\left\{\begin{array}{cc} (a_{111}, a_{222}) & (a_{112}, a_{222}) \\ (a_{111}, a_{221}) & (a_{112}, a_{221}) \\ (a_{121}, a_{211}) & (a_{112}, a_{121}) \end{array}\right\} = \left\{\begin{array}{cc} \underbrace{\text{(SO,DI:II)}}_{\text{inflection source}} & \underbrace{\text{(SI,SO)}}_{\text{saddle}} \\ \underbrace{\text{(SO,II:DI)}}_{\text{inflection source}} & \underbrace{\text{(SI,SI)}}_{\text{sink}} \\ \underbrace{\text{(UP,US)}}_{\text{up-parabola upper-saddle}} & \underbrace{\text{(UP:DP,pF)}}_{\text{hyperbolic-secant flow (+)}} \end{array}\right\}$$

for $a_{110} < 0, a_{220} > 0$,

(5.68)

$$\left\{\begin{array}{cc} (a_{111}, a_{222}) & (a_{112}, a_{222}) \\ (a_{111}, a_{221}) & (a_{112}, a_{221}) \\ (a_{121}, a_{211}) & (a_{112}, a_{121}) \end{array}\right\} = \left\{\begin{array}{cc} \underbrace{\text{(SI,DI:II)}}_{\text{inflection sink}} & \underbrace{\text{(SO,SI)}}_{\text{ssaddle}} \\ \underbrace{\text{(SI,II:DI)}}_{\text{inflection sink}} & \underbrace{\text{(SO,SO)}}_{\text{source}} \\ \underbrace{\text{(UP,LS)}}_{\text{up-parabola lower-saddle}} & \underbrace{\text{(UP:DP,nF)}}_{\text{hyperbolic-secant flow (-)}} \end{array}\right\}$$

for $a_{110} > 0, a_{220} < 0$,

(5.69)

$$\left\{\begin{array}{ll}(a_{111},a_{222}) & (a_{112},a_{222})\\(a_{111},a_{221}) & (a_{112},a_{221})\\(a_{121},a_{211}) & (a_{112},a_{121})\end{array}\right\}=\left\{\begin{array}{cc}\underbrace{(\text{SO,II:DI})}_{\text{inflection source}} & \underbrace{(\text{SI,SI})}_{\text{sink}}\\ \underbrace{(\text{SO,DI:II})}_{\text{inflection source}} & \underbrace{(\text{SI,SO})}_{\text{saddle}}\\ \underbrace{(\text{DP,US})}_{\text{down-parabola upper-saddle}} & \underbrace{(\text{DP:UP,nF})}_{\text{hyperbolic flow }(-)}\end{array}\right\} \tag{5.70}$$

for $a_{110} < 0, a_{220} < 0$.

Such infinite-equilibriums of $x_1^* = a_{111} = a_{211}$ with $a_{121} < a_{221}$ for such a dynamical system are presented in Fig. 5.16. The inflection-sink and source infinite-equilibriums are listed in Table 5.7. The infinite-equilibriums are the switching bifurcation for $a_{211} < a_{111}$ and $a_{211} \in (a_{111}, a_{112})$ with $a_{121} < a_{221}$.

(ii) The infinite equilibriums of $x_1^* = a_{211} = a_{111}$ with $a_{222} < a_{121}$ have the following properties:

$$\left\{\begin{array}{ll}(a_{121},a_{211}) & (a_{112},a_{121})\\(a_{211},a_{222}) & (a_{112},a_{222})\\(a_{211},a_{221}) & (a_{112},a_{221})\end{array}\right\}=\left\{\begin{array}{cc}\underbrace{(\text{DP,LS})}_{\text{down-parabola lower-saddle}} & \underbrace{(\text{DP:UP,pF})}_{\text{hyperbolic flow }(+)}\\ \underbrace{(\text{SO,DI:II})}_{\text{inflection source}} & \underbrace{(\text{SI,SO})}_{\text{saddle}}\\ \underbrace{(\text{SO,II:DI})}_{\text{inflection source}} & \underbrace{(\text{SI,SI})}_{\text{sink}}\end{array}\right\} \tag{5.71}$$

for $a_{110} > 0, a_{220} > 0$,

$$\left\{\begin{array}{ll}(a_{121},a_{211}) & (a_{112},a_{121})\\(a_{211},a_{222}) & (a_{112},a_{222})\\(a_{211},a_{221}) & (a_{112},a_{221})\end{array}\right\}=\left\{\begin{array}{cc}\underbrace{(\text{UP,US})}_{\text{up-parabola upper-saddle}} & \underbrace{(\text{UP:DP,pF})}_{\text{hyperbolic-secant flow }(+)}\\ \underbrace{(\text{SI,II:DI})}_{\text{inflection source}} & \underbrace{(\text{SO,SO})}_{\text{source}}\\ \underbrace{(\text{SI,DI:II})}_{\text{inflection source}} & \underbrace{(\text{SO,SI})}_{\text{saddle}}\end{array}\right\}$$

for $a_{110} < 0, a_{220} > 0$,

$$\tag{5.72}$$

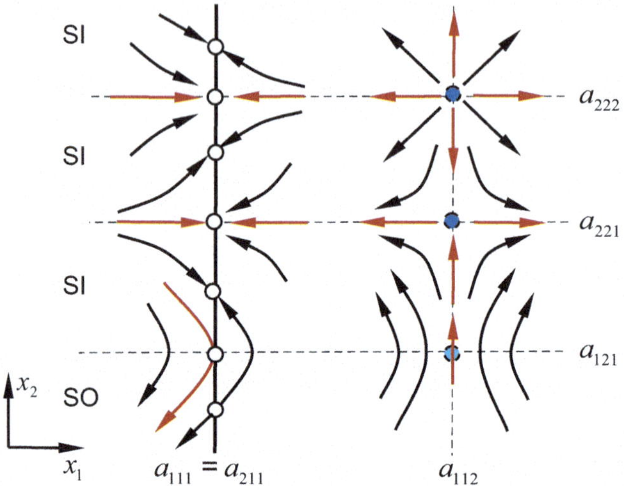

(a)

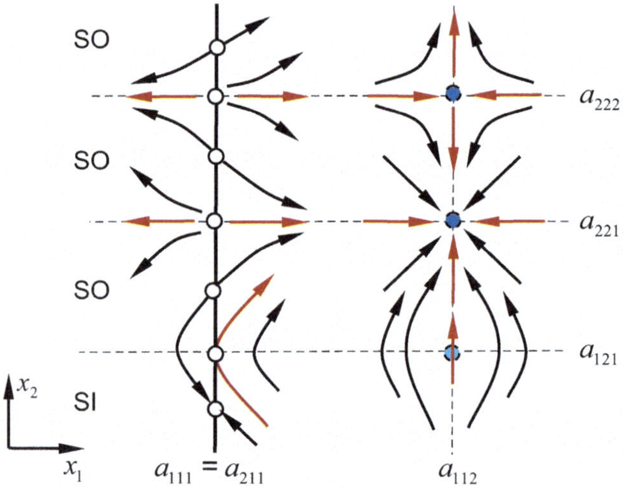

(b)

Fig. 5.16 Phase portraits ($a_{211} = a_{111}$ and $a_{121} < a_{221}$) for two-dimensional systems on the x_1-direction with $x_1^* = a_{111}, a_{211}, a_{112}$ and on the x_2-direction with $x_2^* = a_{121}, a_{221}, a_{222}$. The inflection-sink and source infinite-equilibrium with sink, (source) and saddle plus a hyperbolic flow. (**a**) ($a_{110} > 0, a_{220} > 0$), (**b**) ($a_{110} < 0, a_{220} > 0$), (**c**) ($a_{110} > 0, a_{220} < 0$), and (**d**) ($a_{110} < 0, a_{220} < 0$)

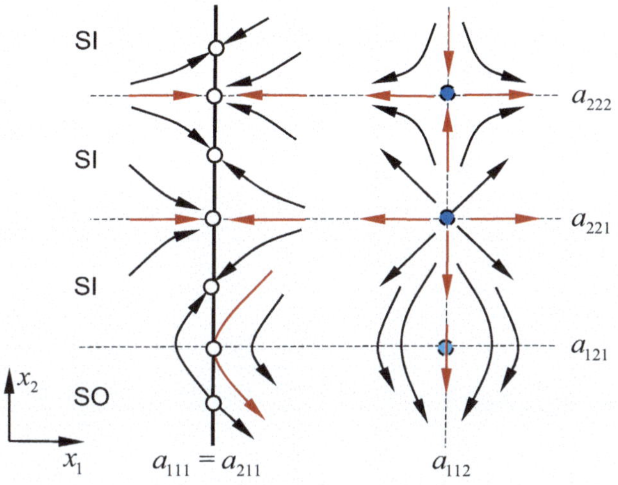

(c)

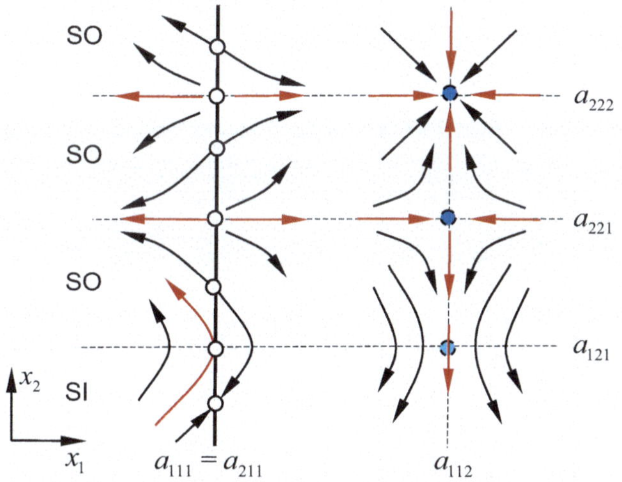

(d)

Fig. 5.16 (continued)

Table 5.7 Inflection-sink and source infinite-equilibriums of $x_1^* = a_{211} = a_{111}$ ($a_{211} = a_{111}$ and $a_{121} < a_{221}$)

	$(a_{111}, \bar{x}_2) \equiv (a_{211}, \bar{x}_2)$			
	$\bar{x}_2$			
(a_{110}, a_{220})	$(-\infty, a_{121})$	(a_{121}, a_{221})	(a_{221}, a_{222})	(a_{222}, ∞)
(+, +)	(SO,II)	(SI,DI)	(SI,II)	(SI,DI)
(−, +)	(SI,DI)	(SO,II)	(SO,DI)	(SO,II)
(+, −)	(SO,DI)	(SI,II)	(SI,DI)	(SI,II)
(−, −)	(SI,II)	(SO,DI)	(SO,II)	(SO,DI)

$$\left\{\begin{array}{cc}(a_{121},a_{211}) & (a_{112},a_{121})\\ (a_{211},a_{222}) & (a_{112},a_{222})\\ (a_{211},a_{221}) & (a_{112},a_{221})\end{array}\right\}=\left\{\begin{array}{cc}\underbrace{(\text{UP,LS})}_{\text{up-parabola lower-saddle}} & \underbrace{(\text{UP:DP,nF})}_{\text{hyperbolic-secant flow }(-)}\\ \underbrace{(\text{SO,II:DI})}_{\text{inflection source}} & \underbrace{(\text{SI,SI})}_{\text{sink}}\\ \underbrace{(\text{SO,DI:II})}_{\text{inflection source}} & \underbrace{(\text{SI,SO})}_{\text{saddle}}\end{array}\right\}$$

$$\text{for } a_{110}>0, a_{220}<0, \tag{5.73}$$

$$\left\{\begin{array}{cc}(a_{121},a_{211}) & (a_{112},a_{121})\\ (a_{211},a_{222}) & (a_{112},a_{222})\\ (a_{211},a_{221}) & (a_{112},a_{221})\end{array}\right\}=\left\{\begin{array}{cc}\underbrace{(\text{DP,US})}_{\text{down-parabola upper-saddle}} & \underbrace{(\text{DP:UP,nF})}_{\text{hyperbolic flow }(-)}\\ \underbrace{(\text{SI,DI:II})}_{\text{inflection sink}} & \underbrace{(\text{SO,SI})}_{\text{saddle}}\\ \underbrace{(\text{SI,II:DI})}_{\text{inflection sink}} & \underbrace{(\text{SO,SO})}_{\text{source}}\end{array}\right\} \tag{5.74}$$

$$\text{for } a_{110}<0, a_{220}<0.$$

Such infinite-equilibriums of $x_1^* = a_{111} = a_{211}$ with $a_{121} > a_{222}$ for such a dynamical system are presented in Fig. 5.17. The inflection-sink and source infinite-equilibriums are listed in Table 5.8. The infinite-equilibriums are the switching bifurcation for $a_{211} < a_{111}$ and $a_{211} \in (a_{111}, a_{112})$ with $a_{121} > a_{222}$.

(iii) The infinite-equilibriums of $x_1^* = a_{211} = a_{111}$ with $a_{121} \in (a_{221}, a_{222})$ have the following properties:

$$\left\{\begin{array}{cc}(a_{211},a_{222}) & (a_{112},a_{222})\\ (a_{121},a_{211}) & (a_{112},a_{121})\\ (a_{211},a_{221}) & (a_{112},a_{221})\end{array}\right\}=\left\{\begin{array}{cc}\underbrace{(\text{SI,II:DI})}_{\text{inflection sink}} & \underbrace{(\text{SO,SO})}_{\text{source}}\\ \underbrace{(\text{UP,LS})}_{\text{iup-parabola lower-saddle}} & \underbrace{(\text{UP:DP,nF})}_{\text{hyperbolic-secant flow }(-)}\\ \underbrace{(\text{SO,II:DI})}_{\text{inflection source}} & \underbrace{(\text{SI,SI})}_{\text{sink}}\end{array}\right\}$$

$$\text{for } a_{110}>0, a_{220}>0, \tag{5.75}$$

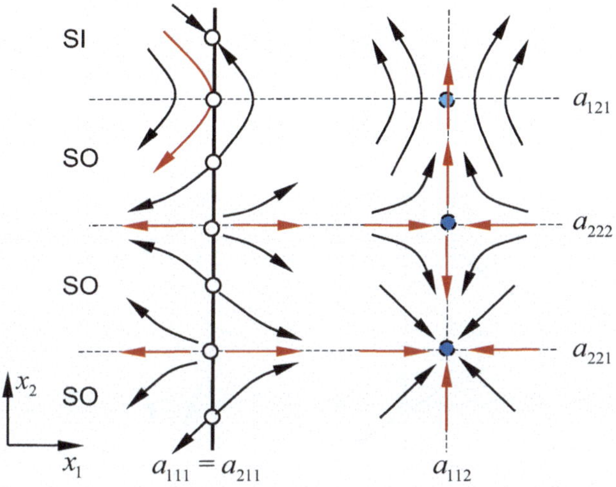

(a)

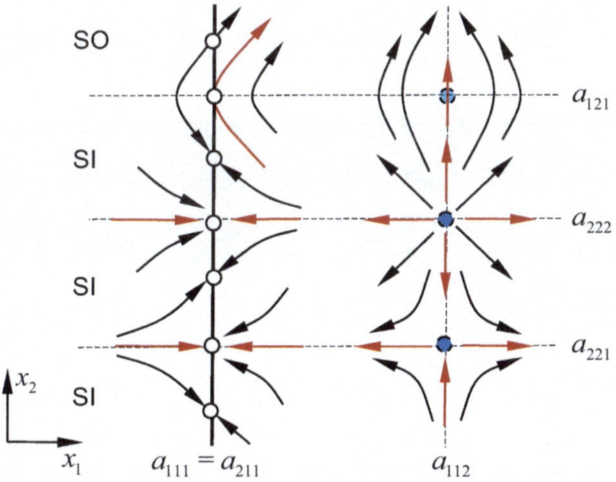

(b)

Fig. 5.17 Phase portraits ($a_{211} = a_{111}$ and $a_{222} < a_{121}$) for two-dimensional systems on the x_1-direction with $x_1^* = a_{111}, a_{211}, a_{112}$ and on the x_2-direction with $x_2^* = a_{121}, a_{221}, a_{222}$. The inflection-sink and source infinite-equilibriums with sink, source, and saddle plus a hyperbolic flow: (**a**) ($a_{110} > 0, a_{220} > 0$), (**b**) ($a_{110} < 0, a_{220} > 0$), (**c**) ($a_{110} > 0, a_{220} < 0$), and (**d**) ($a_{110} < 0, a_{220} < 0$)

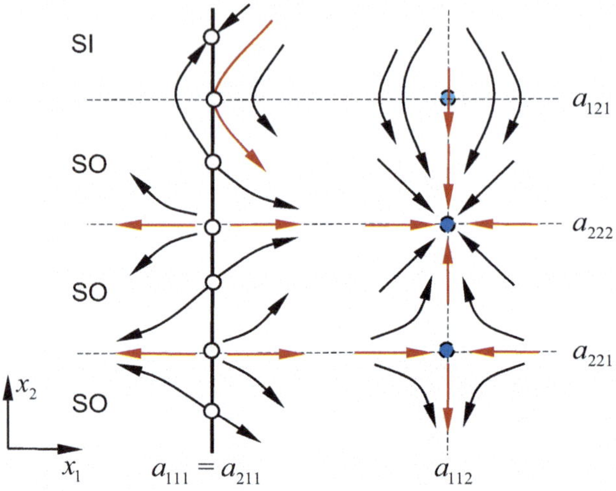

(c)

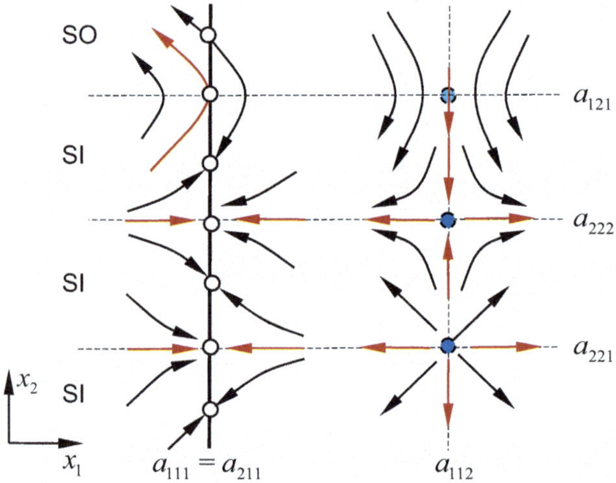

(d)

Fig. 5.17 (continued)

Table 5.8 Inflection-sink and source-infinite equilibriums of $x_1^* = a_{211} = a_{111}$ ($a_{211} = a_{111}$ and $a_{121} > a_{222}$)

	$(a_{111}, \bar{x}_2) \equiv (a_{211}, \bar{x}_2)$			
	$\bar{x}_2$			
(a_{110}, a_{220})	$(-\infty, a_{121})$	(a_{121}, a_{221})	(a_{221}, a_{222})	(a_{222}, ∞)
$(+, +)$	(SO,II)	(SO,DI)	(SO,II)	(SI,DI)
$(-, +)$	(SI,DI)	(SI,II)	(SI,DI)	(SO,II)
$(+, -)$	(SO,DI)	(SO,II)	(SO,DI)	(SI,II)
$(-, -)$	(SI,II)	(SI,DI)	(SI,II)	(SO,DI)

$$\left\{\begin{array}{ll}(a_{211},a_{222}) & (a_{112},a_{222})\\ (a_{121},a_{211}) & (a_{112},a_{121})\\ (a_{211},a_{221}) & (a_{112},a_{221})\end{array}\right\}=\left\{\begin{array}{cc}\underbrace{(\text{SO,DI:II})}_{\text{inflection source}} & \underbrace{(\text{SI,SO})}_{\text{saddle}}\\ \underbrace{(\text{DP,US})}_{\text{down-parabola upper-saddle}} & \underbrace{(\text{DP:UP,nF})}_{\text{hyperbolic flow }(-)}\\ \underbrace{(\text{SI,DI:II})}_{\text{inflection sink}} & \underbrace{(\text{SO,SI})}_{\text{saddle}}\end{array}\right\} \tag{5.76}$$

for $a_{110} < 0, a_{220} > 0$,

$$\left\{\begin{array}{ll}(a_{211},a_{222}) & (a_{112},a_{222})\\ (a_{121},a_{211}) & (a_{112},a_{121})\\ (a_{211},a_{221}) & (a_{112},a_{221})\end{array}\right\}=\left\{\begin{array}{cc}\underbrace{(\text{SI,DI:II})}_{\text{inflection sink}} & \underbrace{(\text{SO,SI})}_{\text{saddle}}\\ \underbrace{(\text{DP,LS})}_{\text{down-parabola lower-saddle}} & \underbrace{(\text{DP:UP,pF})}_{\text{hyperbolic flow }(+)}\\ \underbrace{(\text{SO,DI:II})}_{\text{inflection source}} & \underbrace{(\text{SI,SO})}_{\text{saddle}}\end{array}\right\} \tag{5.77}$$

for $a_{110} > 0, a_{220} < 0$,

$$\left\{\begin{array}{ll}(a_{211},a_{222}) & (a_{112},a_{222})\\ (a_{121},a_{211}) & (a_{112},a_{121})\\ (a_{211},a_{221}) & (a_{112},a_{221})\end{array}\right\}=\left\{\begin{array}{cc}\underbrace{(\text{SO,II:DI})}_{\text{inflection source}} & \underbrace{(\text{SI,SI})}_{\text{sink}}\\ \underbrace{(\text{UP,US})}_{\text{up-parabola upper-saddle}} & \underbrace{(\text{UP:DP,pF})}_{\text{hyperbolic-secant flow }(+)}\\ \underbrace{(\text{SI,II:DI})}_{\text{inflection sink}} & \underbrace{(\text{SO,SO})}_{\text{source}}\end{array}\right\}$$

for $a_{110} < 0, a_{220} < 0$.

(5.78)

Such infinite-equilibriums of $x_1^* = a_{111} = a_{211}$ with $a_{121} \in (a_{221}, a_{222})$ for such a dynamical system are presented in Fig. 5.18. The inflection-sink and source infinite-equilibriums are listed in Table 5.9. The infinite-equilibriums are the switching bifurcation for $a_{211} < a_{111}$ and $a_{211} \in (a_{111}, a_{112})$ with $a_{121} \in (a_{221}, a_{222})$.

(iv) The infinite-equilibriums of $x_1^* = a_{211} = a_{112}$ with $a_{121} < a_{221}$ have the following properties:

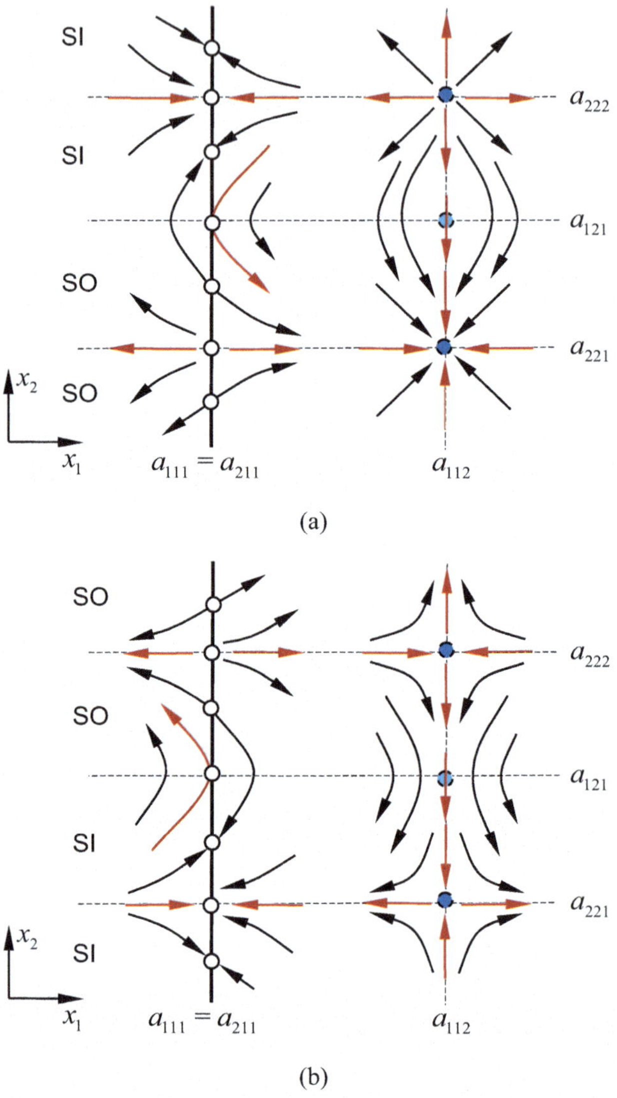

Fig. 5.18 Phase portraits ($a_{211} < a_{111}$ and $a_{121} \in (a_{221}, a_{222})$) for two-dimensional systems on the x_1-direction with $x_1^* = a_{111}, a_{211}, a_{112}$ and on the x_2-direction with $x_2^* = a_{121}, a_{221}, a_{222}$. The inflection-sink and source infinite-equilibriums with sink, source, and saddle plus a hyperbolic flow: (**a**) ($a_{110} > 0, a_{220} > 0$), (**b**) ($a_{110} < 0, a_{220} > 0$), (**c**) ($a_{110} > 0, a_{220} < 0$), and (**d**) ($a_{110} < 0, a_{220} < 0$)

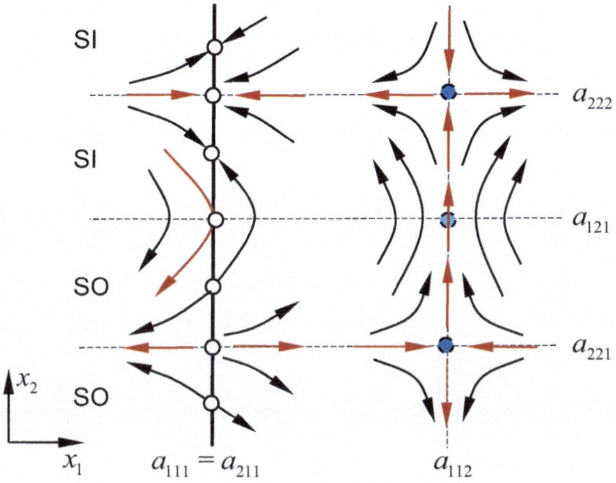

(c)

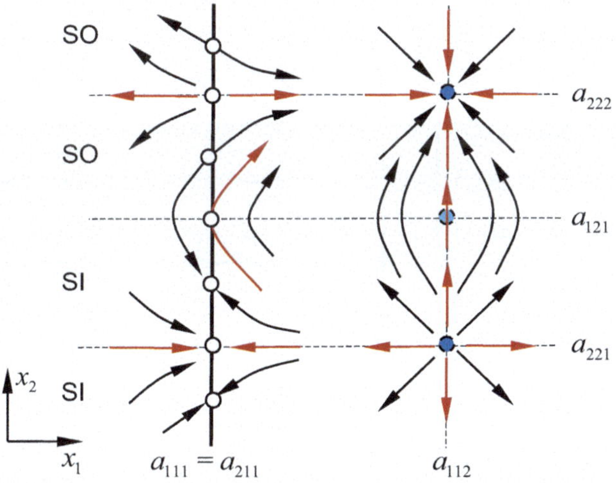

(d)

Fig. 5.18 (continued)

Table 5.9 Inflection-sink and source infinite-equilibriums of $x_1^* = a_{211} = a_{111}$ ($a_{211} = a_{111}$ and $a_{121} \in (a_{221}, a_{222})$)

	$(a_{111}, \bar{x}_2) \equiv (a_{211}, \bar{x}_2)$			
	$\bar{x}_2$			
(a_{110}, a_{220})	$(-\infty, a_{121})$	(a_{121}, a_{221})	(a_{221}, a_{222})	(a_{222}, ∞)
$(+, +)$	(SO,II)	(SO,DI)	(SI,II)	(SI,DI)
$(-, +)$	(SI,DI)	(SI,II)	(SO,DI)	(SO,II)
$(+, -)$	(SO,DI)	(SO,II)	(SI,DI)	(SI,II)
$(-, -)$	(SI,II)	(SI,DI)	(SO,II)	(SO,DI)

$$\left\{\begin{array}{cc}(a_{111},a_{222}) & (a_{112},a_{222})\\ (a_{111},a_{221}) & (a_{112},a_{221})\\ (a_{111},a_{121}) & (a_{121},a_{211})\end{array}\right\}=\left\{\begin{array}{cc}\underbrace{\text{(SI,SI)}}_{\text{sink}} & \underbrace{\text{(SO,DI:II)}}_{\text{inflection source}}\\ \underbrace{\text{(SI,SO)}}_{\text{saddle}} & \underbrace{\text{(SO,II:DI)}}_{\text{inflection source}}\\ \underbrace{\text{(DP:UP,nF)}}_{\text{hyperbolic flow (-)}} & \underbrace{\text{(UP,US)}}_{\text{up-parabola upper-saddle}}\end{array}\right\} \tag{5.79}$$

for $a_{110} > 0, a_{220} > 0$,

$$\left\{\begin{array}{cc}(a_{111},a_{222}) & (a_{112},a_{222})\\ (a_{111},a_{221}) & (a_{112},a_{221})\\ (a_{111},a_{121}) & (a_{121},a_{211})\end{array}\right\}=\left\{\begin{array}{cc}\underbrace{\text{(SO,SI)}}_{\text{saddle}} & \underbrace{\text{(SI,II:DI)}}_{\text{inflection source}}\\ \underbrace{\text{(SO,SO)}}_{\text{source}} & \underbrace{\text{(SI,DI:II)}}_{\text{inflection source}}\\ \underbrace{\text{(UP:DP,nF)}}_{\text{hyperbolic-secant flow (-)}} & \underbrace{\text{(DP,LS)}}_{\text{down-parabola lower-saddle}}\end{array}\right\}$$

for $a_{110} < 0, a_{220} > 0$,

(5.80)

$$\left\{\begin{array}{cc}(a_{111},a_{222}) & (a_{112},a_{222})\\ (a_{111},a_{221}) & (a_{112},a_{221})\\ (a_{111},a_{121}) & (a_{121},a_{211})\end{array}\right\}=\left\{\begin{array}{cc}\underbrace{\text{(SI,SO)}}_{\text{saddle}} & \underbrace{\text{(SO,II:DI)}}_{\text{inflection source}}\\ \underbrace{\text{(SI,SI)}}_{\text{sink}} & \underbrace{\text{(SO,DI:II)}}_{\text{inflection source}}\\ \underbrace{\text{(UP:DP,pF)}}_{\text{hyperbolic-secant flow (+)}} & \underbrace{\text{(DP,US)}}_{\text{down-parabola upper-saddle}}\end{array}\right\}$$

for $a_{110} > 0, a_{220} < 0$;

(5.81)

$$\left\{\begin{array}{cc}(a_{111},a_{222}) & (a_{112},a_{222})\\ (a_{111},a_{221}) & (a_{112},a_{221})\\ (a_{111},a_{121}) & (a_{121},a_{211})\end{array}\right\}=\left\{\begin{array}{cc}\underbrace{\text{(SO,SO)}}_{\text{source}} & \underbrace{\text{(SI,DI:II)}}_{\text{inflection sink}}\\ \underbrace{\text{(SO,SI)}}_{\text{ssaddle}} & \underbrace{\text{(SI,II:DI)}}_{\text{inflection sink}}\\ \underbrace{\text{(DP:UP,pF)}}_{\text{hyperbolic-secant flow (-)}} & \underbrace{\text{(UP,LS)}}_{\text{up-parabola lower-saddle}}\end{array}\right\}$$

for $a_{110} < 0, a_{220} < 0$,

(5.82)

Such infinite-equilibriums of $x_1^* = a_{111} = a_{211}$ with $a_{121} < a_{221}$ for such a dynamical system are presented in Fig. 5.19. The inflection-sink and source infinite-equilibriums are listed in Table 5.10. The infinite-equilibriums are the switching bifurcation for $a_{112} < a_{211}$ and $a_{211} \in (a_{111}, a_{112})$ with $a_{121} < a_{221}$.

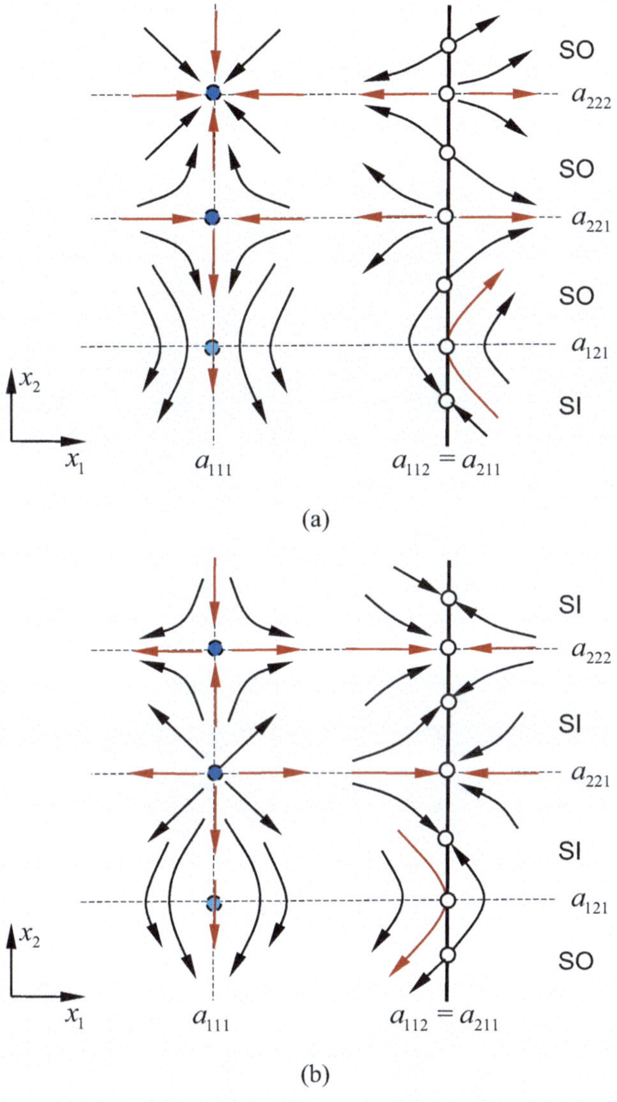

Fig. 5.19 Phase portraits ($a_{211} = a_{112}$ and $a_{121} < a_{221}$) for two-dimensional systems on the x_1-direction with $x_1^* = a_{111}, a_{112}, a_{211}$ and on the x_2-direction with $x_2^* = a_{121}, a_{221}, a_{222}$. The inflection-sink and source infinite-equilibriums with sink, source, and saddle plus a hyperbolic flow: (**a**) ($a_{110} > 0$, $a_{220} > 0$), (**b**) ($a_{110} < 0$, $a_{220} > 0$), (**c**) ($a_{110} > 0$, $a_{220} < 0$), and (**d**) ($a_{110} < 0$, $a_{220} < 0$)

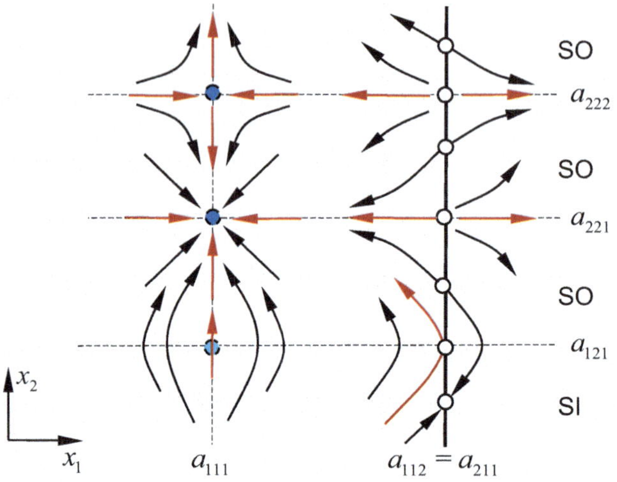

(c)

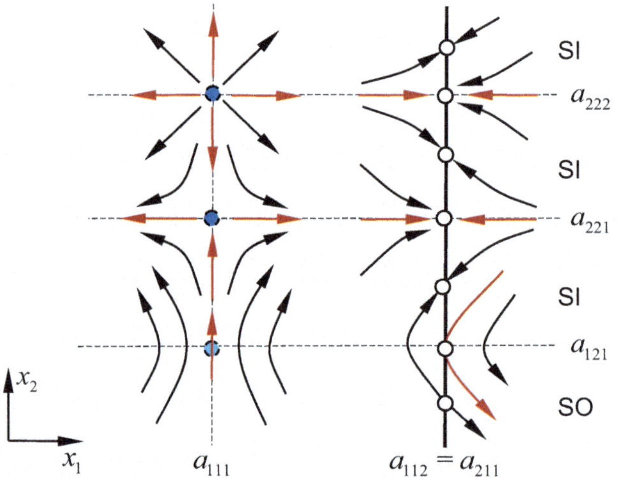

(d)

Fig. 5.19 (continued)

Table 5.10 Inflection-sink and source infinite-equilibriums of $x_1^* = a_{211} = a_{112}$ ($a_{211} = a_{112}$ and $a_{121} < a_{221}$)

	$(a_{111}, \bar{x}_2) \equiv (a_{211}, \bar{x}_2)$			
	$\bar{x}_2$			
(a_{110}, a_{220})	$(-\infty, a_{121})$	(a_{121}, a_{221})	(a_{221}, a_{222})	(a_{222}, ∞)
$(+,+)$	(SI,DI)	(SO,II)	(SO,DI)	(SO,II)
$(-,+)$	(SO,II)	(SI,DI)	(SI,II)	(SI,DI)
$(+,-)$	(SI,II)	(SO,DI)	(SO,II)	(SO,DI)
$(-,-)$	(SO,DI)	(SI,II)	(SI,DI)	(SI,II)

(v) The infinite-equilibriums of $x_1^* = a_{211} = a_{111}$ with $a_{221} < a_{121}$ have the following properties:

$$\left\{\begin{array}{cc}(a_{111},a_{121}) & (a_{121},a_{211})\\ (a_{111},a_{222}) & (a_{112},a_{222})\\ (a_{111},a_{221}) & (a_{112},a_{221})\end{array}\right\} = \left\{\begin{array}{cc}\underbrace{(\text{DP:UP,nF})}_{\text{hyperbolic flow }(-)} & \underbrace{(\text{UP,US})}_{\text{up-parabola upper-saddle}}\\ \underbrace{(\text{SO,SI})}_{\text{saddle}} & \underbrace{(\text{SI,II:DI})}_{\text{inflection source}}\\ \underbrace{(\text{SO,SO})}_{\text{source}} & \underbrace{(\text{SI,DI:II})}_{\text{inflection source}}\end{array}\right\} \tag{5.83}$$

for $a_{110} > 0, a_{220} > 0$,

$$\left\{\begin{array}{cc}(a_{111},a_{121}) & (a_{121},a_{211})\\ (a_{111},a_{222}) & (a_{112},a_{222})\\ (a_{111},a_{221}) & (a_{112},a_{221})\end{array}\right\} = \left\{\begin{array}{cc}\underbrace{(\text{UP:DP,nF})}_{\text{hyperbolic-secant flow }(-)} & \underbrace{(\text{DP,LS})}_{\text{down-parabola lower-saddle}}\\ \underbrace{(\text{SI,SI})}_{\text{sink}} & \underbrace{(\text{SO,DI:II})}_{\text{inflection source}}\\ \underbrace{(\text{SI,SO})}_{\text{saddle}} & \underbrace{(\text{SO,II:DI})}_{\text{inflection source}}\end{array}\right\}$$

for $a_{110} < 0, a_{220} > 0$,

(5.84)

$$\left\{\begin{array}{cc}(a_{111},a_{121}) & (a_{121},a_{211})\\ (a_{111},a_{222}) & (a_{112},a_{222})\\ (a_{111},a_{221}) & (a_{112},a_{221})\end{array}\right\} = \left\{\begin{array}{cc}\underbrace{(\text{UP:DP,pF})}_{\text{hyperbolic-secant flow }(+)} & \underbrace{(\text{DP,US})}_{\text{down-parabola upper-saddle}}\\ \underbrace{(\text{SO,SO})}_{\text{source}} & \underbrace{(\text{SI,DI:II})}_{\text{inflection sink}}\\ \underbrace{(\text{SO,SI})}_{\text{saddle}} & \underbrace{(\text{SI,II:DI})}_{\text{inflection sink}}\end{array}\right\}$$

for $a_{110} > 0, a_{220} < 0$;

(5.85)

$$\left\{\begin{array}{cc}(a_{111},a_{121}) & (a_{121},a_{211})\\ (a_{111},a_{222}) & (a_{112},a_{222})\\ (a_{111},a_{221}) & (a_{112},a_{221})\end{array}\right\} = \left\{\begin{array}{cc}\underbrace{(\text{DP:UP,pF})}_{\text{hyperbolic flow }(+)} & \underbrace{(\text{UP,LS})}_{\text{up-parabola lower-saddle}}\\ \underbrace{(\text{SI,SO})}_{\text{saddle}} & \underbrace{(\text{SO,II:DI})}_{\text{inflection source}}\\ \underbrace{(\text{SI,SI})}_{\text{sink}} & \underbrace{(\text{SO,DI:II})}_{\text{inflection source}}\end{array}\right\} \tag{5.86}$$

for $a_{110} < 0, a_{220} < 0$;

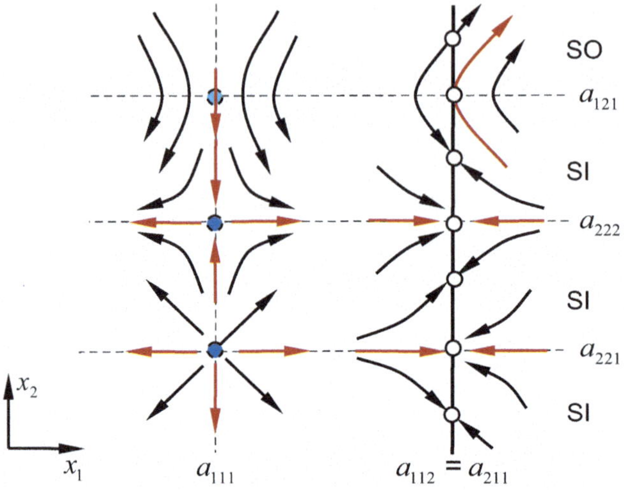

(a)

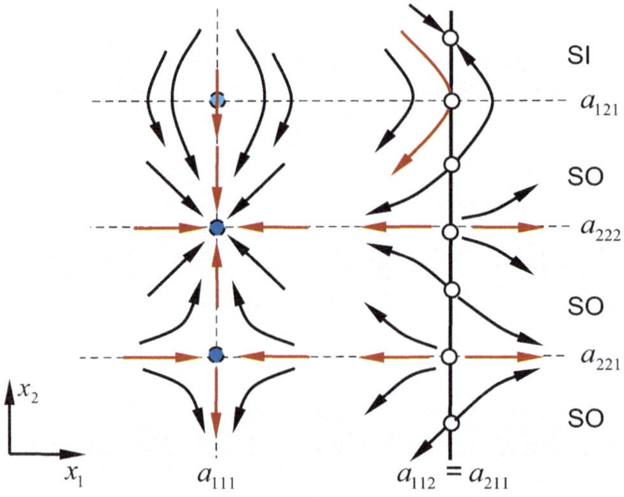

(b)

Fig. 5.20 Phase portraits ($a_{211} = a_{112}$ and $a_{222} < a_{121}$) for two-dimensional systems on the x_1-direction with $x_1^* = a_{111}, a_{112}, a_{211}$ and on the x_2-direction with $x_2^* = a_{121}, a_{221}, a_{222}$. The inflection-sink and source infinite-equilibriums with sink, source, and saddle plus a hyperbolic flow: (**a**) ($a_{110} > 0, a_{220} > 0$), (**b**) ($a_{110} < 0, a_{220} > 0$), (**c**) ($a_{110} > 0, a_{220} < 0$), and (**d**) ($a_{110} < 0, a_{220} < 0$)

Such infinite-equilibriums of $x_1^* = a_{111} = a_{211}$ with $a_{121} > a_{222}$ for such a dynamical system are presented in Fig. 5.20. The inflection-sink and source infinite-equilibriums are listed in Table 5.11. The infinite-equilibriums are the switching bifurcation for $a_{112} < a_{211}$ and $a_{211} \in (a_{111}, a_{112})$ with $a_{121} > a_{222}$.

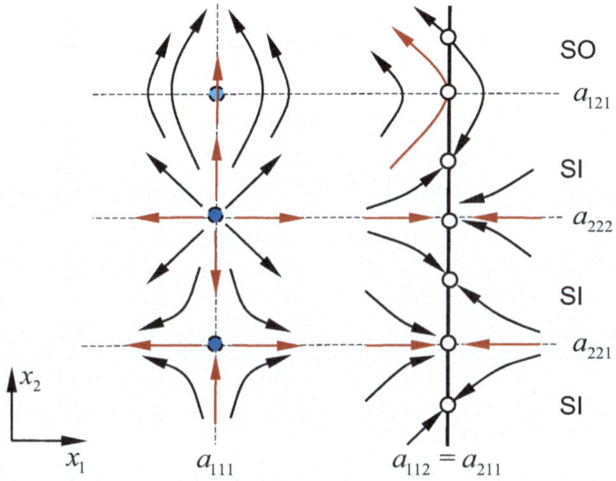

(c)

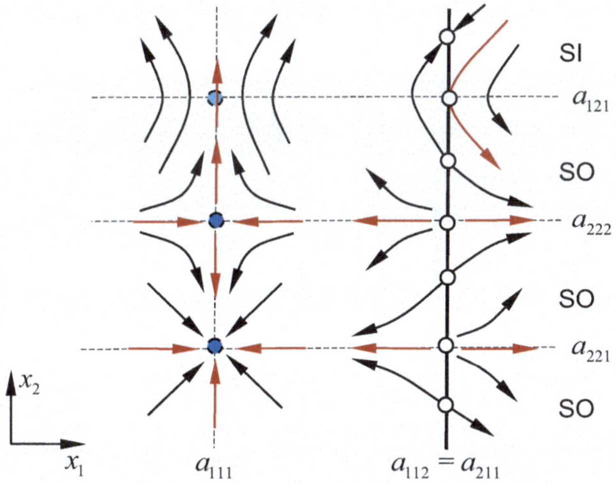

(d)

Fig. 5.20 (continued)

Table 5.11 Inflection-sink and source infinite-equilibriums of $x_1^* = a_{211} = a_{112}$ ($a_{211} = a_{112}$ and $a_{121} > a_{222}$)

	$(a_{111}, \bar{x}_2) \equiv (a_{211}, \bar{x}_2)$			
	$\bar{x}_2$			
(a_{110}, a_{220})	$(-\infty, a_{121})$	(a_{121}, a_{221})	(a_{221}, a_{222})	(a_{222}, ∞)
(+, +)	(SI,DI)	(SI,II)	(SI,DI)	(SO,II)
(−, +)	(SO,II)	(SO,DI)	(SO,II)	(SI,DI)
(+, −)	(SI,II)	(SI,DI)	(SI,II)	(SO,DI)
(−, −)	(SO,DI)	(SO,II)	(SO,DI)	(SI,II)

(vi) The infinite-equilibriums of $x_1^* = a_{211} = a_{111}$ with $a_{121} \in (a_{221}, a_{222})$ have the following properties:

$$\left\{\begin{matrix} (a_{111}, a_{222}) & (a_{112}, a_{222}) \\ (a_{111}, a_{121}) & (a_{121}, a_{211}) \\ (a_{111}, a_{221}) & (a_{112}, a_{221}) \end{matrix}\right\} = \left\{\begin{matrix} \underbrace{(\text{SI,SI})}_{\text{sink}} & \underbrace{(\text{SO,DI:II})}_{\text{inflection source}} \\ \underbrace{(\text{UP:DP,pF})}_{\text{hyperbolic-secant flow }(+)} & \underbrace{(\text{DP,US})}_{\text{down-parabola upper-saddle}} \\ \underbrace{(\text{SO,SO})}_{\text{source}} & \underbrace{(\text{SI,DI:II})}_{\text{inflection sink}} \end{matrix}\right\}$$

for $a_{110} > 0, a_{220} > 0$,

$$(5.87)$$

$$\left\{\begin{matrix} (a_{111}, a_{222}) & (a_{112}, a_{222}) \\ (a_{111}, a_{121}) & (a_{121}, a_{211}) \\ (a_{111}, a_{221}) & (a_{112}, a_{221}) \end{matrix}\right\} = \left\{\begin{matrix} \underbrace{(\text{SO,SI})}_{\text{saddle}} & \underbrace{(\text{SI,II:DI})}_{\text{inflection sink}} \\ \underbrace{(\text{DP:UP,pF})}_{\text{hyperbolic flow }(+)} & \underbrace{(\text{UP,LS})}_{\text{up-parabola lower-saddle}} \\ \underbrace{(\text{SI,SO})}_{\text{saddle}} & \underbrace{(\text{SO,II:DI})}_{\text{inflection source}} \end{matrix}\right\} \tag{5.88}$$

for $a_{110} < 0, a_{220} > 0$,

$$\left\{\begin{matrix} (a_{111}, a_{222}) & (a_{112}, a_{222}) \\ (a_{111}, a_{121}) & (a_{121}, a_{211}) \\ (a_{111}, a_{221}) & (a_{112}, a_{221}) \end{matrix}\right\} = \left\{\begin{matrix} \underbrace{(\text{SI,SO})}_{\text{saddle}} & \underbrace{(\text{SO,II:DI})}_{\text{inflection source}} \\ \underbrace{(\text{DP:UP,nF})}_{\text{hyperbolic flow }(-)} & \underbrace{(\text{UP,US})}_{\text{up-parabola upper-saddle}} \\ \underbrace{(\text{SO,SI})}_{\text{saddle}} & \underbrace{(\text{SI,II:DI})}_{\text{inflection sink}} \end{matrix}\right\} \tag{5.89}$$

for $a_{110} > 0, a_{220} < 0$;

$$\left\{\begin{matrix} (a_{111}, a_{222}) & (a_{112}, a_{222}) \\ (a_{111}, a_{121}) & (a_{121}, a_{211}) \\ (a_{111}, a_{221}) & (a_{112}, a_{221}) \end{matrix}\right\} = \left\{\begin{matrix} \underbrace{(\text{SO,SO})}_{\text{sink}} & \underbrace{(\text{SI,DI:II})}_{\text{inflection sink}} \\ \underbrace{(\text{UP:DP,nF})}_{\text{hyperbolic-secant flow }(-)} & \underbrace{(\text{DP,LS})}_{\text{down-parabola lower-saddle}} \\ \underbrace{(\text{SI,SI})}_{\text{sink}} & \underbrace{(\text{SO,DI:II})}_{\text{inflection source}} \end{matrix}\right\}$$

for $a_{110} < 0, a_{220} < 0$.

$$(5.90)$$

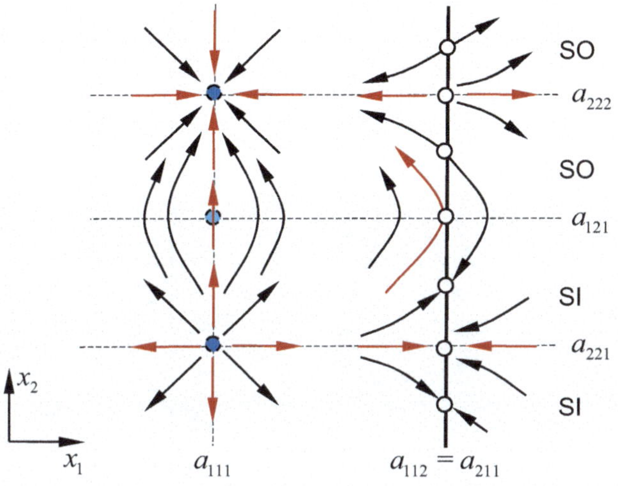

(a)

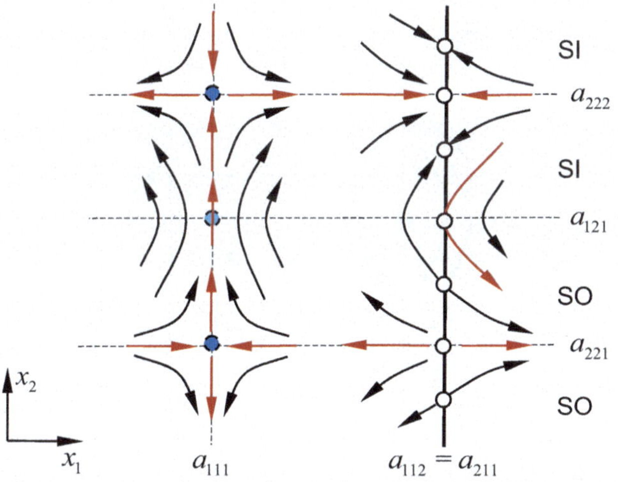

(b)

Fig. 5.21 Phase portraits ($a_{211} = a_{112}$ and $a_{121} < a_{221}$) for two-dimensional systems on the x_1-direction with $x_1^* = a_{111}$ and on the x_2-direction with $x_2^* = a_{121}, a_{221}, a_{222}$. The inflection-sink and source infinite-equilibriums with sink, source, and saddle plus a hyperbolic flow: (**a**) ($a_{110} > 0$, $a_{220} > 0$), (**b**) ($a_{110} < 0$, $a_{220} > 0$), (**c**) ($a_{110} > 0$, $a_{220} < 0$), and (**d**) ($a_{110} < 0$, $a_{220} < 0$)

Such infinite-equilibriums of $x_1^* = a_{111} = a_{211}$ with $a_{121} \in (a_{221}, a_{222})$ for such a dynamical system are presented in Fig. 5.21. The inflection-sink and source infinite-equilibriums are listed in Table 5.12. The infinite-equilibriums are the switching bifurcation for $a_{112} < a_{211}$ and $a_{211} \in (a_{111}, a_{112})$ with $a_{121} \in (a_{221}, a_{222})$.

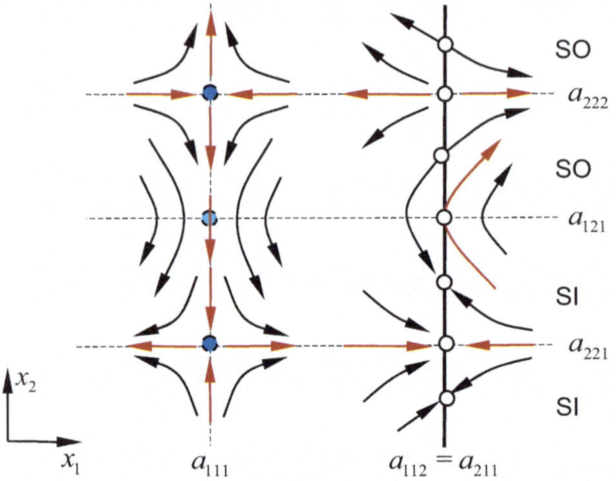

(c)

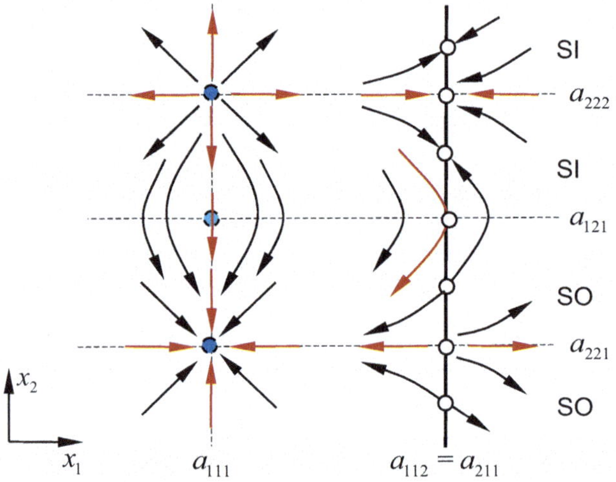

(d)

Fig. 5.21 (continued)

Table 5.12 Inflection-sink and source infinite-equilibriums of $x_1^* = a_{211} = a_{112}$ ($a_{211} = a_{112}$ and $a_{121} \in (a_{221}, a_{222})$)

	$(a_{111}, \bar{x}_2) \equiv (a_{211}, \bar{x}_2)$			
	$\bar{x}_2$			
(a_{110}, a_{220})	$(-\infty, a_{121})$	(a_{121}, a_{221})	(a_{221}, a_{222})	(a_{222}, ∞)
$(+,+)$	(SI,DI)	(SI,II)	(SO,DI)	(SO,II)
$(-,+)$	(SO,II)	(SO,DI)	(SI,II)	(SI,DI)
$(+,-)$	(SI,II)	(SI,DI)	(SO,II)	(SO,DI)
$(-,-)$	(SO,DI)	(SO,II)	(SI,DI)	(SI,II)

5.2.3 *Inflection-Diagonal-Saddles*

In this section, the diagonal-inflection upper-saddle and lower-saddle infinite-equilibriums are discussed, which are for the two switching bifurcations to occur simultaneously.

Consider a dynamical system for $a_{121}=a_{22l_1}$ $(l_1,\ l_2\in\{1,2\},\ l_1\neq l_2)$ and $a_{211}=a_{11s_1}$ $(s_1,\ s_2\in\{1,2\},\ s_1\neq s_2)$ as

$$\begin{aligned}\dot{x}_1&=a_{110}(x_1-a_{111})(x_1-a_{112})(x_2-a_{121}),\\ \dot{x}_2&=a_{220}(x_1-a_{211})(x_2-a_{221})(x_2-a_{222}),\end{aligned}\tag{5.91}$$

and the corresponding first integral manifold is

$$\ln\frac{|x_1-a_{11s_2}|}{|x_{10}-a_{11s_2}|}=\frac{a_{110}}{a_{220}}\ln\frac{|x_2-a_{22l_2}|}{|x_{20}-a_{22l_2}|}.\tag{5.92}$$

(i) The infinite-equilibriums of $x_2^*=a_{121}=a_{221}$ with $x_1^*=a_{211}=a_{111}$ have the following properties:

$$\left\{\begin{matrix}(a_{111},a_{222}) & (a_{112},a_{222})\\ (a_{111},a_{221}) & (a_{112},a_{221})\end{matrix}\right\}=\left\{\begin{matrix}\underbrace{(\text{SI},\text{II:DI})}_{\text{inflection sink}} & \underbrace{(\text{SO},\text{SO})}_{\text{source}}\\ \underbrace{({}_{\text{SO:SI}}\text{II},\ {}_{\text{SO:SI}}\text{II})}_{\text{increasing-inflection diagonal-saddle}} & \underbrace{(\text{II:DI},\text{SI})}_{\text{inflection sink}}\end{matrix}\right\}$$
$$\text{for } a_{110}>0,\ a_{220}>0,\tag{5.93}$$

$$\left\{\begin{matrix}(a_{111},a_{222}) & (a_{112},a_{222})\\ (a_{111},a_{221}) & (a_{112},a_{221})\end{matrix}\right\}=\left\{\begin{matrix}\underbrace{(\text{SO},\text{DI:II})}_{\text{inflection source}} & \underbrace{(\text{SI},\text{SO})}_{\text{saddle}}\\ \underbrace{({}_{\text{SO:SI}}\text{DI},\ {}_{\text{SI:SO}}\text{DI})}_{\text{decreasing-inflection diagonal-saddle}} & \underbrace{(\text{DI:II},\text{SI})}_{\text{inflection sink}}\end{matrix}\right\}$$
$$\text{for } a_{110}<0,\ a_{220}>0,\tag{5.94}$$

$$\left\{\begin{matrix}(a_{111},a_{222}) & (a_{112},a_{222})\\ (a_{111},a_{221}) & (a_{112},a_{221})\end{matrix}\right\}=\left\{\begin{matrix}\underbrace{(\text{SI},\text{DI:II})}_{\text{inflection sink}} & \underbrace{(\text{SO},\text{SI})}_{\text{saddle}}\\ \underbrace{({}_{\text{SI:SO}}\text{DI},\ {}_{\text{SO:SI}}\text{DI})}_{\text{decreasing-inflection diagonal-saddle}} & \underbrace{(\text{DI:II},\text{SO})}_{\text{inflection source}}\end{matrix}\right\}$$
$$\text{for } a_{110}>0,\ a_{220}<0,\tag{5.95}$$

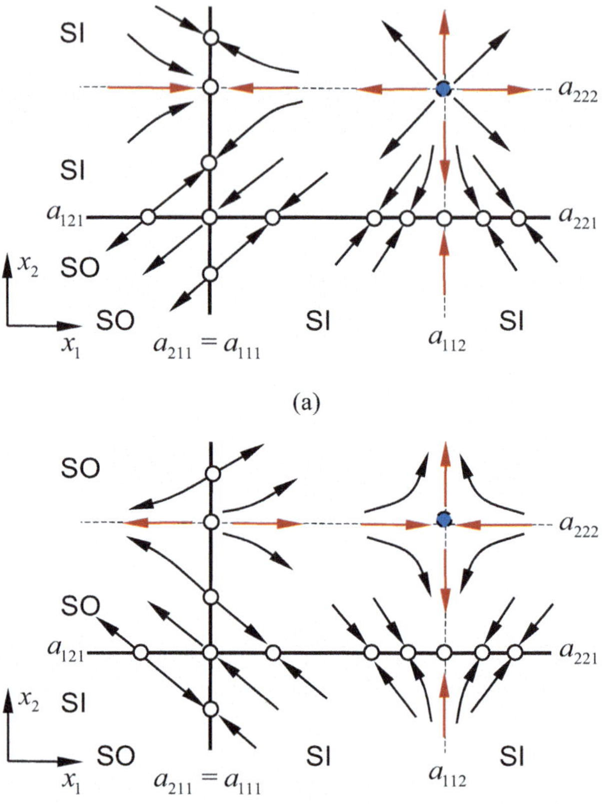

Fig. 5.22 Phase portraits ($a_{121} = a_{221}$ and $a_{111} = a_{211}$) for two-dimensional systems on the x_1-direction with $x_1^* = a_{111}, a_{112}, a_{211}$ and on the x_2-direction with $x_2^* = a_{121}, a_{221}, a_{222}$. The inflection diagonal infinite-equilibriums: (**a**) ($a_{110} > 0$, $a_{220} > 0$), (**b**) ($a_{110} < 0$, $a_{220} > 0$), (**c**) ($a_{110} > 0$, $a_{220} < 0$), and (**d**) ($a_{110} < 0$, $a_{220} < 0$)

$$\left\{\begin{matrix} (a_{111}, a_{222}) & (a_{112}, a_{222}) \\ (a_{111}, a_{221}) & (a_{112}, a_{221}) \end{matrix}\right\} = \left\{\begin{matrix} \underbrace{(\text{SO,II:DI})}_{\text{inflection source}} & \underbrace{(\text{SI,SI})}_{\text{sink}} \\ \underbrace{(\text{SI:SO}^{\text{II}}, \text{SI:SO}^{\text{II}})}_{\text{increasing-inflection diagonal-saddle}} & \underbrace{(\text{II:DI,SO})}_{\text{inflection source}} \end{matrix}\right\}$$

for $a_{110} < 0, a_{220} < 0$.

(5.96)

Such infinite-equilibriums of $x_2^* = a_{121} = a_{221}$ with $x_1^* = a_{211} = a_{111}$ for such a dynamical system are presented in Fig. 5.22. The inflection-sink and source infinite-equilibriums are listed in Table 5.13. The infinite-equilibriums are the switching bifurcation for $a_{121} < a_{221}$ and $a_{121} \in (a_{221}, a_{222})$ with the switching bifurcation of $a_{211} < a_{111}$ and $a_{211} \in (a_{111}, a_{112})$.

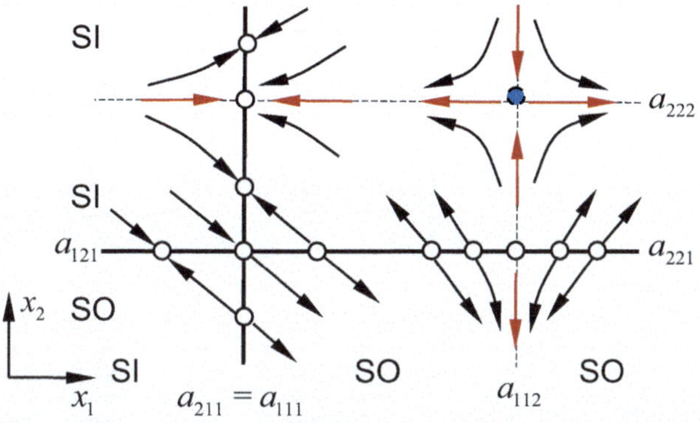

(c)

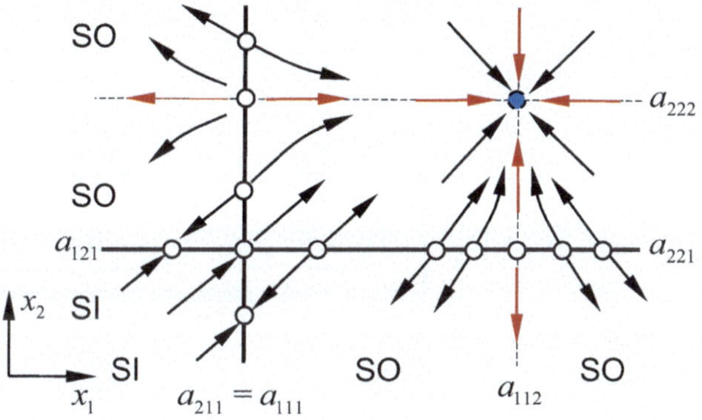

(d)

Fig. 5.22 (continued)

Table 5.13 Inflection-sink and source infinite-equilibriums ($a_{211} = a_{111}$ and $a_{121} = a_{221}$)

	$(\bar{x}_1, a_{221}) \equiv (\bar{x}_1, a_{121})$			$(a_{211}, \bar{x}_2) \equiv (a_{111}, \bar{x}_2)$		
	$\bar{x}_1$			$\bar{x}_2$		
(a_{110}, a_{220})	$(-\infty, a_{111})$	(a_{111}, a_{112})	(a_{112}, ∞)	$(-\infty, a_{221})$	(a_{221}, a_{222})	(a_{222}, ∞)
$(+,+)$	(II,SO)	(II,SI)	(DI,SI)	(SO,II)	(SI,II)	(SI,DI)
$(-,+)$	(DI,SO)	(DI,SI)	(II,SI)	(SI,DI)	(SO,DI)	(SO,II)
$(+,-)$	(DI,SI)	(DI,SO)	(II,SO)	(SO,DI)	(SI,DI)	(SI,II)
$(-,-)$	(II,SI)	(II,SO)	(DI,SO)	(SI,II)	(SO,II)	(SO,DI)

(ii) The infinite-equilibriums of $x_2^* = a_{121} = a_{221}$ with $x_1^* = a_{211} = a_{112}$ have the following properties:

$$
\left\{ \begin{matrix} (a_{111}, a_{222}) & (a_{112}, a_{222}) \\ (a_{111}, a_{221}) & (a_{112}, a_{221}) \end{matrix} \right\} = \left\{ \begin{matrix} \underbrace{(\mathrm{SI,SI})}_{\text{sink}} & \underbrace{(\mathrm{SO,DI:II})}_{\text{inflection source}} \\ \underbrace{(\mathrm{II:DI,SO})}_{\text{inflection source}} & \underbrace{({}_{\mathrm{SO:SI}}\mathrm{DI}, {}_{\mathrm{SI:SO}}\mathrm{DI})}_{\text{decreasing-inflection diagonal-saddle}} \end{matrix} \right\}
$$
$$
\text{for } a_{110} > 0, a_{220} > 0. \tag{5.97}
$$

$$
\left\{ \begin{matrix} (a_{111}, a_{222}) & (a_{112}, a_{222}) \\ (a_{111}, a_{221}) & (a_{112}, a_{221}) \end{matrix} \right\} = \left\{ \begin{matrix} \underbrace{(\mathrm{SO,SI})}_{\text{saddle}} & \underbrace{(\mathrm{SI,II:DI})}_{\text{inflection sink}} \\ \underbrace{(\mathrm{DI:II,SO})}_{\text{inflection source}} & \underbrace{({}_{\mathrm{SO:SI}}\mathrm{II}, {}_{\mathrm{SO:SI}}\mathrm{II})}_{\text{increasing-inflection diagonal-saddle}} \end{matrix} \right\}
$$
$$
\text{for } a_{110} < 0, a_{220} > 0, \tag{5.98}
$$

$$
\left\{ \begin{matrix} (a_{112}, a_{221}) & (a_{112}, a_{222}) \\ (a_{111}, a_{221}) & (a_{111}, a_{222}) \end{matrix} \right\} = \left\{ \begin{matrix} \underbrace{(\mathrm{SI,SO})}_{\text{saddle}} & \underbrace{(\mathrm{SO,II:DI})}_{\text{inflection source}} \\ \underbrace{(\mathrm{DI:II,SI})}_{\text{inflection sink}} & \underbrace{({}_{\mathrm{SI:SO}}\mathrm{II}, {}_{\mathrm{SI:SO}}\mathrm{II})}_{\text{increasing-inflection diagonal-saddle}} \end{matrix} \right\}
$$
$$
\text{for } a_{110} > 0, a_{220} < 0, \tag{5.99}
$$

$$
\left\{ \begin{matrix} (a_{112}, a_{221}) & (a_{112}, a_{222}) \\ (a_{111}, a_{221}) & (a_{111}, a_{222}) \end{matrix} \right\} = \left\{ \begin{matrix} \underbrace{(\mathrm{SO,SO})}_{\text{source}} & \underbrace{(\mathrm{SI,DI:II})}_{\text{inflection sink}} \\ \underbrace{(\mathrm{II:DI,SI})}_{\text{inflection sink}} & \underbrace{({}_{\mathrm{SI:SO}}\mathrm{DI}, {}_{\mathrm{SO:SI}}\mathrm{DI})}_{\text{decreasing-inflection diagonal-saddle}} \end{matrix} \right\}
$$
$$
\text{for } a_{110} < 0, a_{220} < 0, \tag{5.100}
$$

Such infinite-equilibriums of $x_2^* = a_{121} = a_{221}$ with $x_1^* = a_{211} = a_{112}$ for such a dynamical system are presented in Fig. 5.23. The inflection-sink and source infinite-equilibriums are listed in Table 5.14. The infinite-equilibriums are the switching bifurcation for $a_{121} < a_{221}$ and $a_{121} \in (a_{221}, a_{222})$ with the switching bifurcation of $a_{211} > a_{112}$ and $a_{211} \in (a_{111}, a_{112})$.

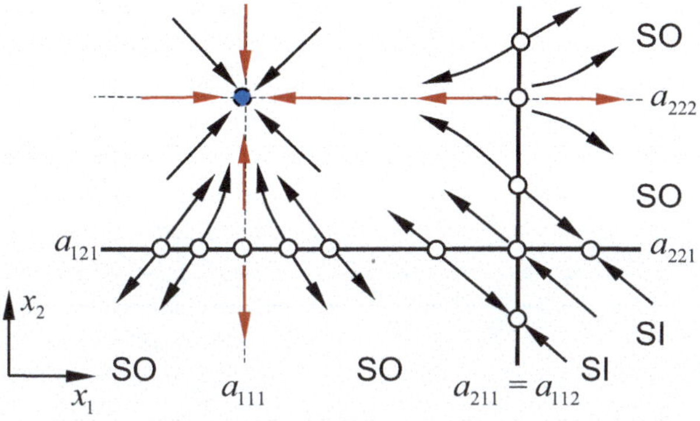

(a)

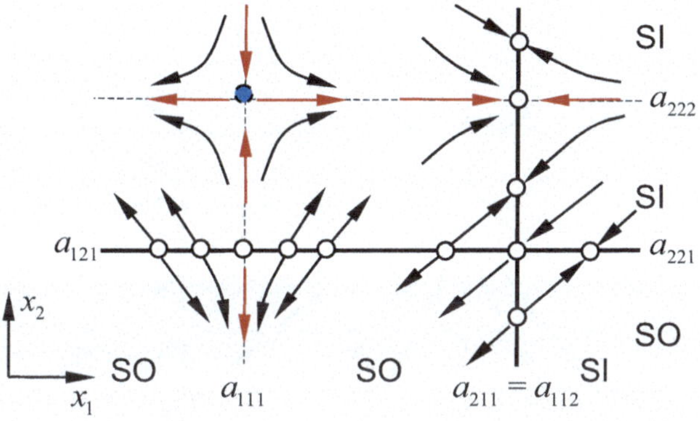

(b)

Fig. 5.23 Phase portraits ($a_{121} = a_{221}$ and $a_{112} = a_{211}$) for two-dimensional systems on the x_1-direction with $x_1^* = a_{111}, a_{112}, a_{211}$ and on the x_2-direction with $x_2^* = a_{121}, a_{221}, a_{222}$. The inflection diagonal infinite-equilibriums: (**a**) ($a_{110} > 0$, $a_{220} > 0$), (**b**) ($a_{110} < 0$, $a_{220} > 0$), (**c**) ($a_{110} > 0$, $a_{220} < 0$), and (**d**) ($a_{110} < 0$, $a_{220} < 0$)

(iii) The infinite equilibriums of $x_2^* = a_{121} = a_{222}$ with $x_1^* = a_{211} = a_{111}$ have the following properties:

$$\left\{\begin{matrix} (a_{111}, a_{222}) & (a_{112}, a_{222}) \\ (a_{111}, a_{221}) & (a_{112}, a_{221}) \end{matrix}\right\} = \left\{\begin{matrix} \underbrace{({}_{\mathrm{SI:SO}}\mathrm{DI}, {}_{\mathrm{SO:SI}}\mathrm{DI})}_{\text{decreasing-inflection diagonal-saddle}} & \underbrace{(\mathrm{DI:II}, \mathrm{SO})}_{\text{inflection source}} \\ \underbrace{(\mathrm{SO}, \mathrm{II:DI})}_{\text{inflection source}} & \underbrace{(\mathrm{SI}, \mathrm{SI})}_{\text{sink}} \end{matrix}\right\}$$

for $a_{110} > 0, a_{220} > 0,$

(5.101)

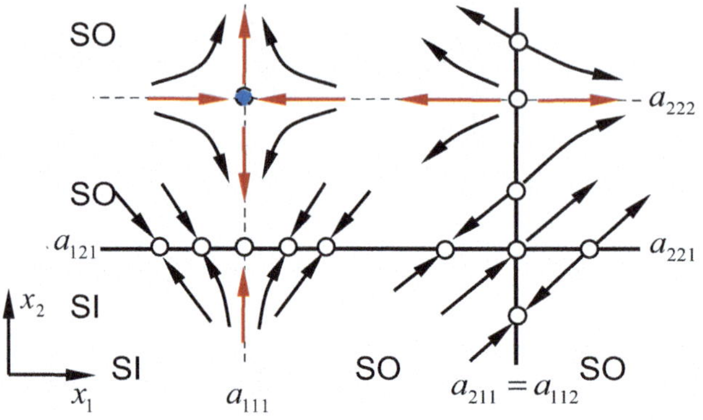

(c)

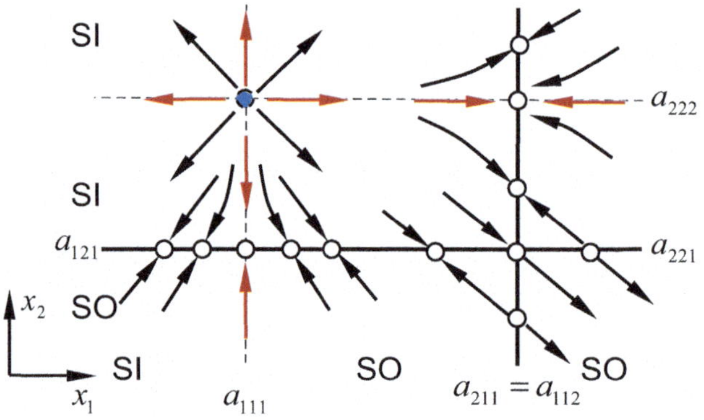

(d)

Fig. 5.23 (continued)

Table 5.14 Inflection-sink and source infinite-equilibriums ($a_{211} = a_{112}$ and $a_{121} = a_{221}$)

	$(\bar{x}_1, a_{221}) \equiv (\bar{x}_1, a_{121})$			$(a_{211}, \bar{x}_2) \equiv (a_{111}, \bar{x}_2)$		
	$\bar{x}_1$			$\bar{x}_2$		
(a_{110}, a_{220})	$(-\infty, a_{111})$	(a_{111}, a_{112})	(a_{112}, ∞)	$(-\infty, a_{221})$	(a_{221}, a_{222})	(a_{222}, ∞)
(+, +)	(II,SO)	(DI,SO)	(DI,SI)	(SI,DI)	(SO,DI)	(SO,II)
(−, +)	(DI,SO)	(II,SO)	(II,SI)	(SO,II)	(SI,II)	(SI,DI)
(+, −)	(DI,SI)	(II,SI)	(II,SO)	(SI,II)	(SO,II)	(SO,DI)
(−, −)	(II,SI)	(DI,SI)	(DI,SO)	(SO,DI)	(SI,DI)	(SI,II)

$$\left\{\begin{array}{ll}(a_{111},a_{222}) & (a_{112},a_{222})\\ (a_{111},a_{221}) & (a_{112},a_{221})\end{array}\right\}=\left\{\begin{array}{ll}\underbrace{({}_{\text{SI:SO}}\text{II},{}_{\text{SI:SO}}\text{II})}_{\text{increasing-inflection diagonal-saddle}} & \underbrace{(\text{II:DI,SO})}_{\text{inflection sink}}\\ \underbrace{(\text{SI,DI:II})}_{\text{inflection sink}} & \underbrace{(\text{SO,SI})}_{\text{saddle}}\end{array}\right\}$$
$$\text{for } a_{110}<0, a_{220}>0, \tag{5.102}$$

$$\left\{\begin{array}{ll}(a_{111},a_{222}) & (a_{112},a_{222})\\ (a_{111},a_{221}) & (a_{112},a_{221})\end{array}\right\}=\left\{\begin{array}{ll}\underbrace{({}_{\text{SO:SI}}\text{II},{}_{\text{SO:SI}}\text{II})}_{\text{increasing-inflection diagonal-saddle}} & \underbrace{(\text{II:DI,SI})}_{\text{inflection sink}}\\ \underbrace{(\text{SO,DI:II})}_{\text{inflection source}} & \underbrace{(\text{SI,SO})}_{\text{saddle}}\end{array}\right\}$$
$$\text{for } a_{110}>0, a_{220}<0, \tag{5.103}$$

$$\left\{\begin{array}{ll}(a_{111},a_{222}) & (a_{112},a_{222})\\ (a_{111},a_{221}) & (a_{112},a_{221})\end{array}\right\}=\left\{\begin{array}{ll}\underbrace{({}_{\text{SO:SI}}\text{DI},{}_{\text{SI:SO}}\text{DI})}_{\text{decreasing-inflection diagonal-saddle}} & \underbrace{(\text{DI:II,SI})}_{\text{inflection sink}}\\ \underbrace{(\text{SI,II:DI})}_{\text{inflection sink}} & \underbrace{(\text{SO,SO})}_{\text{source}}\end{array}\right\}$$
$$\text{for } a_{110}<0, a_{220}<0. \tag{5.104}$$

Such infinite-equilibriums of $x_2^*=a_{121}=a_{222}$ with $x_1^*=a_{211}=a_{111}$ for such a dynamical system are presented in Fig. 5.24. The inflection-sink and source infinite-equilibriums are listed in Table 5.15. The infinite-equilibriums are the switching bifurcation for $a_{121}>a_{222}$ and $a_{121}\in(a_{221},a_{222})$ with the switching bifurcation of $a_{211}<a_{111}$ and $a_{211}\in(a_{111},a_{112})$.

(iv) The infinite-equilibriums of $x_2^*=a_{121}=a_{222}$ with $x_1^*=a_{211}=a_{112}$ have the following properties:

$$\left\{\begin{array}{ll}(a_{111},a_{222}) & (a_{112},a_{222})\\ (a_{111},a_{221}) & (a_{112},a_{221})\end{array}\right\}=\left\{\begin{array}{ll}\underbrace{(\text{DI:II,SI})}_{\text{inflection sink}} & \underbrace{({}_{\text{SI:SO}}\text{II},{}_{\text{SI:SO}}\text{II})}_{\text{increasing-inflection diagonal-saddle}}\\ \underbrace{(\text{SO,SO})}_{\text{source}} & \underbrace{(\text{SI,DI:II})}_{\text{inflection sink}}\end{array}\right\}$$
$$\text{for } a_{110}>0, a_{220}>0, \tag{5.105}$$

$$\left\{\begin{array}{ll}(a_{111},a_{222}) & (a_{112},a_{222})\\ (a_{111},a_{221}) & (a_{112},a_{221})\end{array}\right\}=\left\{\begin{array}{ll}\underbrace{(\text{II:DI,SI})}_{\text{inflection sink}} & \underbrace{({}_{\text{SI:SO}}\text{DI},{}_{\text{SO:SI}}\text{DI})}_{\text{decreasing-inflection diagonal-saddle}}\\ \underbrace{(\text{SI,SO})}_{\text{saddle}} & \underbrace{(\text{SO,II:DI})}_{\text{inflection source}}\end{array}\right\}$$
$$\text{for } a_{110}<0, a_{220}>0, \tag{5.106}$$

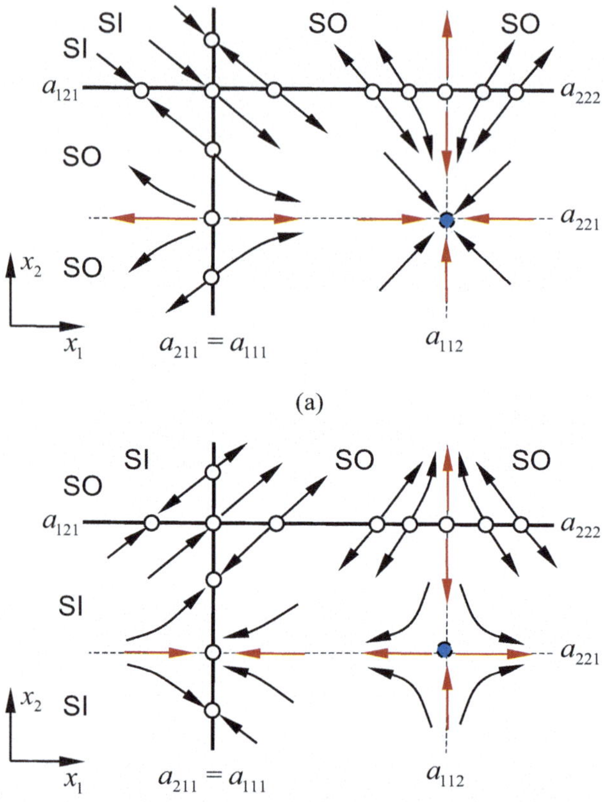

(a)

(b)

Fig. 5.24 Phase portraits ($a_{121} = a_{222}$ and $a_{111} = a_{211}$) for two-dimensional systems on the x_1-direction with $x_1^* = a_{111}, a_{112}, a_{211}$ and on the x_2-direction with $x_2^* = a_{121}, a_{221}, a_{222}$. The inflection diagonal infinite-equilibriums: (**a**) ($a_{110} > 0$, $a_{220} > 0$), (**b**) ($a_{110} < 0$, $a_{220} > 0$), (**c**) ($a_{110} > 0$, $a_{220} < 0$), and (**d**) ($a_{110} < 0$, $a_{220} < 0$)

$$\left\{\begin{array}{ll}(a_{111}, a_{222}) & (a_{112}, a_{222})\\ (a_{111}, a_{221}) & (a_{112}, a_{221})\end{array}\right\} = \left\{\begin{array}{cc}\underbrace{(\mathrm{II{:}DI, SO})}_{\text{inflection source}} & \underbrace{({}_{\mathrm{SO:SI}}\mathrm{DI}, {}_{\mathrm{SI:SO}}\mathrm{DI})}_{\text{decreasing-inflection diagonal-saddle}}\\ \underbrace{(\mathrm{SO,SI})}_{\text{saddle}} & \underbrace{(\mathrm{SI, II{:}DI})}_{\text{inflection sink}}\end{array}\right\}$$
$$\text{for } a_{110} > 0, a_{220} < 0, \tag{5.107}$$

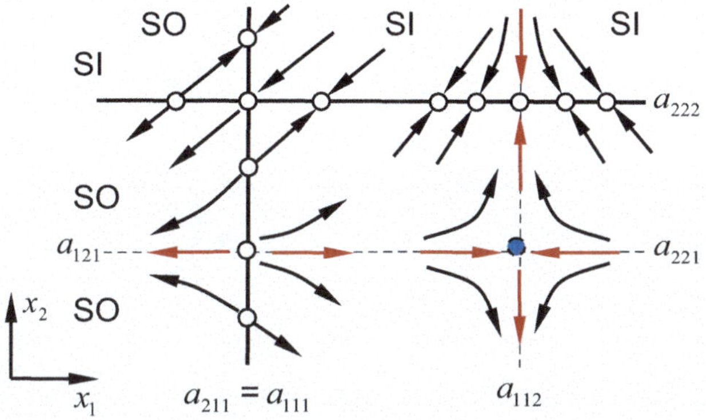

(c)

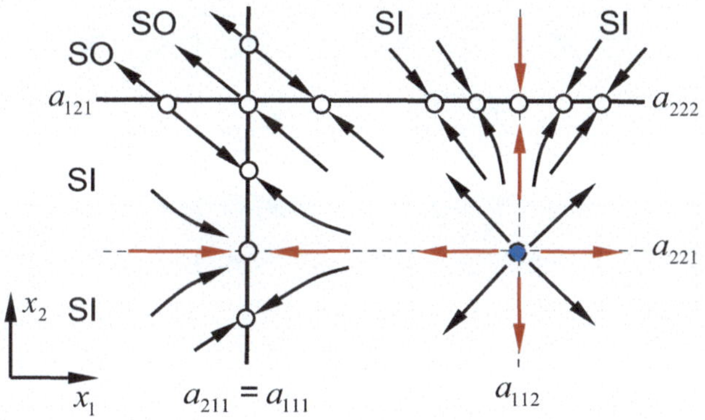

(d)

Fig. 5.24 (continued)

Table 5.15 Inflection-sink and source infinite-equilibriums ($a_{211} = a_{111}$ and $a_{121} = a_{222}$)

	$(\bar{x}_1, a_{221}) \equiv (\bar{x}_1, a_{121})$			$(a_{211}, \bar{x}_2) \equiv (a_{111}, \bar{x}_2)$		
	$\bar{x}_1$			$\bar{x}_2$		
(a_{110}, a_{220})	$(-\infty, a_{111})$	(a_{111}, a_{112})	(a_{112}, ∞)	$(-\infty, a_{221})$	(a_{221}, a_{222})	(a_{222}, ∞)
(+, +)	(DI,SI)	(DI,SO)	(II,SO)	(SO,II)	(SO,DI)	(SI,DI)
(−, +)	(II,SI)	(II,SO)	(DI,SO)	(SI,DI)	(SI,II)	(SO,II)
(+, −)	(II,SO)	(II,SI)	(DI,SI)	(SO,DI)	(SO,II)	(SI,II)
(−, −)	(DI,SO)	(DI,SI)	(II,SI)	(SI,II)	(SI,DI)	(SO,DI)

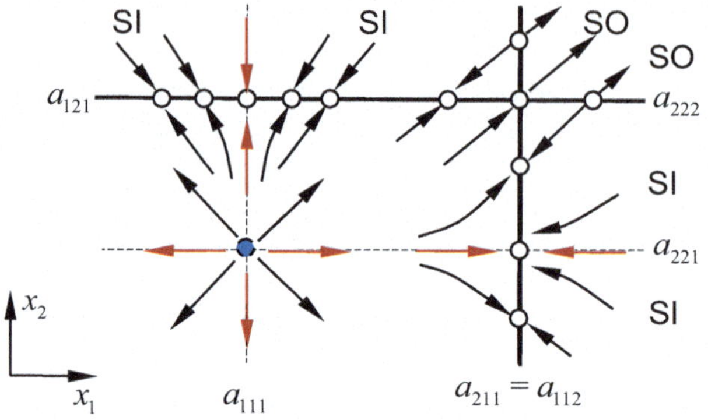

(a)

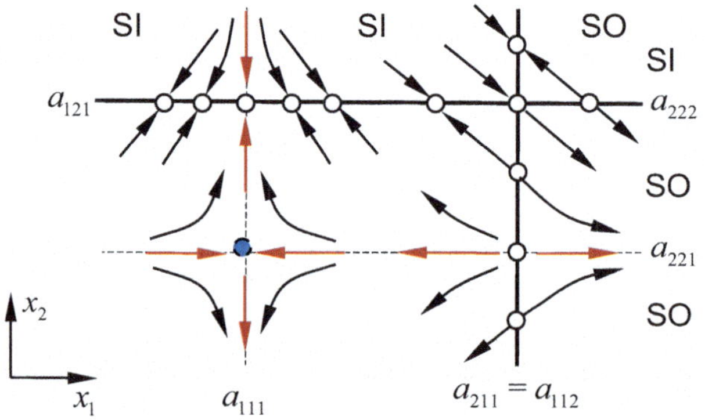

(b)

Fig. 5.25 Phase portraits ($a_{121} = a_{222}$ and $a_{112} = a_{211}$) for two-dimensional systems on the x_1-direction with $x_1^* = a_{111}, a_{112}, a_{211}$ and on the x_2-direction with $x_2^* = a_{121}, a_{221}, a_{222}$. The inflection diagonal infinite-equilibriums: (**a**) ($a_{110} > 0$, $a_{220} > 0$), (**b**) ($a_{110} < 0$, $a_{220} > 0$), (**c**) ($a_{110} > 0$, $a_{220} < 0$), and (**d**) ($a_{110} < 0$, $a_{220} < 0$)

$$\left\{\begin{matrix} (a_{111}, a_{222}) & (a_{112}, a_{222}) \\ (a_{111}, a_{221}) & (a_{112}, a_{221}) \end{matrix}\right\} = \left\{\begin{matrix} \underbrace{(\text{DI:II}, \text{SO})}_{\text{inflection source}} & \underbrace{({}_{\text{SO:SI}}\text{II}, {}_{\text{SO:SI}}\text{II})}_{\text{increasing-inflection diagonal-saddle}} \\ \underbrace{(\text{SI}, \text{SI})}_{\text{sink}} & \underbrace{(\text{SO}, \text{DI:II})}_{\text{inflection source}} \end{matrix}\right\}$$

for $a_{110} < 0, a_{220} < 0$.

(5.108)

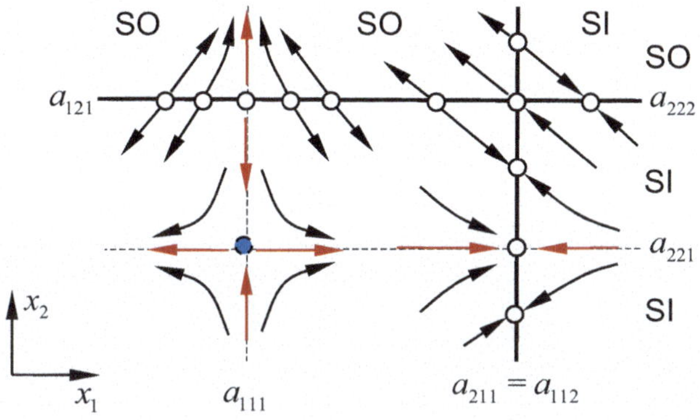

(c)

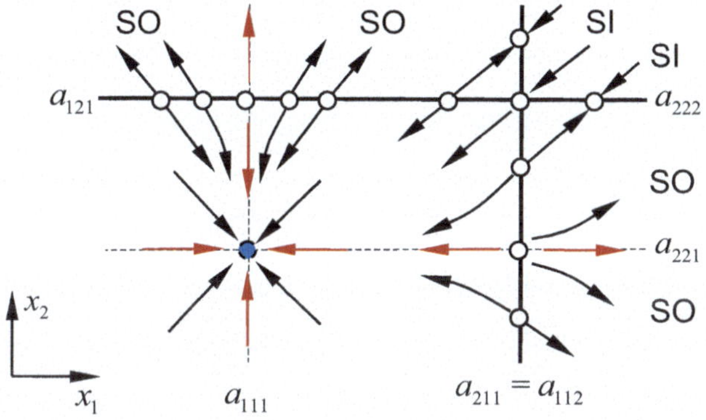

(d)

Fig. 5.25 (continued)

Table 5.16 Inflection-sink and source infinite-equilibriums ($a_{211} = a_{112}$ and $a_{121} = a_{222}$)

	$(\bar{x}_1, a_{221}) \equiv (\bar{x}_1, a_{121})$			$(a_{211}, \bar{x}_2) \equiv (a_{111}, \bar{x}_2)$		
	$\bar{x}_1$			$\bar{x}_2$		
(a_{110}, a_{220})	$(-\infty, a_{111})$	(a_{111}, a_{112})	(a_{112}, ∞)	$(-\infty, a_{221})$	(a_{221}, a_{222})	(a_{222}, ∞)
(+, +)	(DI,SI)	(II,SI)	(II,SO)	(SI,DI)	(SI,II)	(SO,II)
(−, +)	(II,SI)	(DI,SI)	(DI,SO)	(SO,II)	(SO,DI)	(SI,DI)
(+, −)	(II,SO)	(DI,SO)	(DI,SI)	(SI,II)	(SI,DI)	(SO,DI)
(−, −)	(DI,SO)	(II,SO)	(II,SI)	(SO,DI)	(SO,II)	(SI,II)

Such infinite-equilibriums of $x_2^* = a_{121} = a_{222}$ with $x_1^* = a_{211} = a_{112}$ for such a dynamical system are presented in Fig. 5.25. The inflection-sink and source infinite-equilibriums are listed in Table 5.16. The infinite-equilibriums are the switching bifurcation for $a_{121} > a_{222}$ and $a_{121} \in (a_{221}, a_{222})$ with the switching bifurcation of $a_{211} > a_{112}$ and $a_{211} \in (a_{111}, a_{112})$.

Correction to: Two-Dimensional Two-Product Cubic Systems, Vol. II

Albert C. J. Luo

Correction to:
A. C. J. Luo, Two-Dimensional Two-Product Cubic Systems, Vol. II,
https://doi.org/10.1007/978-3-031-48491-9

We have received extensive corrections from the author after the book has been published. The book was updated with all changes mentioned.

The updated version of this book can be found at
https://doi.org/10.1007/978-3-031-48491-9

A. C. J. Luo, *Two-Dimensional Two-Product Cubic Systems, Vol. II*,
https://doi.org/10.1007/978-3-031-48491-9_6

Index

A. C. J. Luo, *Two-Dimensional Two-Product Cubic Systems, Vol. II*,
https://doi.org/10.1007/978-3-031-48491-9

The manufacturer's authorised representative in the EU is Springer Nature Customer Service Centre GmbH, Europaplatz 3, 69115 Heidelberg, Germany. If you have any concerns regarding our products, please contact ProductSafety@springernature.com

Printed and bound by CPI Group (UK) Ltd, Croydon, CR0 4YY

01/10/2026

02225965-0009